Studies in Logic

Volume 54

Proof Theory of N4-related Paraconsistent Logics

Volume 43
Logic, Truth and Inquiry
Mark Weinstein

Volume 44
Meta-logical Investigations in Argumentation Networks
Dov M. Gabbay

Volume 45
Errors of Reasoning. Naturalizing the Logic of Inference
John Woods

Volume 46
Questions, Inferences, and Scenarios
Andrzej Wiśniewski

Volume 47
Logic Across the University: Foundations and Applications. Proceedings of the Tsinghua Logic Conference, Beijing, 2013
Johan van Benthem and Fenrong Liu, eds.

Volume 48
Trends in Belief Revision and Argumentation Dynamics
Eduardo L. Fermé, Dov M. Gabbay, and Guillermo R. Simari

Volume 49
Introduction to Propositional Satisfiability
Victor Marek

Volume 50
Intuitionistic Set Theory
John L. Bell

Volume 51
Metalogical Contributions to the Nonmonotonic Theory of Abstract Argumentation
Ringo Baumann

Volume 52
Inconsistency Robustness
Carl Hewitt and John Woods, eds.

Volume 53
Aristotle's Earlier Logic
John Woods

Volume 54
Proof Theory of N4-related Paraconsistent Logics
Norihiro Kamide and Heinrich Wansing

Studies in Logic Series Editor
Dov Gabbay dov.gabbay@kcl.ac.uk

Proof Theory of N4-related Paraconsistent Logics

Norihiro Kamide
Heinrich Wansing

ISBN 978-1-84890-167-4

College Publications
Scientific Director: Dov Gabbay
Managing Director: Jane Spurr

http://www.collegepublications.co.uk

Original cover design by Orchid Creative www.orchidcreative.co.uk
Printed by Lightning Source, Milton Keynes, UK

Preface

This book is the result of our shared interest in the structural proof theory of paraconsistent logics and it brings together a number of papers which we have written separately or jointly on various systems of inconsistency-tolerant logic. Whereas these papers are scattered over several journals and volumes devoted to different areas ranging from theoretical computer science to philosophical logic, they display a thematic coherence to such an extent that we considered it useful to put this material together to constitute the first monograph ever with a central focus on the proof theory of paraconsistent logics.

It should be emphasized, though, that we do not aim at a comprehensive survey of *all kinds* of proof systems for *all* kinds of paraconsistent logics or forms of paraconsistent reasoning. We do not deal with, for instance, the axiomatic proof theory of da Costa's paraconsistent logics, nor do we contribute to the dynamic proof theory of inconsistency-adaptive logics. Instead, at center stage of our investigations there is the four-valued, constructive paraconsistent logic N4 by David Nelson. This by now well-known logic may be seen as an expressively rich system of departure not only for several closely related paraconsistent systems in the vocabulary of N4 and fragments of this language, but also for numerous modal extensions, such as certain constructive description logics and temporal logics based on N4. The negation, conjunction, disjunction fragment of N4, for example, coincides with Belnap and Dunn's famous so-called useful four-valued logic, and there are also close relations to Arieli and Avron's bilattice logics.

Although the structural proof-theory of N4 and N4-related logics is the central theme of the present monograph, it should come as no surprise that models and model-theoretic semantics also play a role in the presentation. It is not only that methods from model-theoretic semantics are useful for purely technical purposes, such as, for example, the semantical proofs of cut-elimination in this book. The relational, Kripke-style models we shall be dealing with also provide a motivating and intuitively appealing insight into the logics with respect to which they are shown to be complete. Nevertheless, the emphasis is on Gentzen-style proof systems —in particular sequent calculi of a standard and less standard kind— for paraconsistent logics, and cut-elimination and its consequences are a central topic throughout. A unifying element of the presentation is the repeated application of embedding theorems in order to transfer results from other logics such as intuitionistic logic to the paraconsistent case.

We would like to thank various institutions and departments for their support and providing excellent working conditions:

- Waseda Institute for Advanced Study, Waseda University, Japan,
- Department of Human Information Systems, Faculty of Science and Engineering, Teikyo University, Japan,
- Department of Information and Electronic Engineering, Faculty of Science and Engineering, Teikyo University, Japan,
- Japanese Ministry of Education, Culture, Sports, Science and Technology (Grant-in-Aid for Young Scientists (B) 20700015, Grant-in-Aid for Scientific Research (C) 26330263),
- Grant-in-Aid for Okawa Foundation for Information and Telecommunications, Japan,
- Faculty of Information Technology and Business, Cyber University, Japan,
- Institute of Philosophy, Dresden University of Technology, Germany,
- Department of Philosophy II, Ruhr-University Bochum, Germany,
- German Research Foundation, DFG (grant WA 936/6-1).

In particular, we would like to express our gratitude to the Alexander von Humboldt Foundation for conferring a Humboldt Research Fellowship to Norihiro Kamide in 2008/2009. Morever, we would like to thank Prof. Dov Gabbay for supporting the publication of this book in *Studies in Logic* and Mrs Jane Spurr, the managing director of College Publications, for her extremely efficient and kind co-operation.

Norihiro Kamide and Heinrich Wansing
Utsunomiya and Bochum
December 2014

Contents

1. Introduction

1.1. Nelson's paraconsistent logic: A common basis for various paraconsistent logics

Inconsistency handling is of growing importance in Computer Science since inconsistencies may frequently occur in knowledge-based and intelligent information systems [9, 10, 25, 44]. Paraconsistent (or inconsistency-tolerant) logics [154, 157] have been widely studied to cope with such inconsistencies [12, 10, 57, 180]. Moreover, paraconsistent logics are of a broader interest because paraconsistency handling is also relevant to the philosophical analysis of information processing in general and to our understanding of everyday as well as scientific reasoning. One of the aims of this book is to obtain a theoretical foundation of automated inconsistency-tolerant (or paraconsistent) reasoning based on a basic paraconsistent logic. Although there exists a broad variety of paraconsistent logics, *Nelson's paraconsistent (four-valued) logic* N4 (also referred to in the literature as N^-) [6] is known to be one of the most important and basic paraconsistent logics in Computer Science and in Philosophical Logic [32, 51, 140, 142, 148, 194, 196, 197]. In particular, it is known that N4 is a common basis for various extended and useful paraconsistent logics [12, 10, 57, 98, 100, 102, 140, 142, 180]. N4 is also known to be a paraconsistent variant of *Nelson's constructive (three-valued) logic* N3 (also referred to as N) [134], which has been studied by several mathematical logicians.

Gentzen-type sequent calculi for Nelson's logics have been investigated [119, 148, 196], and Kripke semantics for Nelson's logics have also been studied [170, 186, 196]. A translation of N3 into intuitionistic logic has been proposed and studied by Gurevich [73], Rautenberg [164] and Vorob'ev [193]. A similar translation for N4 into LJ, Gentzen's sequent calculus for positive intuitionistic propositional logic, can also be obtained.

A Hilbert-style system HN4 for Nelson's four-valued logic is obtained by adding the following axiom schemes for a paraconsistent negation connective $\sim$:

1. $\sim\sim\alpha \leftrightarrow \alpha$,

2. $\sim(\alpha \wedge \beta) \leftrightarrow (\sim\alpha \vee \sim\beta)$,

3. $\sim(\alpha \vee \beta) \leftrightarrow (\sim\alpha \wedge \sim\beta)$,

4. $\sim(\alpha \rightarrow \beta) \leftrightarrow (\alpha \wedge \sim\beta)$.

to modus ponens and the following axiom schemes of positive intuitionistic logic :

5. $\alpha \rightarrow (\beta \rightarrow \alpha)$,

6. $(\alpha \rightarrow (\beta \rightarrow \gamma)) \rightarrow ((\alpha \rightarrow \beta) \rightarrow (\alpha \rightarrow \gamma))$,

7. $(\alpha \wedge \beta) \rightarrow \alpha$,

8. $(\alpha \wedge \beta) \rightarrow \beta$,

9. $(\alpha \rightarrow \beta) \rightarrow ((\alpha \rightarrow \gamma) \rightarrow (\alpha \rightarrow (\beta \wedge \gamma)))$,

10. $\alpha \rightarrow (\alpha \vee \beta)$,

11. $\beta \rightarrow (\alpha \vee \beta)$,

12. $(\alpha \rightarrow \gamma) \rightarrow ((\beta \rightarrow \gamma) \rightarrow ((\alpha \vee \beta) \rightarrow \gamma))$.

A Hilbert-style system HN3 for N3 is obtained from that of N4 by adding the axiom scheme $(\alpha \wedge \sim\alpha) \rightarrow \beta$. It was shown by Odintsov [136] that N3 is embeddable into N4. The lattices of extensions of N4 and its augmentation by a falsity constant, the system $\mathrm{N4}^{\perp}$, are comprehensively studied in [138].

1.2. Useful four-valued logic, bi-intuitionistic logic and dual-intuitionistic logic

A *useful four-valued logic* (or equivalently *first-degree entailment* logic) was introduced by Belnap [21] and Dunn [47], and some Gentzen-type sequent calculi and their completeness and cut-elimination have been studied (see e.g., [21, 59, 158]). It is known that Belnap and Dunn's four-valued logic and the $\{\wedge, \vee, \sim\}$-fragment of N4 are the same logic. A number of generalized or extended versions of Belnap and Dunn's logic have also been widely studied as some *bilattice* and *trilattice logics* (see e.g., [12, 13, 51, 99, 179, 180, 181, 212, 210]). A similarity between Nelson's logics and certain extensions of Belnap and Dunn's logic was also pointed out by Gargov [62] in a historical account.

N4 is closely related to *Rauszer's* H-B *(Heyting-Brouwer) logic* (or equivalently *bi-intuitionistic logic*) and *dual-intuitionistic logic* (sometimes called *falsification logic*). H-B logic was introduced by Rauszer [161, 163] in order to extend intuitionistic logic by the connective of co-implication. H-B logic can represent a certain notion of "falsification" and the concept of "verification" simultaneously. H-B logic has been investigated using semantic methods such as algebraic and Kripke-type semantics, and has also been extended to some modal versions (see, e.g., [121, 217]).

Dual-intuitionistic logic and certain variants thereof were proposed and studied by Czermak [39], Goodman [71], Urbas [189], Goŕe [68] and Shramko [178]. The relation between H-B logic and variants of dual-intuitionistic logic was pointed out by Goŕe [68]. Dual-intuitionistic logic and its variants have the same connective of co-implication as in H-B logic, and therefore can express a certain notion of falsification.

1.3. Trilattice logics

We shall also consider certain logics that are related to the trilattice $SIXTEEN_3$ of generalized truth values. At first sight, this investigation might appear to be devoted to problems that are rather peripheral to mainstream formal logic. At closer inspection, however, this it not the case. The first impression might be due to the fact that the trilattice $SIXTEEN_3$ has been introduced into the literature only recently, namely in [180], and that the logics under consideration in the present book have been defined even more recently, namely in [139]. We cannot unfold the motivation for studying $SIXTEEN_3$ in any detail here, because this would have to include a historical and philosophical discussion of the notion of a generalized truth value, generalized truth values being subsets of an already given set of truth values. Such a general discussion and motivation can be found in [182]. We may, however, point out that the logical study of trilattices builds upon a mature mathematical theory, namely the theory of bilattices, see, for example, [11, 10, 14, 15, 56, 57, 58, 62, 88, 172]. In particular, the trilattice $SIXTEEN_3$ is a natural and straightforward generalization of the smallest non-trivial bilattice $FOUR_2$. This bilattice is defined on the powerset of the set of classical truth values T and F. The four-valued logic of $FOUR_2$ with $\{T\}$ and $\{T, F\}$ as designated values is known as Dunn and Belnap's (useful) four-valued logic or as first-degree entailment logic, FDE. FDE has found many applications in, for example, many-valued symbolic model-checking, the semantics of logic programs, intelligent tutoring systems, inconsistency-tolerant description logics, and generalizations of algebras of commuting processes.

We will consider two logics related to $SIXTEEN_3$, namely the axiom systems L_B and L_T presented by Odintsov [139]. Since the two axiom systems L_B and L_T can be dealt with in a completely similar way, we shall focus our attention just on the system L_B. The logics L_B and L_T are of interest, among other things because they combine a set of positive connectives (related to truth) and a set of negative connectives (related to falsity), a combination which emerges naturally in the context of $SIXTEEN_3$, see [180, 181, 183, 179].[1] We define a sequent calculus L_{16} for L_B, which differs from the sequent calculus GL_B for L_B defined in [99]. The sequent calculus L_{16} is such that it can be conveniently shown to be strongly complete with respect to a variant of the co-ordinate valuations semantics introduced by Odintsov [139].

Classical propositional logic has both an algebraic semantics in terms of Boolean algebras and a non-algebraic semantics in terms of truth-value assignments. The trilattice logics considered in [180, 181] are non-classical propositional logics with an algebraic semantics. One route to obtain a semantics of *quantified* trilattice logics would consist of defining a suitable notion of cylindric algebras for them. It may be seen as an advantage of the co-ordinate valuations semantics that it admits a simple and straightforward extension to first-order logic. The present chapter gives the first result on first-order trilattice logics. We present a cut-free, sound and complete sequent calculus F_{16} for a first-order extension of the logic L_B.

As noted previously, the trilattice $SIXTEEN_3$ introduced in [180] is a natural generalization of the famous bilattice $FOUR_2$. Whereas in $FOUR_2$, in addition to the information order, there is only one logical order used to define semantical consequence, in $SIXTEEN_3$ there are two logical orders, a *truth* order and a *falsity* order. Truth and falsity are thereby treated on a par as independent notions in their own right. Each of the two logical orders induces a set of logical operations and an entailment relation. The relation $\models_t$ is defined with respect to the truth order and the relation $\models_f$ is defined with respect to the falsity order. In [180] an axiomatization is presented for $\models_t$ in the language based on the truth order and for $\models_f$ in the language based on the falsity order. Moreover, the relations $\models_t$ and $\models_f$ are axiomatized in the positive language extended by falsity negation and in the negative language extended by truth negation, respectively. It was left as an open problem,

[1] In [183] the paper [103] is referred to as an unpublished manuscript with the title "Alternative semantics for trilattice logics".

however, to axiomatize the relations $\models_t$ and $\models_f$ in the full vocabulary containing both the truth and the falsity connectives. Sergei Odintsov [139] presented a sound and complete axiomatization of $\models_t$ based on the full language extended by an implication operation. The implication is interpreted as the residuum of truth conjunction with respect to the truth order.

For a detailed motivation of logics emerging from trilattices of generalized truth values we refer to [180, 181, 183]. The relation of these logics to many-valued logics defined from sets of designated truth values is discussed in [213]; proof-theoretic aspects are investigated in [89, 99, 139, 210, 212].

1.4. Overview and summary of this book

The contents of this book may be summarized as follows.

Part I: Variations of Nelson's paraconsistent logic.

Chapter 2: Nelson's paraconsistent logic. The aim of this chapter is to obtain a theoretical foundation of inconsistency-tolerant (or paraconsistent) reasoning by presenting a uniform perspective on several familiar as well as new proof systems for paraconsistent reasoning. Inconsistency handling is of growing importance in Computer Science since inconsistencies may frequently occur in knowledge-based and intelligent information systems. Paraconsistent, inconsistency-tolerant logics have been studied to cope with such inconsistencies. In this chapter, proof systems for Nelson's paraconsistent logic N4 are comprehensively studied. The logic N4 is a fundamental system and known to be a common basis for various extended and useful paraconsistent logics. Some basic theorems including cut-elimination, normalization and completeness are uniformly proved using various embedding theorems. A variety of sequent calculi and natural deduction systems for N4 and some closely related systems are presented and compared.

Chapter 3: Paraconsistent logics based on trilattices. A sequent calculus L_{16} for Odintsov's Hilbert-style axiomatization L_B of a logic related to the trilattice $SIXTEEN_3$ of generalized truth values is introduced. The completeness theorem w.r.t. a simple semantics for L_{16} is proved using Maehara's decomposition method that simultaneously derives the cut-elimination theorem for L_{16}. A first-order extension F_{16} of L_{16} and its semantics are also introduced. The completeness and cut-elimination theorems for F_{16} are proved using Schütte's method.

Chapter 4: Generalized paraconsistent logics. New propositional and first-order paraconsistent logics (called L_ω and FL_ω, respectively) are introduced as Gentzen-type sequent calculi with classical and paraconsistent negations. Embedding theorems of L_ω and FL_ω into propositional (first-order, respectively) classical logic are shown, and completeness theorems with respect to simple semantics for L_ω and FL_ω are proved. The cut-elimination theorems for L_ω and FL_ω are shown using both syntactical ways via the embedding theorems and semantical ways via the completeness theorems.

Part II: Variations of bi-intuitionistic logic.

Chapter 5: Paraconsistent logics extending bi-intuitionistic logic. In this chapter, a family of propositional logics with constructive implication, constructive co-implication and three kinds of negation is introduced. The logics extend bi-intuitionistic propositional logic, in which both intuitionistic negation and dual intuitionistic co-negation can be defined, by a primitive paraconsistent strong negation connective. A relational possible worlds semantics as well as sound and complete display sequent calculi for the logics under consideration are presented. Moreover, an extended inferentialist semantics is presented. This semantics generalizes the familiar Brouwer-Heyting-Kolmogorov (BHK) interpretation of intuitionistic logic in terms of canonical proofs and the slightly less well-known interpretation of N4 in terms of canonical proofs and disproofs. The logics under consideration are shown to be sound with respect to the inferentialist semantics. We will first present motivating considerations that focus on the notion of constructiveness and we shall later present motivating ideas based on considering the speech act of denial.

Chapter 6: Summetric and dual paraconsistent logics. Two new systems of first-order paraconsistent logic with De Morgan-type negations and co-implication, called symmetric paraconsistent logic (SPL) and dual paraconsistent logic (DPL), are introduced as Gentzen-type sequent calculi. The logic SPL is symmetric in the sense that the rule of contraposition is admissible in cut-free SPL. By using this symmetry property, a simpler cut-free sequent calculus for SPL is obtained. The logic DPL is not symmetric, but it has the duality principle known from classical logic. Simple semantics for SPL and DPL are introduced, and the completeness theorems with respect to these semantics are proved. The cut-elimination theorems for SPL and DPL are proved in two ways: One is a syntactical way which is based on the embedding theorems of SPL and DPL into Gentzen's LK, and the other is a semantical way which is based on the completeness theorems.

Part III: Paraconsistent temporal logics.

Chapter 7: Paraconsistent intuitionistic temporal logics. It is known that linear-time temporal logic (LTL), which is an extension of classical logic, is useful for expressing temporal reasoning as investigated in computer science. In this chapter, two constructive and bounded versions of LTL, which are extensions of intuitionistic logic or Nelson's paraconsistent logic, are introduced as Gentzen-type sequent calculi. These logics, IB[l] and PB[l], are intended to provide a useful theoretical basis for representing not only temporal (linear-time), but also constructive, and paraconsistent (inconsistency-tolerant) reasoning. The time domain of the proposed logics is bounded by a fixed positive integer. Despite the restriction on the time domain, the logics can derive almost all the typical temporal axioms of LTL. As a merit of bounding time, faithful embeddings into intuitionistic logic and Nelson's paraconsistent logic are shown for IB[l] and PB[l], respectively. Completeness (with respect to Kripke semantics), cut-elimination, normalization (with respect to natural deduction), and decidability theorems for the newly defined logics are proved as the main results of this chapter. Moreover, we present sound and complete display calculi for IB[l] and PB[l].

In [124] it has been emphasized that intuitionistic linear-time logic (ILTL) admits an elegant characterization of safety and liveness properties. The system ILTL, however, has been presented only in an algebraic setting. The present chapter is the first semantical *and* proof-theoretical study of bounded constructive linear-time temporal logics containing either intuitionistic or strong negation.

Chapter 8: Paraconsistent classical temporal logics. Inconsistency-tolerant reasoning and paraconsistent logic are of growing importance not only in Knowledge Representation, AI and other areas of Computer Science, but also in Philosophical Logic. In this chapter, a new logic, paraconsistent linear-time temporal logic (PLTL), is obtained semantically from the linear-time temporal logic LTL by adding a paraconsistent negation. Some theorems for embedding PLTL into LTL are proved, and PLTL is shown to be decidable. A Gentzen-type sequent calculus PLT_ω for PLTL is introduced, and the completeness and cut-elimination theorems for this calculus are proved. In addition, a display calculus $\delta\mathrm{PLT}_\omega$ for PLTL is defined.

Part IV: Paraconsistent substructural logics.

Chapter 9: Paraconsistent classical substructural logics. A general Gentzen-style framework for handling both bilattice (or strong) negation and usual negation is introduced based on the characterization of negation by a modal-like operator. This framework is regarded as an extension, generalization or refinement of not only bilattice logics and logics with strong negation, but also traditional logics including classical logic LK, classical modal logic S4 and classical linear logic CL. Cut-elimination theorems are proved for a variety of proposed sequent calculi including CLS (a conservative extension of CL) and $\mathrm{CLS_{cw}}$ (a conservative extension of some bilattice logics, LK and S4). Completeness theorems are given for these calculi with respect to phase semantics, for SLK (a conservative extension and fragment of LK and $\mathrm{CLS_{cw}}$, respectively) with respect to a classical-like semantics, and for SS4 (a conservative extension and fragment of S4 and $\mathrm{CLS_{cw}}$, respectively) with respect to a Kripke-type semantics. The proposed framework allows for an embedding of the proposed calculi into LK, S4 and CL.

Chapter 10: Paraconsistent intuitionistic linear logic. The extended intuitionistic linear logic with strong negation, investigated in [196, 197] and referred to as WILL in [82] and elsewhere, may be regarded as a resource-conscious refinement of Nelson's constructive logics with strong negation. In this chapter, (1) the completeness theorem with respect to phase semantics is proved for WILL using a method that simultaneously allows one to derive the cut-elimination theorem, (2) a simple correspondence between the class of Petri nets with inhibitor arcs and a fragment of WILL is obtained using a Kripke semantics, (3) a cut-free sequent calculus for WILL, called twist calculus, is presented, and (4) new applications of WILL in medical diagnosis and electric circuit theory are proposed. Strong negation in WILL is found to be expressible as a resource-conscious operation of refutability, and is shown to correspond to inhibitor arcs in Petri net theory.

Chapter 11: Paraconsistent logics based on involutive quantales. A new logic, quantized intuitionistic linear logic (QILL), is introduced. This logic is closely related to the logic which corresponds to Mulvey and Pelletier's (commutative) involutive quantales. Some cut-free sequent calculi with a new property "quantization principle" and some complete semantics such as an involutive quantale model and a quantale model are obtained for QILL. The relationship between QILL and the extended intuitionistic linear logic with strong negation WILL is also observed using such syntactical and semantical frameworks.

Chapter 12: Paraconsistent Lambek logics. In non-commutative substructural logics and in Categorial Grammar, a distinction is drawn between two implication connectives, often denoted by $\backslash$ and $/$, reflecting the linear order of syntactic expressions in most formal and natural languages. These implications satisfy two directional residuation laws with respect to a non-commutative, conjunction connective called "fusion" or "multiplicative conjunction". In this chapter, we introduce negated types into Categorial Grammar and obtain a form of *strong*, non-classical negation that turns the functor-type forming directional implications of Categorial Grammar into *connexive implications.* The introduction of negation into Categorial Grammar thereby provides additional motivation for systems of connexive logic presented along the lines of [204], [206]. The starting point of our considerations are powerset residuated groupoids and semigroups.

Part I.

Variations of Nelson's paraconsistent logic

2. Nelson's paraconsistent logic

The aim of this chapter is to obtain a theoretical foundation of inconsistency-tolerant (or paraconsistent) reasoning by presenting a uniform perspective on several familiar as well as new proof systems for paraconsistent reasoning. Inconsistency handling is of growing importance in Computer Science since inconsistencies may frequently occur in knowledge-based and intelligent information systems. Paraconsistent, inconsistency-tolerant logics have been studied to cope with such inconsistencies. In this chapter, proof systems for Nelson's paraconsistent logic N4 are comprehensively studied. The logic N4 is a fundamental system and known to be a common basis for various extended and useful paraconsistent logics. Some basic theorems including cut-elimination, normalization and completeness are uniformly proved using various embedding theorems. A variety of sequent calculi and natural deduction systems for N4 and some closely related systems are presented and compared.[1]

2.1. Introduction

2.1.1. Basic results

Some basic results on N4 are addressed in Section 2.2. The contents of Section 2.2 are summarized as follows.

Firstly, N4 and LJ are defined as some standard Gentzen-type sequent calculi, and some standard Kripke semantics for N4 and LJ are given. Two theorems for syntactically and semantically embedding N4 into LJ are presented based on the sequent calculi and the Kripke semantics, respectively. By using these embedding theorems, the cut-elimination and completeness theorems for N4 are proved uniformly. The embedding-based cut-elimination and completeness proof is a new contribution.

Secondly, relationships among N4, Rauszer's H-B logic and dual-intuitionistic logics are clarified using some sequent calculus based embedding theorems. To consider this issue, the logical connectives used are restricted to $\{\wedge, \vee\}$ or $\{\wedge, \vee, \sim\}$, i.e., the discussion is focused on the corresponding fragments of these logics. A sequent calculus DJ^-,

[1] This chapter includes some reformulated and refined results of some parts of the papers [90, 96, 97, 104, 209] and some standard, established material.

which is the $\{\wedge, \vee\}$-fragment of dual-intuitionistic logic, is presented, and a sequent calculus HB^-, which is theorem-equivalent to the $\{\wedge, \vee\}$-fragment of Rauszer's sequent calculus G1 [161] for H-B logic, is defined as $\mathrm{HB}^- = \mathrm{LJ}^- + \mathrm{DJ}^-$ where LJ^- is the $\{\wedge, \vee\}$-fragment of LJ. Some theorems for embedding $\mathrm{N4}^-$, the $\{\wedge, \vee, \sim\}$-fragment of N4, into DJ^- and for embedding HB^- into $\mathrm{N4}^-$ are proved.

2.1.2. Sequent calculi

Varieties of alternative sequent calculi for N4 and its fragments are presented in Section 2.3. The contents of Section 2.3 can be summarized as follows.

Firstly, a contraction-free system G4np for N4 is introduced by extending the positive fragment of the contraction-free system G4ip [188] for intuitionistic logic, and the structural rule elimination theorem for G4np is shown. The equivalence between G4np and N4 is also derived using this structural rule elimination theorem. Some backgrounds for contraction-free systems are briefly explained below. It is known that there are many cut-free sequent calculi for intuitionistic logic, such as Gentzen's LJ (containing also rules for intuitionistic negation) and its variants G1ip, G2ip, G3ip, G4ip and G5ip [188]. In particular, the contraction-free system G4ip has the useful feature that bottom-up proof search terminates without any loop-detection, and hence G4ip is known to be a convenient basis for automated theorem proving. A direct simple proof of the structural rule elimination theorem for G4ip was given by Dyckhoff and Negri [52].

Secondly, a resolution system Rnp for N4 is introduced by modifying a resolution system Rip for intuitionistic logic, and the equivalence between Rnp and an auxiliary system G5np is proved by using Troelstra and Schwichtenberg's method [188]. The system Rip was introduced by Troelstra and Schwichtenberg, and, as mentioned in [188], it can be regarded as a modification of Mints' original system RIp [127]. The system G5np presented here may be regarded as a modified extension of Troelstra and Schwichtenberg's system G5ip for intuitionistic logic.

Thirdly, a subformula calculus Sn4 and a dual calculus Dn4 are introduced. Sn4 has the subformula property, and Dn4 has dual-context sequents. The equivalence among Sn4, Dn4 and N4 is presented, and the cut-elimination theorems for Sn4 and Dn4 are shown.

Fourthly, a display calculus $\delta\mathrm{N4}^{\perp}$ for the logic $\mathrm{N4}^{\perp}$, which conservatively extends N4 by a falsity constant $\perp$, and a display calculus $\delta\mathrm{HB}$ for Heyting-Brouwer logic are presented.

Finally, some sequent calculi for the $\rightarrow$-free fragment of N4, which is equivalent to Belnap and Dunn's four-valued logic, and a sequent calculus for Arieli and Avron's bilattice logic, which is an extension of N4, are reviewed.

2.1.3. Natural deduction

Varieties of natural deduction systems for N4 are presented in Section 2.4. The contents of Section 2.4 may be summarized as follows.

Firstly, a standard type natural deduction system N_{N4} for N4 is introduced based on the idea of Priest [154]. N_{N4} is obtained from Priest's system for the logic E_{fde} of first-degree entailment by adding the inference rules that correspond to some axiom schemes with respect to the implication and negation connectives. N_{N4} is also obtained from the well-known natural deduction system for positive intuitionistic logic by adding some inference rules for $\sim$. The (weak) normalization theorem for N_{N4} is shown via a correspondence between N_{N4} and an alternative cut-free sequent calculus S_{N4} for N4.

Secondly, a sequent calculus L_{N4} in natural deduction style and a natural deduction system G_{N4} in sequent calculus style are introduced based on the framework by Negri and von Plato [132]. The (general) normalization theorem for G_{N4} is shown. G_{N4} is an extension of the original system by Negri and von Plato for (positive) intuitionistic logic. Schroeder-Heister's idea [173] to use some general elimination rules for strongly negated formulas is applied in the negation part of G_{N4}. The framework by Negri and von Plato is known to be an alternative intermediate framework between natural deduction and sequent calculus.

Thirdly, the following systems are addressed: Prawitz's system [153], here called P_{N4}, a special case of Schroeder-Heister's formulation, here called H_{N4}, two extensions of Negri and von Plato's uniform calculi [133], called here U_{N4} and V_{N4}, and an extension of Tamminga and Tanaka's system [184] for E_{fde}, here called T_{N4}.

Some historical backgrounds for natural deduction systems for N4 are briefly explained below. A typed λ-calculus λ^c for N4 was introduced by Wansing in [196, 195], where a Curry-Howard correspondence (with respect to λ^c and N4) and the completeness theorem (for λ^c) with respect to an extended version of Friedman's full type structures over infinite sets is proved. It is known that Prawitz's natural deduction system P_{N4} [153] has some inference rules with respect to the paraconsistent negation axioms which correspond to the axiom schemes $\sim(\alpha \rightarrow \beta) \leftrightarrow \alpha \wedge \sim\beta$ and $\sim(\alpha \wedge \beta) \leftrightarrow \sim\alpha \vee \sim\beta$.

For the sake of surveyability, the various proof systems to be dealt with are listed in Table 2.1.

Section 1.1.1	HN4	axiom system for Nelson's four-valued logic
Section 1.2.1	N4	sequent calculus for Nelson's four-valued logic
	LJ	sequent calculus for positive intuitionistic logic
Section 1.2.3	DJ^-	sequent calculus for the $\{\wedge, \vee\}$-fragment of dual intuitionistic logic
	LJ^-	$\{\wedge, \vee\}$-fragment of LJ
	HB^-	$\text{DJ}^- + \text{IJ}^-$
	N4^-	$\{\wedge, \vee, \sim\}$-fragment of N4
Section 1.3.1	G4np	contraction free sequent calculus for N4
	$\text{G4ip}^{\perp}$	$\perp$-free fragment of the intuitionistic sequent calculus G4ip
Section 1.3.2	G5np	auxiliary sequent calculus for N4
	Rnp	resolution calculus for N4
Section 1.3.3	Sn4	subformula sequent calculus for N4
	Dn4	dual sequent calculus for N4
Section 1.3.4	$\text{HN4}^{\perp}$	axiom system for N4 extended by $\perp$
	$\delta\text{N4}^{\perp}$	display calculus for $\text{N4}^{\perp}$
	G1	non-cut-free sequent calculus for H-B
	δHB	cut-free display sequent calculus for H-B
Section 1.3.5	$\mathcal{G}_B$	sequent calculus for first-degree entailment logic
	$\mathcal{G}_{FV}$	sequent calculus for first-degree entailment logic
	LE_{fde1}	sequent calculus for first-degree entailment logic
	LE_{fde2}	sequent calculus for first-degree entailment logic
	BL	sequent calculus for bilattice logic
Section 1.4.1	N_{N4}	natural deduction system for N4
	S_{N4}	alternative sequent calculus for N4
Section 1.4.2	L_{N4}	natural deduction style sequent calculus for N4
	L_{LJ}	natural deduction style sequent calculus for positive intuitionistic logic
	G_{N4}	sequent style natural deduction calculus for N4
Section 1.4.3	P_{N4}	Prawitz's natural deduction system for N4
	H_{N4}	another natural deduction system for N4
	U_{N4}	another natural deduction system for N4
	V_{N4}	another natural deduction system for N4
	T_{N4}	another natural deduction system for N4

Table 2.1.: A list of proof systems.

2.2. Basic results

2.2.1. Sequent calculus and cut-elimination

The usual propositional language with $\rightarrow$ (implication), $\wedge$ (conjunction), $\vee$ (disjunction) and $\sim$ (paraconsistent negation) is used in this section. Lower-case letters $p, q, r, ...$ are used to denote propositional variables, Greek lower-case letters $\alpha, \beta, \gamma, ...$ are used to denote formulas, and Greek capital letters $\Gamma, \Delta, ...$ are used to represent finite (possibly empty) multisets of formulas. We use $\alpha \leftrightarrow \beta$ as an abbreviation of $(\alpha \to \beta) \wedge (\beta \to \alpha)$. A *sequent* is an expression of the form $\Gamma \Rightarrow \gamma$. If a sequent S is provable in a sequent calculus L, then such a fact is denoted as $L \vdash S$ or $\vdash S$. A rule of inference R is said to be *admissible* in a sequent calculus L if the following condition is satisfied: for any instance

$$\frac{S_1 \cdots S_n}{S}$$

of R, if $L \vdash S_i$ for all i, then $L \vdash S$.

Definition 2.2.1 (N4 [119, 148, 196]) *The initial sequents of* N4 *are of the form: for any propositional variable* p,

$$p \Rightarrow p \qquad\qquad {\sim}p \Rightarrow {\sim}p.$$

The structural inference rules of N4 *are of the form:*

$$\frac{\Gamma \Rightarrow \alpha \quad \alpha, \Sigma \Rightarrow \gamma}{\Gamma, \Sigma \Rightarrow \gamma}\ (\text{cut}) \quad \frac{\Gamma \Rightarrow \gamma}{\alpha, \Gamma \Rightarrow \gamma}\ (\text{we}) \quad \frac{\alpha, \alpha, \Gamma \Rightarrow \gamma}{\alpha, \Gamma \Rightarrow \gamma}\ (\text{co}).$$

The logical inference rules of N4 *are of the form:*

$$\frac{\Gamma \Rightarrow \alpha \quad \beta, \Delta \Rightarrow \gamma}{\alpha{\to}\beta, \Gamma, \Delta \Rightarrow \gamma}\ ({\to}\text{l}) \quad \frac{\alpha, \Gamma \Rightarrow \beta}{\Gamma \Rightarrow \alpha{\to}\beta}\ ({\to}\text{r})$$

$$\frac{\alpha, \beta, \Gamma \Rightarrow \gamma}{\alpha \wedge \beta, \Gamma \Rightarrow \gamma}\ (\wedge\text{l}) \quad \frac{\Gamma \Rightarrow \alpha \quad \Gamma \Rightarrow \beta}{\Gamma \Rightarrow \alpha \wedge \beta}\ (\wedge\text{r})$$

$$\frac{\alpha, \Gamma \Rightarrow \gamma \quad \beta, \Gamma \Rightarrow \gamma}{\alpha \vee \beta, \Gamma \Rightarrow \gamma}\ (\vee\text{l}) \quad \frac{\Gamma \Rightarrow \alpha}{\Gamma \Rightarrow \alpha \vee \beta}\ (\vee\text{r1}) \quad \frac{\Gamma \Rightarrow \beta}{\Gamma \Rightarrow \alpha \vee \beta}\ (\vee\text{r2})$$

$$\frac{\alpha, \Gamma \Rightarrow \gamma}{{\sim}{\sim}\alpha, \Gamma \Rightarrow \gamma}\ ({\sim}{\sim}\text{l}) \quad \frac{\Gamma \Rightarrow \alpha}{\Gamma \Rightarrow {\sim}{\sim}\alpha}\ ({\sim}{\sim}\text{r})$$

$$\frac{\alpha, {\sim}\beta, \Gamma \Rightarrow \gamma}{{\sim}(\alpha{\to}\beta), \Gamma \Rightarrow \gamma}\ ({\sim}{\to}\text{l}) \quad \frac{\Gamma \Rightarrow \alpha \quad \Gamma \Rightarrow {\sim}\beta}{\Gamma \Rightarrow {\sim}(\alpha{\to}\beta)}\ ({\sim}{\to}\text{r})$$

$$\frac{{\sim}\alpha, \Gamma \Rightarrow \gamma \quad {\sim}\beta, \Gamma \Rightarrow \gamma}{{\sim}(\alpha \wedge \beta), \Gamma \Rightarrow \gamma}\ ({\sim}\wedge\text{l})$$

$$\frac{\Gamma \Rightarrow {\sim}\alpha}{\Gamma \Rightarrow {\sim}(\alpha \wedge \beta)}\ ({\sim}\wedge\text{r1}) \quad \frac{\Gamma \Rightarrow {\sim}\beta}{\Gamma \Rightarrow {\sim}(\alpha \wedge \beta)}\ ({\sim}\wedge\text{r2})$$

$$\frac{{\sim}\alpha, {\sim}\beta, \Gamma \Rightarrow \gamma}{{\sim}(\alpha \vee \beta), \Gamma \Rightarrow \gamma}\ ({\sim}\vee\text{l1}) \quad \frac{\Gamma \Rightarrow {\sim}\alpha \quad \Gamma \Rightarrow {\sim}\beta}{\Gamma \Rightarrow {\sim}(\alpha \vee \beta)}\ ({\sim}\vee\text{r}).$$

Definition 2.2.2 (LJ) *A sequent calculus* LJ *for positive intuitionistic logic is defined as the* $\sim$*-free fragment of* N4, *i.e.,* LJ *is obtained from* N4 *by deleting all the initial sequents of the form* ${\sim}p \Rightarrow {\sim}p$ *for any propositional variable* p *and deleting all the logical inference rules concerning* $\sim$.

The sequents of the form $\alpha \Rightarrow \alpha$ for any formula α are provable in cut-free LJ and cut-free N4. This fact is proved by induction on the complexity of α.

Definition 2.2.3 *We fix a set Φ of propositional variables and define the set $\Phi' := \{p' \mid p \in \Phi\}$ of propositional variables. The language $\mathcal{L}_{\mathrm{N4}}$ of* N4 *is defined using Φ, $\rightarrow, \wedge, \vee$ and $\sim$. The language $\mathcal{L}_{\mathrm{LJ}}$ of* LJ *is obtained from $\mathcal{L}_{\mathrm{N4}}$ by adding Φ' and deleting $\sim$.*

A mapping f from $\mathcal{L}_{\mathrm{N4}}$ to $\mathcal{L}_{\mathrm{LJ}}$ is defined inductively by:

1. *for any $p \in \Phi$, $f(p) := p$ and $f(\sim p) := p' \in \Phi'$,*
2. *$f(\alpha \circ \beta) := f(\alpha) \circ f(\beta)$ where $\circ \in \{\rightarrow, \wedge, \vee\}$,*
3. *$f(\sim\sim\alpha) := f(\alpha)$,*
4. *$f(\sim(\alpha\rightarrow\beta)) := f(\alpha) \wedge f(\sim\beta)$,*
5. *$f(\sim(\alpha \wedge \beta)) := f(\sim\alpha) \vee f(\sim\beta)$,*
6. *$f(\sim(\alpha \vee \beta)) := f(\sim\alpha) \wedge f(\sim\beta)$.*

An expression $f(\Gamma)$ denotes the result of replacing every occurrence of a formula α in Γ by an occurrence of $f(\alpha)$.

Theorem 2.2.4 (Syntactical embedding) *Let Γ be a multiset of formulas in $\mathcal{L}_{\mathrm{N4}}$, γ be a formula in $\mathcal{L}_{\mathrm{N4}}$, and f be the mapping defined in Definition 2.2.3. Then:*

1. *$\mathrm{N4} \vdash \Gamma \Rightarrow \gamma$ iff $\mathrm{LJ} \vdash f(\Gamma) \Rightarrow f(\gamma)$,*
2. *$\mathrm{N4} - (\mathrm{cut}) \vdash \Gamma \Rightarrow \gamma$ iff $\mathrm{LJ} - (\mathrm{cut}) \vdash f(\Gamma) \Rightarrow f(\gamma)$.*

Proof. Since the claim (2) can be obtained from a part of the proof of (1), we show only (1) in the following.

• ($\Longrightarrow$) : By induction on the proofs P of $\Gamma \Rightarrow \gamma$ in N4. We distinguish the cases according to the last inference of P, and show some cases.

Case ($\sim p \Rightarrow \sim p$): The last inference of P is of the form: $\sim p \Rightarrow \sim p$ for a propositional variable p. In this case, we obtain the required fact LJ $\vdash f(\sim p) \Rightarrow f(\sim p)$ since $f(\sim p)$ coincides with $p' \in \Phi'$ by the definition of f.

Case ($\sim\rightarrow$r): The last inference of P is of the form:

$$\frac{\Gamma \Rightarrow \alpha \quad \Gamma \Rightarrow \sim\beta}{\Gamma \Rightarrow \sim(\alpha\rightarrow\beta)}\ (\sim\rightarrow\mathrm{r}).$$

By induction hypothesis, we have: LJ $\vdash f(\Gamma) \Rightarrow f(\alpha)$ and LJ $\vdash f(\Gamma) \Rightarrow f({\sim}\beta)$. Then, we obtain the required fact:

$$\dfrac{\begin{array}{c}\vdots\\ f(\Gamma) \Rightarrow f(\alpha)\end{array} \quad \begin{array}{c}\vdots\\ f(\Gamma) \Rightarrow f({\sim}\beta)\end{array}}{f(\Gamma) \Rightarrow f(\alpha) \wedge f({\sim}\beta)}\ (\wedge\mathrm{r})$$

where $f(\alpha) \wedge f({\sim}\beta)$ coincides with $f({\sim}(\alpha{\to}\beta))$ by the definition of f.

Case $({\sim}{\sim}\mathrm{l})$: The last inference of P is of the form:

$$\frac{\alpha, \Sigma \Rightarrow \gamma}{{\sim}{\sim}\alpha, \Sigma \Rightarrow \gamma}\ ({\sim}{\sim}\mathrm{l}).$$

By induction hypothesis, we have: LJ $\vdash f(\alpha), f(\Sigma) \Rightarrow f(\gamma)$, and hence obtain: LJ $\vdash f({\sim}{\sim}\alpha), f(\Sigma) \Rightarrow f(\gamma)$ where $f({\sim}{\sim}\alpha)$ coincides with $f(\alpha)$ by the definition of f.

Case $({\sim}\wedge\mathrm{l})$: The last inference of P is of the form:

$$\frac{{\sim}\alpha, \Sigma \Rightarrow \gamma \quad {\sim}\beta, \Sigma \Rightarrow \gamma}{{\sim}(\alpha \wedge \beta), \Sigma \Rightarrow \gamma}\ ({\sim} \wedge \mathrm{l}).$$

By induction hypothesis, we have: LJ $\vdash f({\sim}\alpha), f(\Sigma) \Rightarrow f(\gamma)$ and LJ $\vdash f({\sim}\beta), f(\Sigma) \Rightarrow f(\gamma)$. Thus, we obtain the required fact:

$$\dfrac{\begin{array}{c}\vdots\\ f({\sim}\alpha), f(\Sigma) \Rightarrow f(\gamma)\end{array} \quad \begin{array}{c}\vdots\\ f({\sim}\beta), f(\Sigma) \Rightarrow f(\gamma)\end{array}}{f({\sim}\alpha) \vee f({\sim}\beta), f(\Sigma) \Rightarrow f(\gamma)}\ (\vee\mathrm{l})$$

where $f({\sim}\alpha) \vee f({\sim}\alpha)$ coincides with $f({\sim}(\alpha \wedge \beta))$ by the definition of f.

• $(\Longleftarrow)$: By induction on the proofs Q of $f(\Gamma) \Rightarrow f(\gamma)$ in LJ. We distinguish the cases according to the last inference of Q, and show some cases.

Case (cut): The last inference of Q is of the form:

$$\frac{f(\Gamma_1) \Rightarrow \beta \quad \beta, f(\Gamma_2) \Rightarrow f(\gamma)}{f(\Gamma_1), f(\Gamma_2) \Rightarrow f(\gamma)}\ (\mathrm{cut}).$$

Since β is in $\mathcal{L}_{\mathrm{LJ}}$, we have the fact $\beta = f(\beta)$. This fact can be shown by induction on β. Then, by induction hypothesis, we have: N4 $\vdash \Gamma_1 \Rightarrow \beta$ and N4 $\vdash \beta, \Gamma_2 \Rightarrow \gamma$. We then obtain the required fact: N4 $\vdash \Gamma_1, \Gamma_2 \Rightarrow \gamma$ by using (cut) in N4.

Case $(\vee\mathrm{l})$: The last inference rule of Q is of the form:

$$\frac{f({\sim}\alpha), f(\Sigma) \Rightarrow f(\gamma) \quad f({\sim}\beta), f(\Sigma) \Rightarrow f(\gamma)}{f({\sim}\alpha) \vee f({\sim}\beta), f(\Sigma) \Rightarrow f(\gamma)}\ (\vee\mathrm{l})$$

where $f(\sim\alpha) \vee f(\sim\beta)$ coincides with $f(\sim(\alpha \wedge \beta))$ by the definition of f. By induction hypothesis, we have: N4 $\vdash \sim\alpha, \Sigma \Rightarrow \gamma$ and N4 $\vdash \sim\beta, \Sigma \Rightarrow \gamma$. Thus, we obtain the required fact:

$$\dfrac{\begin{array}{c}\vdots \\ \sim\alpha, \Sigma \Rightarrow \gamma\end{array} \quad \begin{array}{c}\vdots \\ \sim\beta, \Sigma \Rightarrow \gamma\end{array}}{\sim(\alpha \wedge \beta), \Sigma \Rightarrow \gamma}\ (\sim\wedge 1).$$

∎

Theorem 2.2.5 (Cut-elimination) *The rule* (cut) *is admissible in cut-free* N4.

Proof. Suppose N4 $\vdash \Gamma \Rightarrow \gamma$. Then, we have: LJ $\vdash f(\Gamma) \Rightarrow f(\gamma)$ by Theorem 2.2.4 (1), and hence obtain: LJ $-$ (cut) $\vdash f(\Gamma) \Rightarrow f(\gamma)$ by the well-known cut-elimination theorem for LJ. By Theorem 2.2.4 (2), we obtain the required fact: N4 $-$ (cut) $\vdash \Gamma \Rightarrow \gamma$. ∎

Theorem 2.2.6 (Decidability) N4 *is decidable.*

Proof. By the decidability of LJ, for each α, it is possible to decide whether $f(\alpha)$ is provable in LJ. Then, by Theorem 2.2.4, N4 is decidable. ∎

By Theorem 2.2.5, we can obtain the following properties of constructible falsity and paraconsistency.

Proposition 2.2.7 (Constructible falsity) *For any formulas α and β, if* N4 $\vdash\ \Rightarrow \sim(\alpha \wedge \beta)$*, then either* N4 $\vdash\ \Rightarrow \sim\alpha$ *or* N4 $\vdash\ \Rightarrow \sim\beta$.

Definition 2.2.8 *Let $\sharp$ be a unary connective and L be a sequent calculus L. L is called* explosive *with respect to $\sharp$ if $L \vdash \alpha, \sharp\alpha \Rightarrow \beta$ for any formulas α and β. L is called* paraconsistent *with respect to $\sharp$ if it is not explosive with respect to $\sharp$.*

Proposition 2.2.9 (Paraconsistency) N4 *is paraconsistent with respect to $\sim$.*

2.2.2. Kripke semantics and completeness

Definition 2.2.10 *A* Kripke frame *is a structure $\langle M, R\rangle$ satisfying the following conditions:*

1. *M is a nonempty set,*

2. *R is a reflexive and transitive binary relation on M.*

Definition 2.2.11 *A* valuation $\models$ *on a Kripke frame* $\langle M, R\rangle$ *is a mapping from the set* Φ *of propositional variables to the power set* 2^M *of* M *such that for any* $p \in \Phi$ *and any* $x, y \in M$, *if* $x \in \models (p)$ *and* xRy, *then* $y \in \models (p)$. *We will write* $x \models p$ *for* $x \in \models (p)$. *This valuation* $\models$ *is extended to a mapping from the set of all formulas to* 2^M *by:*

1. $x \models \alpha \to \beta$ *iff* $\forall y \in M$ $[xRy$ *and* $y \models \alpha$ *imply* $y \models \beta]$,

2. $x \models \alpha \wedge \beta$ *iff* $x \models \alpha$ *and* $x \models \beta$,

3. $x \models \alpha \vee \beta$ *iff* $x \models \alpha$ *or* $x \models \beta$.

The following *heredity condition* holds for $\models$: for any formula α and any $x, y \in M$, if $x \models \alpha$ and xRy, then $y \models \alpha$. This is proved by induction on α.

Definition 2.2.12 *A* Kripke model *is a structure* $\langle M, R, \models\rangle$ *such that*

1. $\langle M, R\rangle$ *is a Kripke frame,*

2. $\models$ *is a valuation on* $\langle M, R\rangle$.

A formula α *is* true *in a Kripke model* $\langle M, R, \models\rangle$ *if* $x \models \alpha$ *for any* $x \in M$, *and is* LJ-valid *in a Kripke frame* $\langle M, R\rangle$ *if it is true for every valuation* $\models$ *on the Kripke frame.*

Definition 2.2.13 Paraconsistent valuations $\models^+$ *and* $\models^-$ *on a Kripke frame* $\langle M, R\rangle$ *are mappings from the set* Φ *of propositional variables to the power set* 2^M *of* M *such that for any* $\star \in \{+, -\}$, *any* $p \in \Phi$ *and any* $x, y \in M$, *if* $x \in \models^\star (p)$ *and* xRy, *then* $y \in \models^\star (p)$. *We will write* $x \models^\star p$ *for* $x \in \models^\star (p)$. *These paraconsistent valuations* $\models^+$ *and* $\models^-$ *are extended to mappings from the set of all formulas to* 2^M *by:*

1. $x \models^+ \alpha \to \beta$ *iff* $\forall y \in M$ $[xRy$ *and* $y \models^+ \alpha$ *imply* $y \models^+ \beta]$,

2. $x \models^+ \alpha \wedge \beta$ *iff* $x \models^+ \alpha$ *and* $x \models^+ \beta$,

3. $x \models^+ \alpha \vee \beta$ *iff* $x \models^+ \alpha$ *or* $x \models^+ \beta$,

4. $x \models^+ {\sim}\alpha$ *iff* $x \models^- \alpha$,

5. $x \models^- {\sim}\alpha$ *iff* $x \models^+ \alpha$,

6. $x \models^- \alpha \rightarrow \beta$ *iff* $x \models^+ \alpha$ *and* $x \models^- \beta$,

7. $x \models^- \alpha \wedge \beta$ *iff* $x \models^- \alpha$ *or* $x \models^- \beta$,

8. $x \models^- \alpha \vee \beta$ *iff* $x \models^- \alpha$ *and* $x \models^- \beta$.

The heredity condition holds for $\models^+$ and $\models^-$.

Definition 2.2.14 *A* paraconsistent Kripke model *is a structure* $\langle M, R, \models^+, \models^- \rangle$ *such that*

1. $\langle M, R \rangle$ *is a Kripke frame,*

2. $\models^+$ *and* $\models^-$ *are paraconsistent valuations on* $\langle M, R \rangle$.

A formula α *is* true *in a paraconsistent Kripke model* $\langle M, R, \models^+, \models^- \rangle$ *if* $x \models^+ \alpha$ *for any* $x \in M$*, and is* N4-valid *in a Kripke frame* $\langle M, R \rangle$ *if it is true for every paraconsistent valuations* $\models^+$ *and* $\models^-$ *on the Kripke frame.*

Lemma 2.2.15 *Let* f *be the mapping defined in Definition 2.2.3. For any paraconsistent Kripke model* $\langle M, R, \models^+, \models^- \rangle$*, we can construct a Kripke model* $\langle M, R, \models \rangle$ *such that for any formula* $\alpha \in \mathcal{L}_{\mathrm{N4}}$ *and any* $x \in M$*,*

1. $x \models^+ \alpha$ *iff* $x \models f(\alpha)$,

2. $x \models^- \alpha$ *iff* $x \models f({\sim}\alpha)$.

Proof. Let Φ be a set of propositional variables and Φ' be the set $\{p' \mid p \in \Phi\}$ of propositional variables. Suppose that $\langle M, R, \models^+, \models^- \rangle$ is a paraconsistent Kripke model where $\models^+$ and $\models^-$ are mappings from Φ to the power set 2^M of M, and that the heredity condition w.r.t. $p \in \Phi$ holds for $\models^+$ and $\models^-$. Suppose that $\langle M, R, \models \rangle$ is a Kripke model where $\models$ is a mapping from $\Phi \cup \Phi'$ to 2^M, and that the heredity condition w.r.t. $p \in \Phi \cup \Phi'$ holds for $\models$. Suppose moreover that these models satisfy the following conditions: for any $x \in M$ and any $p \in \Phi$,

1. $x \models^+ p$ iff $x \models p$,

2. $x \models^- p$ iff $x \models p'$.

Then, the claim is proved by (simultaneous) induction on the complexity of α. We use the symbol $\equiv$ to denote syntactical identity.

- Base step:

Case $\alpha \equiv p \in \Phi$: For (1), we obtain: $x \models^{+} p$ iff $x \models p$ iff $x \models f(p)$ (by the definition of f). For (2), we obtain: $x \models^{-} p$ iff $x \models p'$ iff $x \models f({\sim}p)$ (by the definition of f).

• Induction step:

Case $\alpha \equiv \beta \wedge \gamma$: For (1), we obtain: $x \models^{+} \beta \wedge \gamma$ iff $x \models^{+} \beta$ and $x \models^{+} \gamma$ iff $x \models f(\beta)$ and $x \models f(\gamma)$ (by induction hypothesis for 1) iff $x \models f(\beta) \wedge f(\gamma)$ iff $x \models f(\beta \wedge \gamma)$ (by the definition of f). For (2), we obtain: $x \models^{-} \beta \wedge \gamma$ iff $x \models^{-} \beta$ or $x \models^{-} \gamma$ iff $x \models f({\sim}\beta)$ or $x \models f({\sim}\gamma)$ (by induction hypothesis for 2) iff $x \models f({\sim}\beta) \vee f({\sim}\gamma)$ iff $x \models f({\sim}(\beta \wedge \gamma))$ (by the definition of f).

Case $\alpha \equiv \beta \vee \gamma$: Similar to the above case.

Case $\alpha \equiv \beta \rightarrow \gamma$: For (1), we obtain: $x \models^{+} \beta \rightarrow \gamma$ iff $\forall y \in M[xRy$ and $y \models^{+} \beta$ imply $y \models^{+} \gamma]$ iff $\forall y \in M[xRy$ and $y \models f(\beta)$ imply $y \models f(\gamma)]$ (by induction hypothesis for 1) iff $x \models f(\beta) \rightarrow f(\gamma)$ iff $x \models f(\beta \rightarrow \gamma)$ (by the definition of f). For (2), we obtain: $x \models^{-} \beta \rightarrow \gamma$ iff $x \models^{+} \beta$ and $x \models^{-} \gamma$ iff $x \models f(\beta)$ and $x \models f({\sim}\gamma)$ (by induction hypothesis for 1 and 2) iff $x \models f(\beta) \wedge f({\sim}\gamma)$ iff $x \models f({\sim}(\beta \rightarrow \gamma))$ (by the definition of f).

Case $\alpha \equiv {\sim}\beta$: For (1), we obtain: $x \models^{+} {\sim}\beta$ iff $x \models^{-} \beta$ iff $x \models f({\sim}\beta)$ (by induction hypothesis for 2). For (2), we obtain: $x \models^{-} {\sim}\beta$ iff $x \models^{+} \beta$ iff $x \models f(\beta)$ (by induction hypothesis for 1) iff $x \models f({\sim}{\sim}\beta)$ (by the definition of f). ∎

Lemma 2.2.16 *Let f be the mapping defined in Definition 2.2.3. For any Kripke model $\langle M, R, \models \rangle$, we can construct a paraconsistent Kripke model $\langle M, R, \models^{+}, \models^{-} \rangle$ such that for any formula α and any $x \in M$,*

1. $x \models f(\alpha)$ *iff* $x \models^{+} \alpha$,
2. $x \models f({\sim}\alpha)$ *iff* $x \models^{-} \alpha$.

Proof. Similar to the proof of Lemma 2.2.15. ∎

Theorem 2.2.17 (Semantical embedding) *Let f be the mapping defined in Definition 2.2.3. For any formula α,*

α *is N4-valid iff* $f(\alpha)$ *is LJ-valid.*

Proof. By Lemmas 2.2.15 and 2.2.16. ∎

Theorem 2.2.18 (Completeness) *For any formula α,*

N4 $\vdash \Rightarrow \alpha$ *iff* α *is N4-valid.*

Proof. N4 $\vdash \Rightarrow \alpha$ iff LJ $\vdash \Rightarrow f(\alpha)$ (by Theorem 2.2.4) iff $f(\alpha)$ is LJ-valid (by the well-known Kripke completeness theorem for LJ) iff α is N4-valid (by Theorem 2.2.17). ∎

2.2.3. Relation to other logics

The logical connectives used below are restricted to $\{\wedge, \vee\}$ or $\{\wedge, \vee, \sim\}$. The following discussion is focused on the corresponding fragments of N4, dual intuitionistic logic and H-B logic.

A sequent calculus DJ^- for the $\{\wedge, \vee\}$-fragment of dual-intuitionistic logic is introduced below. A sequent of DJ^- is an expression of the form $\gamma \Rightarrow \Gamma$.

Definition 2.2.19 ($\mathbf{DJ}^-$) *The initial sequents of* DJ^- *are of the form: for any propositional variable p,*

$$p \Rightarrow p.$$

The structural rules of DJ^- *are of the form:*

$$\frac{\gamma \Rightarrow \Gamma, \alpha \quad \alpha \Rightarrow \Delta}{\gamma \Rightarrow \Gamma, \Delta}\ (\text{cut-d}) \qquad \frac{\gamma \Rightarrow \Gamma, \alpha, \alpha}{\gamma \Rightarrow \Gamma, \alpha}\ (\text{co-d}) \qquad \frac{\gamma \Rightarrow \Gamma}{\gamma \Rightarrow \Gamma, \alpha}\ (\text{wk-d}).$$

The logical inference rules of DJ^- *are of the form:*

$$\frac{\alpha \Rightarrow \Gamma}{\alpha \wedge \beta \Rightarrow \Gamma}\ (\wedge\text{l1-d}) \qquad \frac{\beta \Rightarrow \Gamma}{\alpha \wedge \beta \Rightarrow \Gamma}\ (\wedge\text{l2-d}) \qquad \frac{\gamma \Rightarrow \Gamma, \alpha \quad \gamma \Rightarrow \Gamma, \beta}{\gamma \Rightarrow \Gamma, \alpha \wedge \beta}\ (\wedge\text{r-d})$$

$$\frac{\alpha \Rightarrow \Gamma \quad \beta \Rightarrow \Gamma}{\alpha \vee \beta \Rightarrow \Gamma}\ (\vee\text{l-d}) \qquad \frac{\gamma \Rightarrow \Gamma, \alpha}{\gamma \Rightarrow \Gamma, \alpha \vee \beta}\ (\vee\text{r1-d}) \qquad \frac{\gamma \Rightarrow \Gamma, \beta}{\gamma \Rightarrow \Gamma, \alpha \vee \beta}\ (\vee\text{r2-d}).$$

Definition 2.2.20 ($\mathbf{N4}^-$) *The sequent calculus* $\mathrm{N4}^-$ *is defined as the* $\{\wedge, \vee, \sim\}$*-fragment of* N4.

Definition 2.2.21 *We fix a set* Φ *of propositional variables and define the set* $\Phi' := \{p' \mid p \in \Phi\}$ *of propositional variables. The language* $\mathcal{L}_{\mathrm{N4}^-}$ *of* $\mathrm{N4}^-$ *is defined using* Φ, $\wedge, \vee$ *and* $\sim$. *The language* $\mathcal{L}_{\mathrm{DJ}^-}$ *of* DJ^- *is obtained from* $\mathcal{L}_{\mathrm{N4}^-}$ *by adding* Φ' *and deleting* $\sim$.

A mapping e from $\mathcal{L}_{\mathrm{N4}^-}$ *to* $\mathcal{L}_{\mathrm{DJ}^-}$ *is defined as follows.*

1. $e(p) := p$ *and* $e({\sim}p) := p' \in \Phi'$ *for any* $p \in \Phi$,
2. $e(\alpha \wedge \beta) := e(\alpha) \vee e(\beta)$,
3. $e(\alpha \vee \beta) := e(\alpha) \wedge e(\beta)$,
4. $e({\sim}{\sim}\alpha) := e(\alpha)$,
5. $e({\sim}(\alpha \circ \beta)) := e({\sim}\alpha) \circ e({\sim}\beta)$ *where* $\circ \in \{\wedge, \vee\}$.

Theorem 2.2.22 (Syntactical embedding) *Let* Γ *be a multiset of formulas in* $\mathcal{L}_{\mathrm{N4}^-}$, γ *be a formula in* $\mathcal{L}_{\mathrm{N4}^-}$, *and* e *be the mapping defined in Definition 2.2.21. Then:*

1. $\mathrm{N4}^- \vdash \Gamma \Rightarrow \gamma$ *iff* $\mathrm{DJ}^- \vdash e(\gamma) \Rightarrow e(\Gamma)$,
2. $\mathrm{N4}^- - (\mathrm{cut}) \vdash \Gamma \Rightarrow \gamma$ *iff* $\mathrm{DJ}^- - (\mathrm{cut\text{-}d}) \vdash e(\gamma) \Rightarrow e(\Gamma)$.

Proof. Similar to the proof of Theorem 2.2.4. ∎

Theorem 2.2.23 (Cut-elimination) *The rule* (cut-d) *is admissible in cut-free* DJ^-.

Proof. By Theorem 2.2.22 and the cut-elimination theorem for $\mathrm{N4}^-$. ∎

A sequent calculus HB^- for the $\{\wedge, \vee\}$-fragment of H-B logic is introduced below. A sequent of HB^- is an expression of the form $\Sigma \Rightarrow \Delta$ where Σ and Δ cannot have more than one element simultaneously, or equivalently, are of the form either $\Sigma \Rightarrow \delta$ or $\sigma \Rightarrow \Delta$.

Definition 2.2.24 (H-B logic) *Let* LJ^- *be the* $\{\wedge, \vee\}$*-fragment of* LJ. *Then,* HB^- *is defined as*

$$\mathrm{HB}^- = \mathrm{DJ}^- + \mathrm{LJ}^-.$$

Note that the original sequent calculus G1 [161] for H-B logic adopts the one-element restriction of sequents, and uses the restricted versions of $\wedge$-right and $\vee$-left introduction rules:

$$\frac{\Gamma \Rightarrow \Delta, \alpha \quad \Gamma \Rightarrow \Delta, \beta}{\Gamma \Rightarrow \Delta, \alpha \wedge \beta}\ (\wedge \mathrm{r}') \qquad \frac{\alpha, \Gamma \Rightarrow \Delta \quad \beta, \Gamma \Rightarrow \Delta}{\alpha \vee \beta, \Gamma \Rightarrow \Delta}\ (\vee \mathrm{l}')$$

where the antecedent and the succedent of the sequents of these rules cannot be multisets of more than one-element simultaneously. Also note that G1 adopts the corresponding restricted right-weakening rule with the one-element condition. In HB^-, the original version $(\wedge\mathrm{r}')$ (and $(\vee\mathrm{l}')$) is divided into the present versions $(\wedge\mathrm{r})$ and $(\wedge\mathrm{r\text{-}d})$ ($(\vee\mathrm{l})$ and $(\vee\mathrm{l\text{-}d})$, respectively). The one-element condition for $(\wedge\mathrm{r}')$ and $(\vee\mathrm{l}')$ gives the same result as using the two sequent expressions $\Gamma \Rightarrow \gamma$ and $\gamma \Rightarrow \Gamma$.

Definition 2.2.25 *Let* $\mathcal{F}_{\mathrm{HB}^-}$ *and* $\mathcal{F}_{\mathrm{N4}^-}$ *(*$\mathcal{S}_{\mathrm{HB}^-}$ *and* $\mathcal{S}_{\mathrm{N4}^-}$*) be the sets of formulas (the sets of sequents) of* HB^- *and* $\mathrm{N4}^-$, *respectively. Let* $\mathcal{T}_{\mathrm{HB}^-}$ *and* $\mathcal{T}_{\mathrm{N4}^-}$ *be the sets of multisets consisting of the formulas in* $\mathcal{F}_{\mathrm{HB}^-}$ *and* $\mathcal{F}_{\mathrm{N4}^-}$, *respectively.*

A mapping g *from* $\mathcal{F}_{\mathrm{HB}^-}$ *to* $\mathcal{F}_{\mathrm{N4}^-}$, *from* $\mathcal{T}_{\mathrm{HB}^-}$ *to* $\mathcal{T}_{\mathrm{N4}^-}$ *and from* $\mathcal{S}_{\mathrm{HB}^-}$ *to* $\mathcal{S}_{\mathrm{N4}^-}$ *is defined as follows.*

1. $g(p) := p$ *for any* $p \in \Phi$,

2. $g(\alpha \circ \beta) := g(\alpha) \circ g(\beta)$ *where* $\circ \in \{\wedge, \vee\}$,

3. $g(\{\gamma_1, ..., \gamma_n\}) := \{g(\gamma_1), ..., g(\gamma_n)\}$,

4. $g(\Gamma \Rightarrow \gamma) := g(\Gamma) \Rightarrow g(\gamma)$,

5. $g(\gamma \Rightarrow \Gamma) := {\sim}g(\Gamma) \Rightarrow {\sim}g(\gamma)$.

Theorem 2.2.26 (Syntactical embedding) *Let g be the mapping defined in Definition 2.2.25, and S be a sequent in* $\mathcal{S}_{\mathrm{HB}^-}$. *Then:*

1. $\mathrm{HB}^- \vdash S$ *iff* $\mathrm{N4}^- \vdash g(S)$,

2. $\mathrm{HB}^- - \{(\mathrm{cut}), (\mathrm{cut\text{-}d})\} \vdash S$ *iff* $\mathrm{N4}^- - (\mathrm{cut}) \vdash g(S)$.

Proof. We only show (1).

• $(\Longrightarrow)$: By induction on the proofs P of S in HB^-. We distinguish the cases according to the last inference of P. We only show the following case.

Case ($\wedge$r-d): The last inference of P is of the form:

$$\frac{\gamma \Rightarrow \Gamma, \alpha \quad \gamma \Rightarrow \Gamma, \beta}{\gamma \Rightarrow \Gamma, \alpha \wedge \beta} \ (\wedge\text{r-d})$$

where $S \equiv \gamma \Rightarrow \Gamma, \alpha \wedge \beta$. By induction hypothesis, we have: $\mathrm{N4}^- \vdash g(\gamma \Rightarrow \Gamma, \alpha)$, i.e., $\mathrm{N4}^- \vdash {\sim}g(\Gamma), {\sim}g(\alpha) \Rightarrow {\sim}g(\gamma)$ by the definition of g, and $\mathrm{N4}^- \vdash g(\gamma \Rightarrow \Gamma, \beta)$, i.e., $\mathrm{N4}^- \vdash {\sim}g(\Gamma), {\sim}g(\beta) \Rightarrow {\sim}g(\gamma)$ by the definition of g. Thus, we obtain: $\mathrm{N4}^- \vdash {\sim}g(\Gamma), {\sim}(g(\alpha) \wedge g(\beta)) \Rightarrow {\sim}g(\gamma)$ using (${\sim}\wedge$l). This means the required fact: $\mathrm{N4}^- \vdash g(\gamma \Rightarrow \Gamma, \alpha \wedge \beta)$ by the definition of g.

• $(\Longleftarrow)$: By induction on the proofs P of $g(S)$ in $\mathrm{N4}^-$. We distinguish the cases according to the last inference of P. Note that there are no cases for (${\sim}$r) and (${\sim}$l), because by the definition of g, $g(S)$ does not allow a nested negation expression such as ${\sim}g(\Gamma) \Rightarrow {\sim}{\sim}g(\alpha)$. We only show the following case.

Case (${\sim}\wedge$l): The last inference of P is of the form:

$$\frac{{\sim}g(\alpha), {\sim}g(\Gamma) \Rightarrow {\sim}g(\gamma) \quad {\sim}g(\beta), {\sim}g(\Gamma) \Rightarrow {\sim}g(\gamma)}{{\sim}(g(\alpha) \wedge g(\beta)), {\sim}g(\Gamma) \Rightarrow {\sim}g(\gamma)} \ ({\sim}\wedge\mathrm{l})$$

where $g(S) \equiv g(\gamma \Rightarrow \alpha \wedge \beta, \Gamma)$ coincides with ${\sim}(g(\alpha) \wedge g(\beta)), {\sim}g(\Gamma) \Rightarrow {\sim}g(\gamma)$ by the definition of g. By induction hypothesis, we have: $\mathrm{HB}^- \vdash \gamma \Rightarrow \alpha, \Gamma$ and $\mathrm{HB}^- \vdash \gamma \Rightarrow \beta, \Gamma$. We thus obtain the required fact: $\mathrm{HB}^- \vdash \gamma \Rightarrow \alpha \wedge \beta, \Gamma$ by using ($\wedge$r-d). ∎

Theorem 2.2.27 (Cut-elimination) *The rules* (cut) *and* (cut-d) *are admissible in cut-free* HB$^-$.

Proof. By Theorem 2.2.26 and the cut-elimination theorem for N4$^-$. ∎

2.3. Sequent calculi

2.3.1. Contraction-free system

In this section we present the contraction-free, cut-free and weakening-free sequent calculus G4np for Nelson's four-valued logic.

Definition 2.3.1 (G4np) *The initial sequents of* G4np *are of the form: for any propositional variable* p*,*

$$p, \Gamma \Rightarrow p \qquad\qquad {\sim}p, \Gamma \Rightarrow {\sim}p.$$

The logical inference rules of G4np *are* ($\to$r), ($\wedge$l), ($\wedge$r), ($\vee$l), ($\wedge$r1), ($\wedge$r2), ($\sim$l), ($\sim$r), ($\sim\wedge$l), ($\sim\wedge$r1), ($\sim\wedge$r2), ($\sim\vee$l), ($\sim\vee$r), ($\sim\to$l), ($\sim\to$r) *from Definition 2.2.1, and the rules of the form:*

$$\frac{p, \beta, \Gamma \Rightarrow \gamma}{p, p{\to}\beta, \Gamma \Rightarrow \gamma}\ ({\to}\mathrm{l0}) \qquad \frac{{\sim}p, \beta, \Gamma \Rightarrow \gamma}{{\sim}p, {\sim}p{\to}\beta, \Gamma \Rightarrow \gamma}\ ({\sim}{\to}\mathrm{l0})$$

$$\frac{\alpha_1{\to}(\alpha_2{\to}\beta), \Gamma \Rightarrow \gamma}{(\alpha_1 \wedge \alpha_2){\to}\beta, \Gamma \Rightarrow \gamma}\ (\wedge{\to}\mathrm{l}) \qquad \frac{\alpha_1{\to}\beta, \alpha_2{\to}\beta, \Gamma \Rightarrow \gamma}{(\alpha_1 \vee \alpha_2){\to}\beta, \Gamma \Rightarrow \gamma}\ (\vee{\to}\mathrm{l})$$

$$\frac{\alpha_1, \alpha_2{\to}\beta, \Gamma \Rightarrow \alpha_2 \quad \beta, \Gamma \Rightarrow \gamma}{(\alpha_1{\to}\alpha_2){\to}\beta, \Gamma \Rightarrow \gamma}\ ({\to}{\to}\mathrm{l})$$

$$\frac{{\sim}\alpha_1{\to}\beta, {\sim}\alpha_2{\to}\beta, \Gamma \Rightarrow \gamma}{{\sim}(\alpha_1 \wedge \alpha_2){\to}\beta, \Gamma \Rightarrow \gamma}\ ({\sim} \wedge {\to}\mathrm{l})$$

$$\frac{{\sim}\alpha_1{\to}({\sim}\alpha_2{\to}\beta), \Gamma \Rightarrow \gamma}{{\sim}(\alpha_1 \vee \alpha_2){\to}\beta, \Gamma \Rightarrow \gamma}\ ({\sim} \vee {\to}\mathrm{l}) \qquad \frac{\alpha_1{\to}({\sim}\alpha_2{\to}\beta), \Gamma \Rightarrow \gamma}{{\sim}(\alpha_1{\to}\alpha_2){\to}\beta, \Gamma \Rightarrow \gamma}\ ({\sim}{\to}{\to}\mathrm{l}).$$

The $\sim$*-free part of* G4np *is called here* G4ip$^\perp$*, which is the* $\perp$*-free fragment of* G4ip [188].

Roughly speaking, the rules ($\to$l0), ($\sim\to$l0), ($\wedge\to$l), ($\vee\to$l), ($\to\to$l), ($\sim\wedge\to$l), ($\sim\vee\to$l) and ($\sim\to\to$l) can be regarded as some divided versions of the rule:

$$\frac{\alpha{\to}\beta, \Gamma \Rightarrow \alpha \quad \beta, \Gamma \Rightarrow \gamma}{\alpha{\to}\beta, \Gamma \Rightarrow \gamma}$$

where α is divided into $p, {\sim}p$, $\alpha_1 \wedge \alpha_2$, $\alpha_1 \vee \alpha_2$, $\alpha_1{\to}\alpha_2$, ${\sim}(\alpha_1 \wedge \alpha_2)$, ${\sim}(\alpha_1 \vee \alpha_2)$ and ${\sim}(\alpha_1{\to}\alpha_2)$. In the rule just displayed above, the principal formula $\alpha{\to}\beta$ appears twice, i.e., in one of the upper sequents and in the lower sequent. Such occurrences of $\alpha{\to}\beta$ derive some inefficient proof search procedures with *loops*. Since G4np is loop-free, it is regarded as efficient.

Lemma 2.3.2 *Let* Γ *be a multiset of formulas in* $\mathcal{L}_{\mathrm{N4}}$, γ *be a formula in* $\mathcal{L}_{\mathrm{N4}}$, R *be the set* $\{(\mathrm{cut}), (\mathrm{co}), (\mathrm{we})\}$, *and* f *be the mapping defined in Definition 2.2.3. Then:*

1. *if* $\mathrm{G4np} + \mathrm{R} \vdash \Gamma \Rightarrow \gamma$, *then* $\mathrm{G4ip}^{\perp} + \mathrm{R} \vdash f(\Gamma) \Rightarrow f(\gamma)$,
2. *if* $\mathrm{G4ip}^{\perp} \vdash f(\Gamma) \Rightarrow f(\gamma)$, *then* $\mathrm{G4np} \vdash \Gamma \Rightarrow \gamma$.

Proof. Similar to the proof of Theorem 2.2.4. ∎

Theorem 2.3.3 (Structural rule elimination) *The rules* (cut), (co) *and* (we) *are admissible in* G4np.

Proof. Suppose $\mathrm{G4np} + \{(\mathrm{cut}), (\mathrm{co}), (\mathrm{we})\} \vdash \Gamma \Rightarrow \gamma$. Then, we have $\mathrm{G4ip}^{\perp} + \{(\mathrm{cut}), (\mathrm{co}), (\mathrm{we})\} \vdash f(\Gamma) \Rightarrow f(\gamma)$ by Lemma 2.3.2 (1), and hence $\mathrm{G4ip}^{\perp} \vdash f(\Gamma) \Rightarrow f(\gamma)$ by the structural rule elimination theorem for $\mathrm{G4ip}^{\perp}$, which was directly proved by Dyckhoff and Negri [52]. By Lemma 2.3.2 (2), we obtain the required fact: $\mathrm{G4np} \vdash \Gamma \Rightarrow \gamma$. ∎

Theorem 2.3.4 (Equivalence between Gn4 and G4np)
For any sequent S,

$$\mathrm{G4np} \vdash S \;\; \textit{iff} \;\; \mathrm{N4} \vdash S.$$

Proof. • ($\Longrightarrow$): We can straightforwardly prove that if a sequent S is provable in G4np then it is provable in N4. This is proved by induction on the proofs P of S in G4np. We distinguish the cases according to the last inference of P. We only illustrate the case that the last inference of P is of the form:

$$\dfrac{\alpha_1, \alpha_2{\to}\beta, \Gamma \Rightarrow \alpha_2 \quad \beta, \Gamma \Rightarrow \gamma}{(\alpha_1{\to}\alpha_2){\to}\beta, \Gamma \Rightarrow \gamma}\ ({\to}{\to}\mathrm{l}).$$

By induction hypothesis, both $\alpha_1, \alpha_2{\to}\beta, \Gamma \Rightarrow \alpha_2$ and $\beta, \Gamma \Rightarrow \gamma$ are provable in N4. Then, we have the required proof:

$$\dfrac{\dfrac{\dfrac{\dfrac{\dfrac{\dfrac{\alpha_2 \Rightarrow \alpha_2}{\alpha_1, \alpha_2 \Rightarrow \alpha_2}}{\alpha_2 \Rightarrow \alpha_1{\to}\alpha_2} \quad \dfrac{\beta \Rightarrow \beta}{\beta, \alpha_2 \Rightarrow \beta}}{(\alpha_1{\to}\alpha_2){\to}\beta, \alpha_2 \Rightarrow \beta}}{(\alpha_1{\to}\alpha_2){\to}\beta \Rightarrow \alpha_2{\to}\beta} \quad \dfrac{\alpha_1, \alpha_2{\to}\beta, \Gamma \Rightarrow \alpha_2}{\alpha_2{\to}\beta, \Gamma \Rightarrow \alpha_1{\to}\alpha_2}}{(\alpha_1{\to}\alpha_2){\to}\beta, \Gamma \Rightarrow \alpha_1{\to}\alpha_2}\,(\mathrm{cut}) \quad \dfrac{\beta, \Gamma \Rightarrow \gamma}{\beta, (\alpha_1{\to}\alpha_2){\to}\beta, \Gamma \Rightarrow \gamma}}{\dfrac{(\alpha_1{\to}\alpha_2){\to}\beta, (\alpha_1{\to}\alpha_2){\to}\beta, \Gamma \Rightarrow \gamma}{(\alpha_1{\to}\alpha_2){\to}\beta, \Gamma \Rightarrow \gamma}\,(\mathrm{co}).}$$

• ($\Longleftarrow$): We prove that if a sequent S is provable in N4 then it is provable in G4np. This is proved by induction on the proofs P of S in N4. We distinguish the cases according to the last inference of P. We show some cases. The cases that the last inference of P is (cut), (co) or (we) can be shown by Theorem 2.3.3. The case that the last inference of P is ($\to$l) can be proved using the fact that ($\to$l) is admissible in G4np. This fact can be shown in the same way as in the proof of Theorem 2.3.3: We use the result by Dyckhoff and Negri [52][2] that ($\to$l) is admissible in G4ip, and prove a similar lemma as Lemma 2.3.2 with respect to ($\to$l). ∎

2.3.2. Auxiliary system and resolution system

On the way to defining a resolution calculus for Nelson's four-valued logic, we first introduce the auxiliary sequent system G5np.

For the multiset with multiplicity one which is obtained from a multiset Γ, we write (Γ), i.e., the multiset (Γ) contains the formulas of Γ with multiplicity one. For example, if Γ is the multiset $\{\alpha, \alpha, \beta\}$, then (Γ) is the multiset $\{\alpha, \beta\}$.

Definition 2.3.5 (G5np) *The initial sequents of* G5np *are of the form: for any propositional variable p,*

$$p, \Gamma \Rightarrow p \qquad\qquad {\sim}p, \Gamma \Rightarrow {\sim}p.$$

The inference rules of G5np *are of the form:*

$$\frac{\Gamma \Rightarrow \alpha \quad \alpha, \Delta \Rightarrow \gamma}{(\Gamma, \Delta) \Rightarrow \gamma}\ (\text{cut}*)$$

$$\frac{\alpha, \Gamma \Rightarrow \gamma}{(\alpha \wedge \beta, \Gamma) \Rightarrow \gamma}\ (\wedge\text{l1}*) \qquad \frac{\beta, \Gamma \Rightarrow \gamma}{(\alpha \wedge \beta, \Gamma) \Rightarrow \gamma}\ (\wedge\text{l2}*)$$

$$\frac{\alpha, \beta, \Gamma \Rightarrow \gamma}{(\alpha \wedge \beta, \Gamma) \Rightarrow \gamma}\ (\wedge\text{l3}*) \qquad \frac{\Gamma \Rightarrow \alpha \quad \Delta \Rightarrow \beta}{(\Gamma, \Delta) \Rightarrow \alpha \wedge \beta}\ (\wedge\text{r}*)$$

$$\frac{\alpha, \Gamma \Rightarrow \gamma \quad \beta, \Delta \Rightarrow \gamma}{(\alpha \vee \beta, \Gamma, \Delta) \Rightarrow \gamma}\ (\vee\text{l}*) \qquad \frac{\Gamma \Rightarrow \alpha}{\Gamma \Rightarrow \alpha \vee \beta}\ (\vee\text{r1}*) \qquad \frac{\Gamma \Rightarrow \beta}{\Gamma \Rightarrow \alpha \vee \beta}\ (\vee\text{r2}*)$$

$$\frac{\Gamma \Rightarrow \alpha \quad \beta, \Delta \Rightarrow \gamma}{(\alpha \to \beta, \Gamma, \Delta) \Rightarrow \gamma}\ (\to\text{l}*) \qquad \frac{\alpha, \Gamma \Rightarrow \beta}{\Gamma \Rightarrow \alpha \to \beta}\ (\to\text{r1}*) \qquad \frac{\Gamma \Rightarrow \beta}{\Gamma \Rightarrow \alpha \to \beta}\ (\to\text{r2}*)$$

$$\frac{\alpha, \Gamma \Rightarrow \gamma}{({\sim}{\sim}\alpha, \Gamma) \Rightarrow \gamma}\ ({\sim}\text{l}*) \qquad \frac{\Gamma \Rightarrow \alpha}{\Gamma \Rightarrow {\sim}{\sim}\alpha}\ ({\sim}\text{r}*)$$

[2] Lemma 4.1 on page 1503 in [52].

$$\frac{{\sim}\alpha, \Gamma \Rightarrow \gamma \quad {\sim}\beta, \Delta \Rightarrow \gamma}{({\sim}(\alpha \wedge \beta), \Gamma, \Delta) \Rightarrow \gamma} \ ({\sim} \wedge \mathrm{l}{*})$$

$$\frac{\Gamma \Rightarrow {\sim}\alpha}{\Gamma \Rightarrow {\sim}(\alpha \wedge \beta)} \ ({\sim} \wedge \mathrm{r1}{*}) \qquad \frac{\Gamma \Rightarrow {\sim}\beta}{\Gamma \Rightarrow {\sim}(\alpha \wedge \beta)} \ ({\sim} \wedge \mathrm{r2}{*})$$

$$\frac{{\sim}\alpha, \Gamma \Rightarrow \gamma}{({\sim}(\alpha \vee \beta), \Gamma) \Rightarrow \gamma} \ ({\sim} \vee \mathrm{l1}{*}) \qquad \frac{{\sim}\beta, \Gamma \Rightarrow \gamma}{({\sim}(\alpha \vee \beta), \Gamma) \Rightarrow \gamma} \ ({\sim} \vee \mathrm{l2}{*})$$

$$\frac{{\sim}\alpha, {\sim}\beta, \Gamma \Rightarrow \gamma}{({\sim}(\alpha \vee \beta), \Gamma) \Rightarrow \gamma} \ ({\sim} \vee \mathrm{l3}{*}) \qquad \frac{\Gamma \Rightarrow {\sim}\alpha \quad \Delta \Rightarrow {\sim}\beta}{(\Gamma, \Delta) \Rightarrow {\sim}(\alpha \vee \beta)} \ ({\sim} \vee \mathrm{r}{*})$$

$$\frac{\alpha, \Gamma \Rightarrow \gamma}{({\sim}(\alpha {\rightarrow} \beta), \Gamma) \Rightarrow \gamma} \ ({\sim}{\rightarrow}\mathrm{l1}{*}) \qquad \frac{{\sim}\beta, \Gamma \Rightarrow \gamma}{({\sim}(\alpha {\rightarrow} \beta), \Gamma) \Rightarrow \gamma} \ ({\sim}{\rightarrow}\mathrm{l2}{*})$$

$$\frac{\alpha, {\sim}\beta, \Gamma \Rightarrow \gamma}{({\sim}(\alpha {\rightarrow} \beta), \Gamma) \Rightarrow \gamma} \ ({\sim}{\rightarrow}\mathrm{l3}{*}) \qquad \frac{\Gamma \Rightarrow \alpha \quad \Delta \Rightarrow {\sim}\beta}{(\Gamma, \Delta) \Rightarrow {\sim}(\alpha {\rightarrow} \beta)} \ ({\sim}{\rightarrow}\mathrm{r}{*}).$$

The sequents of the form $\alpha, \Gamma \Rightarrow \alpha$ for any formula α are provable in G5np, and hence these sequents are also regarded as initial sequents.

Note that G5np is a modified extension of the system G5ip which is introduced by Troelstra and Schwichtenberg [188] in order to prove the equivalence between Rip (see p. 14) and G5ip. We also remark that the initial sequents of G5np are different from G5ip: G5ip has initial sequents of the forms $\alpha \Rightarrow \alpha$ and $\bot \Rightarrow \alpha$.

Proposition 2.3.6 (Equivalence between G5np and G4np) *For any sequent* $\Gamma \Rightarrow \gamma$,

1. *if* G5np $\vdash \Gamma \Rightarrow \gamma$, *then* G4np $\vdash \Gamma \Rightarrow \gamma$,
2. *if* G4np $\vdash \Gamma \Rightarrow \gamma$, *then* G5np $\vdash \Gamma' \Rightarrow \gamma$ *for some* $\Gamma' \subseteq \Gamma$.

Next, we introduce Rnp based on the notion of intuitionistic clauses.

Definition 2.3.7 *A formula is called a* literal *if it is a propositional variable or a negated propositional variable. A sequent is called an* intuitionistic clause *if it is one of the following forms:*

$$(P{\rightarrow}Q) \Rightarrow R, \qquad P \Rightarrow (Q \vee R), \qquad P_1, ..., P_n \Rightarrow Q$$

where $P, Q, R, P_1, ..., P_n$ *represent literals, and* n *can be* 0.

Definition 2.3.8 (Rnp) *Let* P, Q, R, S *be literals and all the sequents of* Rnp *be intuitionistic clauses.*

The axioms of Rnp *are of the form:*

$$P, \Delta \Rightarrow P.$$

The inference rules of Rnp *are of the form:*

$$\frac{\Gamma \Rightarrow P \quad P, \Delta \Rightarrow Q}{(\Gamma, \Delta) \Rightarrow Q} \text{ (resol)}$$

$$\frac{P \Rightarrow Q \vee R \quad \Gamma \Rightarrow P \quad Q, \Delta \Rightarrow S \quad R, \Sigma \Rightarrow S}{(\Gamma, \Delta, \Sigma) \Rightarrow S} \text{ ($\vee$resol)}$$

$$\frac{(P \rightarrow Q) \Rightarrow R \quad [P], \Delta \Rightarrow Q}{\Delta \Rightarrow R} \text{ ($\rightarrow$resol)}$$

where $[P]$ *represents* P *or the empty multiset.*

Note that Rnp is a modification of the system Rip (for intuitionistic logic) introduced in [188]. In Rip, the formulas P, Q, R, S, which are used as literals in Rnp, are propositional variables. As mentioned in [188], Rip is based on Mints' framework in [127]. Also note that the axioms in Rip are of the forms $P \Rightarrow P$ and $\bot \Rightarrow P$.

The expression l_α, called the label of a formula α, stands for a literal which corresponds to a formula α. For a multiset $\Gamma \equiv \{\gamma_1, ..., \gamma_n\}$ of formulas, l_Γ means the multiset $\{l_{\gamma_1}, ..., l_{\gamma_n}\}$. For any propositional variable p, l_p and $l_{\sim p}$ are defined by p and $\sim p$, respectively. For any compound formula γ, the interpretation of l_γ is considered below.

Definition 2.3.9 *For any formulas* α *and* β, *we define:*

1. $C_{\alpha \wedge \beta} = \{\, l_{\alpha \wedge \beta} \Rightarrow l_\alpha; \quad l_{\alpha \wedge \beta} \Rightarrow l_\beta; \quad l_\alpha, l_\beta \Rightarrow l_{\alpha \wedge \beta}\}$,
2. $C_{\alpha \vee \beta} = \{\, l_{\alpha \vee \beta} \Rightarrow l_\alpha \vee l_\beta; \quad l_\alpha \Rightarrow l_{\alpha \vee \beta}; \quad l_\beta \Rightarrow l_{\alpha \vee \beta}\}$,
3. $C_{\alpha \rightarrow \beta} = \{\, l_{\alpha \rightarrow \beta}, l_\alpha \Rightarrow l_\beta; \quad (l_\alpha \rightarrow l_\beta) \Rightarrow l_{\alpha \rightarrow \beta}\}$,
4. $C_{\sim\sim\alpha} = \{\, l_{\sim\sim\alpha} \Rightarrow l_\alpha; \quad l_\alpha \Rightarrow l_{\sim\sim\alpha}\}$,
5. $C_{\sim(\alpha \wedge \beta)} = \{\, l_{\sim(\alpha \wedge \beta)} \Rightarrow l_{\sim\alpha} \vee l_{\sim\beta}; \quad l_{\sim\alpha} \Rightarrow l_{\sim(\alpha \wedge \beta)};$
   $l_{\sim\beta} \Rightarrow l_{\sim(\alpha \wedge \beta)}\}$,
6. $C_{\sim(\alpha \vee \beta)} = \{\, l_{\sim(\alpha \vee \beta)} \Rightarrow l_{\sim\alpha}; \quad l_{\sim(\alpha \vee \beta)} \Rightarrow l_{\sim\beta};$
   $l_{\sim\alpha}, l_{\sim\beta} \Rightarrow l_{\sim(\alpha \vee \beta)}\}$,
7. $C_{\sim(\alpha \rightarrow \beta)} = \{\, l_{\sim(\alpha \rightarrow \beta)} \Rightarrow l_\alpha; \quad l_{\sim(\alpha \rightarrow \beta)} \Rightarrow l_{\sim\beta};$
   $l_\alpha, l_{\sim\beta} \Rightarrow l_{\sim(\alpha \rightarrow \beta)}\}$.

The expression $Nsub(\alpha)$ *represents the set of all non-literal subformulas and non-literal negated-subformulas of a formula* α. *Let* γ *be an arbitrary non-literal formula. Then, we define:*

$$Cl(\gamma) = \bigcup\{C_\delta \mid \delta \in Nsub(\gamma)\}.$$

Let Γ *be a set* $\{\gamma_1, ..., \gamma_n\}$ $(n \geq 2)$ *of non-literal formulas. Then, we define:*

$$Cl(\Gamma) = Cl(\gamma_1) \cup \cdots \cup Cl(\gamma_n).$$

Note that $Cl(\gamma)$ is a set of intuitionistic clauses. Suppose that an expression $\alpha \leftrightarrow \beta$ means $\vdash \alpha \Rightarrow \beta$ and $\vdash \beta \Rightarrow \alpha$. $Cl(\gamma)$ is intended to address the interpretation of the label expression l_γ as $l_p \leftrightarrow p$, $l_{\sim p} \leftrightarrow {\sim}p$, $l_{\alpha\wedge\beta} \leftrightarrow (l_\alpha \wedge l_\beta)$, $l_{\alpha\vee\beta} \leftrightarrow (l_\alpha \vee l_\beta)$, $l_{\alpha\rightarrow\beta} \leftrightarrow (l_\alpha \rightarrow l_\beta)$, $l_{\sim\sim\alpha} \leftrightarrow l_\alpha$, $l_{\sim(\alpha\wedge\beta)} \leftrightarrow (l_{\sim\alpha} \vee l_{\sim\beta})$, $l_{\sim(\alpha\vee\beta)} \leftrightarrow (l_{\sim\alpha} \wedge l_{\sim\beta})$ and $l_{\sim(\alpha\rightarrow\beta)} \leftrightarrow (l_\alpha \wedge l_{\sim\beta})$. Such an interpretation is regarded as the label version of the mapping f defined in Definition 2.2.3.

Lemma 2.3.10 *If* G5np + $Cl(\gamma) \vdash \Rightarrow l_\gamma$, *then* G5np $\vdash \Rightarrow \gamma$.

Proof. Suppose G5np + $Cl(\gamma) \vdash \Rightarrow l_\gamma$. Let P be a proof of $\Rightarrow l_\gamma$ in G5np + $Cl(\gamma)$. If we substitute α for all the labels l_α everywhere in P, then l_γ becomes γ and all the sequents in $Cl(\gamma)$ appearing in P become G5np-provable sequents. Therefore G5np $\vdash \Rightarrow \gamma$. ∎

Lemma 2.3.11 *If* G5np $\vdash \Gamma \Rightarrow \gamma$, *then* Rnp + $Cl(\Gamma, \gamma) \vdash l_\Gamma \Rightarrow l_\gamma$.

Proof. By induction on the proofs P of $\Gamma \Rightarrow \gamma$ in G5np. We distinguish the cases according to the last inference of P. We show some cases.

Case (${\sim}\vee$l3$*$): The last inference of P is of the form:

$$\frac{{\sim}\alpha, {\sim}\beta, \Sigma \Rightarrow \gamma}{({\sim}(\alpha \vee \beta), \Sigma) \Rightarrow \gamma}\ ({\sim}\vee \text{l}3*).$$

By induction hypothesis, we have:
Rnp + $Cl({\sim}\alpha, {\sim}\beta, \Sigma, \gamma) \vdash l_{\sim\alpha}, l_{\sim\beta}, l_\Sigma \Rightarrow l_\gamma$. We then obtain:

$$\frac{l_{\sim(\alpha\vee\beta)} \Rightarrow l_{\sim\beta} \qquad \dfrac{l_{\sim(\alpha\vee\beta)} \Rightarrow l_{\sim\alpha} \quad l_{\sim\alpha}, l_{\sim\beta}, l_\Sigma \Rightarrow l_\gamma}{(l_{\sim(\alpha\vee\beta)}, l_{\sim\beta}, l_\Sigma) \Rightarrow l_\gamma}\ (\text{resol})}{(l_{\sim(\alpha\vee\beta)}, l_\Sigma) \Rightarrow l_\gamma}\ (\text{resol}).$$

Since $l_{\sim(\alpha\vee\beta)} \Rightarrow l_{\sim\alpha}$, $l_{\sim(\alpha\vee\beta)} \Rightarrow l_{\sim\beta} \in Nsub({\sim}(\alpha \vee \beta))$, it holds that Rnp + $Cl({\sim}(\alpha \vee \beta), \Sigma, \gamma) \vdash l_{\sim(\alpha\vee\beta)}, l_\Sigma \Rightarrow l_\gamma$.

Case (${\sim}\vee$r$*$): The last inference of P is of the form:

$$\frac{\Sigma \Rightarrow {\sim}\alpha \quad \Delta \Rightarrow {\sim}\beta}{(\Sigma, \Delta) \Rightarrow {\sim}(\alpha \vee \beta)}\ ({\sim}\vee \text{r}*).$$

By induction hypothesis, we have: Rnp + $Cl(\Sigma, {\sim}\alpha) \vdash l_{\Sigma} \Rightarrow l_{\sim\alpha}$ and Rnp + $Cl(\Delta, {\sim}\beta) \vdash l_{\Delta} \Rightarrow l_{\sim\beta}$. We then obtain:

$$\cfrac{l_{\Delta} \Rightarrow l_{\sim\beta} \quad \cfrac{l_{\Sigma} \Rightarrow l_{\sim\alpha} \quad l_{\sim\alpha}, l_{\sim\beta} \Rightarrow l_{\sim(\alpha\vee\beta)}}{(l_{\Sigma}, l_{\sim\beta}) \Rightarrow l_{\sim(\alpha\vee\beta)}}\ \text{(resol)}}{(l_{\Sigma}, l_{\Delta}) \Rightarrow l_{\sim(\alpha\vee\beta)}}\ \text{(resol)}.$$

Case (${\sim}\wedge$l$*$): The last inference of P is of the form:

$$\frac{{\sim}\alpha, \Sigma \Rightarrow \gamma \quad {\sim}\beta, \Delta \Rightarrow \gamma}{({\sim}(\alpha \wedge \beta), \Sigma, \Delta) \Rightarrow \gamma}\ ({\sim} \wedge \text{l}*).$$

By induction hypothesis, we have: Rnp + $Cl({\sim}\alpha, \Sigma, \gamma) \vdash l_{\Sigma}, l_{\sim\alpha} \Rightarrow l_{\gamma}$ and Rnp + $Cl({\sim}\beta, \Delta, \gamma) \vdash l_{\Delta}, l_{\sim\beta} \Rightarrow l_{\gamma}$. We then obtain by an application of ($\vee$resol):

$$\frac{l_{\sim(\alpha\wedge\beta)} \Rightarrow l_{\sim\alpha} \vee l_{\sim\beta} \quad l_{\sim(\alpha\wedge\beta)} \Rightarrow l_{\sim(\alpha\wedge\beta)} \quad l_{\Sigma}, l_{\sim\alpha} \Rightarrow l_{\gamma} \quad l_{\Delta}, l_{\sim\beta} \Rightarrow l_{\gamma}}{(l_{\sim(\alpha\wedge\beta)}, l_{\Sigma}, l_{\Delta}) \Rightarrow l_{\gamma}}$$

Case ($\rightarrow$r1$*$): The last inference of P is of the form:

$$\frac{\alpha, \Delta \Rightarrow \beta}{\Delta \Rightarrow \alpha\rightarrow\beta}\ (\rightarrow\text{r1}*).$$

By induction hypothesis, we have: Rnp + $Cl(\alpha, \Delta, \beta) \vdash l_{\alpha}, l_{\Delta} \Rightarrow l_{\beta}$. We then obtain:

$$\frac{l_{\alpha} \rightarrow l_{\beta} \Rightarrow l_{\alpha\rightarrow\beta} \quad l_{\alpha}, l_{\Delta} \Rightarrow l_{\beta}}{l_{\Delta} \Rightarrow l_{\alpha\rightarrow\beta}}\ (\rightarrow\text{resol}).$$

Case ($\rightarrow$l$*$): The last inference of P is of the form:

$$\frac{\Sigma \Rightarrow \alpha \quad \beta, \Delta \Rightarrow \gamma}{(\alpha\rightarrow\beta, \Sigma, \Delta) \Rightarrow \gamma}\ (\rightarrow\text{l}*).$$

By induction hypothesis, we have: Rnp + $Cl(\Sigma, \alpha) \vdash l_{\Sigma} \Rightarrow l_{\alpha}$ and Rnp + $Cl(\beta, \Delta, \gamma) \vdash l_{\beta}, l_{\Delta} \Rightarrow l_{\gamma}$. We then obtain:

$$\cfrac{l_{\Sigma} \Rightarrow l_{\alpha} \quad \cfrac{l_{\alpha\rightarrow\beta}, l_{\alpha} \Rightarrow l_{\beta} \quad l_{\beta}, l_{\Delta} \Rightarrow l_{\gamma}}{(l_{\alpha\rightarrow\beta}, l_{\alpha}, l_{\Delta}) \Rightarrow l_{\gamma}}\ \text{(resol)}}{(l_{\alpha\rightarrow\beta}, l_{\Sigma}, l_{\Delta}) \Rightarrow l_{\gamma}}\ \text{(resol)}.$$

∎

Lemma 2.3.12 *Let C be a set of intuitionistic clauses. For any intuitionistic clause $\Gamma \Rightarrow P$, if* $\mathrm{Rnp} + C \vdash \Gamma \Rightarrow P$, *then* $\mathrm{G5np} + C \vdash \Gamma \Rightarrow P$.

Proof. By induction on the proofs of $\Gamma \Rightarrow P$ in $\mathrm{Rnp} + C$. ∎

Theorem 2.3.13 (Equivalence between Rnp and G5np) *For any formula γ,*

$$\mathrm{G5np} \vdash \Rightarrow \gamma \quad \textit{iff} \quad \mathrm{Rnp} + Cl(\gamma) \vdash \Rightarrow l_\gamma.$$

Proof. ($\Longrightarrow$): By Lemma 2.3.11. ($\Longleftarrow$): Suppose $\mathrm{Rnp} + Cl(\gamma) \vdash \Rightarrow l_\gamma$. By Lemma 2.3.12, we have $\mathrm{G5np} + Cl(\gamma) \vdash \Rightarrow l_\gamma$. We thus obtain G5np $\vdash \Rightarrow \gamma$ by Lemma 2.3.10. ∎

2.3.3. Subformula calculus and dual calculus

A subformula calculus Sn4, which has the subformula property, is introduced below. The sequents of Sn4 are of the form $\Gamma : \Delta \Rightarrow \varnothing : \gamma$ and $\Gamma : \Delta \Rightarrow \gamma : \varnothing$ where γ is a formula, and Γ and Δ are multisets of formulas. We call Γ (Δ) in $\Gamma : \Delta \Rightarrow \varnothing : \gamma$ and $\Gamma : \Delta \Rightarrow \gamma : \varnothing$ a *negative (positive) context*, and also call the left/right-conclusion γ in the sequents a *negative/positive conclusion.* The sequents of the form:

$$\gamma_1, ..., \gamma_m : \delta_1, ..., \delta_n \Rightarrow \varnothing : \gamma \qquad \gamma_1, ..., \gamma_m : \delta_1, ..., \delta_n \Rightarrow \gamma : \varnothing$$

$(0 \leq m, n)$ in Sn4 correspond to the sequents of the form:

$$\sim\gamma_1, ..., \sim\gamma_m, \delta_1, ..., \delta_n \Rightarrow \gamma \qquad \sim\gamma_1, ..., \sim\gamma_m, \delta_1, ..., \delta_n \Rightarrow \sim\gamma$$

in N4. In the following definitions, C stands for $\varnothing : \gamma$ or $\gamma : \varnothing$.

Definition 2.3.14 (Sn4) *The initial sequents of* Sn4 *are of the form:*

$$\varnothing : \alpha \Rightarrow \varnothing : \alpha \qquad \alpha : \varnothing \Rightarrow \alpha : \varnothing.$$

The specific inference rules of Sn4 *are of the form:*

$$\frac{\Gamma : \Delta \Rightarrow \alpha : \varnothing}{\Gamma : \Delta \Rightarrow \varnothing : \sim\alpha}\ (\sim\mathrm{r}+) \quad \frac{\Gamma : \Delta \Rightarrow \varnothing : \alpha}{\Gamma : \Delta \Rightarrow \sim\alpha : \varnothing}\ (\sim\mathrm{r}-)$$

$$\frac{\alpha, \Gamma : \Delta \Rightarrow C}{\Gamma : \sim\alpha, \Delta \Rightarrow C}\ (\sim\mathrm{l}+) \quad \frac{\Gamma : \alpha, \Delta \Rightarrow C}{\sim\alpha, \Gamma : \Delta \Rightarrow C}\ (\sim\mathrm{l}-).$$

The structural rules of Sn4 *are of the form:*

$$\frac{\Gamma_1 : \Delta_1 \Rightarrow \alpha : \varnothing \quad \alpha, \Gamma_2 : \Delta_2 \Rightarrow C}{\Gamma_1, \Gamma_2 : \Delta_1, \Delta_2 \Rightarrow C}\ (\mathrm{cut}-)$$

$$\frac{\Gamma_1 : \Delta_1 \Rightarrow \varnothing : \alpha \quad \Gamma_2 : \alpha, \Delta_2 \Rightarrow C}{\Gamma_1, \Gamma_2 : \Delta_1, \Delta_2 \Rightarrow C}\ (\text{cut}+)$$

$$\frac{\alpha, \alpha, \Gamma : \Delta \Rightarrow C}{\alpha, \Gamma : \Delta \Rightarrow C}\ (\text{co}-) \qquad \frac{\Gamma : \alpha, \alpha, \Delta \Rightarrow C}{\Gamma : \alpha, \Delta \Rightarrow C}\ (\text{co}+)$$

$$\frac{\Gamma : \Delta \Rightarrow C}{\alpha, \Gamma : \Delta \Rightarrow C}\ (\text{we}-) \qquad \frac{\Gamma : \Delta \Rightarrow C}{\Gamma : \alpha, \Delta \Rightarrow C}\ (\text{we}+).$$

The positive inference rules of Sn4 *are of the form:*

$$\frac{\Gamma_1 : \Delta_1 \Rightarrow \varnothing : \alpha \quad \Gamma_2 : \beta, \Delta_2 \Rightarrow C}{\Gamma_1, \Gamma_2 : \alpha {\to} \beta, \Delta_1, \Delta_2 \Rightarrow C}\ ({\to}\text{l}+) \qquad \frac{\Gamma : \alpha, \Delta \Rightarrow \varnothing : \beta}{\Gamma : \Delta \Rightarrow \varnothing : \alpha {\to} \beta}\ ({\to}\text{r}+)$$

$$\frac{\Gamma : \alpha, \Delta \Rightarrow C}{\Gamma : \alpha \wedge \beta, \Delta \Rightarrow C}\ (\wedge\text{l1}+) \qquad \frac{\Gamma : \beta, \Delta \Rightarrow C}{\Gamma : \alpha \wedge \beta, \Delta \Rightarrow C}\ (\wedge\text{l2}+)$$

$$\frac{\Gamma : \Delta \Rightarrow \varnothing : \alpha \quad \Gamma : \Delta \Rightarrow \varnothing : \beta}{\Gamma : \Delta \Rightarrow \varnothing : \alpha \wedge \beta}\ (\wedge\text{r1}+)$$

$$\frac{\Gamma : \alpha, \Delta \Rightarrow C \quad \Gamma : \beta, \Delta \Rightarrow C}{\Gamma : \alpha \vee \beta, \Delta \Rightarrow C}\ (\vee\text{l1}+)$$

$$\frac{\Gamma : \Delta \Rightarrow \varnothing : \alpha}{\Gamma : \Delta \Rightarrow \varnothing : \alpha \vee \beta}\ (\vee\text{r1}+) \qquad \frac{\Gamma : \Delta \Rightarrow \varnothing : \beta}{\Gamma : \Delta \Rightarrow \varnothing : \alpha \vee \beta}\ (\vee\text{r2}+).$$

The negative inference rules of Sn4 *are of the form:*

$$\frac{\beta, \Gamma : \alpha, \Delta \Rightarrow C}{\alpha {\to} \beta, \Gamma : \Delta \Rightarrow C}\ ({\to}\text{l}-) \qquad \frac{\Gamma_1 : \Delta_1 \Rightarrow \beta : \varnothing \quad \Gamma_2 : \Delta_2 \Rightarrow \varnothing : \alpha}{\Gamma_1, \Gamma_2 : \Delta_1, \Delta_2 \Rightarrow \alpha {\to} \beta : \varnothing}\ ({\to}\text{r}-)$$

$$\frac{\alpha, \Gamma : \Delta \Rightarrow C \quad \beta, \Gamma : \Delta \Rightarrow C}{\alpha \wedge \beta, \Gamma : \Delta \Rightarrow C}\ (\wedge\text{l}-)$$

$$\frac{\Gamma : \Delta \Rightarrow \alpha : \varnothing}{\Gamma : \Delta \Rightarrow \alpha \wedge \beta : \varnothing}\ (\wedge\text{r1}-) \qquad \frac{\Gamma : \Delta \Rightarrow \beta : \varnothing}{\Gamma : \Delta \Rightarrow \alpha \wedge \beta : \varnothing}\ (\wedge\text{r2}-)$$

$$\frac{\alpha, \Gamma : \Delta \Rightarrow C}{\alpha \vee \beta, \Gamma : \Delta \Rightarrow C}\ (\vee\text{l1}-) \qquad \frac{\beta, \Gamma : \Delta \Rightarrow C}{\alpha \vee \beta, \Gamma : \Delta \Rightarrow C}\ (\vee\text{l2}-)$$

$$\frac{\Gamma : \Delta \Rightarrow \alpha : \varnothing \quad \Gamma : \Delta \Rightarrow \beta : \varnothing}{\Gamma : \Delta \Rightarrow \alpha \vee \beta : \varnothing}\ (\vee\text{r}-).$$

Theorem 2.3.15 (Equivalence between Sn4 and N4) *Let* Γ *and* Δ *be multisets of formulas, and* γ *be a formula. Then:*

1. *if* Sn4 $\vdash \Gamma : \Delta \Rightarrow C$ *where* C *is either* $\varnothing : \gamma$ *or* $\gamma : \varnothing$, *then* N4 $\vdash$ ${\sim}\Gamma, \Delta \Rightarrow C'$ *where* $C' \equiv \gamma$ *if* $C \equiv \varnothing : \gamma$ *and* $C' \equiv {\sim}\gamma$ *if* $C \equiv \gamma : \varnothing$,

2. *if* N4 $-$ (cut) $\vdash {\sim}\Gamma, \Delta \Rightarrow C'$ *where* C' *is either* γ *or* ${\sim}\gamma$, *then* Sn4 $-$ {(cut$-$), (cut$+$)} $\vdash \Gamma : \Delta \Rightarrow C$ *where* $C \equiv \varnothing : \gamma$ *if* $C' \equiv \gamma$ *and* $C \equiv \gamma : \varnothing$ *if* $C' \equiv {\sim}\gamma$.

Theorem 2.3.16 (Cut-elimination) *The rules* (cut+) *and* (cut−) *are admissible in cut-free* Sn4.

Proof. Suppose that a sequent $\Gamma : \Delta \Rightarrow C$ is provable in Sn4. Then, the sequent $\sim\Gamma, \Delta \Rightarrow C'$ is provable in N4 by Theorem 2.3.15 (1), and hence the sequent $\sim\Gamma, \Delta \Rightarrow C'$ is provable in cut-free N4 by the cut-elimination theorem for N4. Therefore $\Gamma : \Delta \Rightarrow C$ is provable in cut-free Sn4 by Theorem 2.3.15 (2). ∎

Corollary 2.3.17 (Subformula property) Sn4 *has the subformula property, i.e., if a sequent S is provable in* Sn4, *then there is a proof P of S such that all formulas appearing in P are subformulas of some formula in S.*

Next, a dual calculus Dn4 is introduced. A sequent of the form $\Gamma \Rightarrow^+ \gamma$ is called a *positive sequent*, and a sequent of the form $\Gamma \Rightarrow^- \gamma$ is called a *negative sequent.* In the following definitions, γ in an expression $\Gamma \Rightarrow^+ \gamma$ or $\Gamma \Rightarrow^- \gamma$ for any multiset Γ of formulas stands for a single formula, and for any multiset Δ of $\mathcal{L}_{\mathrm{N4}}$ formulas, $\sim\Delta$ is defined as $\{\neg\alpha : \alpha \in \Delta\}$.

Definition 2.3.18 (Dn4) *The initial sequents of* Dn4 *are of the form:*

$$\alpha \Rightarrow^+ \alpha \qquad\qquad \alpha \Rightarrow^- \alpha.$$

The specific inference rules of Dn4 *are of the form:*

$$\frac{\sim\Gamma, \Delta \Rightarrow^- \gamma}{\Gamma, \sim\Delta \Rightarrow^+ \sim\gamma}\ (-/+1) \quad \frac{\sim\Gamma, \Delta \Rightarrow^- \sim\gamma}{\Gamma, \sim\Delta \Rightarrow^+ \gamma}\ (-/+2)$$

$$\frac{\sim\Gamma, \Delta \Rightarrow^+ \gamma}{\Gamma, \sim\Delta \Rightarrow^- \sim\gamma}\ (+/-1) \quad \frac{\sim\Gamma, \Delta \Rightarrow^+ \sim\gamma}{\Gamma, \sim\Delta \Rightarrow^- \gamma}\ (+/-2).$$

The structural inference rules of Dn4 *are of the form:*

$$\frac{\Gamma \Rightarrow^+ \alpha \quad \alpha, \Sigma \Rightarrow^+ \gamma}{\Gamma, \Sigma \Rightarrow^+ \gamma}\ (+\mathrm{cut}) \quad \frac{\Gamma \Rightarrow^- \alpha \quad \alpha, \Sigma \Rightarrow^- \gamma}{\Gamma, \Sigma \Rightarrow^- \gamma}\ (-\mathrm{cut})$$

$$\frac{\alpha, \alpha, \Gamma \Rightarrow^+ \gamma}{\alpha, \Gamma \Rightarrow^+ \gamma}\ (+\mathrm{co}) \quad \frac{\alpha, \alpha, \Gamma \Rightarrow^- \gamma}{\alpha, \Gamma \Rightarrow^- \gamma}\ (-\mathrm{co})$$

$$\frac{\Gamma \Rightarrow^+ \gamma}{\alpha, \Gamma \Rightarrow^+ \gamma}\ (+\mathrm{we}) \quad \frac{\Gamma \Rightarrow^- \gamma}{\alpha, \Gamma \Rightarrow^- \gamma}\ (-\mathrm{we}).$$

The positive inference rules of Dn4 *are of the form:*

$$\frac{\Gamma \Rightarrow^+ \alpha \quad \beta, \Sigma \Rightarrow^+ \gamma}{\alpha{\to}\beta, \Gamma, \Sigma \Rightarrow^+ \gamma}\ (+{\to}\mathrm{l}) \quad \frac{\Gamma, \alpha \Rightarrow^+ \beta}{\Gamma \Rightarrow^+ \alpha{\to}\beta}\ (+{\to}\mathrm{r})$$

$$\frac{\alpha, \Delta \Rightarrow^+ \gamma}{\alpha \wedge \beta, \Delta \Rightarrow^+ \gamma}\ (+\wedge \mathrm{l1}) \quad \frac{\beta, \Delta \Rightarrow^+ \gamma}{\alpha \wedge \beta, \Delta \Rightarrow^+ \gamma}\ (+\wedge \mathrm{l2})$$

$$\frac{\Gamma \Rightarrow^+ \alpha \quad \Gamma \Rightarrow^+ \beta}{\Gamma \Rightarrow^+ \alpha \wedge \beta}\ (+\wedge \mathrm{r}) \quad \frac{\alpha, \Delta \Rightarrow^+ \gamma \quad \beta, \Delta \Rightarrow^+ \gamma}{\alpha \vee \beta, \Delta \Rightarrow^+ \gamma}\ (+\vee \mathrm{l})$$

$$\frac{\Gamma \Rightarrow^+ \alpha}{\Gamma \Rightarrow^+ \alpha \vee \beta}\ (+\vee \mathrm{r1}) \quad \frac{\Gamma \Rightarrow^+ \beta}{\Gamma \Rightarrow^+ \alpha \vee \beta}\ (+\vee \mathrm{r2}).$$

The negative inference rules of Dn4 *are of the form:*

$$\frac{\beta, {\sim}\alpha, \Delta \Rightarrow^- \gamma}{\alpha{\to}\beta, \Delta \Rightarrow^- \gamma}\ (-{\to}\mathrm{l}) \quad \frac{\Gamma \Rightarrow^- \beta \quad \Delta \Rightarrow^- {\sim}\alpha}{\Gamma, \Delta \Rightarrow^- \alpha{\to}\beta}\ (-{\to}\mathrm{r})$$

$$\frac{\alpha, \Delta \Rightarrow^- \gamma \quad \beta, \Delta \Rightarrow^- \gamma}{\alpha \wedge \beta, \Delta \Rightarrow^- \gamma}\ (-\wedge \mathrm{l})$$

$$\frac{\Gamma \Rightarrow^- \alpha}{\Gamma \Rightarrow^- \alpha \wedge \beta}\ (-\wedge \mathrm{r1}) \quad \frac{\Gamma \Rightarrow^- \beta}{\Gamma \Rightarrow^- \alpha \wedge \beta}\ (-\wedge \mathrm{r2})$$

$$\frac{\alpha, \Delta \Rightarrow^- \gamma}{\alpha \vee \beta, \Delta \Rightarrow^- \gamma}\ (-\vee \mathrm{l1}) \quad \frac{\beta, \Delta \Rightarrow^- \gamma}{\alpha \vee \beta, \Delta \Rightarrow^- \gamma}\ (-\vee \mathrm{l2})$$

$$\frac{\Gamma \Rightarrow^- \alpha \quad \Gamma \Rightarrow^- \beta}{\Gamma \Rightarrow^- \alpha \vee \beta}\ (-\vee \mathrm{r}).$$

Let L be Dn4 $-$ {($-$cut), (+cut)}. Then, the following facts hold: for any multisets Γ, Δ, and any formulas α, γ,

1. $\mathrm{L} \vdash \Gamma \Rightarrow^+ \gamma$ iff $\mathrm{L} \vdash \Gamma \Rightarrow^+ {\sim\sim}\gamma$,
2. $\mathrm{L} \vdash \Gamma \Rightarrow^- \gamma$ iff $\mathrm{L} \vdash \Gamma \Rightarrow^- {\sim\sim}\gamma$,
3. $\mathrm{L} \vdash \alpha, \Gamma \Rightarrow^+ \gamma$ iff $\mathrm{L} \vdash {\sim\sim}\alpha, \Gamma \Rightarrow^+ \gamma$,
4. $\mathrm{L} \vdash \alpha, \Gamma \Rightarrow^- \gamma$ iff $\mathrm{L} \vdash {\sim\sim}\alpha, \Gamma \Rightarrow^- \gamma$.

Theorem 2.3.19 (Equivalence between Dn4 and N4) *Let* Γ *be a multiset of formulas, and* γ *be a formula. Then:*

1. *if* Dn4 $\vdash \Gamma \Rightarrow^* \gamma$ $(* \in \{+, -\})$, *then* N4 $\vdash \Gamma \Rightarrow \gamma$ *if* $* = +$, *or* N4 $\vdash {\sim}\Gamma \Rightarrow {\sim}\gamma$ *if* $* = -$,
2. *if* N4 $-$ (cut) $\vdash \Gamma \Rightarrow \gamma$, *then* Dn4 $-$ {($-$cut), (+cut)} $\vdash \Gamma \Rightarrow^+ \gamma$.

Proof. • (1): By induction on the proofs P of $\Gamma \Rightarrow^* \gamma$ ($* \in \{+, -\}$) in Dn4. We distinguish the cases according to the last inference of P. We only show the following cases.

Case $(+/-1)$: The last inference of P is of the form:

$$\frac{{\sim}\Gamma, \Delta \Rightarrow^+ \gamma}{\Gamma, {\sim}\Delta \Rightarrow^- {\sim}\gamma}\ (+/-1).$$

By induction hypothesis, we have: N4 $\vdash {\sim}\Gamma, \Delta \Rightarrow \gamma$. Then we obtain the required fact: N4 $\vdash {\sim}\Gamma, {\sim}{\sim}\Delta \Rightarrow {\sim}{\sim}\gamma$ by using $({\sim}\mathrm{l})$ and $({\sim}\mathrm{r})$.

Case $(-{\rightarrow}\mathrm{l})$: The last inference of P is of the form:

$$\frac{\beta, {\sim}\alpha, \Delta \Rightarrow^- \gamma}{\alpha{\rightarrow}\beta, \Delta \Rightarrow^- \gamma}\ (-{\rightarrow}\mathrm{l}).$$

By induction hypothesis, we have: N4 $\vdash {\sim}\beta, {\sim}{\sim}\alpha, {\sim}\Delta \Rightarrow {\sim}\gamma$. Then we obtain the required fact:

$$\cfrac{\cfrac{\cfrac{\alpha \Rightarrow \alpha}{\alpha \Rightarrow {\sim}{\sim}\alpha}\ ({\sim}\mathrm{r}) \qquad \begin{array}{c}\vdots\\ {\sim}\beta, {\sim}{\sim}\alpha, {\sim}\Delta \Rightarrow {\sim}\gamma\end{array}}{{\sim}\beta, \alpha, {\sim}\Delta \Rightarrow {\sim}\gamma}\ (\mathrm{cut})}{{\sim}(\alpha{\rightarrow}\beta), {\sim}\Delta \Rightarrow {\sim}\gamma}\ ({\sim}{\rightarrow}\mathrm{l}).$$

• (2): By induction on the cut-free proofs P of $\Gamma \Rightarrow \gamma$ in N4. We distinguish cases according to the last inference of P. We only show the following case.

Case $({\sim}{\rightarrow}\mathrm{l})$: The last inference of P is of the form:

$$\frac{{\sim}\beta, \alpha, \Delta \Rightarrow \gamma}{{\sim}(\alpha{\rightarrow}\beta), \Delta \Rightarrow \gamma}\ ({\sim}{\rightarrow}\mathrm{l}).$$

By induction hypothesis, we have:
Dn4 $- \{(-\mathrm{cut}), (+\mathrm{cut})\} \vdash {\sim}\beta, \alpha, \Delta \Rightarrow^+ \gamma$. Then we have the required fact:

$$\cfrac{\cfrac{\cfrac{\begin{array}{c}\vdots\\ {\sim}\beta, \alpha, \Delta \Rightarrow^+ \gamma\end{array}}{\beta, {\sim}\alpha, {\sim}\Delta \Rightarrow^- {\sim}\gamma}\ (+/-1)}{\alpha{\rightarrow}\beta, {\sim}\Delta \Rightarrow^- {\sim}\gamma}\ (-{\rightarrow}\mathrm{l})}{{\sim}(\alpha{\rightarrow}\beta), \Delta \Rightarrow^+ \gamma}\ (-/+2).$$

∎

Theorem 2.3.20 (Cut-elimination) *The rules* (+cut) *and* (−cut) *are admissible in cut-free* Dn4.

Proof. Suppose that a sequent $\Gamma \Rightarrow^* \gamma$ ($* \in \{+, -\}$) is provable in Dn4. Then, by Theorem 2.3.19 (1), the sequent $\Gamma \Rightarrow \gamma$ is provable in N4 if $* = +$, or $\sim\Gamma \Rightarrow \sim\gamma$ is provable in N4 if $* = -$. Hence the sequent $\Gamma \Rightarrow \gamma$ or $\sim\Gamma \Rightarrow \sim\gamma$ is provable in cut-free N4 by the cut-elimination theorem for N4. If $\Gamma \Rightarrow \gamma$ is provable in cut-free N4, then $\Gamma \Rightarrow^+ \gamma$ is provable in cut-free Dn4 by Theorem 2.3.19 (2). If $\sim\Gamma \Rightarrow \sim\gamma$ is provable in cut-free N4, then $\sim\Gamma \Rightarrow^+ \sim\gamma$ is provable in cut-free Dn4 by Theorem 2.3.19 (2), and hence $\Gamma \Rightarrow^- \gamma$ is provable in cut-free Dn4. ∎

2.3.4. Display calculus

Display calculi have been developed by Nuel Belnap [23] as a flexible extension of Gentzen's sequent calculi. They use a language of sequents that enables the formulation of structural rules in terms of structure connectives and permit the combination of structure operations having a different inferential behaviour. Display sequent systems use their structure connectives as context sensitive or context-shift operations. Like Gentzen's comma, a structure operation typically has an antecedent interpretation as a certain object language connective and a succedent interpretation as another object language connective. These object language connectives may be said to be Gentzen duals of each other and often they form a residuated pair. The term "display logic" emphasizes the fact that the structure connectives are regulated by rules (the so-called "display postulates") that allow a single formula or a more complex structure of a sequent to be displayed as the entire antecedent or succedent of a structurally equivalent sequent.

A display sequent calculus for Nelson's four-valued logic may be found in [209]. This system extends rules for a display calculus for positive intuitionistic logic by rules for strongly negated formulas. We here present the display calculus $\delta\text{N4}^\perp$ for the logic $\text{N4}^\perp$ from [138].[3] The system $\delta\text{N4}^\perp$ has separate rules for each connective. The logic $\text{N4}^\perp$ conservatively extends N4 by a falsity constant $\perp$. An axiom system $\text{HN4}^\perp$ for $\text{N4}^\perp$ is obtained from HN4 by adding the axiom schemes

$$\perp \to \alpha \quad \text{and} \quad \alpha \to \sim\perp.$$

The system δN4 makes use of a binary structure connective $\circ$, a unary shift operation $*$, and the structure constant $\mathbf{I}$ ("the empty structure").

Definition 2.3.21 *The set of structures* Struc$(\mathcal{L}_{\text{N4}\cup\{\perp\}})$ *generated*

[3] This presentation modifies and amends the presentation in [200].

from $\mathcal{L}_{\text{N4}} \cup \{\perp\}$ *is defined as follows:*

$$\begin{array}{rl} \textit{formulas:} & \alpha \in \mathcal{L}_{\text{N4}} \cup \{\perp\} \\ \textit{structures:} & X \in \text{Struc}(\mathcal{L}_{\text{N4}}) \\ X ::= & \alpha \mid \mathbf{I} \mid (X \circ X). \end{array}$$

The sequents of δN4 are expressions

$$X_1 \mid X_2 \Rightarrow X_3 \mid X_4,$$

where each X_i ($i \in \{1, 2, 3, 4\}$) is a structure.

Definition 2.3.22 (Display calculus δN4$^\perp$) *The initial sequents of* δN4$^\perp$ *are of the form: for any propositional variable* p,

$$p \mid \mathbf{I} \Rightarrow p \mid \mathbf{I} \qquad \mathbf{I} \mid p \Rightarrow \mathbf{I} \mid p.$$

The display postulates of δN4$^\perp$ *are of the form:*

$$\begin{array}{c} X_1 \mid X_2 \circ *Y \Rightarrow X_3 \mid X_4 \\ \hline\hline X_1 \circ Y \mid X_2 \Rightarrow X_3 \mid X_4 \\ \hline\hline Y \mid *X_1 \circ X_2 \Rightarrow X_3 \mid X_4 \end{array} \qquad \begin{array}{c} X_1 \circ *Y \mid X_2 \Rightarrow X_3 \mid X_4 \\ \hline\hline X_1 \mid X_2 \circ Y \Rightarrow X_3 \mid X_4 \\ \hline\hline *X_2 \circ X_1 \mid Y \Rightarrow X_3 \mid X_4 \end{array}$$

$$\begin{array}{c} X_1 \mid X_2 \Rightarrow X_3 \mid X_4 \circ *Y \\ \hline\hline X_1 \mid X_2 \Rightarrow X_3 \circ Y \mid X_4 \\ \hline\hline X_1 \mid X_2 \Rightarrow Y \mid *X_3 \circ X_4 \end{array} \qquad \begin{array}{c} X_1 \mid X_2 \Rightarrow X_3 \circ *Y \mid X_4 \\ \hline\hline X_1 \mid X_2 \Rightarrow X_3 \mid X_4 \circ Y \\ \hline\hline X_1 \mid X_2 \Rightarrow *X_4 \circ X_3 \mid Y \end{array}$$

where two double-line rules are combined into a single one at a time, and the double-line rules indicate that the rules may be applied both downwards and upwards.

The additional structural rules of δN4$^\perp$ *are of the form:*

$$\frac{X_1 \mid X_2 \Rightarrow \alpha \mid X_4 \quad \alpha \mid X_2' \Rightarrow X_3 \mid X_4'}{X_1 \mid X_2 \circ X_2' \Rightarrow X_3 \mid X_4 \circ X_4'} \ \textit{(odd cut)}$$

$$\frac{X_1 \mid X_2 \Rightarrow X_3 \mid \alpha \quad X_1' \mid \alpha \Rightarrow X_3' \mid X_4}{X_1 \circ X_1' \mid X_2 \Rightarrow X_3 \circ X_3' \mid X_4} \ \textit{(even cut)}$$

$$\begin{array}{c} \mathbf{I} \circ X_1 \mid X_2 \Rightarrow X_3 \mid X_4 \\ \hline\hline X_1 \mid X_2 \Rightarrow X_3 \mid X_4 \end{array} \qquad \begin{array}{c} X_1 \mid \mathbf{I} \circ X_2 \Rightarrow X_3 \mid X_4 \\ \hline\hline X_1 \mid X_2 \Rightarrow X_3 \mid X_4 \end{array}$$

$$\begin{array}{c} X_1 \mid X_2 \Rightarrow \mathbf{I} \circ X_3 \mid X_4 \\ \hline\hline X_1 \mid X_2 \Rightarrow X_3 \mid X_4 \end{array} \qquad \begin{array}{c} X_1 \mid X_2 \Rightarrow X_3 \mid \mathbf{I} \circ X_4 \\ \hline\hline X_1 \mid X_2 \Rightarrow X_3 \mid X_4 \end{array}$$

$$\frac{(X \circ Y) \circ Z \mid X_2 \Rightarrow X_3 \mid X_4}{X \circ (Y \circ Z) \mid X_2 \Rightarrow X_3 \mid X_4} \quad \frac{X_1 \mid X_2 \Rightarrow (X \circ Y) \circ Z \mid X_4}{X_1 \mid X_2 \Rightarrow X \circ (Y \circ Z) \mid X_4}$$

$$\frac{Y \circ X \mid X_2 \Rightarrow X_3 \mid X_4}{X \circ Y \mid X_2 \Rightarrow X_3 \mid X_4} \quad \frac{X_1 \mid X_2 \Rightarrow Y \circ X \mid X_4}{X_1 \mid X_2 \Rightarrow X \circ Y \mid X_4}$$

$$\frac{X \circ X \mid X_2 \Rightarrow X_3 \mid X_4}{X \mid X_2 \Rightarrow X_3 \mid X_4} \quad \frac{X_1 \mid X_2 \Rightarrow X \circ X \mid X_4}{X_1 \mid X_2 \Rightarrow X \mid X_4}$$

$$\frac{X_1 \mid X_2 \Rightarrow X_3 \mid X_4}{X_1 \circ Y \mid X_2 \Rightarrow X_3 \mid X_4} \quad \frac{X_1 \mid X_2 \Rightarrow X_3 \mid X_4}{X_1 \mid X_2 \Rightarrow X_3 \circ Y \mid X_4}$$

The introduction rules of $\delta\mathrm{N4}^{\perp}$ *are of the form:*

$$\frac{X_1 \mid \mathrm{X}_2 \Rightarrow \mathbf{I} \mid X_4}{X_1 \mid \mathrm{X}_2 \Rightarrow \perp \mid X_4} \quad X_1 \mid \mathrm{X}_2 \Rightarrow X_3 \mid \perp$$

$$\perp \mid \mathrm{X}_2 \Rightarrow X_3 \mid X_4 \quad \frac{X_1 \mid \mathbf{I} \Rightarrow X_3 \mid X_4}{X_1 \mid \perp \Rightarrow X_3 \mid X_4}$$

$$\frac{X_1 \mid \mathrm{X}_2 \Rightarrow *\alpha \mid X_4}{X_1 \mid \mathrm{X}_2 \Rightarrow {\sim}\alpha \mid X_4} \quad \frac{X_1 \mid \mathrm{X}_2 \Rightarrow X_3 \mid *\alpha}{X_1 \mid \mathrm{X}_2 \Rightarrow X_3 \mid {\sim}\alpha}$$

$$\frac{*\alpha \mid \mathrm{X}_2 \Rightarrow X_3 \mid X_4}{{\sim}\alpha \mid \mathrm{X}_2 \Rightarrow X_3 \mid X_4} \quad \frac{X_1 \mid *\alpha \Rightarrow X_3 \mid X_4}{X_1 \mid {\sim}\alpha \Rightarrow X_3 \mid X_4}$$

$$\frac{X_1 \mid \mathrm{X}_2 \Rightarrow \alpha \mid X_4 \quad X_1 \mid \mathrm{X}_2 \Rightarrow \beta \mid X_4}{X_1 \mid \mathrm{X}_2 \Rightarrow \alpha \wedge \beta \mid X_4} \quad \frac{X_1 \mid \mathrm{X}_2 \Rightarrow X_3 \mid \alpha \circ \beta}{X_1 \mid \mathrm{X}_2 \Rightarrow X_3 \mid \alpha \wedge \beta}$$

$$\frac{\alpha \circ \beta \mid \mathrm{X}_2 \Rightarrow X_3 \mid X_4 \quad X_1 \mid \alpha \Rightarrow X_3 \mid X_4 \quad X_1 \mid \beta \Rightarrow X_3 \mid X_4}{\alpha \wedge \beta \mid \mathrm{X}_2 \Rightarrow X_3 \mid X_4 \quad X_1 \mid \alpha \wedge \beta \Rightarrow X_3 \mid X_4}$$

$$\frac{X_1 \mid X_2 \Rightarrow \alpha \circ \beta \mid X_4}{X_1 \mid X_2 \Rightarrow \alpha \vee \beta \mid X_4} \quad \frac{X_1 \mid X_2 \Rightarrow X_3 \mid \alpha \quad X_1 \mid X_2 \Rightarrow X_3 \mid \beta}{X_1 \mid X_2 \Rightarrow X_3 \mid \alpha \vee \beta}$$

$$\frac{\alpha \mid \mathrm{X}_2 \Rightarrow X_3 \mid X_4 \quad \beta \mid \mathrm{X}_2 \Rightarrow X_3 \mid X_4}{\alpha \vee \beta \mid \mathrm{X}_2 \Rightarrow X_3 \mid X_4} \quad \frac{X_1 \mid \alpha \circ \beta \Rightarrow X_3 \mid X_4}{X_1 \mid \alpha \vee \beta \Rightarrow X_3 \mid X_4}$$

$$\frac{X_1 \circ \alpha \mid \mathrm{X}_2 \Rightarrow \beta \mid \mathbf{I}}{X_1 \mid \mathrm{X}_2 \Rightarrow \alpha \to \beta \mid \mathbf{I}} \quad \frac{X \mid X_2 \Rightarrow \alpha \mid \mathbf{I} \quad Y \mid X_2 \Rightarrow \mathbf{I} \mid \beta}{X \circ Y \mid X_2 \Rightarrow \mathbf{I} \mid \alpha \to \beta}$$

$$\frac{X_1 \mid \mathrm{X}_2 \Rightarrow \alpha \mid X_4 \quad \beta \mid \mathrm{X}_2 \Rightarrow X_3 \mid X_4}{\alpha \to \beta \mid \mathrm{X}_2 \circ *X_1 \Rightarrow X_3 \mid X_4} \quad \frac{\alpha \circ *\beta \mid *X_1 \Rightarrow X_3 \mid X_4}{X_1 \mid \alpha \to \beta \Rightarrow X_3 \mid X_4}$$

Observation 2.3.23 *For any formula α from $\mathcal{L}_{\mathrm{N4}} \cup \{\bot\}$: $\delta\mathrm{N4}^{\bot} \vdash \mathbf{I} \mid \alpha \Rightarrow \mathbf{I} \mid \alpha$ and $\delta\mathrm{N4}^{\bot} \vdash \alpha \mid \mathbf{I} \Rightarrow \alpha \mid \mathbf{I}$.*

Proof. By induction on the structure of α. ∎

Two sequents are said to be structurally equivalent if they are interderivable by means of display postulates only. It is characteristic of display calculi that any substructure of a given sequent may be displayed as the entire antecedent or succedent of a structurally equivalent sequent. In the case of $\delta\mathrm{N4}^{\bot}$, a distinction is drawn between odd and even positions of the antecedent and the succedent. An occurrence of a substructure in a given structure is said to be positive (negative) iff it is in the scope of an even (uneven) number of $*$'s. An o-antecedent (e-antecedent) part of a sequent $X_1 \mid X_2 \Rightarrow X_3 \mid X_4$ is a positive occurrence of a substructure of X_1 or a negative occurrence of substructure of X_2 (a positive occurrence of a substructure of X_2 or a negative occurrence of substructure of X_1). An o-succedent (e-succedent) part of $X_1 \mid X_2 \Rightarrow X_3 \mid X_4$ is a positive occurrence of a substructure of X_3 or a negative occurrence of substructure of X_4 (a positive occurrence of a substructure of X_4 or a negative occurrence of substructure of X_3).

Theorem 2.3.24 ([23]) *For every sequent* s *and every o-antecedent (e-antecedent) part X of* s*, there exists a sequent* s′ *structurally equivalent to* s *such that X is the entire o-antecedent (e-antecedent) of* s′*. For every sequent* s *and every o-succedent (e-succedent) part X of* s*, there exists a sequent* s′ *structurally equivalent to* s *such that X is the entire o-succedent (e-succedent) of* s′*.*

We use $\top$ as an abbreviation of $p \to p$ for some propositional variable p and define a translation τ from sequents into formulas of $\delta\mathrm{N4}^{\bot}$ as follows:

$$\tau(X_1 \mid X_2 \Rightarrow X_3 \mid X_4) := (t_1(X_1) \wedge \tau_2(X_2)) \to (\tau_3(X_3) \vee \tau_4(X_4)),$$

where t_i ($i \in \{1,2,3,4\}$ is defined as follows:

$$\begin{array}{rcl} \tau_1(\alpha) & := & \tau_3(\alpha) & = & \alpha, \\ \tau_2(\alpha) & := & \tau_4(\alpha) & = & {\sim}\alpha, \\ \tau_1(\mathbf{I}) & := & \tau_2(\mathbf{I}) & = & \top, \\ \tau_3(\mathbf{I}) & := & \tau_4(\mathbf{I}) & = & \bot, \\ \tau_1(*X) & := & \tau_2(X), \\ \tau_2(*X) & := & \tau_1(X), \\ \tau_3(*X) & := & \tau_4(X), \\ \tau_4(*X) & := & \tau_3(X), \end{array}$$

$$\begin{array}{rcl}
\tau_1(X \circ Y) & := & t_1(X) \wedge t_1(Y), \\
\tau_2(X \circ Y) & := & t_2(X) \wedge t_2(Y), \\
\tau_3(X \circ Y) & := & t_3(X) \vee t_3(Y), \\
\tau_4(X \circ Y) & := & t_4(X) \vee t_4(Y).
\end{array}$$

Theorem 2.3.25 *For any formula* α *from* $\mathrm{HN4}^{\perp}$*:*

(i) $\mathrm{HN4}^{\perp} \vdash \alpha$ *implies* $\delta\mathrm{N4}^{\perp} \vdash \mathbf{I} \mid \mathbf{I} \Rightarrow \alpha \mid \mathbf{I}$.

(ii) $\delta\mathrm{N4}^{\perp} \vdash X_1 \mid X_2 \Rightarrow X_3 \mid X_4$ *implies* $\mathrm{HN4}^{\perp} \vdash \tau(X_1 \mid X_2 \Rightarrow X_3 \mid X_4)$.

Proof. (i): By induction on proofs in $\mathrm{HN4}^{\perp}$. We present some cases and use vertical dots to indicate a series of applications of display postulates and other structural rules.

Axiom schemes $\perp \to \alpha$ and $\alpha \to {\sim}\perp$:

$$\dfrac{\dfrac{\perp \mid \mathbf{I} \Rightarrow \alpha \mid \mathbf{I}}{\mathbf{I} \circ \perp \mid \mathbf{I} \Rightarrow \alpha \mid \mathbf{I}}}{\mathbf{I} \mid \mathbf{I} \Rightarrow \perp \to \alpha \mid \mathbf{I}} \qquad \dfrac{\dfrac{\dfrac{\mathbf{I} \circ \alpha \mid \mathbf{I} \Rightarrow *\mathbf{I} \mid \perp}{\vdots}}{\mathbf{I} \circ \alpha \mid \mathbf{I} \Rightarrow *\perp \mid \mathbf{I}}}{\dfrac{\mathbf{I} \circ \alpha \mid \mathbf{I} \Rightarrow {\sim}\perp \mid \mathbf{I}}{\mathbf{I} \mid \mathbf{I} \Rightarrow \alpha \to {\sim}\perp \mid \mathbf{I}}}$$

Axiom scheme ${\sim}(\alpha \vee \beta) \leftrightarrow {\sim}\alpha \wedge {\sim}\beta$:

$$\dfrac{\dfrac{\dfrac{\dfrac{\dfrac{\dfrac{\dfrac{\mathbf{I} \mid \alpha \Rightarrow \mathbf{I} \mid \alpha}{\mathbf{I} \mid \alpha \circ \beta \Rightarrow \mathbf{I} \mid \alpha}}{\mathbf{I} \mid \alpha \vee \beta \Rightarrow \mathbf{I} \mid \alpha}}{\vdots}}{*(\alpha \vee \beta) \mid \mathbf{I} \Rightarrow *\alpha \mid \mathbf{I}}}{*(\alpha \vee \beta) \mid \mathbf{I} \Rightarrow {\sim}\alpha \mid \mathbf{I}}}{{\sim}(\alpha \vee \beta) \mid \mathbf{I} \Rightarrow {\sim}\alpha \mid \mathbf{I}} \quad \dfrac{\dfrac{\dfrac{\dfrac{\dfrac{\dfrac{\mathbf{I} \mid \beta \Rightarrow \mathbf{I} \mid \beta}{\mathbf{I} \mid \alpha \circ \beta \Rightarrow \mathbf{I} \mid \beta}}{\mathbf{I} \mid \alpha \vee \beta \Rightarrow \mathbf{I} \mid \beta}}{\vdots}}{*(\alpha \vee \beta) \mid \mathbf{I} \Rightarrow *\beta \mid \mathbf{I}}}{*(\alpha \vee \beta) \mid \mathbf{I} \Rightarrow {\sim}\beta \mid \mathbf{I}}}{{\sim}(\alpha \vee \beta) \mid \mathbf{I} \Rightarrow {\sim}\beta \mid \mathbf{I}}}{\dfrac{{\sim}(\alpha \vee \beta) \circ {\sim}(\alpha \vee \beta) \mid \mathbf{I} \Rightarrow {\sim}\alpha \wedge {\sim}\beta \mid \mathbf{I}}{\dfrac{{\sim}(\alpha \vee \beta) \mid \mathbf{I} \Rightarrow {\sim}\alpha \wedge {\sim}\beta \mid \mathbf{I}}{\dfrac{\mathbf{I} \circ {\sim}(\alpha \vee \beta) \mid \mathbf{I} \Rightarrow {\sim}\alpha \wedge {\sim}\beta \mid \mathbf{I}}{\mathbf{I} \mid \mathbf{I} \Rightarrow {\sim}(\alpha \vee \beta) \to ({\sim}\alpha \wedge {\sim}\beta) \mid \mathbf{I}}}}}$$

$$
\begin{array}{c}
\mathbf{I} \mid \alpha \Rightarrow \mathbf{I} \mid \alpha \quad \mathbf{I} \mid \beta \Rightarrow \mathbf{I} \mid \beta \\
\hline
\mathbf{I} \mid \alpha \circ \beta \Rightarrow \mathbf{I} \mid \alpha \vee \beta \\
\hline
\vdots \\
\hline
*\beta \mid \alpha \Rightarrow \mathbf{I} \mid \alpha \vee \beta \\
\hline
{\sim}\beta \mid \alpha \Rightarrow \mathbf{I} \mid \alpha \vee \beta \\
\hline
\vdots \\
\hline
*\alpha \mid *{\sim}\beta \Rightarrow \mathbf{I} \mid \alpha \vee \beta \\
\hline
{\sim}\alpha \mid *{\sim}\beta \Rightarrow \mathbf{I} \mid \alpha \vee \beta \\
\hline
\vdots \\
\hline
{\sim}\alpha \circ {\sim}\beta \mid \mathbf{I} \Rightarrow *(\alpha \vee \beta) \mid \mathbf{I} \\
\hline
{\sim}\alpha \wedge {\sim}\beta \mid \mathbf{I} \Rightarrow *(\alpha \vee \beta) \mid \mathbf{I} \\
\hline
{\sim}\alpha \wedge {\sim}\beta \mid \mathbf{I} \Rightarrow {\sim}(\alpha \vee \beta) \mid \mathbf{I} \\
\hline
\mathbf{I} \circ ({\sim}\alpha \wedge {\sim}\beta) \mid \mathbf{I} \Rightarrow {\sim}(\alpha \vee \beta) \mid \mathbf{I} \\
\hline
\mathbf{I} \mid \mathbf{I} \Rightarrow ({\sim}\alpha \wedge {\sim}\beta) \to {\sim}(\alpha \vee \beta) \mid \mathbf{I}
\end{array}
$$

Axiom scheme $(\alpha \to (\beta \to \gamma)) \to ((\alpha \to \beta) \to (\alpha \to \gamma))$:

$$
\begin{array}{c}
\dfrac{
\begin{array}{c}
\alpha \mid \mathbf{I} \Rightarrow \alpha \mid \mathbf{I} \\
\hline
\alpha \mid \mathbf{I} \circ (*(\alpha \to \beta) \circ *\alpha) \Rightarrow \alpha \mid \mathbf{I} \\
\hline
\alpha \mid *(\alpha \to \beta) \circ *\alpha \Rightarrow \alpha \mid \mathbf{I}
\end{array}
\qquad
\begin{array}{c}
\dfrac{
\begin{array}{c}
\alpha \mid \mathbf{I} \Rightarrow \alpha \mid \mathbf{I} \\
\hline
\alpha \mid \mathbf{I} \circ *(\beta \to \gamma) \Rightarrow \alpha \mid \mathbf{I} \\
\hline
\alpha \mid *(\beta \to \gamma) \Rightarrow \alpha \mid \mathbf{I}
\end{array}
\quad
\begin{array}{c}
\beta \mid \mathbf{I} \Rightarrow \beta \mid \mathbf{I} \quad \gamma \mid \mathbf{I} \Rightarrow \gamma \mid \mathbf{I} \\
\hline
\beta \to \gamma \mid \mathbf{I} \circ *\beta \Rightarrow \gamma \mid \mathbf{I} \\
\hline
\vdots \\
\hline
\beta \mid *(\beta \to \gamma) \Rightarrow \gamma \mid \mathbf{I}
\end{array}
}{\alpha \to \beta \mid *(\beta \to \gamma) \circ *\alpha \Rightarrow \gamma \mid \mathbf{I}} \\
\hline
\vdots \\
\hline
\beta \to \gamma \mid *(\alpha \to \beta) \circ *\alpha \Rightarrow \gamma \mid \mathbf{I}
\end{array}
}{\alpha \to (\beta \to \gamma) \mid (*(\alpha \to \beta)) \circ *\alpha) \circ *\alpha \Rightarrow \gamma \mid \mathbf{I}} \\
\hline
\vdots \\
\hline
(((\alpha \to (\beta \to \gamma)) \circ (\alpha \to \beta)) \circ \alpha) \circ \alpha \mid \mathbf{I} \Rightarrow \gamma \mid \mathbf{I} \\
\hline
((\alpha \to (\beta \to \gamma)) \circ (\alpha \to \beta)) \circ \alpha \mid \mathbf{I} \Rightarrow \gamma \mid \mathbf{I} \\
\hline
(\alpha \to (\beta \to \gamma)) \circ (\alpha \to \beta) \mid \mathbf{I} \Rightarrow \alpha \to \gamma \mid \mathbf{I} \\
\hline
(\alpha \to (\beta \to \gamma)) \mid \mathbf{I} \Rightarrow (\alpha \to \beta) \to (\alpha \to \gamma) \mid \mathbf{I} \\
\hline
\mathbf{I} \circ (\alpha \to (\beta \to \gamma)) \mid \mathbf{I} \Rightarrow (\alpha \to \beta) \to (\alpha \to \gamma) \mid \mathbf{I} \\
\hline
\mathbf{I} \mid \mathbf{I} \Rightarrow (\alpha \to (\beta \to \gamma)) \to ((\alpha \to \beta) \to (\alpha \to \gamma)) \mid \mathbf{I}
\end{array}
$$

Axiom scheme $(\alpha \to \beta) \to ((\alpha \to \gamma) \to (\alpha \to (\beta \wedge \gamma)))$:

$$\dfrac{\dfrac{\dfrac{\dfrac{\dfrac{\begin{array}{c}\dfrac{\alpha \mid \mathbf{I} \Rightarrow \alpha \mid \mathbf{I} \quad \beta \mid \mathbf{I} \Rightarrow \beta \mid \mathbf{I}}{\alpha \to \beta \mid \mathbf{I} \circ *\alpha \Rightarrow \beta \mid \mathbf{I}} \\ \vdots \end{array}}{((\alpha \to \beta) \circ (\alpha \to \gamma)) \circ \alpha \mid \mathbf{I} \Rightarrow \beta \mid \mathbf{I}} \quad \dfrac{\begin{array}{c}\dfrac{\alpha \mid \mathbf{I} \Rightarrow \alpha \mid \mathbf{I} \quad \gamma \mid \mathbf{I} \Rightarrow \gamma \mid \mathbf{I}}{\alpha \to \gamma \mid \mathbf{I} \circ *\alpha \Rightarrow \gamma \mid \mathbf{I}} \\ \vdots \end{array}}{((\alpha \to \beta) \circ (\alpha \to \gamma)) \circ \alpha \mid \mathbf{I} \Rightarrow \gamma \mid \mathbf{I}}}{\dfrac{(((\alpha \to \beta) \circ (\alpha \to \gamma)) \circ \alpha) \circ (((\alpha \to \beta) \circ (\alpha \to \gamma)) \circ \alpha) \mid \mathbf{I} \Rightarrow \beta \wedge \gamma \mid \mathbf{I}}{((\alpha \to \beta) \circ (\alpha \to \gamma)) \circ \alpha \mid \mathbf{I} \Rightarrow \beta \wedge \gamma \mid \mathbf{I}}}}{(\alpha \to \beta) \circ (\alpha \to \gamma) \mid \mathbf{I} \Rightarrow \alpha \to (\beta \wedge \gamma) \mid \mathbf{I}}}{\dfrac{\alpha \to \beta \mid \mathbf{I} \Rightarrow ((\alpha \to \gamma) \to (\alpha \to (\beta \wedge \gamma))) \mid \mathbf{I}}{\mathbf{I} \circ (\alpha \to \beta) \mid \mathbf{I} \Rightarrow ((\alpha \to \gamma) \to (\alpha \to (\beta \wedge \gamma))) \mid \mathbf{I}}}}{\mathbf{I} \mid \mathbf{I} \Rightarrow (\alpha \to \beta) \to ((\alpha \to \gamma) \to (\alpha \to (\beta \wedge \gamma))) \mid \mathbf{I}}$$

Axiom scheme $(\alpha \to \gamma) \to ((\beta \to \gamma) \to ((\alpha \vee \beta) \to \gamma))$:

$$\dfrac{\begin{array}{c}\dfrac{\dfrac{\dfrac{\begin{array}{c}\dfrac{\alpha \mid \mathbf{I} \Rightarrow \alpha \mid \mathbf{I} \quad \gamma \mid \mathbf{I} \Rightarrow \gamma \mid \mathbf{I}}{\alpha \to \gamma \mid \mathbf{I} \circ *\alpha \Rightarrow \gamma \mid \mathbf{I}} \\ \vdots \end{array}}{\alpha \mid *(\alpha \to \gamma) \Rightarrow \gamma \mid \mathbf{I}}}{\alpha \mid *(\alpha \to \gamma) \circ *(\beta \to \gamma) \Rightarrow \gamma \mid \mathbf{I}} \quad \dfrac{\dfrac{\begin{array}{c}\dfrac{\beta \mid \mathbf{I} \Rightarrow \beta \mid \mathbf{I} \quad \gamma \mid \mathbf{I} \Rightarrow \gamma \mid \mathbf{I}}{\beta \to \gamma \mid \mathbf{I} \circ *\beta \Rightarrow \gamma \mid \mathbf{I}} \\ \vdots \end{array}}{\beta \mid *(\beta \to \gamma) \Rightarrow \gamma \mid \mathbf{I}}}{\beta \mid *(\alpha \to \gamma) \circ *(\beta \to \gamma) \Rightarrow \gamma \mid \mathbf{I}}}{\alpha \vee \beta \mid *(\alpha \to \gamma) \circ *(\beta \to \gamma) \Rightarrow \gamma \circ \gamma \mid \mathbf{I}} \\ \vdots \end{array}}{\dfrac{((\alpha \to \gamma) \circ (\beta \to \gamma)) \circ (\alpha \vee \beta) \mid \mathbf{I} \Rightarrow \gamma \circ \gamma \mid \mathbf{I}}{\dfrac{((\alpha \to \gamma) \circ (\beta \to \gamma)) \circ (\alpha \vee \beta) \mid \mathbf{I} \Rightarrow \gamma \mid \mathbf{I}}{\dfrac{(\alpha \to \gamma) \circ (\beta \to \gamma) \mid \mathbf{I} \Rightarrow (\alpha \vee \beta) \to \gamma \mid \mathbf{I}}{\dfrac{\alpha \to \gamma \mid \mathbf{I} \Rightarrow (\beta \to \gamma) \to ((\alpha \vee \beta) \to \gamma) \mid \mathbf{I}}{\dfrac{\mathbf{I} \circ (\alpha \to \gamma) \mid \mathbf{I} \Rightarrow (\beta \to \gamma) \to ((\alpha \vee \beta) \to \gamma) \mid \mathbf{I}}{\mathbf{I} \mid \mathbf{I} \Rightarrow (\alpha \to \gamma) \to ((\beta \to \gamma) \to ((\alpha \vee \beta) \to \gamma)) \mid \mathbf{I}}}}}}}$$

Modus ponens: By the induction hypothesis, $\delta\mathrm{N4}^{\perp} \vdash \mathbf{I} \mid \mathbf{I} \Rightarrow \alpha \mid \mathbf{I}$ and $\delta\mathrm{N4}^{\perp} \vdash \mathbf{I} \mid \mathbf{I} \Rightarrow \alpha \to \beta \mid \mathbf{I}$. We have:

$$\dfrac{\dfrac{\dfrac{\dfrac{\mathbf{I} \mid \mathbf{I} \Rightarrow \alpha \mid \mathbf{I} \quad \mathbf{I} \mid \mathbf{I} \Rightarrow \alpha \to \beta \mid \mathbf{I}}{\mathbf{I} \circ \mathbf{I} \mid \mathbf{I} \Rightarrow \alpha \wedge (\alpha \to \beta) \mid \mathbf{I}}}{\mathbf{I} \mid \mathbf{I} \Rightarrow \alpha \wedge (\alpha \to \beta) \mid \mathbf{I}} \quad \dfrac{\dfrac{\dfrac{\alpha \mid \mathbf{I} \Rightarrow \alpha \mid \mathbf{I} \quad \beta \mid \mathbf{I} \Rightarrow \beta \mid \mathbf{I}}{\alpha \to \beta \mid \mathbf{I} \circ *\alpha \Rightarrow \beta \mid \mathbf{I}}}{\alpha \circ (\alpha \to \beta) \mid \mathbf{I} \Rightarrow \beta \mid \mathbf{I}}}{\alpha \wedge (\alpha \to \beta) \mid \mathbf{I} \Rightarrow \beta \mid \mathbf{I}}}{\dfrac{\mathbf{I} \mid \mathbf{I} \circ \mathbf{I} \Rightarrow \beta \mid \mathbf{I} \circ \mathbf{I}}{\dfrac{\mathbf{I} \mid \mathbf{I} \Rightarrow \beta \mid \mathbf{I} \circ \mathbf{I}}{\mathbf{I} \mid \mathbf{I} \Rightarrow \beta \mid \mathbf{I}}}}$$

(ii): It is enough to show by induction on proofs in $\delta\mathrm{N4}^\perp$ that $\delta\mathrm{N4}^\perp$ $\vdash X_1 \mid X_2 \Rightarrow X_3 \mid X_4$ implies the validity of the formula $\tau(X_1 \mid X_2 \Rightarrow X_3 \mid X_4)$ in every Kripke model of Section 2.2.2 with the additional clauses

$$[\text{not } x \models^+ \bot] \quad \text{and} \quad x \models^- \bot.$$

The completeness of $\mathrm{HN4}^\perp$ with respect to the Kripke semantics follows from a generalization of Theorem 2.2.17 by extending the mapping f as follows: $f(\bot) := \bot$ and $f(\sim\bot) := \top$.

Initial sequents: The τ-translations $(p \wedge \top) \to (p \vee \bot)$ and $(\top\wedge \sim p) \to (\bot\vee \sim p)$ of initial sequents are clearly valid.

Display postulates:
Consider $X_1 \mid X_2 \Rightarrow X_3 \mid X_4 \circ Y \vdash X_1 \mid X_2 \Rightarrow X_3 \circ *Y \mid X_4$ and suppose $(\tau_1(X_1) \wedge \tau_2(X_2)) \Rightarrow (\tau_3(X_3) \vee \tau_4(X_4) \vee \tau_4(Y))$ is valid. It is enough to note that $\tau_3(*Y) = \tau_4(Y)$.

Case $\{X_1 \mid \alpha \Rightarrow X_3 \mid X_4,\ X_1 \mid \beta \Rightarrow X_3 \mid X_4\} \vdash X_1 \mid \alpha \wedge \beta \Rightarrow X_3 \mid X_4$. Let (1) $(\tau_1(X_1) \wedge \sim\alpha) \to (\tau_3(X_3) \vee \tau_4(X_4))$ be valid, let (2) $(\tau_1(X_1) \wedge \sim\beta) \to (\tau_3(X_3) \vee \tau_4(X_4))$ be valid, and let $\langle M, R, \models^+, \models^- \rangle$ be a model and $x \in M$. Suppose that $x \models^+ \tau_1(X_1)$ and $x \models^+ \sim(\alpha \wedge \beta)$. Then $x \models^+ \sim\alpha$ or $x \models^+ \sim\beta$. In both cases, either by (1) or by (2), $x \models^+ \tau_3(X_3) \vee \tau_4(X_4)$.

Case $\{X_1 \mid X_2 \Rightarrow \alpha \mid X_4,\ \beta \mid X_2 \Rightarrow X_3 \mid X_4\} \vdash \alpha \to \beta \mid X_2 \circ *X_1 \Rightarrow X_3 \mid X_4$. Let (1) $(\tau_1(X_1) \wedge \tau_2(X)) \to (\alpha \vee \tau_4(X_4))$ be valid, let (2) $(\beta \wedge \tau_2(X_2)) \to (\tau_3(X_3) \vee \tau_4(X_4))$ be valid, and let $\langle M, R, \models^+, \models^- \rangle$ be a model and $x \in M$. Suppose that (3) $x \models^+ \alpha \to \beta$ $x \models^+ \tau_2(X_2)$ and $x \models^+ \tau_2(*X_1)$. Then $x \models^+ \tau_1(X_1)$. By (1), $x \models^+ \alpha$ or $x \models^+ \tau_4(X_4)$. In the latter case, $x \models^+ \tau_3(X_3) \vee \tau_4(X_4)$. In the former case, $x \models^+ \beta$, by (3). It follows from (2) that $x \models^+ \tau_3(X_3) \vee \tau_4(X_4)$. ∎

Corollary 2.3.26 *For any formula α from* $\mathrm{HN4}^\perp$*:* $\mathrm{HN4}^\perp \vdash \alpha$ *iff* $\delta\mathrm{N4}^\perp$ $\vdash \mathbf{I} \mid \mathbf{I} \Rightarrow \alpha \mid \mathbf{I}$.

Belnap [23] presents a very general cut-elimination theorem covering all so-called properly displayable logics, which are logics satisfying a number of conditions (C1)–(C8). The conditions (C1)–(C8) are stated for binary sequents of the shape $X \Rightarrow Y$ but can easily be restated for sequents $X_1 \mid X_2 \Rightarrow X_3 \mid X_4$. Conditions (C6) and (C7), for example, stipulate that each rule is closed under simultaneous substitution of arbitrary structures for congruent formulas which are succedent parts, respectively antecedent parts. Instead of these requirements it has to be postulated that each rule is closed under simultaneous substitution of arbitrary structures for congruent formulas which are o-succedent parts, e-succedent parts, o-antecedent parts, and e-antecedent parts, respectively. Analogous adjustments have to be made to Conditions (C4) and (C5). Conditions (C1)–(C7) can be checked by mere inspection of the rules. (C8) is the requirement of eliminability of principal cuts, i.e., applications of (*cut*) in the form

$$\frac{X \Rightarrow \alpha \quad \alpha \Rightarrow Y}{X \Rightarrow Y}$$

or of (odd cut) and (even cut) in which the two premise sequents have been obtained by introducing the main connective of the cut-formula α.

Theorem 2.3.27 *If* $\delta\mathrm{N4}^{\bot} \vdash X1 \mid X_2 \Rightarrow X_3 \mid X_4$*, then we have* $\delta\mathrm{N4}^{\bot} - \{$*(odd cut), (even cut)*$\} \vdash X1 \mid X_2 \Rightarrow X_3 \mid X_4$.

Proof. It is enough to note that $\delta\mathrm{N4}^{\bot}$ satisfies conditions (C1)-(C8). For (C8) we have, for example:

$$\dfrac{\dfrac{X_1 \mid X_2 \Rightarrow \mathbf{I} \mid X_4}{X_1 \mid X_2 \Rightarrow \bot \mid X_4} \quad \bot \mid X_2' \Rightarrow X_3 \mid X_4'}{X_1 \mid X_2 \circ X_2' \Rightarrow X_3 \mid X_4 \circ X_4'}$$

is replaced by

$$\dfrac{\dfrac{\dfrac{\dfrac{\dfrac{X_1 \mid X_2 \Rightarrow \mathbf{I} \mid X_4}{X_1 \circ * X_2' \mid X_2 \Rightarrow \mathbf{I} \mid X_4}}{X_1 \mid X_2 \circ X_2' \Rightarrow \mathbf{I} \mid X_4}}{X_1 \mid X_2 \circ X_2' \Rightarrow \mathbf{I} \circ (X_3 \circ * X_4') \mid X_4}}{X_1 \mid X_2 \circ X_2' \Rightarrow X_3 \circ * X_4' \mid X_4}}{X_1 \mid X_2 \circ X_2' \Rightarrow X_3 \mid X_4 \circ X_4'}$$

$$\dfrac{\dfrac{X_1' \circ \alpha \mid X_2' \Rightarrow \beta \mid \mathbf{I}}{X_1' \mid X_2' \Rightarrow \alpha \to \beta \mid \mathbf{I}} \quad \dfrac{X_1 \mid X_2 \Rightarrow \alpha \mid X_4 \quad \beta \mid X_2 \Rightarrow X_3 \mid X_4}{\alpha \to \beta \mid X_2 \circ * X_1 \Rightarrow X_3 \mid X_4}}{X_1' \mid X_2' \circ (X_2 \circ * X_1) \Rightarrow X_3 \mid \mathbf{I} \circ X_4}$$

is replaced by

$$\dfrac{X_1 \mid X_2 \Rightarrow \alpha \mid X_4 \qquad \dfrac{\dfrac{X_1' \circ \alpha \mid X_2' \Rightarrow \beta \mid \mathbf{I} \quad \beta \mid X_2 \Rightarrow X_3 \mid X_4}{X_1' \circ \alpha \mid (X_2' \circ X_2) \Rightarrow X_3 \mid \mathbf{I} \circ X_4}}{\alpha \mid (X_2' \circ X_2) \circ *X_1' \Rightarrow X_3 \mid \mathbf{I} \circ X_4}}{X_1 \mid X_2 \circ ((X_2' \circ X_2) \circ *X_1') \Rightarrow X_3 \mid X_4 \circ (\mathbf{I} \circ X_4)}$$

$$\vdots$$

$$\dfrac{\dfrac{\overline{*X_2 \circ *X_2 \mid (X_2' \circ *X_1') \circ *X_1) \Rightarrow *X_4 \circ *X_4 \mid \mathbf{I} \circ *X_3}}{*X_2 \mid (X_2' \circ *X_1') \circ *X_1) \Rightarrow *X_4 \circ *X_4 \mid \mathbf{I} \circ *X_3}}{*X_2 \mid (X_2' \circ *X_1') \circ *X_1) \Rightarrow *X_4 \mid \mathbf{I} \circ *X_3}$$

$$\vdots$$

$$\overline{X_1' \mid X_2' \circ (X_2 \circ *X_1) \Rightarrow X_3 \mid \mathbf{I} \circ X_4}$$

∎

In Section ?? (on page 3) we mentioned Heyting-Brouwer logic H-B as a logic closely related to Nelson's four-valued logic. Rauszer's sequent calculus G1 [161] for H-B logic is presented below. The language of G1 includes $\rightarrow$, $\prec$ (the dual of $\rightarrow$), $\neg$ and $-$(co-negation, the dual of $\neg$). Note that in the presence of both $\rightarrow$ and $\prec$, the two negations are definable as follows:

$$\neg\alpha := \alpha \rightarrow \bot, \quad -\alpha := \top \prec \alpha,$$

where for some arbitrary but fixed propositional variable p, $\bot$ is defined as $p \prec p$ and $\top$ is defined as $p \rightarrow p$. Note also that whereas intuitionistic negation is paracomplete in the sense that $\Rightarrow (\alpha \vee \neg\alpha)$ fails to be provable in G1, co-negation is paraconsistent in the sense that $\Rightarrow -(-\alpha \wedge \alpha)$ fails to be provable in G1.

Definition 2.3.28 (Rauszer's G1 for H-B logic [161]) *An expression* Γ *is used to represent a sequence of formulas. The sequents of* G1 *are of the form* $\Gamma \Rightarrow \Delta$ *with the one-element restriction:* Γ *and* Δ *can not be sequences with more than one-element simultaneously.*

The initial sequents of G1 *are of the form:*

$$\alpha \Rightarrow \alpha.$$

The inference rules of G1 *are of the form:*

$$\dfrac{\Gamma \Rightarrow \Sigma, \alpha \quad \alpha, \Delta \Rightarrow \Pi}{\Gamma, \Delta \Rightarrow \Sigma, \Pi} * \qquad \dfrac{\Gamma, \beta, \alpha, \Delta \Rightarrow \gamma}{\Gamma, \alpha, \beta, \Delta \Rightarrow \gamma} \qquad \dfrac{\alpha, \alpha, \Gamma \Rightarrow \gamma}{\alpha, \Gamma \Rightarrow \gamma} \qquad \dfrac{\Gamma \Rightarrow \Delta}{\alpha, \Gamma \Rightarrow \Delta} *$$

$$\frac{\gamma \Rightarrow \Gamma, \beta, \alpha, \Delta}{\gamma \Rightarrow \Gamma, \alpha, \beta, \Delta} \qquad \frac{\gamma \Rightarrow \Gamma, \alpha, \alpha}{\gamma \Rightarrow \Gamma, \alpha} \qquad \frac{\Gamma \Rightarrow \Delta}{\Gamma \Rightarrow \Delta, \alpha}\,*$$

$$\frac{\Gamma \Rightarrow \Delta, \alpha \quad \Gamma \Rightarrow \Delta, \beta}{\Gamma \Rightarrow \Delta, \alpha \wedge \beta}\,* \qquad \frac{\alpha, \beta, \Gamma \Rightarrow \gamma}{\alpha \wedge \beta, \Gamma \Rightarrow \gamma} \qquad \frac{\alpha \Rightarrow \Delta}{\alpha \wedge \beta \Rightarrow \Delta} \qquad \frac{\beta \Rightarrow \Delta}{\alpha \wedge \beta \Rightarrow \Delta}$$

$$\frac{\Gamma \Rightarrow \alpha}{\Gamma \Rightarrow \alpha \vee \beta} \qquad \frac{\Gamma \Rightarrow \beta}{\Gamma \Rightarrow \alpha \vee \beta} \qquad \frac{\gamma \Rightarrow \Gamma, \alpha, \beta}{\gamma \Rightarrow \Gamma, \alpha \vee \beta} \qquad \frac{\alpha, \Gamma \Rightarrow \Delta \quad \beta, \Gamma \Rightarrow \Delta}{\alpha \vee \beta, \Gamma \Rightarrow \Delta}\,*$$

$$\frac{\alpha, \Gamma \Rightarrow}{\Gamma \Rightarrow \neg\alpha} \qquad \frac{\Gamma \Rightarrow \Delta, \alpha}{\neg\alpha, \Gamma \Rightarrow \Delta}\,* \qquad \frac{\alpha, \Gamma \Rightarrow \Delta}{\Gamma \Rightarrow \Delta, -\alpha}\,* \qquad \frac{\Rightarrow \Delta, \alpha}{-\alpha \Rightarrow \Delta}$$

$$\frac{\alpha, \Gamma \Rightarrow \beta}{\Gamma \Rightarrow \alpha \to \beta} \qquad \frac{\Gamma, \Delta \Rightarrow \alpha \quad \beta, \Gamma, \Delta \Rightarrow \gamma}{\alpha \to \beta, \Gamma, \Delta \Rightarrow \gamma}$$

$$\frac{\gamma \Rightarrow \Sigma, \Delta, \alpha \quad \beta \Rightarrow \Sigma, \Delta}{\gamma \Rightarrow \Sigma, \Delta, \alpha \mathrel{-\!\!<} \beta} \qquad \frac{\alpha \Rightarrow \Delta, \beta}{\alpha \mathrel{-\!\!<} \beta \Rightarrow \Delta}$$

where "" means that the rule has the one-element restriction.*

It is known that the cut-elimination theorem for G1 does not hold, a counterexample due to T. Uustalu, namely the formula $p \to (q \vee (r \to ((p \mathrel{-\!\!<} q) \wedge r))$, is presented in [28]. The sequent calculus for Heyting-Brouwer logic in [38] uses single-conclusion sequents but imposes a singleton-on-the-left restriction on the left introduction rule for co-implication and a singleton-on-the-right constraint on the right introduction rule for implication. Uustalu's counterexample also shows that this calculus is not cut-free. A sequent calculus for Heyting-Brouwer logic that admits cut-elimination can be found in [68], and Buisman and Goré [28] present a non-standard cut-free sequent calculus for H-B. In this proof system, the sequent rule for implications in succedent position of a sequent and the rule for co-implications in antecedent position of a sequent have side conditions on variables for families of sets of formulas. Two other, cut-free sequent calculi for Heyting-Brouwer logic are presented in [70]. The first calculus is intermediate between display calculi and standard sequent systems. From this system a variant is defined, which is amenable to automated proof-search. A cut-elimination admitting display sequent calculus for certain conservative extensions of H-B by several versions of strong negation may be found in [209] and Chapter 12. We will here present the fragment of the latter system equivalent with G1. This display calculus, δHB, is a variant of Goré's system [68]. Whereas Goré [68] treats the pair of commutative operations $\wedge$ and $\vee$ as Gentzen duals and the non-commutative operations $\to$ and $\mathrel{-\!\!<}$, the display sequent calculus δHB uses the residuated pairs ($\wedge$, $\to$) and ($\mathrel{-\!\!<}$, $\vee$) as pairs of Gentzen duals. We will use the binary operations $\circ$ and $\bullet$ as structural connectives. In antecedent position, $\circ$ is to be interpreted as conjunction and in succedent position as implication. In antecedent position, $\bullet$ is to be read as co-implication and in succedent position as disjunction. A sequent is an expression of the form $X \Rightarrow Y$, where X and Y are structures. We also assume the structure constant **I**.

Definition 2.3.29 *Let* $\mathcal{L}_{\text{H-B}}$ *be the set of all* H-B*-formulas. The set of structures* Struc*(*$\mathcal{L}_{\text{H-B}}$*) generated from* $\mathcal{L}_{\text{H-B}}$ *is defined as follows:*

$$\begin{array}{rl} \textit{formulas:} & \alpha \in \mathcal{L}_{\text{H-B}} \\ \textit{structures:} & X \in \text{Struc}(\mathcal{L}_{\text{H-B}}) \\ X ::= & \alpha \mid \mathbf{I} \mid (X \circ X) \mid (X \bullet X). \end{array}$$

Definition 2.3.30 (Display calculus δHB) *The initial sequents of* δHB *are of the form: for any propositional variable p,*

$$p \Rightarrow p.$$

The display postulates of δHB *are of the form:*

$$\dfrac{\dfrac{Y \Rightarrow X \circ Z}{X \circ Y \Rightarrow Z}}{X \Rightarrow Y \circ Z} \qquad \dfrac{\dfrac{X \Rightarrow Y \circ Z}{X \circ Y \Rightarrow Z}}{Y \Rightarrow X \circ Z} \qquad \dfrac{\dfrac{X \bullet Z \Rightarrow Y}{X \Rightarrow Y \bullet Z}}{X \bullet Y \Rightarrow Z} \qquad \dfrac{\dfrac{X \bullet Y \Rightarrow Z}{X \Rightarrow Y \bullet Z}}{X \bullet Z \Rightarrow Y.}$$

where two rules are combined into a single one at a time.

The additional structural rules of δHB *are of the form:*

$$\dfrac{X \Rightarrow \alpha \quad \alpha \Rightarrow Y}{X \Rightarrow Y}\ (cut)$$

$$\dfrac{\dfrac{X \circ \mathbf{I} \Rightarrow Y}{X \Rightarrow Y}}{\mathbf{I} \circ X \Rightarrow Y} \qquad \dfrac{\dfrac{\mathbf{I} \circ X \Rightarrow Y}{X \Rightarrow Y}}{X \circ \mathbf{I} \Rightarrow Y} \qquad \dfrac{\dfrac{X \Rightarrow Y \bullet \mathbf{I}}{X \Rightarrow Y}}{X \Rightarrow \mathbf{I} \bullet Y} \qquad \dfrac{\dfrac{X \Rightarrow \mathbf{I} \bullet Y}{X \Rightarrow Y}}{X \Rightarrow Y \bullet \mathbf{I}}$$

$$\dfrac{X \Rightarrow Y}{X \Rightarrow Y \bullet Z}\ (rm) \qquad \dfrac{X \Rightarrow Y}{X \circ Z \Rightarrow Y}\ (lm)$$

$$\dfrac{X \Rightarrow Y \bullet Z}{X \Rightarrow Z \bullet Y}\ (re) \qquad \dfrac{X \circ Z \Rightarrow Y}{Z \circ X \Rightarrow Y}\ (le)$$

$$\dfrac{X \Rightarrow Y \bullet Y}{X \Rightarrow Y}\ (rc) \qquad \dfrac{X \circ X \Rightarrow Y}{X \Rightarrow Y}\ (lc)$$

$$\dfrac{X \Rightarrow (Y \bullet Z) \bullet X'}{X \Rightarrow Y \bullet (Z \bullet X')}\ (ra) \qquad \dfrac{(X \circ Y) \circ Z \Rightarrow X'}{X \circ (Y \circ Z) \Rightarrow X'}\ (la)$$

The introduction rules of δHB are of the form:

$$\frac{X \Rightarrow \alpha \quad Y \Rightarrow \beta}{X \circ Y \Rightarrow (\alpha \wedge \beta)} (\Rightarrow \wedge) \qquad \frac{\alpha \circ \beta \Rightarrow X}{(\alpha \wedge \beta) \Rightarrow X} (\wedge \Rightarrow)$$

$$\frac{X \Rightarrow \alpha \bullet \beta}{X \Rightarrow (\alpha \vee \beta)} (\Rightarrow \vee) \qquad \frac{\alpha \Rightarrow X \quad \beta \Rightarrow Y}{(\alpha \vee \beta) \Rightarrow X \bullet Y} (\vee \Rightarrow)$$

$$\frac{X \Rightarrow \alpha \circ \beta}{X \Rightarrow (\alpha \to \beta)} (\Rightarrow \to) \qquad \frac{X \Rightarrow \alpha \quad \beta \Rightarrow Y}{(\alpha \to \beta) \Rightarrow X \circ Y} (\to \Rightarrow)$$

$$\frac{X \Rightarrow \beta \quad \alpha \Rightarrow Y}{X \bullet Y \Rightarrow \beta \prec \alpha} (\Rightarrow \prec) \qquad \frac{\beta \bullet \alpha \Rightarrow X}{\beta \prec \alpha \Rightarrow X} (\prec \Rightarrow)$$

Definition 2.3.31 *If* $\mathsf{s} = X \Rightarrow Y$ *is a sequent, then the displayed occurrence of* X *(*Y*) is an antecedent (succedent) part of* s*. If an occurrence of* $(Z \circ W)$ *is an antecedent part of* s*, then the displayed occurrences of* Z *and* W *are antecedent parts of* s*. If an occurrence of* $(Z \bullet W)$ *is an antecedent part of* s*, then the displayed occurrence of* Z *(*W*) is an antecedent (succedent) part of* s*. If an occurrence of* $(Z \circ W)$ *is a succedent part of* s*, then the displayed occurrence of* Z *(*W*) is an antecedent (succedent) part of* s*. If an occurrence of* $(Z \bullet W)$ *is a succedent part of* s*, then the displayed occurrences of* Z *and* W *are succedent parts of* s*.*

Theorem 2.3.32 ([23]) *For every sequent* s *and every antecedent (succedent) part* X *of* s*, there exists a sequent* s' *structurally equivalent to* s *such that* X *is the entire antecedent (succedent) of* s'*.*

By induction on the structure of H-B-formulas α, it can be shown that for every H-B-formula α, $\delta\mathrm{HB} \vdash \alpha \Rightarrow \alpha$.

One can define translations τ_1 and τ_2 from structures into formulas such that these translations reflect the context-sensitive interpretation of the structure connectives: τ_1 translates structures in antecedent position and τ_2 in succedent position.

Definition 2.3.33 *The translations* τ_1 *and* τ_2 *from structures into* H-B*-formulas are inductively defined as follows, where* α *is a formula and* p *is an arbitrary but fixed propositional variable:*

$$\begin{array}{rclrcl} \tau_1(\alpha) & = & \alpha & \tau_2(\alpha) & = & \alpha \\ \tau_1(\mathbf{I}) & = & p \to p & \tau_2(\mathbf{I}) & = & p \prec p \\ \tau_1(X \circ Y) & = & \tau_1(X) \wedge \tau_1(Y) & \tau_2(X \circ Y) & = & \tau_1(X) \to \tau_2(Y) \\ \tau_1(X \bullet Y) & = & \tau_1(X) \prec \tau_2(Y) & \tau_2(X \bullet Y) & = & \tau_2(X) \vee \tau_2(Y) \end{array}$$

Theorem 2.3.34 (Soundness of δHB) *For every sequent $X \Rightarrow Y$, if $\delta\mathrm{HB} \vdash X \Rightarrow Y$, then $\mathrm{G1} \vdash \tau_1(X) \Rightarrow \tau_2(Y)$.*

Proof. By induction on proofs in δHB. For initial sequents the claim is obvious. We here present some of the remaining cases.
Case $X \bullet Y \Rightarrow Z \vdash X \Rightarrow Y \bullet Z$: By the induction hypothesis, G1 $\vdash \tau_1(X \bullet Y) \Rightarrow \tau_2(Z)$, i.e., $\vdash \tau_1(X) -\!\!\!\prec \tau_2(Y) \Rightarrow \tau_2(Z)$. It has to be shown that $\vdash \tau_1(X) \Rightarrow \tau_2(Y \bullet Z)$, i.e., $\vdash \tau_1(X) \Rightarrow \tau_2(Y) \vee \tau_2(Z)$. We have:

$$\dfrac{\dfrac{\dfrac{\dfrac{\tau_1(X) \Rightarrow \tau_1(X)}{\tau_1(X) \Rightarrow \tau_2(Y), \tau_1(X)} \quad \tau_2(Y) \Rightarrow \tau_2(Y)}{\tau_1(X) \Rightarrow \tau_2(Y), \tau_1(X) -\!\!\!\prec \tau_2(Y)} \qquad \tau_1(X) -\!\!\!\prec \tau_2(Y) \Rightarrow \tau_2(Z)}{\tau_1(X) \Rightarrow \tau_2(Y), \tau_2(Z)}}{\tau_1(X) \Rightarrow \tau_2(Y) \vee \tau_2(Z)}$$

Case $X \Rightarrow Y \bullet Z \vdash X \bullet Z \Rightarrow Y$: By the induction hypothesis, G1 $\vdash \tau_1(X) \Rightarrow \tau_2(Y) \vee \tau_2(Z)$. It has to be shown that $\vdash \tau_1(X) -\!\!\!\prec \tau_2(Z) \Rightarrow \tau_2(Y)$. We have:

$$\dfrac{\dfrac{\tau_1(X) \Rightarrow \tau_2(Y) \vee \tau_2(Z) \qquad \dfrac{\dfrac{\tau_2(Y) \Rightarrow \tau_2(Y)}{\tau_2(Y) \Rightarrow \tau_2(Y), \tau_2(Z)} \quad \dfrac{\tau_2(Z) \Rightarrow \tau_2(Z)}{\tau_2(Z) \Rightarrow \tau_2(Y), \tau_2(Z)}}{\tau_2(Y) \vee \tau_2(Z) \Rightarrow \tau_2(Y), \tau_2(Z)}}{\tau_1(X) \Rightarrow \tau_2(Y), \tau_2(Z)}}{\tau_1(X) -\!\!\!\prec \tau_2(Z) \Rightarrow \tau_2(Y)}$$

Case $X \Rightarrow Y \bullet \mathbf{I} \vdash X \Rightarrow Y$: By the induction hypothesis, G1 $\vdash \tau_1(X) \Rightarrow \tau_2(Y) \vee (p -\!\!\!\prec p)$. It has to be shown that $\vdash \tau_1(X) \Rightarrow \tau_2(Y)$. We have:

$$\dfrac{\tau_1(X) \Rightarrow \tau_2(Y) \vee (p -\!\!\!\prec p) \qquad \dfrac{\tau_2(Y) \Rightarrow \tau_2(Y) \quad \dfrac{\dfrac{p \Rightarrow p}{p \Rightarrow \tau_2(Y), p}}{p -\!\!\!\prec p \Rightarrow \tau_2(Y)}}{\tau_2(Y) \vee (p -\!\!\!\prec p) \Rightarrow \tau_2(Y)}}{\tau_1(X) \Rightarrow \tau_2(Y)}$$

Case $(\Rightarrow -\!\!\!\prec)$: By the induction hypothesis, G1 $\vdash \tau_1(X) \Rightarrow \beta$, $\vdash \alpha \Rightarrow \tau_2(Y)$. We have:

$$\dfrac{\dfrac{\tau_1(X) \Rightarrow \beta \quad \alpha \Rightarrow \tau_2(Y)}{\tau_1(X) \bullet \tau_2(Y) \Rightarrow \beta -\!\!\!\prec \alpha}}{\tau_1(X) -\!\!\!\prec \tau_2(Y) \Rightarrow \beta -\!\!\!\prec \alpha}$$

∎

In order to prove the converse of Theorem 2.3.34, we will apply two lemmas.

Lemma 2.3.35 *The sequent* $\neg\alpha \Rightarrow -\alpha$ *is derivable in* δHB.

Proof. The proof follows the pattern of the cut-free derivation in [68], Figure 4. ∎

Definition 2.3.36 *If* $\Gamma \Rightarrow \Delta$ *is a sequent of G1 then* $h(\Gamma \Rightarrow \Delta)$ *is defined as the sequent* $h(\Gamma) \Rightarrow h(\Delta)$ *of* δHB, *where*

$$h(\Gamma) = \begin{cases} \alpha_1 \wedge \ldots \wedge \alpha_n & \text{if } \Gamma \equiv \alpha_1, \ldots, \alpha_n \text{ and } n > 1 \\ \alpha & \text{if } \Gamma \equiv \alpha \\ p \rightarrow p & \text{if } \Gamma \text{ is the empty sequence} \end{cases}$$

$$h(\Delta) = \begin{cases} \beta_1 \vee \ldots \vee \beta_m & \text{if } \Delta \equiv \beta_1, \ldots, \beta_m \text{ and } m > 1 \\ \beta & \text{if } \Gamma \equiv \beta \\ p \prec p & \text{if } \Delta \text{ is the empty sequence.} \end{cases}$$

and p *is the fixed propositional variable from Lemma 2.3.33.*

Lemma 2.3.37 *For every sequent* $\Gamma \Rightarrow \Delta$ *of* G1, *if* G1 $\vdash \Gamma \Rightarrow \Delta$, *then* δHB $\vdash h(\Gamma) \Rightarrow h(\Delta)$.

Proof. By induction on proofs in G1. For initial sequents the claim is obvious. We here present some of the remaining cases.

Case

$$\frac{\Gamma \Rightarrow \Sigma, \alpha \quad \alpha, \Delta \Rightarrow \Pi}{\Gamma, \Delta \Rightarrow \Sigma, \Pi} \; * :$$

Due to the one-element restriction, there are four subcases: (1) both Γ and Δ are the empty sequence, (2) both Γ and Π are the empty sequence, (3) both Σ and Δ are the empty sequence, or (4) both Σ and Π are the empty sequence. We here present Subcase (1). By the induction hypothesis, δHB $\vdash \mathbf{I} \Rightarrow h(\Sigma) \vee \alpha$ and δHB $\vdash \alpha \Rightarrow h(\Pi)$. We have:

$$\dfrac{\dfrac{\dfrac{\dfrac{\mathbf{I} \Rightarrow h(\Sigma) \vee \alpha \quad \dfrac{h(\Sigma) \Rightarrow h(\Sigma) \quad \alpha \Rightarrow \alpha}{h(\Sigma) \vee \alpha \Rightarrow h(\Sigma) \bullet \alpha}}{\mathbf{I} \Rightarrow h(\Sigma) \bullet \alpha}}{\mathbf{I} \bullet h(\Sigma) \Rightarrow \alpha} \quad \alpha \Rightarrow h(\Pi)}{\dfrac{\dfrac{\mathbf{I} \bullet h(\Sigma) \Rightarrow h(\Pi)}{\mathbf{I} \Rightarrow h(\Sigma) \bullet h(\Pi)}}{\mathbf{I} \Rightarrow h(\Sigma) \vee h(\Pi)}}$$

Case $\Rightarrow \Delta, \alpha \vdash -\alpha \Rightarrow \Delta$: By the induction hypothesis, $\delta\mathrm{HB} \vdash p \to p \Rightarrow h(\Delta) \vee \alpha$. It has to be shown that $\delta\mathrm{HB} \vdash (p \to p) \mathbin{-\!\!\!\prec} \alpha \Rightarrow h(\Delta)$. We have:

$$\dfrac{\dfrac{\dfrac{p \to p \Rightarrow h(\Delta) \vee \alpha \qquad \dfrac{h(\Delta) \Rightarrow h(\Delta) \quad \alpha \Rightarrow \alpha}{h(\Delta) \vee \alpha \Rightarrow h(\Delta) \bullet \alpha}}{(p \to p) \Rightarrow h(\Delta) \bullet \alpha}}{(p \to p) \bullet \alpha \Rightarrow h(\Delta)}}{(p \to p) \mathbin{-\!\!\!\prec} \alpha \Rightarrow h(\Delta)}$$

Case $\Gamma \Rightarrow \Delta, \alpha \vdash \neg\alpha, \Gamma \Rightarrow \Delta$ $*$: Due to the one-element restriction, Γ or Δ are the empty sequence. Suppose Γ is the empty sequence. By the induction hypothesis, $\delta\mathrm{HB} \vdash p \to p \Rightarrow h(\Delta) \vee \alpha$. It has to be shown that $\delta\mathrm{HB} \vdash \alpha \to (p \mathbin{-\!\!\!\prec} p) \Rightarrow h(\Delta)$. We have:

$$\dfrac{s \qquad \dfrac{\dfrac{\dfrac{p \to p \Rightarrow h(\Delta) \vee \alpha \qquad \dfrac{h(\Delta) \Rightarrow h(\Delta) \quad \alpha \Rightarrow \alpha}{h(\Delta) \vee \alpha \Rightarrow h(\Delta) \bullet \alpha}}{p \to p \Rightarrow h(\Delta) \bullet \alpha}}{(p \to p) \bullet \alpha \Rightarrow h(\Delta)}}{(p \to p) \mathbin{-\!\!\!\prec} \alpha \Rightarrow h(\Delta)}}{\alpha \to (p \mathbin{-\!\!\!\prec} p) \Rightarrow (h(\Delta)} \; (cut), \text{Lemma 2.3.35}$$

where s is $\alpha \to (p \mathbin{-\!\!\!\prec} p) \Rightarrow (p \to p) \mathbin{-\!\!\!\prec} \alpha$.
Suppose Δ is the empty sequence. By the induction hypothesis, $\delta\mathrm{HB} \vdash h(\Gamma) \Rightarrow \alpha$. It has to be shown that $\delta\mathrm{HB} \vdash (\alpha \to (p \mathbin{-\!\!\!\prec} p)) \wedge h(\Gamma) \Rightarrow p \mathbin{-\!\!\!\prec} p$. We have:

$$\dfrac{\dfrac{\dfrac{h(\Gamma) \Rightarrow \alpha \qquad \dfrac{\dfrac{\dfrac{\alpha \Rightarrow \alpha \quad p \mathbin{-\!\!\!\prec} p \Rightarrow p \mathbin{-\!\!\!\prec} p}{\alpha \to (p \mathbin{-\!\!\!\prec} p) \Rightarrow \alpha \circ (p \mathbin{-\!\!\!\prec} p)}}{\alpha \circ (\alpha \to (p \mathbin{-\!\!\!\prec} p)) \Rightarrow p \mathbin{-\!\!\!\prec} p}}{\alpha \Rightarrow (\alpha \to (p \mathbin{-\!\!\!\prec} p)) \circ (p \mathbin{-\!\!\!\prec} p)}}{h(\Gamma) \Rightarrow (\alpha \to (p \mathbin{-\!\!\!\prec} p)) \circ (p \mathbin{-\!\!\!\prec} p)}}{(\alpha \to (p \mathbin{-\!\!\!\prec} p)) \circ h(\Gamma) \Rightarrow p \mathbin{-\!\!\!\prec} p}}{(\alpha \to (p \mathbin{-\!\!\!\prec} p)) \wedge h(\Gamma) \Rightarrow p \mathbin{-\!\!\!\prec} p}$$

∎

Theorem 2.3.38 (Completeness of δHB) *For every sequent $X \Rightarrow Y$ of δHB, if G1 $\vdash \tau_1(X) \Rightarrow \tau_2(Y)$, then δHB $\vdash X \Rightarrow Y$.*

Proof. Suppose G1 $\vdash \tau_1(X) \Rightarrow \tau_2(Y)$. By Lemma 2.3.37, G1 $\vdash h(\tau_1(X)) \Rightarrow h(\tau_2(Y))$. But $h(\tau_1(X)) \Rightarrow h(\tau_2(Y)) \equiv \tau_1(X) \Rightarrow \tau_2(Y)$. ∎

Corollary 2.3.39 *For every* H-B*-formula* α*,* δHB $\vdash \mathbf{I} \Rightarrow \alpha$ *iff* G1 $\vdash \Rightarrow \alpha$.

Proof. By Theorems 2.3.34 and 2.3.38 and the observation that δHB $\vdash \mathbf{I} \Rightarrow \alpha$ iff δHB $\vdash p \to p \Rightarrow \alpha$ and G1 $\vdash \Rightarrow \alpha$ iff G1 $\vdash p \to p \Rightarrow \alpha$. ∎

Theorem 2.3.40 *If* δHB $\vdash X \Rightarrow Y$*, then* δHB$-$*(cut)* $\vdash X \Rightarrow Y$.

It is enough to note that δHB satisfies Belnap's conditions (C1)–(C8) . The principal cut-elimination step for $\mathrel{-\!\!<}$ is:

$$\dfrac{\dfrac{X \Rightarrow B \quad A \Rightarrow Y}{X \bullet Y \Rightarrow B \mathrel{-\!\!<} A} \qquad \dfrac{B \bullet A \Rightarrow Z}{B \mathrel{-\!\!<} A \Rightarrow Z}}{X \bullet Y \Rightarrow Z}$$

is replaced by

$$\dfrac{\dfrac{\dfrac{\dfrac{\dfrac{X \Rightarrow \beta \quad \dfrac{\beta \bullet \alpha \Rightarrow Z}{\beta \Rightarrow \alpha \bullet Z}}{X \Rightarrow \alpha \bullet Z}}{X \bullet Z \Rightarrow \alpha} \quad \alpha \Rightarrow Y}{X \bullet Z \Rightarrow Y}}{X \Rightarrow Z \Rightarrow Y}}{X \bullet Y \Rightarrow Z}$$

∎

2.3.5. Other systems

As mentioned before, it is known that the following logics are the same logic: the $\to$-free fragment of N4, Belnap and Dunn's four-valued logic and Anderson and Belnap's logic E_{fde} of first-degree entailment. In the following, we present some sequent calculi for the logic. Unless stated otherwise, in this subsection uppercase Greek letters Γ, DE etc. are used to represent finite sequences of formulas.

Definition 2.3.41 (Pynko's $\mathcal{G}_B$ [158]) *The sequents of* $\mathcal{G}_B$ *are of the form:* $\Gamma \Rightarrow \Delta$ *with the non-emptiness restriction: both sequences* Γ *and* Δ *are non-empty.*

The initial sequents of $\mathcal{G}_B$ *are of the form:*

$$\Gamma, \alpha, \Delta \Rightarrow \Sigma, \alpha, \Pi.$$

The inference rules of $\mathcal{G}_B$ *are of the form:*

$$\dfrac{\Gamma, \alpha, \beta, \Delta \Rightarrow \Pi}{\Gamma, \alpha \wedge \beta, \Delta \Rightarrow \Pi} \qquad \dfrac{\Gamma \Rightarrow \Delta, \alpha, \Sigma \quad \Gamma \Rightarrow \Delta, \beta, \Sigma}{\Gamma \Rightarrow \Delta, \alpha \wedge \beta, \Sigma}$$

$$\frac{\Gamma, \alpha, \Delta \Rightarrow \Pi \quad \Gamma, \beta, \Delta \Rightarrow \Pi}{\Gamma, \alpha \vee \beta, \Delta \Rightarrow \Pi} \quad \frac{\Gamma \Rightarrow \Delta, \alpha, \beta, \Sigma}{\Gamma \Rightarrow \Delta, \alpha \vee \beta, \Sigma}$$

$$\frac{\Gamma, \alpha, \Delta \Rightarrow \Pi}{\Gamma, \sim\sim\alpha, \Delta \Rightarrow \Pi} \quad \frac{\Gamma \Rightarrow \Delta, \alpha, \Sigma}{\Gamma \Rightarrow \Delta, \sim\sim\alpha, \Sigma}$$

$$\frac{\Gamma, \sim\alpha, \Delta \Rightarrow \Pi \quad \Gamma, \sim\beta, \Delta \Rightarrow \Pi}{\Gamma, \sim(\alpha \wedge \beta), \Delta \Rightarrow \Pi} \quad \frac{\Gamma \Rightarrow \Delta, \sim\alpha, \sim\beta, \Sigma}{\Gamma \Rightarrow \Delta, \sim(\alpha \wedge \beta), \Sigma}$$

$$\frac{\Gamma, \sim\alpha, \sim\beta, \Delta \Rightarrow \Pi}{\Gamma, \sim(\alpha \vee \beta), \Delta \Rightarrow \Pi} \quad \frac{\Gamma \Rightarrow \Delta, \sim\alpha, \Sigma \quad \Gamma \Rightarrow \Delta, \sim\beta, \Sigma}{\Gamma \Rightarrow \Delta, \sim(\alpha \vee \beta), \Sigma} .$$

It was shown in [158] that all the structural rules are admissible in $\mathcal{G}_B$.

Definition 2.3.42 (LE$_{fde2}$ [7] pp. 179–180.]) *A sequent calculus* LE$_{fde2}$ *for Anderson and Belnap's first-degree entailment* E$_{fde}$ *(or equivalently* R$_{fde}$*) is obtained from* $\mathcal{G}_B$ *in Definition 2.3.41 by deleting the non-empty sequent restriction and by adding the inference rule (for an implication connective) of the form:*

$$\frac{\alpha \Rightarrow \beta}{\Rightarrow \alpha{\to}\beta} \ (\to\text{right}).$$

Definition 2.3.43 (LE$_{fde1}$ [7] pp. 177–178.) *An expression* $\alpha^\star$ *is the result of adding a sign of negation to* α *if the number of outer negation signs on* α *is even (or zero), or the result of removing a sign of negation from* α *if the number is odd. We remark that* $\alpha^{\star\star}$ *is the same formula as* α*. An expression* $\Gamma^\star$ *is the sequence obtained from* Γ *by replacing each member* α *of* Γ *by* $\alpha^\star$*.*

LE$_{fde1}$ *is obtained from the* $\{\wedge, \vee\}$*-fragment of Gentzen's* LK *(with the right weakening rule) for classical logic by adding* ($\to$right) *in Definition 2.3.42 and the inference rules of the form:*

$$\frac{\Gamma \Rightarrow \Delta}{\Delta^\star \Rightarrow \Gamma^\star} \quad \frac{\Gamma, \alpha \Rightarrow \Delta}{\Delta^\star \Rightarrow \Gamma^\star, \sim\alpha} \quad \frac{\Gamma \Rightarrow \Delta, \alpha}{\Delta^\star, \sim\alpha \Rightarrow \Gamma^\star} .$$

It was shown by Dunn in [51] that we can obtain various important extensions of E$_{fde}$ by adding some axiom schemes:

1. Kl$_{fde}$ = E$_{fde}$ + $(\alpha\wedge\sim\alpha)\to\beta$ (the first-degree entailment of Kleene),
2. LP$_{fde}$ = E$_{fde}$ + $\alpha\to(\beta\vee\sim\beta)$ (the first-degree entailment of Priest),
3. RM$_{fde}$ = E$_{fde}$ + $(\alpha \wedge \sim\alpha)\to(\beta \vee \sim\beta)$ (the first-degree entailment of Dunn and McCall's system RM ("R-mingle")).

Definition 2.3.44 (Font and Verdú's $\mathcal{G}_{FV}$ [59]) *The sequents of $\mathcal{G}_{FV}$ are of the form $\Gamma \Rightarrow \gamma$ with the non-emptiness restriction: Γ is a non-empty sequence.*

The initial sequents of $\mathcal{G}_{FV}$ are of the form:

$$\alpha \Rightarrow \alpha.$$

The inference rules of $\mathcal{G}_{FV}$ are of the form:

$$\frac{\Gamma \Rightarrow \alpha \quad \alpha, \Gamma \Rightarrow \gamma}{\Gamma \Rightarrow \gamma} \quad \frac{\Gamma, \beta, \alpha, \Delta \Rightarrow \gamma}{\Gamma, \alpha, \beta, \Delta \Rightarrow \gamma} \quad \frac{\alpha, \alpha, \Gamma \Rightarrow \gamma}{\alpha, \Gamma \Rightarrow \gamma} \quad \frac{\Gamma \Rightarrow \gamma}{\alpha, \Gamma \Rightarrow \gamma}$$

$$\frac{\alpha, \beta, \Gamma \Rightarrow \gamma}{\alpha \wedge \beta, \Gamma \Rightarrow \gamma} \quad \frac{\Gamma \Rightarrow \alpha \quad \Gamma \Rightarrow \beta}{\Gamma \Rightarrow \alpha \wedge \beta}$$

$$\frac{\alpha, \Gamma \Rightarrow \gamma \quad \beta, \Gamma \Rightarrow \gamma}{\alpha \vee \beta, \Gamma \Rightarrow \gamma} \quad \frac{\Gamma \Rightarrow \alpha}{\Gamma \Rightarrow \alpha \vee \beta} \quad \frac{\Gamma \Rightarrow \beta}{\Gamma \Rightarrow \alpha \vee \beta}$$

$$\frac{\alpha, \Gamma \Rightarrow \gamma}{\sim\sim\alpha, \Gamma \Rightarrow \gamma} \quad \frac{\Gamma \Rightarrow \alpha}{\Gamma \Rightarrow \sim\sim\alpha} \quad \frac{\alpha \Rightarrow \beta}{\sim\beta \Rightarrow \sim\alpha}.$$

A sequent calculus (called here BL) for Arieli and Avron's bilattice logic is presented below.

Definition 2.3.45 (Arieli and Avron's bilattice logic [12, 13]) *An expression Γ is used to represent a set of formulas. The sequents of* BL *are of the form $\Gamma \Rightarrow \Delta$.*

The initial sequents of BL *are of the form:*

$$\alpha \Rightarrow \alpha.$$

The inference rules of BL *are of the form:*

$$\frac{\Gamma \Rightarrow \Delta, \alpha \quad \alpha, \Gamma \Rightarrow \Delta}{\Gamma \Rightarrow \Delta} \quad \frac{\Gamma \Rightarrow \Delta}{\alpha, \Gamma \Rightarrow \Delta} \quad \frac{\Gamma \Rightarrow \Delta}{\Gamma \Rightarrow \Delta, \alpha}$$

$$\frac{\Gamma \Rightarrow \Delta, \alpha \quad \beta, \Gamma \Rightarrow \Delta}{\alpha \to \beta, \Gamma \Rightarrow \Delta} \quad \frac{\alpha, \Gamma \Rightarrow \Delta, \beta}{\Gamma \Rightarrow \Delta, \alpha \to \beta} \ (\to\text{R})$$

$$\frac{\alpha, \Gamma \Rightarrow \Delta}{\alpha \wedge \beta, \Gamma \Rightarrow \Delta} \quad \frac{\beta, \Gamma \Rightarrow \Delta}{\alpha \wedge \beta, \Gamma \Rightarrow \Delta} \quad \frac{\Gamma \Rightarrow \Delta, \alpha \quad \Gamma \Rightarrow \Delta, \beta}{\Gamma \Rightarrow \Delta, \alpha \wedge \beta}$$

$$\frac{\alpha, \Gamma \Rightarrow \Delta \quad \beta, \Gamma \Rightarrow \Delta}{\alpha \vee \beta, \Gamma \Rightarrow \Delta} \quad \frac{\Gamma \Rightarrow \Delta, \alpha}{\Gamma \Rightarrow \Delta, \alpha \vee \beta} \quad \frac{\Gamma \Rightarrow \Delta, \beta}{\Gamma \Rightarrow \Delta, \alpha \vee \beta}$$

$$\frac{\alpha, \Gamma \Rightarrow \Delta}{\sim\sim\alpha, \Gamma \Rightarrow \Delta} \ (\sim\text{L}) \quad \frac{\Gamma \Rightarrow \Delta, \alpha}{\Gamma \Rightarrow \Delta, \sim\sim\alpha}$$

$$\frac{\alpha, \Gamma \Rightarrow \Delta}{{\sim}(\alpha{\rightarrow}\beta), \Gamma \Rightarrow \Delta} \quad \frac{{\sim}\beta, \Gamma \Rightarrow \Delta}{{\sim}(\alpha{\rightarrow}\beta), \Gamma \Rightarrow \Delta} \quad \frac{\Gamma \Rightarrow \Delta, \alpha \quad \Gamma \Rightarrow \Delta, {\sim}\beta}{\Gamma \Rightarrow \Delta, {\sim}(\alpha{\rightarrow}\beta)}$$

$$\frac{{\sim}\alpha, \Gamma \Rightarrow \Delta \quad {\sim}\beta, \Gamma \Rightarrow \Delta}{{\sim}(\alpha \wedge \beta), \Gamma \Rightarrow \Delta} \quad \frac{\Gamma \Rightarrow \Delta, {\sim}\alpha}{\Gamma \Rightarrow \Delta, {\sim}(\alpha \wedge \beta)} \quad \frac{\Gamma \Rightarrow \Delta, {\sim}\beta}{\Gamma \Rightarrow \Delta, {\sim}(\alpha \wedge \beta)}$$

$$\frac{{\sim}\alpha, \Gamma \Rightarrow \Delta}{{\sim}(\alpha \vee \beta), \Gamma \Rightarrow \Delta} \quad \frac{{\sim}\beta, \Gamma \Rightarrow \Delta}{{\sim}(\alpha \vee \beta), \Gamma \Rightarrow \Delta} \quad \frac{\Gamma \Rightarrow \Delta, {\sim}\alpha \quad \Gamma \Rightarrow \Delta, {\sim}\beta}{\Gamma \Rightarrow \Delta, {\sim}(\alpha \vee \beta)} .$$

As mentioned in [17], an alternative sequent calculus for N4 is obtained from BL by replacing (→R) of BL with (→r) of N4. A sequent calculus for the paraconsistent logic C_{min} studied by Carnielli and Marcos in [33] is obtained from the ∼-free fragment of BL by adding (∼L) and the rule of the form:

$$\frac{\alpha, \Gamma \Rightarrow \Delta}{\Gamma \Rightarrow \Delta, {\sim}\alpha} .$$

Roughly speaking, an intuitionistic version of C_{min} is Raggio's formulation [160] of da Costa's well-known paraconsistent logic C_ω [40].

2.4. Natural deduction

In this section, we consider a variety of natural deduction proof system for Nelson's four-valued logic.

2.4.1. Natural deduction in standard style

Definition 2.4.1 (Extended version of Priest's system [154], N_{N4}) *The inference rules of* N_{N4} *are of the form:*

$$\frac{\begin{matrix}[\alpha]\\ \vdots \\ \beta\end{matrix}}{\alpha{\rightarrow}\beta}\ ({\rightarrow}I) \qquad \frac{\alpha{\rightarrow}\beta \quad \alpha}{\beta}\ ({\rightarrow}E)$$

$$\frac{\alpha_1 \quad \alpha_2}{\alpha_1 \wedge \alpha_2}\ (\wedge I) \qquad \frac{\alpha_1 \wedge \alpha_2}{\alpha_1}\ (\wedge E1) \qquad \frac{\alpha_1 \wedge \alpha_2}{\alpha_2}\ (\wedge E2)$$

$$\frac{\alpha_1}{\alpha_1 \vee \alpha_2}\ (\vee I1) \qquad \frac{\alpha_2}{\alpha_1 \vee \alpha_2}\ (\vee I2) \qquad \frac{\alpha_1 \vee \alpha_2 \quad \begin{matrix}[\alpha_1]\\ \vdots \\ \beta\end{matrix} \quad \begin{matrix}[\alpha_2]\\ \vdots \\ \beta\end{matrix}}{\beta}\ (\vee E)$$

$$\frac{\alpha}{{\sim}{\sim}\alpha}\ ({\sim}I) \qquad \frac{{\sim}{\sim}\alpha}{\alpha}\ ({\sim}E)$$

$$\frac{\alpha \wedge {\sim}\beta}{{\sim}(\alpha{\rightarrow}\beta)}\ ({\sim}{\rightarrow}I) \qquad \frac{{\sim}(\alpha{\rightarrow}\beta)}{\alpha \wedge {\sim}\beta}\ ({\sim}{\rightarrow}E)$$

$$\frac{{\sim}\alpha \vee {\sim}\beta}{{\sim}(\alpha \wedge \beta)}\ ({\sim}\wedge I) \qquad \frac{{\sim}(\alpha \wedge \beta)}{{\sim}\alpha \vee {\sim}\beta}\ ({\sim}\wedge E)$$

$$\frac{{\sim}\alpha \wedge {\sim}\beta}{{\sim}(\alpha \vee \beta)}\ ({\sim}\vee I) \qquad \frac{{\sim}(\alpha \vee \beta)}{{\sim}\alpha \wedge {\sim}\beta}\ ({\sim}\vee E).$$

The $\rightarrow$-free part of N_{N4} is Priest's system [154] for E_{fde}. Note that the inference rules $({\sim}I)$, $({\sim}E)$, $({\sim}{\rightarrow}I)$, $({\sim}{\rightarrow}E)$, $({\sim}{\wedge}I)$, $({\sim}{\wedge}E)$, $({\sim}{\vee}I)$ and $({\sim}\vee E)$ correspond to the Hilbert-style axiom schemes: ${\sim}{\sim}\alpha \leftrightarrow \alpha$, ${\sim}(\alpha{\rightarrow}\beta) \leftrightarrow \alpha \wedge {\sim}\beta$, ${\sim}(\alpha \wedge \beta) \leftrightarrow {\sim}\alpha \vee {\sim}\beta$ and ${\sim}(\alpha \vee \beta) \leftrightarrow {\sim}\alpha \wedge {\sim}\beta$, and also correspond to the definition of the ${\sim}$-part of the typed λ-terms for λ^c [196]. Note also that the ${\sim}$-free part of N_{N4} is identical to the natural deduction system for positive propositional intuitionistic logic.

The inference rules $({\rightarrow}I)$, $({\wedge}I)$, $({\vee}I1)$, $({\vee}I2)$, $({\sim}I)$, $({\sim}{\rightarrow}I)$, $({\sim}\wedge I)$, and $({\sim}\vee I)$ are called *introduction rules*, and the inference rules $({\rightarrow}E)$, $({\wedge}E1)$, $({\wedge}E2)$, $({\vee}E)$, $({\sim}E)$, $({\sim}{\rightarrow}E)$, $({\sim}\wedge E)$ and $({\sim}\vee E)$ are called *elimination rules*. The usual terminologies of *major or minor premise* of inference rules are used. The notions of *proof* (of N_{N4}), *(open and discharged) assumptions* of a proof, and *end-formula* of a proof are also defined as usual and used. A formula α is said to be *provable* in N_{N4} if there exists a proof of N_{N4} with no open assumption whose end-formula is α. Similar terminologies and notions will also be used in other systems.

Definition 2.4.2 *Let α be a formula occurring in a proof D in N_{N4}. Then, α is called a* maximum formula *in D if α satisfies the following conditions:*

1. *α is the conclusion of an introduction rule or $({\vee}E)$,*
2. *α is the major premise of an elimination rule.*

A proof is said to be normal *if it contains no maximum formula.*

In order to define a reduction relation $\rhd$ on the set of proofs, we assume the usual definition of substitution for proofs (for assumptions). The set of proofs is closed under substitutions.

Definition 2.4.3 *Let γ be a maximum formula in a proof which is the conclusion of an inference rule R. The reduction relation $\rhd$ at γ is defined as follows.*

1. *R is $({\rightarrow}I)$, and γ is $\alpha{\rightarrow}\beta$:*

$$\dfrac{\dfrac{\begin{array}{c}[\alpha]\\ \vdots\ D\\ \beta\end{array}}{\alpha{\rightarrow}\beta}\ R \qquad \begin{array}{c}\vdots\ E\\ \alpha\end{array}}{\beta} \qquad \rhd \qquad \begin{array}{c}\vdots\ E\\ \alpha\\ \vdots\ D\\ \beta\end{array}$$

2. R is $(\wedge I)$, and γ is $\alpha_1 \wedge \alpha_2$:

$$\dfrac{\dfrac{\begin{matrix}\vdots\, D_1\\ \alpha_1\end{matrix}\qquad\begin{matrix}\vdots\, D_2\\ \alpha_2\end{matrix}}{\alpha_1\wedge\alpha_2}\,R}{\alpha_i}\qquad\rhd\qquad\begin{matrix}\vdots\, D_i\\ \alpha_i\end{matrix}$$

where i is 1 or 2.

3. R is $(\vee I1)$ or $(\vee I2)$, and γ is $\alpha_1 \vee \alpha_2$:

$$\dfrac{\dfrac{\begin{matrix}\vdots\, D\\ \alpha_i\end{matrix}}{\alpha_1\vee\alpha_2}\,R\qquad\begin{matrix}[\alpha_1]\\ \vdots\, E_1\\ \beta\end{matrix}\qquad\begin{matrix}[\alpha_2]\\ \vdots\, E_2\\ \beta\end{matrix}}{\beta}\qquad\rhd\qquad\begin{matrix}\vdots\, D\\ \alpha_i\\ \vdots\, E_i\\ \beta\end{matrix}$$

where i is 1 or 2.

4. R is $(\vee E)$:

$$\dfrac{\dfrac{\begin{matrix}\vdots\, D_1\\ \alpha_1\vee\alpha_2\end{matrix}\qquad\begin{matrix}[\alpha_1]\\ \vdots\, D_2\\ \gamma\end{matrix}\qquad\begin{matrix}[\alpha_2]\\ \vdots\, D_3\\ \gamma\end{matrix}}{\gamma}\,R\qquad E_1\quad E_2}{\beta}\,R'$$

$$\rhd\qquad\dfrac{\begin{matrix}\vdots\, D_1\\ \alpha_1\vee\alpha_2\end{matrix}\qquad\dfrac{\begin{matrix}[\alpha_1]\\ \vdots\, D_2\\ \gamma\end{matrix}\quad E_1\quad E_2}{\beta}\qquad\dfrac{\begin{matrix}[\alpha_2]\\ \vdots\, D_3\\ \gamma\end{matrix}\quad E_1\quad E_2}{\beta}}{\beta}$$

where R' is an arbitrary inference rule, and both E_1 and E_2 are proofs of the minor premises of R' if they exist.

5. R is $({\sim}I)$, and γ is ${\sim}{\sim}\alpha$:

$$\dfrac{\dfrac{\begin{matrix}\vdots\, D\\ \alpha\end{matrix}}{{\sim}{\sim}\alpha}\,R}{\alpha}\qquad\rhd\qquad\begin{matrix}\vdots\, D\\ \alpha\end{matrix}$$

6. R is $({\sim}{\to}I)$, and γ is ${\sim}(\alpha{\to}\beta)$:

$$\dfrac{\dfrac{\begin{matrix}\vdots\, D\\ \alpha\wedge{\sim}\beta\end{matrix}}{{\sim}(\alpha{\to}\beta)}\,R}{\alpha\wedge{\sim}\beta}\qquad\rhd\qquad\begin{matrix}\vdots\, D\\ \alpha\wedge{\sim}\beta\end{matrix}$$

7. *R is* $(\sim\wedge I)$, *and* γ *is* $\sim(\alpha \wedge \beta)$:

$$\dfrac{\dfrac{\begin{matrix}\vdots\, D\\ \sim\alpha \vee \sim\beta\end{matrix}}{\sim(\alpha \wedge \beta)}\,R}{\sim\alpha \vee \sim\beta} \quad \rhd \quad \begin{matrix}\vdots\, D\\ \sim\alpha \vee \sim\beta\end{matrix}$$

8. *R is* $(\sim\vee I)$, *and* γ *is* $\sim(\alpha \vee \beta)$:

$$\dfrac{\dfrac{\begin{matrix}\vdots\, D\\ \sim\alpha \wedge \sim\beta\end{matrix}}{\sim(\alpha \vee \beta)}\,R}{\sim\alpha \wedge \sim\beta} \quad \rhd \quad \begin{matrix}\vdots\, D\\ \sim\alpha \wedge \sim\beta\end{matrix}$$

9. *Let* D, D', E, F *be proofs. If* $D \rhd D'$, *then*

$$\frac{D}{\alpha}\,(I) \;\rhd\; \frac{D'}{\alpha}\,(I), \quad \frac{D \quad E}{\alpha}\,(R) \;\rhd\; \frac{D' \quad E}{\alpha}\,(R),$$

$$\frac{E \quad D}{\alpha}\,(R) \;\rhd\; \frac{E \quad D'}{\alpha}\,(R),$$

$$\frac{D \quad E \quad F}{\alpha}\,(\vee E) \;\rhd\; \frac{D' \quad E \quad F}{\alpha}\,(\vee E),$$

$$\frac{E \quad D \quad F}{\alpha}\,(\vee E) \;\rhd\; \frac{E \quad D' \quad F}{\alpha}\,(\vee E),$$

$$\frac{E \quad F \quad D}{\alpha}\,(\vee E) \;\rhd\; \frac{E \quad F \quad D'}{\alpha}\,(\vee E)$$

where $I \in \{\to I, \wedge E1, \wedge E2, \vee I1, \vee I2, \sim I, \sim E, \sim\to I, \sim\to E, \sim\wedge I, \sim\wedge E, \sim\vee I, \sim\vee E\}$ *and* $R \in \{\to E, \wedge I\}$.

In this Definition 2.4.3, the conditions 5–8 are taken from the following conditions in the definition of the reduction relation $\rhd$ with respect to λ^c [196]: $M^{\sim\sim\alpha} \rhd M^{\alpha}$, $M^{\sim(\alpha\to\beta)} \rhd M^{\alpha\wedge\sim\beta}$, $M^{\sim(\alpha\wedge\beta)} \rhd M^{\sim\alpha\vee\sim\beta}$ and $M^{\sim(\alpha\vee\beta)} \rhd M^{\sim\alpha\wedge\sim\beta}$ where M is an arbitrary term and the superscripts are types.

Definition 2.4.4 *A sequence* $D_0, D_1, \ldots$ *of proofs is called a* reduction sequence *if it satisfies the following conditions:*

1. $D_i \rhd D_{i+1}$ *for all* $i \geq 0$,

2. *the last proof in the sequence is normal if the sequence is finite.*

A proof D is called normalizable *if there is a finite reduction sequence starting from D.*

In order to consider a correspondence between N_{N4} and a sequent calculus, we introduce an alternative sequent calculus S_{N4} for N4.

Definition 2.4.5 (S_{N4}) *An expression Γ is used in S_{N4} to represent a finite (possibly empty)* set *of formulas. A* sequent *for S_{N4} is an expression of the form $\Gamma \Rightarrow \gamma$. Then, S_{N4} is obtained from* N4 *by deleting* (co).

S_{N4} enjoys cut-elimination, and is theorem equivalent to N4.

Let P be a natural deduction proof. The expression $\mathrm{oa}(P)$ denotes the set of open assumptions of P, and the expression $\mathrm{end}(P)$ denotes the end-formula of P.

Theorem 2.4.6 (Equivalence between N_{N4} and S_{N4}) *We have the following:*

1. *if P is a proof in N_{N4} such that $\mathrm{oa}(P) = \Gamma$ and $\mathrm{end}(P) = \beta$, then $S_{N4} \vdash \Gamma \Rightarrow \beta$,*

2. *if $S_{N4} - (\mathrm{cut}) \vdash \Gamma \Rightarrow \beta$, then there is a proof Q in N_{N4} which satisfies the following conditions:*

   a) $\mathrm{oa}(Q) = \Gamma$,

   b) $\mathrm{end}(Q) = \beta$,

   c) Q *is normal.*

Proof. (1): By induction on the proofs P of N_{N4} such that $\mathrm{oa}(P) = \Gamma$ and $\mathrm{end}(P) = \beta$. We distinguish the cases according to the last inference of P. We show only the following case.

Case $(\sim\!\to\! I)$: P is of the form:

$$\dfrac{\begin{array}{c}\Gamma \\ \vdots\; P' \\ \alpha \wedge {\sim}\gamma\end{array}}{{\sim}(\alpha \to \gamma)}\ ({\sim}{\to}I)$$

where $\mathrm{oa}(P) = \Gamma$ and $\mathrm{end}(P) = {\sim}(\alpha\to\gamma)$. By the induction hypothesis, we have: $S_{N4} \vdash \Gamma \Rightarrow \alpha \wedge {\sim}\gamma$. Then, we have the required fact:

$$\dfrac{\begin{array}{c}\vdots \\ \Gamma \Rightarrow \alpha \wedge {\sim}\gamma\end{array} \qquad \dfrac{\dfrac{\dfrac{\alpha \Rightarrow \alpha}{\alpha, {\sim}\gamma \Rightarrow \alpha}\,(\mathrm{we}) \quad \dfrac{{\sim}\gamma \Rightarrow {\sim}\gamma}{\alpha, {\sim}\gamma \Rightarrow {\sim}\gamma}\,(\mathrm{we})}{\alpha, {\sim}\gamma \Rightarrow {\sim}(\alpha\to\gamma)}\,({\sim}{\to}\mathrm{r})}{\alpha \wedge {\sim}\gamma \Rightarrow {\sim}(\alpha\to\gamma)}\,({\sim}{\to}\mathrm{l})}{\Gamma \Rightarrow {\sim}(\alpha\to\gamma)}\,(\mathrm{cut}).$$

(2): By induction on the proofs P of $\Gamma \Rightarrow \beta$ in S_{N4} $-$ (cut). We distinguish the cases according to the inference of P. We show only the following case.

Case (∼→right): P is of the form:

$$\dfrac{\begin{matrix}\vdots\, P_1 \\ \Gamma \Rightarrow \alpha\end{matrix} \quad \begin{matrix}\vdots\, P_2 \\ \Gamma \Rightarrow {\sim}\gamma\end{matrix}}{\Gamma \Rightarrow {\sim}(\alpha{\rightarrow}\gamma)}\ ({\sim}{\rightarrow}\text{right}).$$

By the induction hypothesis, there are normal proofs Q_1 and Q_2 in N_{N4} of the form:

$$\begin{matrix}\Gamma \\ \vdots\, Q_1 \\ \alpha,\end{matrix} \qquad \begin{matrix}\Gamma \\ \vdots\, Q_2 \\ {\sim}\gamma\end{matrix}$$

where $\mathrm{oa}(Q_1) = \mathrm{oa}(Q_2) = \Gamma$, $\mathrm{end}(Q_1) = \alpha$ and $\mathrm{end}(Q_2) = {\sim}\gamma$. Then we obtain a normal proof Q:

$$\dfrac{\dfrac{\begin{matrix}\Gamma \\ \vdots\, Q_1 \\ \alpha\end{matrix} \quad \begin{matrix}\Gamma \\ \vdots\, Q_2 \\ {\sim}\gamma\end{matrix}}{\alpha \wedge {\sim}\gamma}\ (\wedge I)}{{\sim}(\alpha{\rightarrow}\gamma)}\ ({\sim}{\rightarrow}I)$$

where $\mathrm{oa}(Q) = \Gamma$ and $\mathrm{end}(Q) = {\sim}(\alpha{\rightarrow}\gamma)$. ∎

Theorem 2.4.7 (Normalization) *All proofs in* N_{N4} *are normalizable. More precisely, if a proof P in* N_{N4} *is given, then there is a normal proof Q such that* $\mathrm{oa}(Q) = \mathrm{oa}(P)$ *and* $\mathrm{end}(Q) = \mathrm{end}(P)$.

Proof. Suppose $\mathrm{oa}(P) = \Gamma$ and $\mathrm{end}(P) = \beta$. By Theorem 2.4.6 (1), the sequent $\Gamma \Rightarrow \beta$ is provable in S_{N4}. By the cut-elimination theorem for S_{N4}, the sequent $\Gamma \Rightarrow \beta$ is also provable in S_{N4} $-$ (cut). Then, by Theorem 2.4.6 (2), there is a normal proof Q in N_{N4} such that $\mathrm{oa}(Q) = \mathrm{oa}(P)$ and $\mathrm{end}(Q) = \mathrm{end}(P)$. ∎

2.4.2. Natural deduction in sequent calculus style

In this subsection, Greek capital letters $\Gamma, \Delta, ...$ are used to represent finite (possibly empty) *multisets* of formulas. A *sequent* is an expression of the form $\Gamma \Rightarrow \gamma$. In a multiset of formulas, n occurrences of a formula α are denoted by α^n.

Definition 2.4.8 *A multiset Δ is called a* multiset reduct *of a multiset Γ if Δ is obtained from Γ by multiplying formulas in Γ, where zero multiplicity is also permitted. An expression Γ^* denotes a multiset reduct of Γ.*

Definition 2.4.9 (Sequent calculus in natural deduction style, L_{N4}) *Let $m, n \geq 0$ be any natural numbers. The initial sequents of* L_{N4} *are of the form:*

$$\alpha \Rightarrow \alpha.$$

The cut rule of L_{N4} *is of the form:*

$$\frac{\Gamma \Rightarrow \alpha \quad \alpha, \Sigma \Rightarrow \gamma}{\Gamma, \Sigma \Rightarrow \gamma} \text{ (cut)}.$$

The logical inference rules of L_{N4} *are of the form:*

$$\frac{\Gamma \Rightarrow \alpha \quad \beta^n, \Sigma \Rightarrow \gamma}{\alpha{\rightarrow}\beta, \Gamma, \Sigma \Rightarrow \gamma} \text{ }({\rightarrow}\text{l}^*) \qquad \frac{\alpha^m, \Gamma \Rightarrow \beta}{\Gamma \Rightarrow \alpha{\rightarrow}\beta} \text{ }({\rightarrow}\text{r}^*)$$

$$\frac{\alpha^m, \beta^n, \Delta \Rightarrow \gamma}{\alpha \wedge \beta, \Delta \Rightarrow \gamma} \text{ }(\wedge\text{l}^*) \qquad \frac{\Gamma \Rightarrow \alpha \quad \Delta \Rightarrow \beta}{\Gamma, \Delta \Rightarrow \alpha \wedge \beta} \text{ }(\wedge\text{r}^*)$$

$$\frac{\alpha^m, \Gamma \Rightarrow \gamma \quad \beta^n, \Delta \Rightarrow \gamma}{\alpha \vee \beta, \Gamma, \Delta \Rightarrow \gamma} \text{ }(\vee\text{l}^*)$$

$$\frac{\Gamma \Rightarrow \alpha}{\Gamma \Rightarrow \alpha \vee \beta} \text{ }(\vee\text{r1}) \qquad \frac{\Gamma \Rightarrow \beta}{\Gamma \Rightarrow \alpha \vee \beta} \text{ }(\vee\text{r2})$$

$$\frac{\alpha, \Delta \Rightarrow \gamma}{{\sim}{\sim}\alpha, \Delta \Rightarrow \gamma} \text{ }({\sim}\text{l}) \qquad \frac{\Gamma \Rightarrow \alpha}{\Gamma \Rightarrow {\sim}{\sim}\alpha} \text{ }({\sim}\text{r})$$

$$\frac{\alpha^m, {\sim}\beta^n, \Delta \Rightarrow \gamma}{{\sim}(\alpha{\rightarrow}\beta), \Delta \Rightarrow \gamma} \text{ }({\sim}{\rightarrow}\text{l}^*) \qquad \frac{\Gamma \Rightarrow \alpha \quad \Delta \Rightarrow {\sim}\beta}{\Gamma, \Delta \Rightarrow {\sim}(\alpha{\rightarrow}\beta)} \text{ }({\sim}{\rightarrow}\text{r}^*)$$

$$\frac{{\sim}\alpha^m, \Gamma \Rightarrow \gamma \quad {\sim}\beta^n, \Delta \Rightarrow \gamma}{{\sim}(\alpha \wedge \beta), \Gamma, \Delta \Rightarrow \gamma} \text{ }({\sim}\wedge\text{l}^*)$$

$$\frac{\Gamma \Rightarrow {\sim}\alpha}{\Gamma \Rightarrow {\sim}(\alpha \wedge \beta)} \text{ }({\sim}\wedge\text{r1}) \qquad \frac{\Gamma \Rightarrow {\sim}\beta}{\Gamma \Rightarrow {\sim}(\alpha \wedge \beta)} \text{ }({\sim}\wedge\text{r2})$$

$$\frac{{\sim}\alpha^m, {\sim}\beta^n, \Delta \Rightarrow \gamma}{{\sim}(\alpha \vee \beta), \Delta \Rightarrow \gamma} \text{ }({\sim}\vee\text{l}^*) \qquad \frac{\Gamma \Rightarrow {\sim}\alpha \quad \Delta \Rightarrow {\sim}\beta}{\Gamma, \Delta \Rightarrow {\sim}(\alpha \vee \beta)} \text{ }({\sim}\vee\text{r}^*).$$

A sequent calculus L_{LJ} *for positive intuitionistic logic is defined as the* $\sim$*-free fragment of* L_{N4}.

The superscript $\cdot^*$ in the inference rules of L_{N4} indicates that their forms differ from those of S_{N4}. It can be shown that a sequent $\Rightarrow \alpha$ is provable in L_{N4} if and only if so is S_{N4}. On the other hand, it cannot be shown that a sequent $\Gamma \Rightarrow \gamma$ is provable in L_{N4} if and only if so is S_{N4}. For example, $p \Rightarrow p \wedge p$ for a propositional variable p is provable in S_{N4}, but it is not provable in L_{N4}.[4] Note that the rule ($\sim$l) is not the multiple occurrence form, but the same as that in S_{N4}.

The following cut-elimination theorem is proved by Negri and von Plato [132]: For any sequent $\Gamma \Rightarrow \gamma$,

> if $L_{LJ} \vdash \Gamma \Rightarrow \gamma$, then L_{LJ} − (cut) $\vdash \Gamma^* \Rightarrow \gamma$ where Γ^* is a multiset reduct of Γ.

Theorem 2.4.10 (Cut-elimination) *For any sequent* $\Gamma \Rightarrow \gamma$*,*

> *if* $L_{N4} \vdash \Gamma \Rightarrow \gamma$*, then* L_{N4} − (cut) $\vdash \Gamma^* \Rightarrow \gamma$ *where* Γ^* *is a multiset reduct of* Γ.

Proof. Similar to the proof of Theorem 2.2.5. We use the cut-elimination theorem for L_{LJ}. ∎

The notation used for L_{N4} is also adopted for a natural deduction system G_{N4} introduced below.

Definition 2.4.11 (Natural deduction system in sequent style, G_{N4}) *Let* m, $n \geq 0$ *be any natural numbers. The axioms of* G_{N4} *are of the form:*

$$\alpha \vdash \alpha.$$

The inference rules of G_{N4} *are of the form:*

$$\frac{\alpha^m, \Gamma \vdash \beta}{\Gamma \vdash \alpha{\to}\beta}\ ({\to}I^G) \qquad \frac{\Pi \vdash \alpha{\to}\beta \quad \Gamma \vdash \alpha \quad \beta^n, \Sigma \vdash \gamma}{\Pi, \Gamma, \Sigma \vdash \gamma}\ ({\to}E^G)$$

$$\frac{\Gamma \vdash \alpha \quad \Delta \vdash \beta}{\Gamma, \Delta \vdash \alpha \wedge \beta}\ (\wedge I^G) \qquad \frac{\Pi \vdash \alpha \wedge \beta \quad \alpha^m, \beta^n, \Delta \vdash \gamma}{\Pi, \Delta \vdash \gamma}\ (\wedge E^G)$$

$$\frac{\Gamma \vdash \alpha}{\Gamma \vdash \alpha \vee \beta}\ (\vee I1^G) \qquad \frac{\Gamma \vdash \beta}{\Gamma \vdash \alpha \vee \beta}\ (\vee I2^G)$$

$$\frac{\Pi \vdash \alpha \vee \beta \quad \alpha^m, \Gamma \vdash \gamma \quad \beta^n, \Delta \vdash \gamma}{\Pi, \Gamma, \Delta \vdash \gamma}\ (\vee E^G)$$

$$\frac{\Gamma \vdash \alpha}{\Gamma \vdash {\sim}{\sim}\alpha}\ ({\sim}I^G) \qquad \frac{\Pi \vdash {\sim}{\sim}\alpha \quad \alpha, \Delta \vdash \gamma}{\Pi, \Delta \vdash \gamma}\ ({\sim}E^G)$$

$$\frac{\Gamma \vdash \alpha \quad \Delta \vdash {\sim}\beta}{\Gamma, \Delta \vdash {\sim}(\alpha{\to}\beta)}\ ({\sim}{\to}I^G) \qquad \frac{\Pi \vdash {\sim}(\alpha{\to}\beta) \quad \alpha^m, {\sim}\beta^n, \Delta \vdash \gamma}{\Pi, \Delta \vdash \gamma}\ ({\sim}{\to}E^G)$$

$$\frac{\Pi \vdash {\sim}(\alpha \wedge \beta) \quad {\sim}\alpha^m, \Gamma \vdash \gamma \quad {\sim}\beta^n, \Delta \vdash \gamma}{\Pi, \Gamma, \Delta \vdash \gamma}\ ({\sim} \wedge E^G)$$

[4] This fact is guaranteed by the cut-elimination theorem for L_{N4}, which will be proved.

$$\frac{\Gamma \vdash {\sim}\alpha}{\Gamma \vdash {\sim}(\alpha \wedge \beta)} \ ({\sim} \wedge I1^G) \qquad \frac{\Gamma \vdash {\sim}\beta}{\Gamma \vdash {\sim}(\alpha \wedge \beta)} \ ({\sim} \wedge I2^G)$$

$$\frac{\Gamma \vdash {\sim}\alpha \quad \Delta \vdash {\sim}\beta}{\Gamma, \Delta \vdash {\sim}(\alpha \vee \beta)} \ ({\sim} \vee I^G)$$

$$\frac{\Pi \vdash {\sim}(\alpha \vee \beta) \quad {\sim}\alpha^m, {\sim}\beta^n, \Delta \vdash \gamma}{\Pi, \Delta \vdash \gamma} \ ({\sim} \vee E^G).$$

The inference rules $(\rightarrow E^G)$, $(\wedge E^G)$, $(\vee E^G)$, $({\sim} E^G)$, $({\sim}{\rightarrow} E^G)$, $({\sim} \wedge E^G)$, and $({\sim} \vee E^G)$ are called *general elimination rules*, and the other inference rules are called *introduction rules*. In general elimination rules, the premise containing the logical connective is called *major premise*. The other premises are called *minor premises*.

We remark that the negation part of G_{N4} is from the idea of Schroeder-Heister [173]. For example, his general elimination rule for the negated implication is of the form:

$$\frac{{\sim}(\alpha{\rightarrow}\beta) \quad \begin{matrix}[\alpha, {\sim}\beta] \\ \vdots \\ \gamma\end{matrix}}{\gamma},$$

and it corresponds to $({\sim}{\rightarrow} E^G)$.

Definition 2.4.12 *A proof in* G_{N4} *is in* general normal form *if all major premises of the general elimination rules in the proof are assumptions.*

In Definition 2.4.12, to distinguish the notion of general normal form from the usual notion of (weak) normalization with respect to reduction, the term "general normal form" in this definition is used.

Theorem 2.4.13 (Equivalence between $\mathbf{G}_{N4}$ and $\mathbf{L}_{N4}$) *We have the following:*

1. *if* $\mathrm{L}_{N4} - (\mathrm{cut}) \vdash \Gamma \Rightarrow \gamma$, *then there is a general normal proof of* $\Gamma \vdash \gamma$ *in* G_{N4},
2. *if there is a proof of* $\Gamma \vdash \gamma$ *in* G_{N4}, *then* $\mathrm{L}_{N4} \vdash \Gamma \Rightarrow \gamma$.

Proof. • (1): By induction on the proofs P of $\Gamma \Rightarrow \gamma$ in $\mathrm{L}_{N4} - (\mathrm{cut})$. We distinguish the cases according to the last inference of P. We show only the following case.

Case (∼→l*): P is of the form:

$$\dfrac{\begin{array}{c}\vdots\ P_1\\ \alpha^m, {\sim}\beta^n, \Delta \Rightarrow \gamma\end{array}}{{\sim}(\alpha{\rightarrow}\beta), \Delta \Rightarrow \gamma}\ ({\sim}{\rightarrow}\mathrm{l}^*).$$

By the induction hypothesis, there is a general normal proof P_1' of $\alpha^m, {\sim}\beta^n, \Delta \vdash \gamma$ in G_{N4}, and hence a required proof is:

$$\dfrac{{\sim}(\alpha{\rightarrow}\beta) \vdash {\sim}(\alpha{\rightarrow}\beta) \quad \begin{array}{c}\vdots\ P_1'\\ \alpha^m, {\sim}\beta^n, \Delta \vdash \gamma\end{array}}{{\sim}(\alpha{\rightarrow}\beta), \Delta \vdash \gamma}\ ({\sim}{\rightarrow}E^G).$$

• (2): By induction on the proofs Q of $\Gamma \vdash \gamma$ in L_{N4}. We distinguish the cases according to the last inference of Q. We show only the following case.

Case (∼→E^G): Q is of the form:

$$\dfrac{\begin{array}{c}\vdots\ Q_1\\ \Pi \vdash {\sim}(\alpha{\rightarrow}\beta)\end{array} \quad \begin{array}{c}\vdots\ Q_2\\ \alpha^m, {\sim}\beta^n, \Delta \vdash \gamma\end{array}}{\Pi, \Delta \vdash \gamma}\ ({\sim}{\rightarrow}E^G).$$

By the induction hypothesis, the sequents $\Pi \Rightarrow {\sim}(\alpha{\rightarrow}\beta)$ and $\alpha^m, {\sim}\beta^n, \Delta \Rightarrow \gamma$ are provable in L_{N4}, and hence a required proof is:

$$\dfrac{\begin{array}{c}\vdots\ Q_1'\\ \Pi \Rightarrow {\sim}(\alpha{\rightarrow}\beta)\end{array} \quad \dfrac{\begin{array}{c}\vdots\ Q_2'\\ \alpha^m, {\sim}\beta^n, \Delta \Rightarrow \gamma\end{array}}{{\sim}(\alpha{\rightarrow}\beta), \Delta \Rightarrow \gamma}\ ({\sim}{\rightarrow}\mathrm{l}^*)}{\Pi, \Delta \Rightarrow \gamma}\ (\mathrm{cut}).$$

∎

Theorem 2.4.14 (Normalization) *Every proof P of $\Gamma \vdash \gamma$ in G_{N4} can be transformed into a general normal proof P' of $\Gamma \vdash \gamma$ in G_{N4}.*

Proof. Let P be a proof of $\Gamma \vdash \gamma$ in G_{N4}. Then, the sequent $\Gamma \Rightarrow \gamma$ is provable in L_{N4} by Theorem 2.4.13 (2), and hence $\Gamma \Rightarrow \gamma$ is provable in cut-free L_{N4} by the cut-elimination theorem for L_{N4}. By Theorem 2.4.13 (1), there is a general normal proof P' of $\Gamma \vdash \gamma$ in G_{N4}. ∎

2.4.3. Other systems

Definition 2.4.15 (Prawitz's system P_{N4} [153]) *Prawitz's natural deduction system P_{N4} for N4 is obtained from N_{N4} by replacing the* *inference rules* $(\sim\to I)$, $(\sim\to E)$, $(\sim\wedge I)$, $(\sim\wedge E)$, $(\sim\vee I)$ *and* $(\sim\vee E)$ *with the inference rules of the form:*

$$\frac{\alpha \quad \sim\beta}{\sim(\alpha\to\beta)}\,(\sim\to I^p) \qquad \frac{\sim(\alpha\to\beta)}{\alpha}\,(\sim\to E1^p) \qquad \frac{\sim(\alpha\to\beta)}{\sim\beta}\,(\sim\to E2^p)$$

$$\frac{\sim\alpha \quad \sim\beta}{\sim(\alpha\vee\beta)}\,(\sim\vee I^p) \qquad \frac{\sim(\alpha\vee\beta)}{\sim\alpha}\,(\sim\vee E1^p) \qquad \frac{\sim(\alpha\vee\beta)}{\sim\beta}\,(\sim\vee E2^p)$$

$$\frac{\sim\alpha}{\sim(\alpha\wedge\beta)}\,(\sim\wedge I1^p) \qquad \frac{\sim\beta}{\sim(\alpha\wedge\beta)}\,(\sim\wedge I2^p)$$

$$\frac{\sim(\alpha\wedge\beta) \quad \begin{matrix}[\sim\alpha]\\ \vdots \\ \gamma\end{matrix} \quad \begin{matrix}[\sim\beta]\\ \vdots \\ \gamma\end{matrix}}{\gamma}\,(\sim\wedge E^p).$$

Proposition 2.4.16 (Equivalence between $\mathbf{P}_{N4}$ and $\mathbf{N}_{N4}$) *For any formula* γ,

γ *is provable in* P_{N4} *iff* γ *is provable in* N_{N4}.

Schroeder-Heister [173] proposed a more general natural deduction framework dealing with some n-ary logical connectives and also with $\sim$. The following system, called H_{N4}, may be regarded as a special case of his framework.

Definition 2.4.17 (Special case of Schroeder-Heister's system [173], $\mathbf{H}_{N4}$) *The inference rules of* H_{N4} *are* $(\to I)$, $(\wedge I)$, $(\vee I1)$, $(\vee I2)$, $(\vee E)$ *and* $(\sim I)$ *in* N_{N4}, *and* $(\sim\to I^p)$, $(\sim\vee I^p)$, $(\sim\wedge I1^p)$, $(\sim\wedge I2^p)$ *and* $(\sim\wedge E^p)$ *in* P_{N4}, *and the inference rules of the form:*

$$\frac{\alpha\to\beta \quad \alpha \quad \begin{matrix}[\beta]\\ \vdots \\ \gamma\end{matrix}}{\gamma}\,(\to E^h) \qquad \frac{\alpha\wedge\beta \quad \begin{matrix}[\alpha,\beta]\\ \vdots \\ \gamma\end{matrix}}{\gamma}\,(\wedge E^h)$$

$$\frac{\sim\sim\alpha \quad \begin{matrix}[\alpha]\\ \vdots \\ \gamma\end{matrix}}{\gamma}\,(\sim E^h) \qquad \frac{\sim(\alpha\to\beta) \quad \begin{matrix}[\alpha,\sim\beta]\\ \vdots \\ \gamma\end{matrix}}{\gamma}\,(\sim\to E^h)$$

$$\frac{\sim(\alpha\vee\beta) \quad \begin{matrix}[\sim\alpha,\sim\beta]\\ \vdots \\ \gamma\end{matrix}}{\gamma}\,(\sim\vee E^h).$$

Note that H_{N4} is the standard-style version of the sequent-style system G_{N4}.

Proposition 2.4.18 (Equivalence between $\mathbf{H}_{N4}$ and $\mathbf{N}_{N4}$) *For any formula γ,*

$$\gamma \textit{ is provable in } \mathrm{H}_{N4} \quad \textit{iff} \quad \gamma \textit{ is provable in } \mathrm{N}_{N4}.$$

We can extend the uniform calculi (for positive intuitionistic logic) introduced by Negri and von Plato [133], which have some *general introduction rules.*

Definition 2.4.19 (Extended version of the systems from [133], $\mathbf{U}_{N4}$) *Let $m \geq 0$ be any natural number. The natural deduction system U_{N4} in sequent calculus style is obtained from G_{N4} (Definition 2.4.11) by replacing ($\rightarrow I^G$), ($\wedge I^G$), ($\vee I1^G$), ($\vee I2^G$), ($\sim I^G$), ($\sim\rightarrow I^G$), ($\sim\wedge I1^G$), ($\sim\wedge I2^G$) and ($\sim\vee I^G$) by the general introduction rules of the form:*

$$\frac{\alpha\rightarrow\beta, \Gamma \vdash \gamma \quad \alpha^m, \Delta \vdash \beta}{\Gamma, \Delta \vdash \gamma}\ (\rightarrow I^U) \qquad \frac{\alpha \wedge \beta, \Gamma \vdash \gamma \quad \Delta \vdash \alpha \quad \Sigma \vdash \beta}{\Gamma, \Delta, \Sigma \vdash \gamma}\ (\wedge I^U)$$

$$\frac{\alpha \vee \beta, \Gamma \vdash \gamma \quad \Delta \vdash \alpha}{\Gamma, \Delta \vdash \gamma}\ (\vee I1^U) \qquad \frac{\alpha \vee \beta, \Gamma \vdash \gamma \quad \Delta \vdash \beta}{\Gamma, \Delta \vdash \gamma}\ (\vee I2^U)$$

$$\frac{\sim\sim\alpha, \Gamma \vdash \gamma \quad \Delta \vdash \alpha}{\Gamma, \Delta \vdash \gamma}\ (\sim I^U)$$

$$\frac{\sim(\alpha\rightarrow\beta), \Gamma \vdash \gamma \quad \Delta \vdash \alpha \quad \Sigma \vdash \sim\beta}{\Gamma, \Delta, \Sigma \vdash \gamma}\ (\sim\rightarrow I^U)$$

$$\frac{\sim(\alpha \wedge \beta), \Gamma \vdash \gamma \quad \Delta \vdash \sim\alpha}{\Gamma, \Delta \vdash \gamma}\ (\sim\wedge I1^U)$$

$$\frac{\sim(\alpha \wedge \beta), \Gamma \vdash \gamma \quad \Delta \vdash \sim\beta}{\Gamma, \Delta \vdash \gamma}\ (\sim\wedge I2^U)$$

$$\frac{\sim(\alpha \vee \beta), \Gamma \vdash \gamma \quad \Delta \vdash \sim\alpha \quad \Sigma \vdash \sim\beta}{\Gamma, \Delta, \Sigma \vdash \gamma}\ (\sim\vee I^U).$$

Proposition 2.4.20 (Equivalence between $\mathbf{U}_{N4}$ and $\mathbf{G}_{N4}$) *For any formula γ,*

$$\vdash \gamma \textit{ is provable in } \mathrm{U}_{N4} \quad \textit{iff} \quad \gamma \textit{ is provable in } \mathrm{G}_{N4}.$$

The following system V_{N4} is the standard-style version of U_{N4}.

Definition 2.4.21 (Extended version of the systems from [133], $\mathbf{V}_{N4}$) *V_{N4} is obtained from H_{N4} (Definition 2.4.17) by replacing $(\to I)$, $(\wedge I)$, $(\vee I1)$, $(\vee I2)$, $({\sim} I)$, $({\sim}{\to} I^p)$, $({\sim} \wedge I1^p)$, $({\sim} \wedge I2^p)$ and $({\sim} \vee I^p)$ by the general introduction rules of the form: Negri and*

$$\dfrac{\begin{matrix}[\alpha\to\beta] \\ \vdots \\ \gamma\end{matrix} \quad \begin{matrix}[\alpha] \\ \vdots \\ \beta\end{matrix}}{\gamma}\ (\to I^V) \qquad \dfrac{\begin{matrix}[\alpha\wedge\beta] \\ \vdots \\ \gamma\end{matrix} \quad \alpha \quad \beta}{\gamma}\ (\wedge I^V) \qquad \dfrac{\begin{matrix}[\alpha\vee\beta] \\ \vdots \\ \gamma\end{matrix} \quad \alpha}{\gamma}\ (\vee I1^V)$$

$$\dfrac{\begin{matrix}[\alpha\vee\beta] \\ \vdots \\ \gamma\end{matrix} \quad \beta}{\gamma}\ (\vee I2^V) \qquad \dfrac{\begin{matrix}[{\sim}{\sim}\alpha] \\ \vdots \\ \gamma\end{matrix} \quad \alpha}{\gamma}\ ({\sim} I^V)$$

$$\dfrac{\begin{matrix}[{\sim}(\alpha\to\beta)] \\ \vdots \\ \gamma\end{matrix} \quad \alpha \quad {\sim}\beta}{\gamma}\ ({\sim}{\to} I^V)$$

$$\dfrac{\begin{matrix}[{\sim}(\alpha\wedge\beta)] \\ \vdots \\ \gamma\end{matrix} \quad {\sim}\alpha}{\gamma}\ ({\sim} \wedge I1^V) \qquad \dfrac{\begin{matrix}[{\sim}(\alpha\wedge\beta)] \\ \vdots \\ \gamma\end{matrix} \quad {\sim}\beta}{\gamma}\ ({\sim} \wedge I2^V)$$

$$\dfrac{\begin{matrix}[{\sim}(\alpha\vee\beta)] \\ \vdots \\ \gamma\end{matrix} \quad {\sim}\alpha \quad {\sim}\beta}{\gamma}\ ({\sim} \vee I^V).$$

Proposition 2.4.22 (Equivalence between $\mathbf{V}_{N4}$ and $\mathbf{N}_{N4}$) *For any formula γ,*

γ is provable in V_{N4} iff γ is provable in N_{N4}.

In the following, we use two expressions $\vdash^+ \alpha$ and $\vdash^- \alpha$, which stand for α and ${\sim}\alpha$, respectively. Using these expressions we introduce a natural deduction system T_{N4} for N4. The $\to$-free part of T_{N4} is just Tamminga and Tanaka's system $\mathrm{ND}_{\mathrm{FDE}}$ [184] for E_{fde}.

Definition 2.4.23 (Extended version of the system from [184], $\mathbf{T}_{N4}$) *Let $\vdash^* \gamma$ be $\vdash^+ \gamma$ or $\vdash^- \gamma$. The inference rules of T_{N4} are of the form:*

$$\dfrac{\begin{matrix}[\vdash^+ \alpha] \\ \vdots \\ \vdash^+ \beta\end{matrix}}{\vdash^+ \alpha\to\beta}\ (\to I^+) \qquad \dfrac{\vdash^+ \alpha\to\beta \quad \vdash^+ \alpha}{\vdash^+ \beta}\ (\to E^+)$$

$$\frac{\vdash^+ \alpha \quad \vdash^+ \beta}{\vdash^+ \alpha \wedge \beta}\ (\wedge I^+) \qquad \frac{\vdash^+ \alpha \wedge \beta}{\vdash^+ \alpha}\ (\wedge E1^+) \qquad \frac{\vdash^+ \alpha \wedge \beta}{\vdash^+ \beta}\ (\wedge E2^+)$$

$$\frac{\vdash^+ \alpha}{\vdash^+ \alpha \vee \beta}\ (\vee I1^+) \qquad \frac{\vdash^+ \beta}{\vdash^+ \alpha \vee \beta}\ (\vee I2^+)$$

$$\frac{\vdash^+ \alpha \vee \beta \quad \begin{array}{c}[\vdash^+ \alpha]\\ \vdots \\ \vdash^* \gamma\end{array} \quad \begin{array}{c}[\vdash^+ \beta]\\ \vdots \\ \vdash^* \gamma\end{array}}{\vdash^* \gamma}\ (\vee E^+)$$

$$\frac{\vdash^- \alpha}{\vdash^+ {\sim}\alpha}\ ({\sim}I^+) \qquad \frac{\vdash^+ {\sim}\alpha}{\vdash^- \alpha}\ ({\sim}E^+)$$

$$\frac{\vdash^+ \alpha}{\vdash^- {\sim}\alpha}\ ({\sim}I^-) \qquad \frac{\vdash^- {\sim}\alpha}{\vdash^+ \alpha}\ ({\sim}E^-)$$

$$\frac{\vdash^+ \alpha \quad \vdash^- \beta}{\vdash^- \alpha{\rightarrow}\beta}\ ({\rightarrow}I^-) \qquad \frac{\vdash^- \alpha{\rightarrow}\beta}{\vdash^+ \alpha}\ ({\rightarrow}E1^-) \qquad \frac{\vdash^- \alpha{\rightarrow}\beta}{\vdash^- \beta}\ ({\rightarrow}E2^-)$$

$$\frac{\vdash^- \alpha \quad \vdash^- \beta}{\vdash^- \alpha \vee \beta}\ (\vee I^-) \qquad \frac{\vdash^- \alpha \vee \beta}{\vdash^- \alpha}\ (\vee E1^-) \qquad \frac{\vdash^- \alpha \vee \beta}{\vdash^- \beta}\ (\vee E2^-)$$

$$\frac{\vdash^- \alpha}{\vdash^- \alpha \wedge \beta}\ (\wedge I1^-) \qquad \frac{\vdash^- \beta}{\vdash^- \alpha \wedge \beta}\ (\wedge I2^-)$$

$$\frac{\vdash^- \alpha \wedge \beta \quad \begin{array}{c}[\vdash^- \alpha]\\ \vdots \\ \vdash^* \gamma\end{array} \quad \begin{array}{c}[\vdash^- \beta]\\ \vdots \\ \vdash^* \gamma\end{array}}{\vdash^* \gamma}\ (\wedge E^-).$$

Proposition 2.4.24 (Equivalence between $\mathbf{T}_{N4}$ and $\mathbf{N}_{N4}$) *For any formula γ,*

1. *if γ is provable in N_{N4}, then $\vdash^+ \gamma$ is provable in T_{N4},*
2. *if $\vdash^+ \gamma$ ($\vdash^- \gamma$) is provable in T_{N4}, then γ (${\sim}\gamma$, respectively) is provable in N_{N4}.*

2.5. Remarks

David Nelson's paraconsistent logic N4 is a common basis for various extended and useful paraconsistent logics which are widely used in Computer Science. Therefore, in order to obtain a proof theoretical foundation for N4 is an important issue for realizing inconsistency-tolerant reasoning with automated theorem-proving. This chapter gives a uniform perspective for such a proof theory.

Various proof systems for N4 and its neighbors were comprehensively studied in this chapter. Some basic results including the completeness, cut-elimination and decidability theorems for N4 were uniformly proved using the syntactical and semantical embedding theorems of N4 into positive intuitionistic logic. The sequent calculi G4np (contraction-free system), G5np (resolution related system), Sn4 (subformula calculus), Dn4 (dual calculus) and some standard and less familiar display sequent calculi were presented and compared. The cut-elimination and equivalence theorems for these calculi were proved. The natural deduction systems N_{N4} (standard-style system), G_{N4} (sequent-style system), U_{N4} (uniform calculus) and some extended traditional natural deduction systems were presented. The normalization and equivalence theorems for these systems were proved.

Some of our recent works on extensions of (a sequent calculus for) N4 are briefly reviewed below. Sequent calculi for some trilattice logics, which may be regarded as extensions of N4 and Arieli and Avron's bilattice logics, were studied in [99], and some sequent and tableaux calculi for some intuitionistic counterparts of the trilattice logics were also studied in [212]. The sequent calculi proposed in [99, 212] were based on some parts of the trilattice $SIXTEEN_3$, i.e., such sequent calculi are not a full system for $SIXTEEN_3$. A full sequent system for $SIXTEEN_3$ was proposed in [210], introducing a new mechanism for representing non-determinism. A paraconsistent constructive linear-time temporal logic, which is obtained from (a sequent calculus for) N4 by adding temporal operators, was studied in [100], and a classical version of this logic was also studied in [102]. Extensions of H-B by various kinds of strong negation are studied in [209], see Chapter 5.

3. Paraconsistent logics based on trilattices

A sequent calculus L_{16} for Odintsov's Hilbert-style axiomatization L_B of a logic related to the trilattice $SIXTEEN_3$ of generalized truth values is introduced. The completeness theorem w.r.t. a simple semantics for L_{16} is proved using Maehara's decomposition method that simultaneously derives the cut-elimination theorem for L_{16}. A first-order extension F_{16} of L_{16} and its semantics are also introduced. The completeness and cut-elimination theorems for F_{16} are proved using Schütte's method.[1]

The present chapter is structured as follows. In Section 3.1, we present Odintsov's axiom systems L_B and L_T and explain how they are related to the trilattice of generalized truth values $SIXTEEN_3$. In Section 3.2, the sequent calculus L_{16} for L_B is defined. Moreover, a variant of Odintsov's co-ordinate valuations semantics is introduced, and Maehara's method is applied to show that L_{16} is strongly sound and complete with respect to this semantics. A semantic proof of cut-elimination follows immediately. Section 3.3 is devoted to extending L_{16} to a first-order sequent calculus F_{16}. Using Schütte's method involving the notion of saturated sequents, F_{16} is shown to be sound and complete with respect to first-order models using four interpretation functions (according to the variation of Odintsov's co-ordinate valuations) and four corresponding satisfaction relations. Again, cut-elimination can be proved semantically.

3.1. Introduction

3.1.1. Trilattices

The structure $SIXTEEN_3$ is the smallest so-called Belnap trilattice [180, 181, 183, 179]. It is based on the 16-element powerset of the powerset of the set of classical truth values $\mathbf{2} = \{T, F\}$, and it is motivated by generalizing N. Belnap's [21, 22] idea of viewing a truth value as

[1] This chapter is based on [103]. Here we take the opportunity to correct some errors in this paper, which were kindly pointed out to us by M. Takano.

information that is told to a computer concerning a given proposition. Whereas the elements of $\mathbf{4} = \mathcal{P}(\mathbf{2})$ can be understood as follows:

$\mathbf{N} = \{\}$ - "told neither true nor false";
$\mathbf{F} = \{F\}$ - "told only false";
$\mathbf{T} = \{T\}$ - "told only true";
$\mathbf{B} = \{T, F\}$ - "told both true and false",

so informed computers themselves may or may not pass Belnap's generalized truth values to other computers, which thereby receive combinations of Belnap's values as information concerning a given proposition. They may, e.g., receive the information that a proposition is "both told neither true nor false and told both true and false" $(\{\varnothing, \{T, F\}\})$. It turns out that on the resulting set of values $\mathcal{P}(\mathbf{4}) = \mathbf{16}$:

1.	$\mathbf{N} = \varnothing$	9.	$\mathbf{FT} = \{\{F\}, \{T\}\}$
2.	$\boldsymbol{N} = \{\varnothing\}$	10.	$\mathbf{FB} = \{\{F\}, \{F, T\}\}$
3.	$\boldsymbol{F} = \{\{F\}\}$	11.	$\mathbf{TB} = \{\{T\}\}, \{F, T\}\}$
4.	$\boldsymbol{T} = \{\{T\}\}$	12.	$\mathbf{NFT} = \{\varnothing, \{F\}, \{T\}\}$
5.	$\boldsymbol{B} = \{\{F, T\}\}$	13.	$\mathbf{NFB} = \{\varnothing, \{F\}, \{F, T\}\}$
6.	$\mathbf{NF} = \{\varnothing, \{F\}\}$	14.	$\mathbf{NTB} = \{\varnothing, \{T\}, \{F, T\}\}$
7.	$\mathbf{NT} = \{\varnothing, \{T\}\}$	15.	$\mathbf{FTB} = \{\{F\}, \{T\}, \{F, T\}\}$
8.	$\mathbf{NB} = \{\varnothing, \{F, T\}\}$	16.	$\mathbf{A} = \{\varnothing, \{T\}, \{F\}, \{F, T\}\}$

in addition to set-inclusion as a natural information order $\leq_i$, a truth order $\leq_t$ and a falsity order $\leq_f$ can be defined (which are not inverses of each other).

Definition 3.1.1 *For every x, y in* $\mathbf{16}$*:*

1. $x \leq_i y$ *iff* $x \subseteq y$*;*
2. $x \leq_t y$ *iff* $x^t \subseteq y^t$ *and* $y^{-t} \subseteq x^{-t}$ *,*
   where $x^t := \{y \in x \mid T \in y\}$ *and* $x^{-t} := \{y \in x \mid T \notin y\}$*;*
3. $x \leq_f y$ *iff* $x^f \subseteq y^f$ *and* $y^{-f} \subseteq x^{-f}$ *,*
   where $x^f := \{y \in x \mid F \in y\}$ *and* $x^{-f} := \{y \in x \mid F \notin y\}$.

Note that the definition of the truth (falsity) order refers only to the classical value T (F) and not to the value F (T).

The three (complete) lattices $(\mathbf{16}, \leq_i)$, $(\mathbf{16}, \leq_t)$, and $(\mathbf{16}, \leq_f)$ can be combined into the *trilattice* $SIXTEEN_3 = (\mathbf{16}, \leq_i, \leq_t, \leq_f)$, see [180, 183]. $SIXTEEN_3$ is depicted as a Hasse diagram in Figure 3.1; alternatively, it may be presented as the algebraic structure $\langle \mathbf{16}, \sqcap_i, \sqcup_i, \sqcap_t, \sqcup_t, \sqcap_f, \sqcup_f \rangle$, where $\sqcap_\sharp$ $(\sqcup_\sharp)$ is the meet (join) with respect to $\leq_\sharp$, $\sharp \in \{i, t, f\}$. Since the "logical" relations $\leq_t$ and $\leq_f$ are treated on a par, the operations

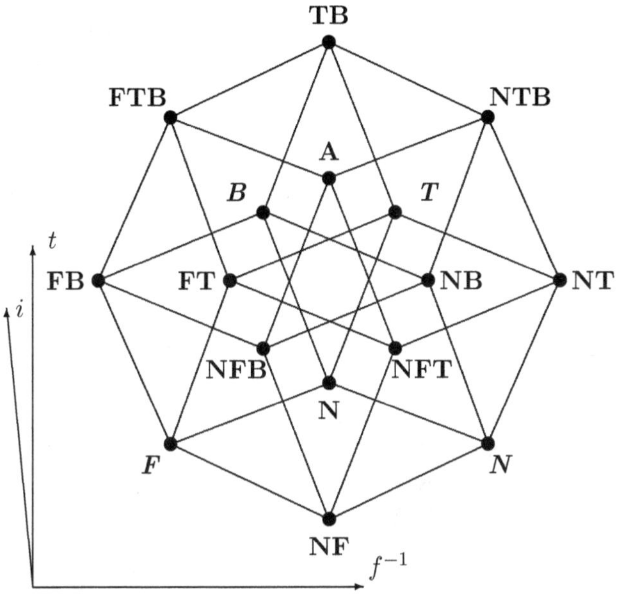

Figure 3.1.: Trilattice $SIXTEEN_3$

$\sqcap_t$ and $\sqcup_t$ are not privileged as interpretations of conjunction and disjunction. The operation $\sqcup_f$ may as well be regarded as a conjunction and $\sqcap_f$ as a disjunction. Therefore, the logical vocabulary can be considered to comprise a positive truth vocabulary together with a negative falsity vocabulary. Also certain unary truth and falsity operations with negation-like properties are available in *SIXTEEN*$_3$.

A unary operation $-_t$ $(-_f)$ on *SIXTEEN*$_3$ is said to be a t-inversion (f-inversion) iff the following conditions are satisfied:

1. t-*inversion* $(-_t)$:	2. f-*inversion* $(-_f)$:
(a) $x \leq_t y \Rightarrow -_t y \leq_t -_t x$;	(a) $x \leq_t y \Rightarrow -_f x \leq_t -_f y$;
(b) $x \leq_f y \Rightarrow -_t x \leq_f -_t y$;	(b) $x \leq_f y \Rightarrow -_f y \leq_f -_f x$;
(c) $x \leq_i y \Rightarrow -_t x \leq_i -_t y$;	(c) $x \leq_i y \Rightarrow -_f x \leq_i -_f y$;
(d) $-_t -_t x = x$.	(d) $-_f -_f x = x$.

A t-inversion (f-inversion) thus inverts the truth (falsity) order, leaves the other orders untouched, and is period-two. In $SIXTEEN_3$ such operations are definable as shown in Table 1. If the condition that an inversion preserves the other orderings is given up, the definition of

t-inversion (f-inversion) refers only to the truth-order (falsity-order). What this suggests is that not only conjunction and disjunction, but also negation comes in two versions, because $-_t$ and $-_f$ are both natural interpretations for a negation connective. Moreover, since $x \sqcap_t y \neq x \sqcup_f y$, $x \sqcup_t y \neq x \sqcap_f y$ and $-_t x \neq -_f x$, the two logical orderings $\leq_t$ and $\leq_f$ indeed give rise to pairs of *distinct* logical operations with the same arity.

x	$-_t x$	$-_f x$	x	$-_t x$	$-_f x$
N	**N**	**N**	**NB**	**FT**	**FT**
N	***T***	***F***	**FB**	**FB**	**NT**
F	***B***	***N***	**TB**	**NF**	**TB**
T	***N***	***B***	**NFT**	**NTB**	**NFB**
B	***F***	***T***	**NFB**	**FTB**	**NFT**
NF	**TB**	**NF**	**NTB**	**NFT**	**FTB**
NT	**NT**	**FB**	**FTB**	**NFB**	**NTB**
FT	**NB**	**NB**	**A**	**A**	**A**

Table 3.1.: t- and f-inversions in $SIXTEEN_3$

3.1.2. Odintsov's axiomatizations

The following list of symbols is adopted for the language used in this chapter: countably many propositional variables $p_0, p_1, ...$, logical connectives $\rightarrow$, $\neg$, $\wedge_t$, $\vee_t$, $\wedge_f$, $\vee_f$, $\sim_t$ and $\sim_f$. The connectives $\rightarrow$, $\neg$, $\wedge_t$ and $\vee_t$ are just the classical implication, negation, conjunction and disjunction, respectively. Note that $\rightarrow$ is denoted as $\rightarrow_t$ in [139] (because it is interpreted as the residuum of the truth order in the trilattice $SIXTEEN_3$). The symbol $\sim_b$ is used to denote $\sim_f\sim_t$ or $\sim_t\sim_f$, and the symbol $\equiv$ is used to denote the equality of sets of symbols. Greek lower-case letters $\alpha, \beta, ...$ are used to denote formulas. An expression $\alpha \leftrightarrow \beta$ is an abbreviation of $(\alpha{\rightarrow}\beta){\wedge_t}(\beta{\rightarrow}\alpha)$.

Odintsov's Hilbert-style axiomatizations L_{base}, L_B and L_T [139] of logics related to the trilattice $SIXTEEN_3$ of generalized truth values in the propositional language based on $\{\sim_t, \sim_f, \wedge_t, \wedge_f, \vee_t, \vee_f, \rightarrow, \neg\}$ are presented below.

Definition 3.1.2 ($\mathbf{L}_{base}$) *The rules of* L_{base} *are of the form:*

$$\frac{\alpha \quad \alpha{\rightarrow}\beta}{\beta} \text{ (mp).}$$

The axiom schemes of L_{base} *are of the form:*

1. $\alpha \to (\beta \to \alpha)$,
2. $(\alpha \to (\beta \to \gamma)) \to ((\alpha \to \beta) \to (\alpha \to \gamma))$,
3. $(\alpha \wedge_t \beta) \to \alpha$,
4. $(\alpha \wedge_t \beta) \to \beta$,
5. $(\alpha \to \beta) \to ((\alpha \to \gamma) \to (\alpha \to (\beta \wedge_t \gamma)))$,
6. $\alpha \to (\alpha \vee_t \beta)$,
7. $\beta \to (\alpha \vee_t \beta)$,
8. $(\alpha \to \gamma) \to ((\beta \to \gamma) \to ((\alpha \vee_t \beta) \to \gamma))$,
9. $\alpha \vee_t (\alpha \to \beta)$,
10. $\alpha \leftrightarrow \sim_t \sim_t \alpha$,
11. $\alpha \leftrightarrow \sim_f \sim_f \alpha$,
12. $\sim_t \sim_f \alpha \leftrightarrow \sim_f \sim_t \alpha$,
13. $\neg \sim_f \alpha \leftrightarrow \sim_f \neg \alpha$,
14. $\neg \sim_t \alpha \leftrightarrow \sim_t \neg \alpha$,
15. $\neg \sim_f \sim_t \alpha \leftrightarrow \sim_f \sim_t \neg \alpha$,
16. $\sim_f \alpha \leftrightarrow \sim_t \sim_f \sim_t \alpha$,
17. $\sim_t \alpha \leftrightarrow \sim_f \sim_t \sim_f \alpha$,
18. $\sim_t (\alpha \wedge_t \beta) \leftrightarrow (\sim_t \alpha \vee_t \sim_t \beta)$,
19. $\sim_t (\alpha \vee_t \beta) \leftrightarrow (\sim_t \alpha \wedge_t \sim_t \beta)$,
20. $\sim_t (\alpha \wedge_f \beta) \leftrightarrow (\sim_t \alpha \wedge_f \sim_t \beta)$,
21. $\sim_t (\alpha \vee_f \beta) \leftrightarrow (\sim_t \alpha \vee_f \sim_t \beta)$,
22. $\sim_f (\alpha \wedge_f \beta) \leftrightarrow (\sim_f \alpha \vee_f \sim_f \beta)$,
23. $\sim_f (\alpha \vee_f \beta) \leftrightarrow (\sim_f \alpha \wedge_f \sim_f \beta)$,
24. $\sim_f (\alpha \wedge_t \beta) \leftrightarrow (\sim_f \alpha \wedge_t \sim_f \beta)$,
25. $\sim_f (\alpha \vee_t \beta) \leftrightarrow (\sim_f \alpha \vee_t \sim_f \beta)$,
26. $\sim_f \sim_t (\alpha \wedge_t \beta) \leftrightarrow (\sim_f \sim_t \alpha \vee_t \sim_f \sim_t \beta)$,

27. $\sim_f\sim_t(\alpha\vee_t\beta) \leftrightarrow (\sim_f\sim_t\alpha\wedge_t\sim_f\sim_t\beta)$,

28. $\sim_f\sim_t(\alpha\wedge_f\beta) \leftrightarrow (\sim_f\sim_t\alpha\vee_f\sim_f\sim_t\beta)$,

29. $\sim_f\sim_t(\alpha\vee_f\beta) \leftrightarrow (\sim_f\sim_t\alpha\wedge_f\sim_f\sim_t\beta)$,

30. $(\alpha\rightarrow\beta) \leftrightarrow (\neg\alpha\vee_t\beta)$,

31. $\sim_t(\alpha\rightarrow\beta) \leftrightarrow (\sim_t\neg\alpha\wedge_t\sim_t\beta)$,

32. $\sim_f(\alpha\rightarrow\beta) \leftrightarrow (\sim_f\alpha\rightarrow\sim_f\beta)$,

33. $\sim_f\sim_t(\alpha\rightarrow\beta) \leftrightarrow (\sim_f\sim_t\neg\alpha\wedge_t\sim_f\sim_t\beta)$.

Definition 3.1.3 ($\mathbf{L}_B$ and $\mathbf{L}_T$) L_B *is obtained from* L_{base} *by adding the*
axiom schemes of the form:

34. $(\alpha\wedge_t\beta) \leftrightarrow (\alpha\vee_f\beta)$,

35. $(\alpha\vee_t\beta) \leftrightarrow (\alpha\wedge_f\beta)$.

L_T *is obtained from* L_{base} *by adding the axiom schemes of the form:*

36. $(\alpha\wedge_t\beta) \leftrightarrow (\alpha\wedge_f\beta)$,

37. $(\alpha\vee_t\beta) \leftrightarrow (\alpha\vee_f\beta)$.

Since we can consider (cut-free) sequent calculi for both L_B and L_T similarly, we only discuss a sequent calculus for L_B and its first-order extension in the following sections. A cut-free sequent calculus for L_{base} has not been obtained yet. Such a calculus may be difficult to construct, see [99].

In the logical language of the sequent calculi L_{base}, L_B, and L_T, negation, conjunction, and disjunction come in two versions. In addition to the truth connectives $\sim_t$, $\wedge_t$, and $\vee_t$, there are also falsity connectives $\sim_f$, $\wedge_f$, and $\vee_f$. This split of the basic propositional vocabulary is induced by the truth order and the falsity order on the set of generalized truth values of $SIXTEEN_3$. The connectives $\wedge_t$ and $\vee_t$ ($\wedge_f$ and $\vee_f$) are interpreted as the lattice meet and lattice join (lattice join and lattice meet) of the truth order $\leq_t$ (the falsity order $\leq_f$), and the negations $\sim_t$ and $\sim_f$ are interpreted as certain truth order and falsity order inversions, respectively. Assignments of generalized truth values from $SIXTEEN_3$ to the propositional variables, i.e., functions from the set of propositional variables into the set **16**, the powerset of the powerset of the set of classical truth values, are thereby homomorphically extended to valuation functions v assigning generalized truth values to arbitrary formulas.

Definition 3.1.4 *The binary entailment relations* $\models_t$ *and* $\models_f$ *on the set of formulas are defined by the following equivalences:*

$$\alpha \models_t \beta \quad \textit{iff} \quad \forall v\,(v(\alpha) \leq_t v(\beta));$$
$$\alpha \models_f \beta \quad \textit{iff} \quad \forall v\,(v(\beta) \leq_f v(\alpha)).$$

Note that $\models_t$ and $\models_f$ are distinct relations.

The basic vocabulary can naturally be extended by a truth implication and a falsity implication, interpreted as the residuum of the truth order and the falsity order, respectively. In [139], Odintsov showed that the relation $\models_t$ can be axiomatized if the truth implication $\rightarrow$ is added to the language. Classical negation $\neg$ can be defined by setting $\neg\alpha := \alpha \rightarrow {\sim_t}(p \rightarrow p)$, for some fixed atom p. Let $\top := p \rightarrow p$ (for some atom p) and $\mathrm{L}^t = \{\alpha \mid \top \models_t \alpha\}$. Odintsov showed that $\mathrm{L}^t = \mathrm{L}_T \cap \mathrm{L}_B$. The intersection of the theorems of L_T and those of L_B coincides with the set formulas $\models_t$-entailed by $\top$, where $\top$ is interpreted as the top-element of the truth order $\leq_t$.

3.2. Propositional case

3.2.1. Sequent calculus

Greek capital letters $\Gamma, \Delta, \ldots$ are used to represent finite (possibly empty) sets of formulas. An expression of the form $\Gamma \Rightarrow \Delta$ is called a *sequent*. An expression $L \vdash S$ (or $\vdash S$) is used to denote the fact that a sequent S is provable in a sequent calculus L.

Definition 3.2.1 (L_{16})
Let ${\sim_b} \in \{{\sim_t}{\sim_f}, {\sim_f}{\sim_t}\}$, ${\sim_d} \in \{{\sim_t}{\sim_t}, {\sim_f}{\sim_f}, {\sim_b}{\sim_b}\}$, *and* ${\sim_e} \in \{{\sim_t}, {\sim_b}\}$.

The initial sequents of L_{16} *are of the form:*

$$\alpha \Rightarrow \alpha \qquad {\sim_f}{\sim_t}\alpha \Rightarrow {\sim_t}{\sim_f}\alpha \qquad {\sim_t}{\sim_f}\alpha \Rightarrow {\sim_f}{\sim_t}\alpha.$$

The structural inference rules of L_{16} *are of the form:*

$$\frac{\Gamma \Rightarrow \Delta, \alpha \quad \alpha, \Sigma \Rightarrow \Pi}{\Gamma, \Sigma \Rightarrow \Delta, \Pi}\ (\text{cut}) \qquad \frac{\Gamma \Rightarrow \Delta}{\alpha, \Gamma \Rightarrow \Delta}\ (\text{w-l}) \qquad \frac{\Gamma \Rightarrow \Delta}{\Gamma \Rightarrow \Delta, \alpha}\ (\text{w-r}).$$

The logical inference rules of L_{16} *are of the form:*

$$\frac{\Gamma \Rightarrow \Sigma, \alpha \quad \beta, \Delta \Rightarrow \Pi}{\alpha{\rightarrow}\beta, \Gamma, \Delta \Rightarrow \Sigma, \Pi}\ ({\rightarrow}\text{l}) \qquad \frac{\alpha, \Gamma \Rightarrow \Delta, \beta}{\Gamma \Rightarrow \Delta, \alpha{\rightarrow}\beta}\ ({\rightarrow}\text{r})$$

$$\frac{\Gamma \Rightarrow \Delta, \alpha}{\neg\alpha, \Gamma \Rightarrow \Delta}\ (\neg\text{l}) \qquad \frac{\alpha, \Gamma \Rightarrow \Delta}{\Gamma \Rightarrow \Delta, \neg\alpha}\ (\neg\text{r})$$

$$\frac{\alpha,\beta,\Gamma\Rightarrow\Delta}{\alpha\wedge_t\beta,\Gamma\Rightarrow\Delta}\ (\wedge_t\mathrm{l})\qquad\frac{\Gamma\Rightarrow\Delta,\alpha\quad\Gamma\Rightarrow\Delta,\beta}{\Gamma\Rightarrow\Delta,\alpha\wedge_t\beta}\ (\wedge_t\mathrm{r})$$

$$\frac{\alpha,\Gamma\Rightarrow\Delta\quad\beta,\Gamma\Rightarrow\Delta}{\alpha\vee_t\beta,\Gamma\Rightarrow\Delta}\ (\vee_t\mathrm{l})\qquad\frac{\Gamma\Rightarrow\Delta,\alpha,\beta}{\Gamma\Rightarrow\Delta,\alpha\vee_t\beta}\ (\vee_t\mathrm{r})$$

$$\frac{\alpha,\Gamma\Rightarrow\Delta\quad\beta,\Gamma\Rightarrow\Delta}{\alpha\wedge_f\beta,\Gamma\Rightarrow\Delta}\ (\wedge_f\mathrm{l})\qquad\frac{\Gamma\Rightarrow\Delta,\alpha,\beta}{\Gamma\Rightarrow\Delta,\alpha\wedge_f\beta}\ (\wedge_f\mathrm{r})$$

$$\frac{\alpha,\beta,\Gamma\Rightarrow\Delta}{\alpha\vee_f\beta,\Gamma\Rightarrow\Delta}\ (\vee_f\mathrm{l})\qquad\frac{\Gamma\Rightarrow\Delta,\alpha\quad\Gamma\Rightarrow\Delta,\beta}{\Gamma\Rightarrow\Delta,\alpha\vee_f\beta}\ (\vee_f\mathrm{r})$$

$$\frac{\alpha,\Gamma\Rightarrow\Delta}{\sim_d\alpha,\Gamma\Rightarrow\Delta}\ (\sim_d\mathrm{l})\qquad\frac{\Gamma\Rightarrow\Delta,\alpha}{\Gamma\Rightarrow\Delta,\sim_d\alpha}\ (\sim_d\mathrm{r})$$

$$\frac{\sim_f\alpha,\Gamma\Rightarrow\Delta}{\sim_t\sim_f\sim_t\alpha,\Gamma\Rightarrow\Delta}\ (\sim_t\sim_f\sim_t\mathrm{l})\qquad\frac{\Gamma\Rightarrow\Delta,\sim_f\alpha}{\Gamma\Rightarrow\Delta,\sim_t\sim_f\sim_t\alpha}\ (\sim_t\sim_f\sim_t\mathrm{r})$$

$$\frac{\sim_t\alpha,\Gamma\Rightarrow\Delta}{\sim_f\sim_t\sim_f\alpha,\Gamma\Rightarrow\Delta}\ (\sim_f\sim_t\sim_f\mathrm{l})\qquad\frac{\Gamma\Rightarrow\Delta,\sim_t\alpha}{\Gamma\Rightarrow\Delta,\sim_f\sim_t\sim_f\alpha}\ (\sim_f\sim_t\sim_f\mathrm{r})$$

$$\frac{\neg\alpha,\Gamma\Rightarrow\Delta}{\sim_t\neg\sim_t\alpha,\Gamma\Rightarrow\Delta}\ (\sim_t\neg\sim_t\mathrm{l})\qquad\frac{\Gamma\Rightarrow\Delta,\neg\alpha}{\Gamma\Rightarrow\Delta,\sim_t\neg\sim_t\alpha}\ (\sim_t\neg\sim_t\mathrm{r})$$

$$\frac{\neg\alpha,\Gamma\Rightarrow\Delta}{\sim_f\neg\sim_f\alpha,\Gamma\Rightarrow\Delta}\ (\sim_f\neg\sim_f\mathrm{l})\qquad\frac{\Gamma\Rightarrow\Delta,\neg\alpha}{\Gamma\Rightarrow\Delta,\sim_f\neg\sim_f\alpha}\ (\sim_f\neg\sim_f\mathrm{r})$$

$$\frac{\sim_f\alpha,\Gamma\Rightarrow\Delta}{\sim_f\sim_t\sim_t\alpha,\Gamma\Rightarrow\Delta}\ (\sim_f\sim_t\sim_t\mathrm{l})\qquad\frac{\Gamma\Rightarrow\Delta,\sim_f\alpha}{\Gamma\Rightarrow\Delta,\sim_f\sim_t\sim_t\alpha}\ (\sim_f\sim_t\sim_t\mathrm{r})$$

$$\frac{\sim_t\alpha,\Gamma\Rightarrow\Delta}{\sim_t\sim_f\sim_f\alpha,\Gamma\Rightarrow\Delta}\ (\sim_t\sim_f\sim_f\mathrm{l})\qquad\frac{\Gamma\Rightarrow\Delta,\sim_t\alpha}{\Gamma\Rightarrow\Delta,\sim_t\sim_f\sim_f\alpha}\ (\sim_t\sim_f\sim_f\mathrm{r})$$

$$\frac{\neg\alpha,\Gamma\Rightarrow\Delta}{\neg\sim_t\sim_t\alpha,\Gamma\Rightarrow\Delta}\ (\neg\sim_t\sim_t\mathrm{l})\qquad\frac{\Gamma\Rightarrow\Delta,\neg\alpha}{\Gamma\Rightarrow\Delta,\neg\sim_t\sim_t\alpha}\ (\neg\sim_t\sim_t\mathrm{r})$$

$$\frac{\neg\alpha,\Gamma\Rightarrow\Delta}{\neg\sim_f\sim_f\alpha,\Gamma\Rightarrow\Delta}\ (\neg\sim_f\sim_f\mathrm{l})\qquad\frac{\Gamma\Rightarrow\Delta,\neg\alpha}{\Gamma\Rightarrow\Delta,\neg\sim_f\sim_f\alpha}\ (\neg\sim_f\sim_f\mathrm{r})$$

$$\frac{\sim_e\neg\alpha,\sim_e\beta,\Gamma\Rightarrow\Delta}{\sim_e(\alpha\to\beta),\Gamma\Rightarrow\Delta}\ (\sim_e\!\to\mathrm{l})\qquad\frac{\Gamma\Rightarrow\Delta,\sim_e\neg\alpha\quad\Gamma\Rightarrow\Delta,\sim_e\beta}{\Gamma\Rightarrow\Delta,\sim_e(\alpha\to\beta)}\ (\sim_e\!\to\mathrm{r})$$

$$\frac{\Gamma\Rightarrow\Delta,\sim_e\alpha}{\sim_e\neg\alpha,\Gamma\Rightarrow\Delta}\ (\sim_e\neg\mathrm{l})\qquad\frac{\sim_e\alpha,\Gamma\Rightarrow\Delta}{\Gamma\Rightarrow\Delta,\sim_e\neg\alpha}\ (\sim_e\neg\mathrm{r})$$

$$\frac{\sim_e\alpha,\Gamma\Rightarrow\Delta\quad\sim_e\beta,\Gamma\Rightarrow\Delta}{\sim_e(\alpha\wedge_t\beta),\Gamma\Rightarrow\Delta}\ (\sim_e\wedge_t\mathrm{l})\qquad\frac{\Gamma\Rightarrow\Delta,\sim_e\alpha,\sim_e\beta}{\Gamma\Rightarrow\Delta,\sim_e(\alpha\wedge_t\beta)}\ (\sim_e\wedge_t\mathrm{r})$$

$$\frac{\sim_e\alpha,\sim_e\beta,\Gamma\Rightarrow\Delta}{\sim_e(\alpha\vee_t\beta),\Gamma\Rightarrow\Delta}\ (\sim_e\vee_t\mathrm{l})\qquad\frac{\Gamma\Rightarrow\Delta,\sim_e\alpha\quad\Gamma\Rightarrow\Delta,\sim_e\beta}{\Gamma\Rightarrow\Delta,\sim_e(\alpha\vee_t\beta)}\ (\sim_e\vee_t\mathrm{r})$$

$$\frac{\sim_t\alpha,\sim_t\beta,\Gamma\Rightarrow\Delta}{\sim_t(\alpha\wedge_f\beta),\Gamma\Rightarrow\Delta}\ (\sim_t\wedge_f\mathrm{l})\qquad\frac{\Gamma\Rightarrow\Delta,\sim_t\alpha\quad\Gamma\Rightarrow\Delta,\sim_t\beta}{\Gamma\Rightarrow\Delta,\sim_t(\alpha\wedge_f\beta)}\ (\sim_t\wedge_f\mathrm{r})$$

$$\frac{{\sim_t}\alpha, \Gamma \Rightarrow \Delta \quad {\sim_t}\beta, \Gamma \Rightarrow \Delta}{{\sim_t}(\alpha \vee_f \beta), \Gamma \Rightarrow \Delta} \ ({\sim_t}{\vee_f}\mathrm{l}) \qquad \frac{\Gamma \Rightarrow \Delta, {\sim_t}\alpha, {\sim_t}\beta}{\Gamma \Rightarrow \Delta, {\sim_t}(\alpha \vee_f \beta)} \ ({\sim_t}{\vee_f}\mathrm{r})$$

$$\frac{{\sim_b}\alpha, \Gamma \Rightarrow \Delta \quad {\sim_b}\beta, \Gamma \Rightarrow \Delta}{{\sim_b}(\alpha \wedge_f \beta), \Gamma \Rightarrow \Delta} \ ({\sim_b}{\wedge_f}\mathrm{l}) \qquad \frac{\Gamma \Rightarrow \Delta, {\sim_b}\alpha, {\sim_b}\beta}{\Gamma \Rightarrow \Delta, {\sim_b}(\alpha \wedge_f \beta)} \ ({\sim_b}{\wedge_f}\mathrm{r})$$

$$\frac{{\sim_b}\alpha, {\sim_b}\beta, \Gamma \Rightarrow \Delta}{{\sim_b}(\alpha \vee_f \beta), \Gamma \Rightarrow \Delta} \ ({\sim_b}{\vee_f}\mathrm{l}) \qquad \frac{\Gamma \Rightarrow \Delta, {\sim_b}\alpha \quad \Gamma \Rightarrow \Delta, {\sim_b}\beta}{\Gamma \Rightarrow \Delta, {\sim_b}(\alpha \vee_f \beta)} \ ({\sim_b}{\vee_f}\mathrm{r})$$

$$\frac{\Gamma \Rightarrow \Sigma, {\sim_f}\alpha \quad {\sim_f}\beta, \Delta \Rightarrow \Pi}{{\sim_f}(\alpha \rightarrow \beta), \Gamma, \Delta \Rightarrow \Sigma, \Pi} \ ({\sim_f}{\rightarrow}\mathrm{l}) \qquad \frac{{\sim_f}\alpha, \Gamma \Rightarrow \Delta, {\sim_f}\beta}{\Gamma \Rightarrow \Delta, {\sim_f}(\alpha \rightarrow \beta)} \ ({\sim_f}{\rightarrow}\mathrm{r})$$

$$\frac{\Gamma \Rightarrow \Delta, {\sim_f}\alpha}{{\sim_f}\neg\alpha, \Gamma \Rightarrow \Delta} \ ({\sim_f}{\neg}\mathrm{l}) \qquad \frac{{\sim_f}\alpha, \Gamma \Rightarrow \Delta}{\Gamma \Rightarrow \Delta, {\sim_f}\neg\alpha} \ ({\sim_f}{\neg}\mathrm{r})$$

$$\frac{{\sim_f}\alpha, {\sim_f}\beta, \Gamma \Rightarrow \Delta}{{\sim_f}(\alpha \wedge_t \beta), \Gamma \Rightarrow \Delta} \ ({\sim_f}{\wedge_t}\mathrm{l}) \qquad \frac{\Gamma \Rightarrow \Delta, {\sim_f}\alpha \quad \Gamma \Rightarrow \Delta, {\sim_f}\beta}{\Gamma \Rightarrow \Delta, {\sim_f}(\alpha \wedge_t \beta)} \ ({\sim_f}{\wedge_t}\mathrm{r})$$

$$\frac{{\sim_f}\alpha, \Gamma \Rightarrow \Delta \quad {\sim_f}\beta, \Gamma \Rightarrow \Delta}{{\sim_f}(\alpha \vee_t \beta), \Gamma \Rightarrow \Delta} \ ({\sim_f}{\vee_t}\mathrm{l}) \qquad \frac{\Gamma \Rightarrow \Delta, {\sim_f}\alpha, {\sim_f}\beta}{\Gamma \Rightarrow \Delta, {\sim_f}(\alpha \vee_t \beta)} \ ({\sim_f}{\vee_t}\mathrm{r})$$

$$\frac{{\sim_f}\alpha, {\sim_f}\beta, \Gamma \Rightarrow \Delta}{{\sim_f}(\alpha \wedge_f \beta), \Gamma \Rightarrow \Delta} \ ({\sim_f}{\wedge_f}\mathrm{l}) \qquad \frac{\Gamma \Rightarrow \Delta, {\sim_f}\alpha \quad \Gamma \Rightarrow \Delta, {\sim_f}\beta}{\Gamma \Rightarrow \Delta, {\sim_f}(\alpha \wedge_f \beta)} \ ({\sim_f}{\wedge_f}\mathrm{r})$$

$$\frac{{\sim_f}\alpha, \Gamma \Rightarrow \Delta \quad {\sim_f}\beta, \Gamma \Rightarrow \Delta}{{\sim_f}(\alpha \vee_f \beta), \Gamma \Rightarrow \Delta} \ ({\sim_f}{\vee_f}\mathrm{l}) \qquad \frac{\Gamma \Rightarrow \Delta, {\sim_f}\alpha, {\sim_f}\beta}{\Gamma \Rightarrow \Delta, {\sim_f}(\alpha \vee_f \beta)} \ ({\sim_f}{\vee_f}\mathrm{r}).$$

Note that the $\{\wedge_t, \vee_t, {\sim_t}\}$-fragment of L_{16} is a sequent calculus for Dunn and Belnap's four-valued logic [21], [47] and that the inference rules $({\sim_t}{\wedge_f}\mathrm{l})$, $({\sim_t}{\wedge_f}\mathrm{r})$, $({\sim_t}{\vee_f}\mathrm{l})$ and $({\sim_t}{\vee_f}\mathrm{r})$ can be found in Arieli and Avron's bilattice logic [10], if $\wedge_f$ and $\vee_f$, respectively, are read as the (multiplicative) conjunction and disjunction connectives $*$ and $+$ used in [10]. Thus, L_{16} may be viewed as a natural extension and generalization of Dunn and Belnap's logic and Arieli and Avron's logic.

We also observe that a sequent calculus L^*_{16} for L_T is obtained from L_{16} by replacing the inference rules $\{(\wedge_f\mathrm{l}), (\wedge_f\mathrm{r}), (\vee_f\mathrm{l}), (\vee_f\mathrm{r})\}$ by the inference rules of the form:

$$\frac{\alpha, \beta, \Gamma \Rightarrow \Delta}{\alpha \wedge_f \beta, \Gamma \Rightarrow \Delta} \ (\wedge_f\mathrm{l}) \qquad \frac{\Gamma \Rightarrow \Delta, \alpha \quad \Gamma \Rightarrow \Delta, \beta}{\Gamma \Rightarrow \Delta, \alpha \wedge_f \beta} \ (\wedge_f\mathrm{r})$$

$$\frac{\alpha, \Gamma \Rightarrow \Delta \quad \beta, \Gamma \Rightarrow \Delta}{\alpha \vee_f \beta, \Gamma \Rightarrow \Delta} \ (\vee_f\mathrm{l}) \qquad \frac{\Gamma \Rightarrow \Delta, \alpha, \beta}{\Gamma \Rightarrow \Delta, \alpha \vee_f \beta} \ (\vee_f\mathrm{r}).$$

Moreover, we note that L^*_{16} is an extension of the $\rightarrow$-free fragment of Arieli and Avron's bilattice logic.

Proposition 3.2.2 *The following rules are derivable in cut-free* L_{16}*:*

$$\frac{\sim_t\alpha, \Gamma \Rightarrow \Delta}{\neg\sim_t\neg\alpha, \Gamma \Rightarrow \Delta}\ (\neg\sim_t\neg\mathrm{l}) \qquad \frac{\Gamma \Rightarrow \Delta, \sim_t\alpha}{\Gamma \Rightarrow \Delta, \neg\sim_t\neg\alpha}\ (\neg\sim_t\neg\mathrm{r})$$

$$\frac{\sim_f\alpha, \Gamma \Rightarrow \Delta}{\neg\sim_f\neg\alpha, \Gamma \Rightarrow \Delta}\ (\neg\sim_f\neg\mathrm{l}) \qquad \frac{\Gamma \Rightarrow \Delta, \sim_f\alpha}{\Gamma \Rightarrow \Delta, \neg\sim_f\neg\alpha}\ (\neg\sim_f\neg\mathrm{r})$$

$$\frac{\sim_t\alpha, \Gamma \Rightarrow \Delta}{\sim_t\neg\neg\alpha, \Gamma \Rightarrow \Delta}\ (\sim_t\neg\neg\mathrm{l}) \qquad \frac{\Gamma \Rightarrow \Delta, \sim_t\alpha}{\Gamma \Rightarrow \Delta, \sim_t\neg\neg\alpha}\ (\sim_t\neg\neg\mathrm{r})$$

$$\frac{\sim_f\alpha, \Gamma \Rightarrow \Delta}{\sim_f\neg\neg\alpha, \Gamma \Rightarrow \Delta}\ (\sim_f\neg\neg\mathrm{l}) \qquad \frac{\Gamma \Rightarrow \Delta, \sim_f\alpha}{\Gamma \Rightarrow \Delta, \sim_f\neg\neg\alpha}\ (\sim_f\neg\neg\mathrm{r}).$$

Proposition 3.2.3 *The following rules are admissible in cut-free* L_{16}*:*

$$\frac{\sim_t\sim_f\alpha, \Gamma \Rightarrow \Delta}{\sim_f\sim_t\alpha, \Gamma \Rightarrow \Delta}\ (\sim_f\sim_t\mathrm{l}) \qquad \frac{\Gamma \Rightarrow \Delta, \sim_t\sim_f\alpha}{\Gamma \Rightarrow \Delta, \sim_f\sim_t\alpha}\ (\sim_f\sim_t\mathrm{r})$$

$$\frac{\sim_f\sim_t\alpha, \Gamma \Rightarrow \Delta}{\sim_t\sim_f\alpha, \Gamma \Rightarrow \Delta}\ (\sim_t\sim_f\mathrm{l}) \qquad \frac{\Gamma \Rightarrow \Delta, \sim_f\sim_t\alpha}{\Gamma \Rightarrow \Delta, \sim_t\sim_f\alpha}\ (\sim_t\sim_f\mathrm{r})$$

$$\frac{\sim_t\neg\alpha, \Gamma \Rightarrow \Delta}{\neg\sim_t\alpha, \Gamma \Rightarrow \Delta}\ (\neg\sim_t\mathrm{l}) \qquad \frac{\Gamma \Rightarrow \Delta, \sim_t\neg\alpha}{\Gamma \Rightarrow \Delta, \neg\sim_t\alpha}\ (\neg\sim_t\mathrm{r})$$

$$\frac{\neg\sim_t\alpha, \Gamma \Rightarrow \Delta}{\sim_t\neg\alpha, \Gamma \Rightarrow \Delta}\ (\sim_t\neg\mathrm{l}) \qquad \frac{\Gamma \Rightarrow \Delta, \neg\sim_t\alpha}{\Gamma \Rightarrow \Delta, \sim_t\neg\alpha}\ (\sim_t\neg\mathrm{r})$$

$$\frac{\neg\sim_f\alpha, \Gamma \Rightarrow \Delta}{\sim_f\neg\alpha, \Gamma \Rightarrow \Delta}\ (\sim_f\neg\mathrm{l}) \qquad \frac{\Gamma \Rightarrow \Delta, \neg\sim_f\alpha}{\Gamma \Rightarrow \Delta, \sim_f\neg\alpha}\ (\sim_f\neg\mathrm{r})$$

$$\frac{\sim_f\neg\alpha, \Gamma \Rightarrow \Delta}{\neg\sim_f\alpha, \Gamma \Rightarrow \Delta}\ (\neg\sim_f\mathrm{l}) \qquad \frac{\Gamma \Rightarrow \Delta, \sim_f\neg\alpha}{\Gamma \Rightarrow \Delta, \neg\sim_f\alpha}\ (\neg\sim_f\mathrm{r})$$

$$\frac{\sim_b\neg\alpha, \Gamma \Rightarrow \Delta}{\neg\sim_b\alpha, \Gamma \Rightarrow \Delta}\ (\neg\sim_b\mathrm{l}) \qquad \frac{\Gamma \Rightarrow \Delta, \sim_b\neg\alpha}{\Gamma \Rightarrow \Delta, \neg\sim_b\alpha}\ (\neg\sim_b\mathrm{r})$$

$$\frac{\neg\sim_b\alpha, \Gamma \Rightarrow \Delta}{\sim_b\neg\alpha, \Gamma \Rightarrow \Delta}\ (\sim_b\neg\mathrm{l}) \qquad \frac{\Gamma \Rightarrow \Delta, \neg\sim_b\alpha}{\Gamma \Rightarrow \Delta, \sim_b\neg\alpha}\ (\sim_b\neg\mathrm{r}).$$

Proof. We show only the claim for $(\sim_f\sim_t\mathrm{l})$ and $(\neg\sim_t\mathrm{r})$.

• $(\sim_f\sim_t\mathrm{l})$: We show that the rule $(\sim_f\sim_t\mathrm{l})$ is admissible in cut-free L_{16}, i.e., we show that $\mathrm{L}_{16} - (\mathrm{cut}) \vdash \sim_t\sim_f\alpha, \Gamma \Rightarrow \Delta$ implies $\mathrm{L}_{16} - (\mathrm{cut}) \vdash \sim_f\sim_t\alpha, \Gamma \Rightarrow \Delta$. This is proved by induction on a cut-free proof P of $\sim_t\sim_f\alpha, \Gamma \Rightarrow \Delta$. We distinguish the cases according to the last inference of P. We show some cases.

Case $(\sim_t\sim_f p \Rightarrow \sim_t\sim_f p)$: P is of the form: $\sim_t\sim_f p \Rightarrow \sim_t\sim_f p$, where p is a propositional variable. Then, $\sim_f\sim_t p \Rightarrow \sim_t\sim_f p$ is an initial sequent of L_{16}.

Case ($\sim_t\sim_f\sim_t$l): The last inference of P is of the form:

$$\frac{\sim_f\beta, \Gamma \Rightarrow \Delta}{\sim_t\sim_f\sim_t\beta, \Gamma \Rightarrow \Delta}\ (\sim_t\sim_f\sim_t\text{l})$$

where $\alpha \equiv \sim_t\beta$. By the hypothesis, we have the required fact:

$$\frac{\sim_f\beta, \Gamma \Rightarrow \Delta}{\sim_f\sim_t\sim_t\beta, \Gamma \Rightarrow \Delta}\ (\sim_f\sim_t\sim_t\text{l}).$$

Case ($\sim_b\wedge_t$l): The last inference of P is of the form:

$$\frac{\sim_t\sim_f\alpha', \Gamma \Rightarrow \Delta \quad \sim_t\sim_f\alpha'', \Gamma \Rightarrow \Delta}{\sim_t\sim_f(\alpha'\wedge_t\alpha''), \Gamma \Rightarrow \Delta}\ (\sim_b\wedge_t\text{l})$$

where $\alpha \equiv \alpha'\wedge_t\alpha''$. By the induction hypothesis, we have $\text{L}_{16} - (\text{cut}) \vdash \sim_f\sim_t\alpha', \Gamma \Rightarrow \Delta$ and $\text{L}_{16} - (\text{cut}) \vdash \sim_f\sim_t\alpha'', \Gamma \Rightarrow \Delta$. Thus, we obtain the required fact:

$$\frac{\sim_f\sim_t\alpha', \Gamma \Rightarrow \Delta \quad \sim_f\sim_t\alpha'', \Gamma \Rightarrow \Delta}{\sim_f\sim_t(\alpha'\wedge_t\alpha''), \Gamma \Rightarrow \Delta}\ (\sim_b\wedge_t\text{l}).$$

• ($\neg\sim_t$r): We show that the rule ($\neg\sim_t$r) is admissible in cut-free L_{16}, i.e., we show that $\text{L}_{16} - (\text{cut}) \vdash \Gamma \Rightarrow \Delta, \sim_t\neg\alpha$ implies $\text{L}_{16} - (\text{cut}) \vdash \Gamma \Rightarrow \Delta, \neg\sim_t\alpha$. This is proved by induction on a cut-free proof P of $\Gamma \Rightarrow \Delta, \sim_t\neg\alpha$. We distinguish the cases according to the last inference of P. We show some cases.

Case ($\sim_t\neg\sim_t$r): The last inference of P is of the form:

$$\frac{\Gamma \Rightarrow \Delta, \neg\beta}{\Gamma \Rightarrow \Delta, \sim_t\neg\sim_t\beta}\ (\sim_t\neg\sim_t\text{r})$$

where $\alpha \equiv \sim_t\beta$. By the hypothesis, we obtain the required fact:

$$\frac{\Gamma \Rightarrow \Delta, \neg\beta}{\Gamma \Rightarrow \Delta, \neg\sim_t\sim_t\beta}\ (\neg\sim_t\sim_t\text{r}).$$

Case ($\sim_t\neg$r): The last inference of P is of the form:

$$\frac{\sim_t\alpha, \Gamma \Rightarrow \Delta}{\Gamma \Rightarrow \Delta, \sim_t\neg\alpha}\ (\sim_t\neg\text{r}).$$

By the hypothesis, we obtain the required fact:

$$\frac{\sim_t\alpha, \Gamma \Rightarrow \Delta}{\Gamma \Rightarrow \Delta, \neg\sim_t\alpha}\ (\neg\text{r}).$$

∎

The rules presented in Proposition 3.2.3 are quite natural with regard to obtaining a sequent system for Odintsov's axiom system L_B and, in fact, the sequent system GL_B for L_B defined in [99] differs from L_{16} inter alia in that it comprises these admissible rules. These rules are not adopted in the definition of L_{16}, because they are not suited for applying Maehara's method, see Section 12.4.

3.2.2. Semantics

The semantics which we will extend to a semantics for first-order trilattice logics makes use of Odintsov's co-ordinate valuations [139]. These valuations are defined on the basis of Odintsov's matrix representation of the following operations on **16**: $\sqcap_t$ (lattice meet of the truth order), $\sqcup_t$ (lattice join of the truth order), $-_t$ (truth order inversion), $\sqcap_f$ (lattice meet of the falsity order), $\sqcup_f$ (lattice join of the falsity order), $-_f$ (falsity order inversion), $\sqsupset_t$ (residuum of $\sqcap_t$ with respect to the truth order), see Section 3.1.1 and [99, 139, 180, 181, 183]. Every element of **16** is a subset of the powerset of the set of classical truth values T (true) and F (*false*), i.e., it is a subset of $\{\mathbf{N} = \varnothing, \mathbf{T} = \{T\}, \mathbf{F} = \{F\}, \mathbf{B} = \{T, F\}\}$. Therefore, every element x of **16** can be represented as a 2×2-matrix of values of characteristic functions:

$$\begin{vmatrix} n & f \\ t & b \end{vmatrix}$$

where each element of the matrix is an element of the set $\{0, 1\}$ and the following equivalences hold:

$$n = 1 \text{ iff } \mathbf{N} \in x; \quad f = 1 \text{ iff } \mathbf{F} \in x;$$
$$t = 1 \text{ iff } \mathbf{T} \in x; \quad b = 1 \text{ iff } \mathbf{B} \in x.$$

Proposition 3.2.4 (Odintsov [139]) *Let $\wedge$ and $\vee$ be the classical truth functions of conjunction and disjunction, let $x, y \in$ **16**, let*

$$x = \begin{vmatrix} n & f \\ t & b \end{vmatrix} \text{ and let } y = \begin{vmatrix} y & f' \\ t' & b' \end{vmatrix}.$$

Then the following equations hold:

$$\begin{vmatrix} n & f \\ t & b \end{vmatrix} \sqcap_t \begin{vmatrix} n' & f' \\ t' & b' \end{vmatrix} = \begin{vmatrix} n \vee n' & f \vee f' \\ t \wedge t' & b \wedge b' \end{vmatrix}$$

$$\begin{vmatrix} n & f \\ t & b \end{vmatrix} \sqcup_t \begin{vmatrix} n' & f' \\ t' & b' \end{vmatrix} = \begin{vmatrix} n \wedge n' & f \wedge f' \\ t \vee t' & b \vee b' \end{vmatrix}$$

$$-_t\begin{vmatrix} n & f \\ t & b \end{vmatrix} = \begin{vmatrix} t & b \\ n & f \end{vmatrix}$$

$$\begin{vmatrix} n & f \\ t & b \end{vmatrix} \sqcap_f \begin{vmatrix} n' & f' \\ t' & b' \end{vmatrix} = \begin{vmatrix} n \wedge n' & f \vee f' \\ t \wedge t' & b \vee b' \end{vmatrix}$$

$$\begin{vmatrix} n & f \\ t & b \end{vmatrix} \sqcup_f \begin{vmatrix} n' & f' \\ t' & b' \end{vmatrix} = \begin{vmatrix} n \vee n' & f \wedge f' \\ t \vee t' & b \wedge b' \end{vmatrix}$$

$$-_f\begin{vmatrix} n & f \\ t & b \end{vmatrix} = \begin{vmatrix} f & n \\ b & t \end{vmatrix}$$

Moreover,

$$\begin{vmatrix} n & f \\ t & b \end{vmatrix} \sqsupset_t \begin{vmatrix} n' & f' \\ t' & b' \end{vmatrix} = \begin{vmatrix} \neg n \wedge n' & \neg f \wedge f' \\ t \rightarrow t' & b \rightarrow b' \end{vmatrix}$$

where $\rightarrow$ and $\neg$ denote the truth functions of Boolean implication and Boolean negation, respectively.

In view of the matrix presentation of elements of **16**, every assignment v from the set of propositional variables into **16** can be associated with co-ordinate valuations v_t, v_f, v_n, and v_b which are classical valuations from the set of propositional variables into $\{0, 1\}$. The co-ordinate valuations are defined by the following equivalence:

$$v_c(p) = 1 \text{ iff } \mathbf{C} \in v(p),\ c \in \{n, f, t, b\}.$$

In this section, we shall introduce valuation functions v^n, v^f, v^t, v^b and use them to prove strong completeness and cut-elimination for L_{16}. Moreover, we explain how these mappings are related to Odintsov's co-ordinate valuations.

Let p be a fixed propositional variable. Suppose Γ is a set $\{\alpha_1, ..., \alpha_m\}$ ($m \geq 0$) of formulas. Then Γ^* is defined as $\alpha_1 \vee_t \cdots \vee_t \alpha_m$ if $m \geq 1$, and $\neg(p \rightarrow p)$ if $m = 0$. Also Γ_* is defined as $\alpha_1 \wedge_t \cdots \wedge_t \alpha_m$ if $m \geq 1$, and $p \rightarrow p$ if $m = 0$. In the following discussion, the commutativity of $\wedge_t$ or $\vee_t$ is assumed. We have the following fact: for any formulas $\alpha_1, ..., \alpha_m, \beta_1, ..., \beta_n$, the sequent $\alpha_1, ..., \alpha_m \Rightarrow \beta_1, ..., \beta_n$ is provable in L_{16} if and only if so is $\alpha_1 \wedge_t \cdots \wedge_t \alpha_m \Rightarrow \beta_1 \vee_t \cdots \vee_t \beta_n$.

Definition 3.2.5 *The valuations v^n, v^t, v^f and v^b are mappings from the set of all propositional variables to the set $\{t, f\}$. The valuations v^n, v^t, v^f and v^b are extended to mappings from the set of all formulas to $\{t, f\}$ by the following clauses. For any $e \in \{t, b\}$, $g \in \{f, n\}$:*

1. $v^g(\alpha{\rightarrow}\beta) = t$ *iff* $v^g(\alpha) = f$ *or* $v^g(\beta) = t$,
2. $v^g(\neg\alpha) = t$ *iff* $v^g(\alpha) = f$,
3. $v^g(\alpha{\wedge_t}\beta) = t$ *iff* $v^g(\alpha) = t$ *and* $v^g(\beta) = t$,
4. $v^g(\alpha{\vee_t}\beta) = t$ *iff* $v^g(\alpha) = t$ *or* $v^g(\beta) = t$,
5. $v^g(\alpha{\wedge_f}\beta) = t$ *iff* $v^g(\alpha) = t$ *or* $v^g(\beta) = t$,
6. $v^g(\alpha{\vee_f}\beta) = t$ *iff* $v^g(\alpha) = t$ *and* $v^g(\beta) = t$,
7. $v^n(\sim_t\alpha) = t$ *iff* $v^t(\alpha) = t$,
8. $v^n(\sim_f\alpha) = t$ *iff* $v^f(\alpha) = t$,
9. $v^n(\sim_b\alpha) = t$ *iff* $v^b(\alpha) = t$,
10. $v^e(\alpha{\rightarrow}\beta) = t$ *iff* $v^e(\alpha) = f$ *and* $v^e(\beta) = t$,
11. $v^e(\neg\alpha) = t$ *iff* $v^e(\alpha) = f$,
12. $v^e(\alpha{\wedge_t}\beta) = t$ *iff* $v^e(\alpha) = t$ *or* $v^e(\beta) = t$,
13. $v^e(\alpha{\vee_t}\beta) = t$ *iff* $v^e(\alpha) = t$ *and* $v^e(\beta) = t$,
14. $v^t(\alpha{\wedge_f}\beta) = t$ *iff* $v^t(\alpha) = t$ *and* $v^t(\beta) = t$,
15. $v^t(\alpha{\vee_f}\beta) = t$ *iff* $v^t(\alpha) = t$ *or* $v^t(\beta) = t$,
16. $v^b(\alpha{\wedge_f}\beta) = t$ *iff* $v^b(\alpha) = t$ *or* $v^b(\beta) = t$,
17. $v^b(\alpha{\vee_f}\beta) = t$ *iff* $v^b(\alpha) = t$ *and* $v^b(\beta) = t$,
18. $v^t(\sim_t\alpha) = t$ *iff* $v^n(\alpha) = t$,
19. $v^t(\sim_f\alpha) = t$ *iff* $v^b(\alpha) = t$,
20. $v^t(\sim_b\alpha) = t$ *iff* $v^f(\alpha) = t$,
21. $v^f(\sim_t\alpha) = t$ *iff* $v^b(\alpha) = t$,
22. $v^f(\sim_f\alpha) = t$ *iff* $v^n(\alpha) = t$,
23. $v^f(\sim_b\alpha) = t$ *iff* $v^t(\alpha) = t$,
24. $v^b(\sim_t\alpha) = t$ *iff* $v^f(\alpha) = t$,
25. $v^b(\sim_f\alpha) = t$ *iff* $v^t(\alpha) = t$,
26. $v^b(\sim_b\alpha) = t$ *iff* $v^n(\alpha) = t$.

A formula α is called a tautology *if $v^n(\alpha) = t$ holds for any valuations v^n, v^t, v^f and v^b. A sequent of the form $\Gamma \Rightarrow \Delta$ is called a tautology if so is the formula $\Gamma_* \to \Delta^*$.*

Note that the valuation v^n behaves classically with respect to the classical connectives $\wedge_t$, $\vee_t$, $\neg$, and $\to$. This may be seen as a justification for defining the notion of a tautology with respect to v^n. Moreover, it is noted that the following conditions hold:

1. $v^t(\sim_f \alpha) = v^f(\sim_t \alpha)$,
2. $v^t(\alpha) = v^n(\sim_t \alpha) = v^f(\sim_b \alpha) = v^b(\sim_f \alpha)$,
3. $v^f(\alpha) = v^n(\sim_f \alpha) = v^t(\sim_b \alpha) = v^b(\sim_t \alpha)$,
4. $v^b(\alpha) = v^n(\sim_b \alpha) = v^t(\sim_f \alpha) = v^f(\sim_t \alpha)$.

Theorem 3.2.6 (Soundness of $\mathbf{L_{16}}$) *For any sequent S, if $\mathrm{L}_{16} \vdash S$, then S is a tautology.*

3.2.3. Completeness and cut-elimination

In the following, we prove the (strong) completeness and cut-elimination theorems for L_{16} by using the method by Maehara presented, for instance, in [146].

Definition 3.2.7 *Let $\sim_d \in \{\sim_t\sim_t, \sim_f\sim_f, \sim_b\sim_b\}$ and $\sim_e \in \{\sim_t, \sim_b\}$. A* decomposition *of a sequent S is defined as having the form S' or $S'; S''$ by*

1. *$\Gamma \Rightarrow \Delta, \alpha$; $\beta, \Gamma \Rightarrow \Delta$ is a decomposition of $\alpha \to \beta, \Gamma \Rightarrow \Delta$,*
2. *$\alpha, \Gamma \Rightarrow \Delta, \beta$ is a decomposition of $\Gamma \Rightarrow \Delta, \alpha \to \beta$,*
3. *$\Gamma \Rightarrow \Delta, \alpha$ is a decomposition of $\neg\alpha, \Gamma \Rightarrow \Delta$,*
4. *$\alpha, \Gamma \Rightarrow \Delta$ is a decomposition of $\Gamma \Rightarrow \Delta, \neg\alpha$,*
5. *$\alpha, \beta, \Gamma \Rightarrow \Delta$ is a decomposition of $\alpha \wedge_t \beta, \Gamma \Rightarrow \Delta$,*
6. *$\Gamma \Rightarrow \Delta, \alpha$; $\Gamma \Rightarrow \Delta, \beta$ is a decomposition of $\Gamma \Rightarrow \Delta, \alpha \wedge_t \beta$,*
7. *$\alpha, \Gamma \Rightarrow \Delta$; $\beta, \Gamma \Rightarrow \Delta$ is a decomposition of $\alpha \vee_t \beta, \Gamma \Rightarrow \Delta$,*
8. *$\Gamma \Rightarrow \Delta, \alpha, \beta$ is a decomposition of $\Gamma \Rightarrow \Delta, \alpha \vee_t \beta$,*
9. *$\alpha, \Gamma \Rightarrow \Delta$; $\beta, \Gamma \Rightarrow \Delta$ is a decomposition of $\alpha \wedge_f \beta, \Gamma \Rightarrow \Delta$,*

10. $\Gamma \Rightarrow \Delta, \alpha, \beta$ *is a decomposition of* $\Gamma \Rightarrow \Delta, \alpha \wedge_f \beta$,

11. $\alpha, \beta, \Gamma \Rightarrow \Delta$ *is a decomposition of* $\alpha \vee_f \beta, \Gamma \Rightarrow \Delta$,

12. $\Gamma \Rightarrow \Delta, \alpha$; $\Gamma \Rightarrow \Delta, \beta$ *is a decomposition of* $\Gamma \Rightarrow \Delta, \alpha \vee_f \beta$,

13. $\alpha, \Gamma \Rightarrow \Delta$ *is a decomposition of* $\sim_d \alpha, \Gamma \Rightarrow \Delta$,

14. $\Gamma \Rightarrow \Delta, \alpha$ *is a decomposition of* $\Gamma \Rightarrow \Delta, \sim_d \alpha$,

15. $\sim_f \alpha, \Gamma \Rightarrow \Delta$ *is a decomposition of* $\sim_t \sim_f \sim_t \alpha, \Gamma \Rightarrow \Delta$,

16. $\Gamma \Rightarrow \Delta, \sim_f \alpha$ *is a decomposition of* $\Gamma \Rightarrow \Delta, \sim_t \sim_f \sim_t \alpha$,

17. $\sim_t \alpha, \Gamma \Rightarrow \Delta$ *is a decomposition of* $\sim_f \sim_t \sim_f \alpha, \Gamma \Rightarrow \Delta$,

18. $\Gamma \Rightarrow \Delta, \sim_t \alpha$ *is a decomposition of* $\Gamma \Rightarrow \Delta, \sim_f \sim_t \sim_f \alpha$,

19. $\neg \alpha, \Gamma \Rightarrow \Delta$ *is a decomposition of* $\sim_t \neg \sim_t \alpha, \Gamma \Rightarrow \Delta$,

20. $\Gamma \Rightarrow \Delta, \neg \alpha$ *is a decomposition of* $\Gamma \Rightarrow \Delta, \sim_t \neg \sim_t \alpha$,

21. $\neg \alpha, \Gamma \Rightarrow \Delta$ *is a decomposition of* $\sim_f \neg \sim_f \alpha, \Gamma \Rightarrow \Delta$,

22. $\Gamma \Rightarrow \Delta, \neg \alpha$ *is a decomposition of* $\Gamma \Rightarrow \Delta, \sim_f \neg \sim_f \alpha$,

23. $\sim_f \alpha, \Gamma \Rightarrow \Delta$ *is a decomposition of* $\sim_f \sim_t \sim_t \alpha, \Gamma \Rightarrow \Delta$,

24. $\Gamma \Rightarrow \Delta, \sim_f \alpha$ *is a decomposition of* $\Gamma \Rightarrow \Delta, \sim_f \sim_t \sim_t \alpha$,

25. $\sim_t \alpha, \Gamma \Rightarrow \Delta$ *is a decomposition of* $\sim_t \sim_f \sim_f \alpha, \Gamma \Rightarrow \Delta$,

26. $\Gamma \Rightarrow \Delta, \sim_t \alpha$ *is a decomposition of* $\Gamma \Rightarrow \Delta, \sim_t \sim_f \sim_f \alpha$,

27. $\neg \alpha, \Gamma \Rightarrow \Delta$ *is a decomposition of* $\neg \sim_t \sim_t \alpha, \Gamma \Rightarrow \Delta$,

28. $\Gamma \Rightarrow \Delta, \neg \alpha$ *is a decomposition of* $\Gamma \Rightarrow \Delta, \neg \sim_t \sim_t \alpha$,

29. $\neg \alpha, \Gamma \Rightarrow \Delta$ *is a decomposition of* $\neg \sim_f \sim_f \alpha, \Gamma \Rightarrow \Delta$,

30. $\Gamma \Rightarrow \Delta, \neg \alpha$ *is a decomposition of* $\Gamma \Rightarrow \Delta, \neg \sim_f \sim_f \alpha$,

31. $\sim_e \neg \alpha, \sim_e \beta, \Gamma \Rightarrow \Delta$ *is a decomposition of* $\sim_e (\alpha \rightarrow \beta), \Gamma \Rightarrow \Delta$,

32. $\Gamma \Rightarrow \Delta, \sim_e \neg \alpha$; $\Gamma \Rightarrow \Delta, \sim_e \beta$ *is a decomposition of* $\Gamma \Rightarrow \Delta, \sim_e (\alpha \rightarrow \beta)$,

33. $\Gamma \Rightarrow \Delta, \sim_e \alpha$ *is a decomposition of* $\sim_e \neg \alpha, \Gamma \Rightarrow \Delta$,

34. $\sim_e \alpha, \Gamma \Rightarrow \Delta$ *is a decomposition of* $\Gamma \Rightarrow \Delta, \sim_e \neg \alpha$,

35. $\sim_e \alpha, \Gamma \Rightarrow \Delta$; $\sim_e \beta, \Gamma \Rightarrow \Delta$ *is a decomposition of* $\sim_e (\alpha \wedge_t \beta), \Gamma \Rightarrow \Delta$,

36. $\Gamma \Rightarrow \Delta, {\sim_e}\alpha, {\sim_e}\beta$ *is a decomposition of* $\Gamma \Rightarrow \Delta, {\sim_e}(\alpha \wedge_t \beta)$,

37. ${\sim_e}\alpha, {\sim_e}\beta, \Gamma \Rightarrow \Delta$ *is a decomposition of* ${\sim_e}(\alpha \vee_t \beta), \Gamma \Rightarrow \Delta$,

38. $\Gamma \Rightarrow \Delta, {\sim_e}\alpha$; $\Gamma \Rightarrow \Delta, {\sim_e}\beta$ *is a decomposition of* $\Gamma \Rightarrow \Delta, {\sim_e}(\alpha \vee_t \beta)$,

39. ${\sim_t}\alpha, {\sim_t}\beta, \Gamma \Rightarrow \Delta$ *is a decomposition of* ${\sim_t}(\alpha \wedge_f \beta), \Gamma \Rightarrow \Delta$,

40. $\Gamma \Rightarrow \Delta, {\sim_t}\alpha$; $\Gamma \Rightarrow \Delta, {\sim_t}\beta$ *is a decomposition of* $\Gamma \Rightarrow \Delta, {\sim_t}(\alpha \wedge_f \beta)$,

41. ${\sim_t}\alpha, \Gamma \Rightarrow \Delta$; ${\sim_t}\beta, \Gamma \Rightarrow \Delta$ *is a decomposition of* ${\sim_t}(\alpha \vee_f \beta), \Gamma \Rightarrow \Delta$,

42. $\Gamma \Rightarrow \Delta, {\sim_t}\alpha, {\sim_t}\beta$ *is a decomposition of* $\Gamma \Rightarrow \Delta, {\sim_t}(\alpha \vee_f \beta)$,

43. ${\sim_b}\alpha, \Gamma \Rightarrow \Delta$; ${\sim_b}\beta, \Gamma \Rightarrow \Delta$ *is a decomposition of* ${\sim_b}(\alpha \wedge_f \beta), \Gamma \Rightarrow \Delta$,

44. $\Gamma \Rightarrow \Delta, {\sim_b}\alpha, {\sim_b}\beta$ *is a decomposition of* $\Gamma \Rightarrow \Delta, {\sim_b}(\alpha \wedge_f \beta)$,

45. ${\sim_b}\alpha, {\sim_b}\beta, \Gamma \Rightarrow \Delta$ *is a decomposition of* ${\sim_b}(\alpha \vee_f \beta), \Gamma \Rightarrow \Delta$,

46. $\Gamma \Rightarrow \Delta, {\sim_b}\alpha$; $\Gamma \Rightarrow \Delta, {\sim_b}\beta$ *is a decomposition of* $\Gamma \Rightarrow \Delta, {\sim_b}(\alpha \vee_f \beta)$,

47. $\Gamma \Rightarrow \Delta, {\sim_f}\alpha$; ${\sim_f}\beta, \Gamma \Rightarrow \Delta$ *is a decomposition of* ${\sim_f}(\alpha \rightarrow \beta), \Gamma \Rightarrow \Delta$,

48. ${\sim_f}\alpha, \Gamma \Rightarrow \Delta, {\sim_f}\beta$ *is a decomposition of* $\Gamma \Rightarrow \Delta, {\sim_f}(\alpha \rightarrow \beta)$,

49. $\Gamma \Rightarrow \Delta, {\sim_f}\alpha$ *is a decomposition of* ${\sim_f}(\neg\alpha), \Gamma \Rightarrow \Delta$,

50. ${\sim_f}\alpha, \Gamma \Rightarrow \Delta$ *is a decomposition of* $\Gamma \Rightarrow \Delta, {\sim_f}(\neg\alpha)$,

51. ${\sim_f}\alpha, {\sim_f}\beta, \Gamma \Rightarrow \Delta$ *is a decomposition of* ${\sim_f}(\alpha \wedge_t \beta), \Gamma \Rightarrow \Delta$,

52. $\Gamma \Rightarrow \Delta, {\sim_f}\alpha$; $\Gamma \Rightarrow \Delta, {\sim_f}\beta$ *is a decomposition of* $\Gamma \Rightarrow \Delta, {\sim_f}(\alpha \wedge_t \beta)$,

53. ${\sim_f}\alpha, \Gamma \Rightarrow \Delta$; ${\sim_f}\beta, \Gamma \Rightarrow \Delta$ *is a decomposition of* ${\sim_f}(\alpha \vee_t \beta), \Gamma \Rightarrow \Delta$,

54. $\Gamma \Rightarrow \Delta, {\sim_f}\alpha, {\sim_f}\beta$ *is a decomposition of* $\Gamma \Rightarrow \Delta, {\sim_f}(\alpha \vee_t \beta)$,

55. ${\sim_f}\alpha, {\sim_f}\beta, \Gamma \Rightarrow \Delta$ *is a decomposition of* ${\sim_f}(\alpha \wedge_f \beta), \Gamma \Rightarrow \Delta$,

56. $\Gamma \Rightarrow \Delta, {\sim_f}\alpha$; $\Gamma \Rightarrow \Delta, {\sim_f}\beta$ *is a decomposition of* $\Gamma \Rightarrow \Delta, {\sim_f}(\alpha \wedge_f \beta)$,

57. ${\sim_f}\alpha, \Gamma \Rightarrow \Delta$; ${\sim_f}\beta, \Gamma \Rightarrow \Delta$ *is a decomposition of* ${\sim_f}(\alpha \vee_f \beta), \Gamma \Rightarrow \Delta$,

58. $\Gamma \Rightarrow \Delta, {\sim_f}\alpha, {\sim_f}\beta$ *is a decomposition of* $\Gamma \Rightarrow \Delta, {\sim_f}(\alpha \vee_f \beta)$.

Note that the clauses in the definition of the decompositions just correspond to the logical inference rules of L_{16}.

Definition 3.2.8 *A* decomposition tree of *a sequent S is a tree which expresses a process of some repeated decomposition of S. In other words, a decomposition tree corresponds to a bottom up proof search tree. A* complete decomposition tree of S *is a decomposition tree of a sequent S in which all the formulas occurring in all the leaves of the tree are of one of the following forms: $p, {\sim_t}p, {\sim_f}p$ and ${\sim_b}p$.*

Lemma 3.2.9 *For any sequent S, there is a complete decomposition tree of S, i.e., every repeated decomposition process terminates.*

Proof. By the definition of decomposition, S_1 and S_2 consist only of some subformulas or negated subformulas of a formula in S. ∎

Note that if the corresponding decomposition rules of the admissible inference rules displayed in Proposition 3.2.3 are adopted, then some repeated decomposition processes do not terminate. This is the reason why we do not use the inference rules in Proposition 3.2.3 in the definition of L_{16}.

Lemma 3.2.10 *Let S_1 or S_1 ; S_2 be a decomposition of S. If S is a tautology, then so are S_1 and S_2.*

Proof. We show some cases.

(15): Suppose that ${\sim_t}{\sim_f}{\sim_t}\alpha{\wedge_t}\Gamma_*{\to}\Delta^*$ is a tautology. We show that the sequent ${\sim_f}\alpha{\wedge_t}\Gamma_*{\to}\Delta^*$ is a tautology. Suppose (1) $v^n({\sim_f}\alpha{\wedge_t}\Gamma_*) = t$. We show $v^n(\Delta^*) = t$. By (1), we have (2) $v^n({\sim_f}\alpha) = v^f(\alpha) = t$ and (3) $v^n(\Gamma_*) = t$. By (2), we obtain (4) $v^n({\sim_t}{\sim_f}{\sim_t}\alpha) = v^n({\sim_t}{\sim_b}\alpha) = v^t({\sim_b}\alpha) = v^f(\alpha) = t$. On the other hand, we have (5) $v^n({\sim_t}{\sim_f}{\sim_t}\alpha{\wedge_t}\Gamma_*{\to}\Delta^*) = t$ by the hypothesis. By (5), (4), and (3), we obtain $v^n(\Delta^*) = t$.

(21): Suppose that ${\sim_f}\neg{\sim_f}\alpha{\wedge_t}\Gamma_*{\to}\Delta^*$ is a tautology. We show that the sequent $\neg\alpha{\wedge_t}\Gamma_*{\to}\Delta^*$ is a tautology. Suppose $v^n(\neg\alpha{\wedge_t}\Gamma_*) = t$, i.e., (1) $v^n(\neg\alpha) = t$ and (2) $v^n(\Gamma_*) = t$. Then, we show $v^n(\Delta^*) = t$. We have (3) $v^n({\sim_f}\neg{\sim_f}\alpha) = v^f(\neg{\sim_f}\alpha) = t$ iff $v^f({\sim_f}\alpha) = f$ iff $v^n(\alpha) = f$ iff $v^n(\neg\alpha) = t$. Thus, we obtain (4) $v^n({\sim_f}\neg{\sim_f}\alpha) = t$ by (3) and (1). On the other hand, we have (5) $v^n({\sim_f}\neg{\sim_f}\alpha{\wedge_t}\Gamma_*{\to}\Delta^*) = t$ by the hypothesis. Thus, we obtain $v^n(\Delta^*) = t$ by (5), (4), and (2).

(32): Suppose that $\Gamma_*{\to}\Delta^*{\vee_t}{\sim_t}(\alpha{\to}\beta)$ is a tautology. First, we show that $\Gamma_* \to \Delta^* {\vee_t}{\sim_t}\neg\alpha$ is a tautology. Suppose that (1) $v^n(\Gamma_*) = t$. We show $v^n(\Delta^*{\vee_t}{\sim_t}\neg\alpha) = t$. If $v^n({\sim_t}\neg\alpha) = t$, then $v^n(\Delta^*{\vee_t}{\sim_t}\neg\alpha) = t$. Thus, suppose $v^n({\sim_t}\neg\alpha) = v^t(\neg\alpha) = f$, i.e., $v^t(\alpha) = t$. Then, $v^t(\alpha{\to}\beta) = f$, since $v^t(\alpha{\to}\beta) = f$ iff $v^t(\alpha) = t$ or $v^t(\beta) = f$. Hence, we have (2) $v^n({\sim_t}(\alpha{\to}\beta)) = f$. On the other hand, we have (3) $v^n(\Gamma_*{\to}\Delta^*{\vee_t}{\sim_t}(\alpha{\to}\beta)) = t$ by the hypothesis. Thus, we obtain $v^n(\Delta^*)$

$= t$ by (1), (2), and (3). Therefore $v^n(\Delta^* \vee_t \sim_t \neg\alpha) = t$. Second, we show that $\Gamma_* \rightarrow \Delta^* \vee_t \sim_t \beta$ is a tautology. This case is similar to the proof above, since we can derive $v^t(\beta) = f$.

(47): Suppose that $\sim_f(\alpha \rightarrow \beta) \wedge_t \Gamma_* \rightarrow \Delta^*$ is a tautology. First, we show that $\Gamma_* \rightarrow \Delta^* \vee_t \sim_f \alpha$ is a tautology. Suppose that (1) $v^n(\Gamma_*) = t$. We show $v^n(\Delta^* \vee_t \sim_f \alpha) = t$. If $v^n(\sim_f \alpha) = t$, then $v^n(\Delta^* \vee_t \sim_f \alpha) = t$. Thus, suppose that $v^n(\sim_f \alpha) = v^f(\alpha) = f$. Then, $v^f(\alpha \rightarrow \beta) = t$, and hence (2) $v^n(\sim_f(\alpha \rightarrow \beta)) = v^f(\alpha \rightarrow \beta) = t$. On the other hand, we have (3) $v^n(\sim_f(\alpha \rightarrow \beta) \wedge_t \Gamma_* \rightarrow \Delta^*) = t$ by the hypothesis. Thus, we obtain $v^n(\Delta^*) = t$ by (1), (2) and (3). Therefore $v^n(\Delta^* \vee_t \sim_f \alpha) = t$. Second, we show that $\sim_f \beta \wedge_t \Gamma_* \rightarrow \Delta^*$ is a tautology. Suppose that (4) $v^n(\sim_f \beta \wedge_t \Gamma_*) = t$. Then, (5) $v^n(\sim_f \beta) = v^f(\beta) = t$ and (6) $v^n(\Gamma_*) = t$. By (5), we have $v^f(\alpha \rightarrow \beta) = t$, and hence (7) $v^n(\sim_f(\alpha \rightarrow \beta)) = v^f(\alpha \rightarrow \beta) = t$. By (3), (6), and (7), we obtain the required fact $v^n(\Delta^*) = t$. ∎

Lemma 3.2.11 *Let $\sim_*$ be $\sim_t$, $\sim_f$, or $\sim_b$.*

1. *Suppose that each α_i or β_j in $\{\alpha_1, ..., \alpha_m, \beta_1, ..., \beta_n\}$ is a propositional variable or a formula of the form $\sim_* \gamma$ where γ is a propositional variable. Then, the sequent $\alpha_1, ..., \alpha_m \Rightarrow \beta_1, ..., \beta_n$ is a tautology if and only if* (a) *there are α_i ($i \leq m$) and β_j ($j \leq n$) such that $\alpha_i \equiv \beta_j$, or* (b) *there are α_i ($i \leq m$) and β_j ($j \leq n$) such that ($\alpha_i \equiv \sim_f \sim_t p$ and $\beta_j \equiv \sim_t \sim_f p$) or ($\alpha_i \equiv \sim_t \sim_f p$ and $\beta_j \equiv \sim_f \sim_t p$) where p is a propositional variable.*
2. *Sequents of the form $(\alpha, \Gamma \Rightarrow \Delta, \alpha)$, $(\sim_f \sim_t \alpha, \Gamma \Rightarrow \Delta, \sim_t \sim_f \alpha)$, and $(\sim_t \sim_f \alpha, \Gamma \Rightarrow \Delta, \sim_f \sim_t \alpha)$ are provable in cut-free* L_{16}.

Proof. We show only (1). Suppose there are no α_i and β_j such that (a) and (b). We specify a valuation v^n as follows: $v^n(\alpha_i) = t$ ($i = 1, ..., m$) and $v^n(\beta_j) = f$ ($j = 1, ..., n$). Then we obtain

$$v^n((\alpha_1 \wedge_t \cdots \wedge_t \alpha_m) \rightarrow (\beta_1 \vee_t \cdots \vee_t \beta_n)) = f,$$

and hence $\alpha_1, ..., \alpha_m \Rightarrow \beta_1, ..., \beta_n$ is not tautology. Conversely, suppose there are α_i and β_j such that (a) and (b). Then $\alpha_1, ..., \alpha_m \Rightarrow \beta_1, ..., \beta_n$ is a tautology. ∎

Lemma 3.2.12 *Let S_1 (or S_1 ; S_2) be a decomposition of S. If S_1 (or S_2) is provable in cut-free* L_{16}*, then so is S.*

Theorem 3.2.13 (Strong completeness for $\mathbf{L}_{16}$) *For any sequent S, if S is a tautology, then* $\mathrm{L}_{16} - (\mathrm{cut}) \vdash S$.

Proof. Suppose that a sequent S is a tautology. We can obtain a complete decomposition tree of S by Lemma 3.2.9. All the leaves of this complete decomposition tree are tautologies by using Lemma 3.2.10 repeatedly. Then, these leaves are provable in cut-free L_{16} by Lemma 3.2.11 (1) and (2). By using Lemma 3.2.12 repeatedly for the complete decomposition tree of S, all the sequents in the tree are provable in cut-free L_{16}. Therefore, in particular, S is provable in cut-free L_{16}. ∎

Theorem 3.2.14 (Cut-elimination for $\mathbf{L_{16}}$) *The rule* (cut) *is admissible in cut-free* L_{16}.

Proof. Suppose $L_{16} \vdash S$. Then, S is a tautology by Theorem 3.2.6. By Theorem 3.2.13, we obtain L_{16} − (cut) $\vdash S$. ∎

Odintsov [139] proved that L_B is the set of all formulas α of the language under consideration such that for every assignment v from the set of propositional variables into **16**, $v_b(\alpha) = 1$. (Also, $L_B = \{\alpha \mid \forall v(v_n(\alpha) = 0\}$.) Note that (*) for the two negations $\sim_t$ and $\sim_f$, the valuations v^n, v^t, v^f, and v^b are defined like Odintsov's v_n, v_t, v_f, and v_b, respectively. Moreover, (**) for the other connectives, the valuations v^n and v^f are defined like Odintsov's v_t and v_b (and v^t, v^b are defined like v_n and v_f). The following theorem shows how the valuation functions v^n, v^f, v^t, and v^b are related to Odintsov's co-ordinate valuations.

Theorem 3.2.15 *Suppose that valuations v^n, v^t, v^f, v^b and v_n, v_t, v_f, v_b satisfy the following conditions: for any propositional variable p,*

1. $v^n(p) = t$ *iff* $v_b(p) = 1$,
2. $v^f(p) = t$ *iff* $v_t(p) = 1$,
3. $v^b(p) = t$ *iff* $v_n(p) = 1$,
4. $v^t(p) = t$ *iff* $v_f(p) = 1$.

Then, we have the following conditions: for any formula α,

1. $v^n(\alpha) = t$ *iff* $v_b(\alpha) = 1$,
2. $v^f(\alpha) = t$ *iff* $v_t(\alpha) = 1$,
3. $v^b(\alpha) = t$ *iff* $v_n(\alpha) = 1$,
4. $v^t(\alpha) = t$ *iff* $v_f(\alpha) = 1$.

Proof. By simultaneous induction on α. For atoms and negated atoms the claims hold trivially. For negated complex formulas, the claims hold by (*) and the four induction hypotheses. We consider here two cases for Claim 1.

$v^n(\sim_f(\alpha \wedge_t \beta)) = t$ iff $v^f(\alpha \wedge_t \beta) = t$ iff ($v^f(\alpha) = t$ and $v^f(\beta) = t$) iff $v_t(\alpha) = 1$ and $v_t(\beta) = 1$) (induction hypothesis for 2) iff $v_t(\alpha \wedge_t \beta) = 1$ iff $v_b(\sim_f(\alpha \wedge_t \beta)) = 1$.

$v^n(\sim_t\sim_f\alpha) = t$ iff $v^t(\sim_f\alpha) = t$ iff $v^b(\alpha) = t$ iff (induction hypothesis for 3) $v_n(\alpha) = 1$ iff $v_f(\sim_f\alpha) = 1$ iff $v_b(\sim_t\sim_f\alpha) = 1$.

For formulas of the form $(\beta \sharp \delta)$, where $\sharp$ is a binary connective, the claims follow by (**) and the respective induction hypothesis. We consider here one case for claim 2: $v^f(\alpha \to \beta) = t$ iff ($v^f(\alpha) = f$ or $v^f(\beta) = t$) iff ($v_t(\alpha) = 0$ or $v_t(\beta) = 1$) iff $v_t(\alpha \to \beta) = 1$. ∎

3.3. First-order case

3.3.1. Sequent calculus

We will extend the sequent system L_{16} to a first-order sequent calculus F_{16}. The notational conventions of the previous section are also adopted in this section. To begin with, we introduce the first-order language $\mathcal{L}$, in which the quantifiers come in two versions, one pair of quantifiers is related to truth, the other pair is related to falsity. For the sake of simplicity of the discussion, $\mathcal{L}$ is a language without individual constants and function symbols. *Formulas* are constructed from countably many predicate symbols $p, q, \ldots$, countably many individual variables $x, y, \ldots$, and the logical connectives $\to$, $\neg$, $\wedge_t$, $\vee_t$, $\wedge_f$, $\vee_f$, $\sim_t$, $\sim_f$, $\forall_t$, $\forall_f$, $\exists_t$, and $\exists_f$. An expression $\alpha[y/x]$ means the formula which is obtained from the formula α by replacing all free occurrences of the individual variable x in α by the individual variable y, but avoiding a clash of variables by a suitable renaming of bound variables. A 0-ary predicate is regarded as a propositional variable.

Definition 3.3.1 ($\mathbf{F}_{16}$) *Let* $\sim_d \in \{\sim_t\sim_t, \sim_f\sim_f, \sim_b\sim_b\}$ *and* $\sim_e \in \{\sim_t, \sim_b\}$. *The sequent calculus* F_{16} *is obtained from* L_{16} *by adding the quantifier inference rules of the form:*

$$\frac{\alpha[y/x], \Gamma \Rightarrow \Delta}{\forall_t x\alpha, \Gamma \Rightarrow \Delta}\ (\forall_t\mathrm{l}) \qquad \frac{\Gamma \Rightarrow \Delta, \alpha[z/x]}{\Gamma \Rightarrow \Delta, \forall_t x\alpha}\ (\forall_t\mathrm{r})$$

$$\frac{\alpha[z/x],\Gamma\Rightarrow\Delta}{\exists_t x\alpha,\Gamma\Rightarrow\Delta}\ (\exists_t\mathrm{l})\qquad\frac{\Gamma\Rightarrow\Delta,\alpha[y/x]}{\Gamma\Rightarrow\Delta,\exists_t x\alpha}\ (\exists_t\mathrm{r})$$

$$\frac{\alpha[z/x],\Gamma\Rightarrow\Delta}{\forall_f x\alpha,\Gamma\Rightarrow\Delta}\ (\forall_f\mathrm{l})\qquad\frac{\Gamma\Rightarrow\Delta,\alpha[y/x]}{\Gamma\Rightarrow\Delta,\forall_f x\alpha}\ (\forall_f\mathrm{r})$$

$$\frac{\alpha[y/x],\Gamma\Rightarrow\Delta}{\exists_f x\alpha,\Gamma\Rightarrow\Delta}\ (\exists_f\mathrm{l})\qquad\frac{\Gamma\Rightarrow\Delta,\alpha[z/x]}{\Gamma\Rightarrow\Delta,\exists_f x\alpha}\ (\exists_f\mathrm{r})$$

$$\frac{\sim_e\alpha[z/x],\Gamma\Rightarrow\Delta}{\sim_e\forall_t x\alpha,\Gamma\Rightarrow\Delta}\ (\sim_e\forall_t\mathrm{l})\qquad\frac{\Gamma\Rightarrow\Delta,\sim_e\alpha[y/x]}{\Gamma\Rightarrow\Delta,\sim_e\forall_t x\alpha}\ (\sim_e\forall_t\mathrm{r})$$

$$\frac{\sim_e\alpha[y/x],\Gamma\Rightarrow\Delta}{\sim_e\exists_t x\alpha,\Gamma\Rightarrow\Delta}\ (\sim_e\exists_t\mathrm{l})\qquad\frac{\Gamma\Rightarrow\Delta,\sim_e\alpha[z/x]}{\Gamma\Rightarrow\Delta,\sim_e\exists_t x\alpha}\ (\sim_e\exists_t\mathrm{r})$$

$$\frac{\sim_t\alpha[y/x],\Gamma\Rightarrow\Delta}{\sim_t\forall_f x\alpha,\Gamma\Rightarrow\Delta}\ (\sim_t\forall_f\mathrm{l})\qquad\frac{\Gamma\Rightarrow\Delta,\sim_t\alpha[z/x]}{\Gamma\Rightarrow\Delta,\sim_t\forall_f x\alpha}\ (\sim_t\forall_f\mathrm{r})$$

$$\frac{\sim_t\alpha[z/x],\Gamma\Rightarrow\Delta}{\sim_t\exists_f x\alpha,\Gamma\Rightarrow\Delta}\ (\sim_t\exists_f\mathrm{l})\qquad\frac{\Gamma\Rightarrow\Delta,\sim_t\alpha[y/x]}{\Gamma\Rightarrow\Delta,\sim_t\exists_f x\alpha}\ (\sim_t\exists_f\mathrm{r})$$

$$\frac{\sim_b\alpha[z/x],\Gamma\Rightarrow\Delta}{\sim_b\forall_f x\alpha,\Gamma\Rightarrow\Delta}\ (\sim_b\forall_f\mathrm{l})\qquad\frac{\Gamma\Rightarrow\Delta,\sim_b\alpha[y/x]}{\Gamma\Rightarrow\Delta,\sim_b\forall_f x\alpha}\ (\sim_b\forall_f\mathrm{r})$$

$$\frac{\sim_b\alpha[y/x],\Gamma\Rightarrow\Delta}{\sim_b\exists_f x\alpha,\Gamma\Rightarrow\Delta}\ (\sim_b\exists_f\mathrm{l})\qquad\frac{\Gamma\Rightarrow\Delta,\sim_b\alpha[z/x]}{\Gamma\Rightarrow\Delta,\sim_b\exists_f x\alpha}\ (\sim_b\exists_f\mathrm{r})$$

$$\frac{\sim_f\alpha[y/x],\Gamma\Rightarrow\Delta}{\sim_f\forall_t x\alpha,\Gamma\Rightarrow\Delta}\ (\sim_f\forall_t\mathrm{l})\qquad\frac{\Gamma\Rightarrow\Delta,\sim_f\alpha[z/x]}{\Gamma\Rightarrow\Delta,\sim_f\forall_t x\alpha}\ (\sim_f\forall_t\mathrm{r})$$

$$\frac{\sim_f\alpha[z/x],\Gamma\Rightarrow\Delta}{\sim_f\exists_t x\alpha,\Gamma\Rightarrow\Delta}\ (\sim_f\exists_t\mathrm{l})\qquad\frac{\Gamma\Rightarrow\Delta,\sim_f\alpha[y/x]}{\Gamma\Rightarrow\Delta,\sim_f\exists_t x\alpha}\ (\sim_f\exists_t\mathrm{r})$$

$$\frac{\sim_f\alpha[y/x],\Gamma\Rightarrow\Delta}{\sim_f\forall_f x\alpha,\Gamma\Rightarrow\Delta}\ (\sim_f\forall_f\mathrm{l})\qquad\frac{\Gamma\Rightarrow\Delta,\sim_f\alpha[z/x]}{\Gamma\Rightarrow\Delta,\sim_f\forall_f x\alpha}\ (\sim_f\forall_f\mathrm{r})$$

$$\frac{\sim_f\alpha[z/x],\Gamma\Rightarrow\Delta}{\sim_f\exists_f x\alpha,\Gamma\Rightarrow\Delta}\ (\sim_f\exists_f\mathrm{l})\qquad\frac{\Gamma\Rightarrow\Delta,\sim_f\alpha[y/x]}{\Gamma\Rightarrow\Delta,\sim_f\exists_f x\alpha}\ (\sim_f\exists_f\mathrm{r})$$

where y is an arbitrary individual variable, and z is an individual variable which has the eigenvariable condition, i.e., z does not occur as a free individual variable in the lower sequent of the rule.

Note that, for example, L_{16} − (cut) $\vdash \exists_f x\, p(x) \Rightarrow \forall_f x\, p(x)$, whereas the eigenvariable condition prevents $\exists_t x\, p(x) \Rightarrow \forall_t x\, p(x)$ from being provable.

Propositions 3.2.2 and 3.2.3 hold for F_{16}.

3.3.2. Semantics

Definition 3.3.2 *A structure $\mathcal{A} := \langle U, I^n, I^t, I^f, I^b\rangle$ is called a* model *if the following conditions hold:*

1. *U is a non-empty set,*
2. *I^n, I^t, I^f and I^b are mappings such that $p^{I^n}, p^{I^t}, p^{I^f}, p^{I^b} \subseteq U^n$ (i.e. $p^{I^n}, p^{I^t}, p^{I^f}$ and p^{I^b} are n-ary relations on U) for an n-ary predicate symbol p.*

We introduce the notation $\underline{u}$ as the name of $u \in U$, and we denote as $\mathcal{L}[\mathcal{A}]$ the language obtained from $\mathcal{L}$ by adding the names of all the elements of U. A formula α is called a closed formula *if α has no free individual variable. A formula of the form $\forall_t x_1 \cdots \forall_t x_m \alpha$ is called the* universal closure *of α if the free variables of α are $x_1, ..., x_m$. We write $cl(\alpha)$ for the universal closure of α.*

Definition 3.3.3 *Let $\mathcal{A} := \langle U, I^n, I^t, I^f, I^b\rangle$ be a model. The satisfaction relations $\mathcal{A} \models^n \alpha$, $\mathcal{A} \models^t \alpha$, $\mathcal{A} \models^f \alpha$, and $\mathcal{A} \models^b \alpha$ for any closed formula α of $\mathcal{L}[\mathcal{A}]$ are defined inductively as follows: for any $* \in \{n, t, f, b\}$, any $e \in \{t, b\}$, and any $g \in \{f, n\}$,*

1. *$[\mathcal{A} \models^* p(\underline{x}_1, ..., \underline{x}_n)$ iff $(x_1, ..., x_n) \in p^{I^*}]$ for any n-ary atomic formula $p(\underline{x}_1, ..., \underline{x}_n)$,*
2. *$\mathcal{A} \models^g \alpha{\to}\beta$ iff not-$(\mathcal{A} \models^g \alpha)$ or $\mathcal{A} \models^g \beta$,*
3. *$\mathcal{A} \models^g \neg\alpha$ iff not-$(\mathcal{A} \models^g \alpha)$,*
4. *$\mathcal{A} \models^g \alpha\wedge_t\beta$ iff $\mathcal{A} \models^g \alpha$ and $\mathcal{A} \models^g \beta$,*
5. *$\mathcal{A} \models^g \alpha\vee_t\beta$ iff $\mathcal{A} \models^g \alpha$ or $\mathcal{A} \models^g \beta$,*
6. *$\mathcal{A} \models^g \alpha\wedge_f\beta$ iff $\mathcal{A} \models^g \alpha$ or $\models^g \beta$,*
7. *$\mathcal{A} \models^g \alpha\vee_f\beta$ iff $\mathcal{A} \models^g \alpha$ and $\mathcal{A} \models^g \beta$,*
8. *$\mathcal{A} \models^g \forall_t x\alpha$ iff $\mathcal{A} \models^g \alpha[\underline{u}/x]$ for all $u \in U$,*
9. *$\mathcal{A} \models^g \exists_t x\alpha$ iff $\mathcal{A} \models^g \alpha[\underline{u}/x]$ for some $u \in U$,*
10. *$\mathcal{A} \models^g \forall_f x\alpha$ iff $\mathcal{A} \models^g \alpha[\underline{u}/x]$ for some $u \in U$,*
11. *$\mathcal{A} \models^g \exists_f x\alpha$ iff $\mathcal{A} \models^g \alpha[\underline{u}/x]$ for all $u \in U$,*
12. *$\mathcal{A} \models^n {\sim_t}\alpha$ iff $\mathcal{A} \models^t \alpha$,*

13. $\mathcal{A} \models^n \sim_f\alpha$ *iff* $\mathcal{A} \models^f \alpha$,

14. $\mathcal{A} \models^n \sim_b\alpha$ *iff* $\mathcal{A} \models^b \alpha$,

15. $\mathcal{A} \models^e \alpha{\rightarrow}\beta$ *iff not-*($\mathcal{A} \models^e \alpha$) *and* $\mathcal{A} \models^e \beta$,

16. $\mathcal{A} \models^e \neg\alpha$ *iff not-*($\mathcal{A} \models^e \alpha$),

17. $\mathcal{A} \models^e \alpha{\wedge_t}\beta$ *iff* $\mathcal{A} \models^e \alpha$ *or* $\mathcal{A} \models^e \beta$,

18. $\mathcal{A} \models^e \alpha{\vee_t}\beta$ *iff* $\mathcal{A} \models^e \alpha$ *and* $\mathcal{A} \models^e \beta$,

19. $\mathcal{A} \models^t \alpha{\wedge_f}\beta$ *iff* $\mathcal{A} \models^t \alpha$ *and* $\mathcal{A} \models^t \beta$,

20. $\mathcal{A} \models^t \alpha{\vee_f}\beta$ *iff* $\mathcal{A} \models^t \alpha$ *or* $\mathcal{A} \models^t \beta$,

21. $\mathcal{A} \models^b \alpha{\wedge_f}\beta$ *iff* $\mathcal{A} \models^b \alpha$ *or* $\mathcal{A} \models^b \beta$,

22. $\mathcal{A} \models^b \alpha{\vee_f}\beta$ *iff* $\mathcal{A} \models^b \alpha$ *and* $\mathcal{A} \models^b \beta$,

23. $\mathcal{A} \models^e \forall_t x\alpha$ *iff* $\mathcal{A} \models^e \alpha[\underline{u}/x]$ *for some* $u \in U$,

24. $\mathcal{A} \models^e \exists_t x\alpha$ *iff* $\mathcal{A} \models^e \alpha[\underline{u}/x]$ *for all* $u \in U$,

25. $\mathcal{A} \models^t \forall_f x\alpha$ *iff* $\mathcal{A} \models^t \alpha[\underline{u}/x]$ *for all* $u \in U$,

26. $\mathcal{A} \models^t \exists_f x\alpha$ *iff* $\mathcal{A} \models^t \alpha[\underline{u}/x]$ *for some* $u \in U$,

27. $\mathcal{A} \models^b \forall_f x\alpha$ *iff* $\mathcal{A} \models^b \alpha[\underline{u}/x]$ *for some* $u \in U$,

28. $\mathcal{A} \models^b \exists_f x\alpha$ *iff* $\mathcal{A} \models^b \alpha[\underline{u}/x]$ *for all* $u \in U$,

29. $\mathcal{A} \models^t \sim_t\alpha$ *iff* $\mathcal{A} \models^n \alpha$,

30. $\mathcal{A} \models^t \sim_f\alpha$ *iff* $\mathcal{A} \models^b \alpha$,

31. $\mathcal{A} \models^t \sim_b\alpha$ *iff* $\mathcal{A} \models^f \alpha$,

32. $\mathcal{A} \models^f \sim_t\alpha$ *iff* $\mathcal{A} \models^b \alpha$,

33. $\mathcal{A} \models^f \sim_f\alpha$ *iff* $\mathcal{A} \models^n \alpha$,

34. $\mathcal{A} \models^f \sim_b\alpha$ *iff* $\mathcal{A} \models^t \alpha$,

35. $\mathcal{A} \models^b \sim_t\alpha$ *iff* $\mathcal{A} \models^f \alpha$,

36. $\mathcal{A} \models^b \sim_f\alpha$ *iff* $\mathcal{A} \models^t \alpha$,

37. $\mathcal{A} \models^b \sim_b\alpha$ *iff* $\mathcal{A} \models^n \alpha$.

The satisfaction relations $\mathcal{A} \models^* \alpha$ $(* \in \{n, t, f, b\})$ *for any formula* α *of* $\mathcal{L}$ *are defined by* $(\mathcal{A} \models^* \alpha$ *iff* $\mathcal{A} \models^* cl(\alpha))$. *A formula* α *of* $\mathcal{L}$ *is called* valid *if* $\mathcal{A} \models^n \alpha$ *holds for any model* $\mathcal{A}$. *A sequent* $\Gamma \Rightarrow \Delta$ *of* $\mathcal{L}$ *is called valid if so is the formula* $\Gamma_* \to \Delta^*$.

Note that the four interpretation functions I^n, I^f, I^t and I^b correspond to the valuation functions v^n, v^f, v^t and v^b, respectively. Moreover, it is noted that the following conditions hold:

1. $\mathcal{A} \models^t \sim_f \alpha$ iff $\mathcal{A} \models^f \sim_t \alpha$,
2. $\mathcal{A} \models^t \alpha$ iff $\mathcal{A} \models^n \sim_t \alpha$ iff $\mathcal{A} \models^f \sim_b \alpha$ iff $\mathcal{A} \models^b \sim_f \alpha$,
3. $\mathcal{A} \models^f \alpha$ iff $\mathcal{A} \models^n \sim_f \alpha$ iff $\mathcal{A} \models^t \sim_b \alpha$ iff $\mathcal{A} \models^b \sim_t \alpha$,
4. $\mathcal{A} \models^b \alpha$ iff $\mathcal{A} \models^n \sim_b \alpha$ iff $\mathcal{A} \models^t \sim_f \alpha$ iff $\mathcal{A} \models^f \sim_t \alpha$.

Theorem 3.3.4 (Soundness of $\mathbf{F}_{16}$) *For any sequent* S, *if* $\mathrm{F}_{16} \vdash S$, *then* S *is valid.*

Proof. By induction on the proof P of S. We distinguish the cases according to the last inference of P. We show only the following case.

Case $(\sim_t \exists_t \mathrm{r})$: The last inference of P is of the form:

$$\frac{\Gamma \Rightarrow \Delta, \sim_t \alpha[z/x]}{\Gamma \Rightarrow \Delta, \sim_t \exists_t x \alpha} \ (\sim_t \exists_t \mathrm{r}).$$

We show that "$\Gamma \Rightarrow \Delta, \sim_t \alpha[z/x]$ is valid" implies "$\Gamma \Rightarrow \Delta, \sim_t \exists_t x \alpha$ is valid." By the hypothesis, (i): $\forall_t z_1 \cdots \forall_t z_n \forall_t z(\Gamma_* \to (\Delta^* \vee_t (\sim_t \alpha[z/x])))$ (where $z_1, ..., z_n$ are the free individual variables occurring in $\Gamma \Rightarrow \Delta, \sim_t \exists_t x \alpha$) is valid. We show that $\mathcal{A} \models^n \forall_t z_1 \cdots \forall_t z_n (\Gamma_* \to (\Delta^* \vee_t (\sim_t \exists_t x \alpha)))$ for any model $\mathcal{A} := \langle U, I^n, I^t, I^f, I^b \rangle$, i.e., we show that for any $u_1, ..., u_n \in U$, $\mathcal{A} \models^n \underline{\Gamma}_* \to (\underline{\Delta}^* \vee_t (\sim_t \exists_t x \underline{\alpha}))$, where $\underline{\Gamma}_*, \underline{\Delta}^*$ and $\underline{\alpha}$ are respectively obtained from Γ_*, Δ^* and α by replacing $z_1, ..., z_n$ by $\underline{u}_1, ..., \underline{u}_n$ Here, we note that $(\sim_t \exists_t x \alpha)[\underline{u}_1/z_1, ..., \underline{u}_n/z_n]$ (the result of the simultaneous substitution of z_i by $\underline{u}_i$ $(1 \leq i \leq n)$) is equivalent to $\sim_t \exists_t x(\alpha[\underline{u}_1/z_1, ..., \underline{u}_n/z_n])$, i.e., $\sim_t \exists_t x \underline{\alpha}$. By (i), we have $\mathcal{A} \models^n (\underline{\Gamma}_* \to (\underline{\Delta}^* \vee_t (\sim_t \underline{\alpha}[z/x])))[\underline{w}/z]$ for any $w \in U$. By the eigenvariable condition, z is not occurring freely in $\underline{\Gamma}_*, \underline{\Delta}^*$ and $\underline{\alpha}$. Thus, $\underline{\Gamma}_*[\underline{w}/z]$ and $\underline{\Delta}^*[\underline{w}/z]$ are equivalent to $\underline{\Gamma}_*$and $\underline{\Delta}^*$, respectively, and $\underline{\alpha}[z/x][\underline{w}/z]$ is equivalent to $\underline{\alpha}[\underline{w}/z][\underline{w}/x]$, i.e., $\underline{\alpha}[\underline{w}/x]$. Therefore, for any $w \in U$, we have that (a): $\mathcal{A} \models^n \underline{\Gamma}_* \to (\underline{\Delta}^* \vee_t \sim_t \underline{\alpha}[\underline{w}/x])$. Suppose that (b): $[\mathcal{A} \models^n \underline{\Gamma}_*$ and not $(\mathcal{A} \models^n \underline{\Delta}^*)]$. Then, by (a), we have that for any $w \in U$, $\mathcal{A} \models^n \sim_t \underline{\alpha}[\underline{w}/x]$, i.e., $\mathcal{A} \models^t \underline{\alpha}[\underline{w}/x]$. Therefore, we obtain (c): $\mathcal{A} \models^t \exists_t x\underline{\alpha}$, and hence $\mathcal{A} \models^n \sim_t\exists_t x\underline{\alpha}$. This means that (b) implies (c), i.e., $\mathcal{A} \models^n \underline{\Gamma}_*$ implies ($\mathcal{A} \models^n \underline{\Delta}^*$ or $\mathcal{A} \models^n \sim_t\exists_t x\underline{\alpha}$). Therefore, we have the required fact that $\mathcal{A} \models^n \underline{\Gamma}_*\to(\underline{\Delta}^*\vee_t(\sim_t\exists_t x\underline{\alpha}))$ for any $u_1, ...u_n \in U$.

∎

3.3.3. Completeness and cut-elimination

In the following, we prove the (strong) completeness and cut-elimination theorems for F_{16} by using Schütte's method [185].

Definition 3.3.5 *Let* $\sim_d \in \{\sim_t\sim_t, \sim_f\sim_f, \sim_b\sim_b\}$ *and* $\sim_e \in \{\sim_t, \sim_b\}$. *A sequent* $\Gamma \Rightarrow \Delta$ *is called* saturated *if for any formulas* α *and* β,

1. $\alpha\to\beta \in \Gamma$ *implies* ($\alpha \in \Delta$ *or* $\beta \in \Gamma$),
2. $\alpha\to\beta \in \Delta$ *implies* ($\alpha \in \Gamma$ *and* $\beta \in \Delta$),
3. $\neg\alpha \in \Gamma$ *implies* $\alpha \in \Delta$,
4. $\neg\alpha \in \Delta$ *implies* $\alpha \in \Gamma$,
5. $\alpha\wedge_t\beta \in \Gamma$ *implies* ($\alpha \in \Gamma$ *and* $\beta \in \Gamma$),
6. $\alpha\wedge_t\beta \in \Delta$ *implies* ($\alpha \in \Delta$ *or* $\beta \in \Delta$),
7. $\alpha\vee_t\beta \in \Gamma$ *implies* ($\alpha \in \Gamma$ *or* $\beta \in \Gamma$),
8. $\alpha\vee_t\beta \in \Delta$ *implies* ($\alpha \in \Delta$ *and* $\beta \in \Delta$),
9. $\alpha\wedge_f\beta \in \Gamma$ *implies* ($\alpha \in \Gamma$ *or* $\beta \in \Gamma$),
10. $\alpha\wedge_f\beta \in \Delta$ *implies* ($\alpha \in \Delta$ *and* $\beta \in \Delta$),
11. $\alpha\vee_f\beta \in \Gamma$ *implies* ($\alpha \in \Gamma$ *and* $\beta \in \Gamma$),
12. $\alpha\vee_f\beta \in \Delta$ *implies* ($\alpha \in \Delta$ *or* $\beta \in \Delta$),
13. $\forall_t x\alpha \in \Gamma$ *implies* ($\alpha[y/x] \in \Gamma$ *for any individual variable* y),
14. $\forall_t x\alpha \in \Delta$ *implies* ($\alpha[z/x] \in \Delta$ *for some individual variable* z),
15. $\exists_t x\alpha \in \Gamma$ *implies* ($\alpha[z/x] \in \Gamma$ *for some individual variable* z),
16. $\exists_t x\alpha \in \Delta$ *implies* ($\alpha[y/x] \in \Delta$ *for any individual variable* y),
17. $\forall_f x\alpha \in \Gamma$ *implies* ($\alpha[z/x] \in \Gamma$ *for some individual variable* z),
18. $\forall_f x\alpha \in \Delta$ *implies* ($\alpha[y/x] \in \Delta$ *for any individual variable* y),
19. $\exists_f x\alpha \in \Gamma$ *implies* ($\alpha[y/x] \in \Gamma$ *for any individual variable* y),
20. $\exists_f x\alpha \in \Delta$ *implies* ($\alpha[z/x] \in \Delta$ *for some individual variable* z),
21. $\sim_d\alpha \in \Gamma$ *implies* $\alpha \in \Gamma$,
22. $\sim_d\alpha \in \Delta$ *implies* $\alpha \in \Delta$,

23. $\sim_t\sim_f\sim_t\alpha \in \Gamma$ *implies* $\sim_f\alpha \in \Gamma$,

24. $\sim_t\sim_f\sim_t\alpha \in \Delta$ *implies* $\sim_f\alpha \in \Delta$,

25. $\sim_f\sim_t\sim_f\alpha \in \Gamma$ *implies* $\sim_t\alpha \in \Gamma$,

26. $\sim_f\sim_t\sim_f\alpha \in \Delta$ *implies* $\sim_t\alpha \in \Delta$,

27. $\sim_t\neg\sim_t\alpha \in \Gamma$ *implies* $\neg\alpha \in \Gamma$,

28. $\sim_t\neg\sim_t\alpha \in \Delta$ *implies* $\neg\alpha \in \Delta$,

29. $\sim_f\neg\sim_f\alpha \in \Gamma$ *implies* $\neg\alpha \in \Gamma$,

30. $\sim_f\neg\sim_f\alpha \in \Delta$ *implies* $\neg\alpha \in \Delta$,

31. $\sim_f\sim_t\sim_t\alpha \in \Gamma$ *implies* $\sim_f\alpha \in \Gamma$,

32. $\sim_f\sim_t\sim_t\alpha \in \Delta$ *implies* $\sim_f\alpha \in \Delta$,

33. $\sim_t\sim_f\sim_f\alpha \in \Gamma$ *implies* $\sim_t\alpha \in \Gamma$,

34. $\sim_t\sim_f\sim_f\alpha \in \Delta$ *implies* $\sim_t\alpha \in \Delta$,

35. $\neg\sim_t\sim_t\alpha \in \Gamma$ *implies* $\neg\alpha \in \Gamma$,

36. $\neg\sim_t\sim_t\alpha \in \Delta$ *implies* $\neg\alpha \in \Delta$,

37. $\neg\sim_f\sim_f\alpha \in \Gamma$ *implies* $\neg\alpha \in \Gamma$,

38. $\neg\sim_f\sim_f\alpha \in \Delta$ *implies* $\neg\alpha \in \Delta$,

39. $\sim_e(\alpha\rightarrow\beta) \in \Gamma$ *implies* ($\sim_e\neg\alpha \in \Gamma$ *and* $\sim_e\beta \in \Gamma$),

40. $\sim_e(\alpha\rightarrow\beta) \in \Delta$ *implies* ($\sim_e\neg\alpha \in \Delta$ *or* $\sim_e\beta \in \Delta$),

41. $\sim_e\neg\alpha \in \Gamma$ *implies* $\sim_e\alpha \in \Delta$,

42. $\sim_e\neg\alpha \in \Delta$ *implies* $\sim_e\alpha \in \Gamma$,

43. $\sim_e(\alpha\wedge_t\beta) \in \Gamma$ *implies* ($\sim_e\alpha \in \Gamma$ *or* $\sim_e\beta \in \Gamma$),

44. $\sim_e(\alpha\wedge_t\beta) \in \Delta$ *implies* ($\sim_e\alpha \in \Delta$ *and* $\sim_e\beta \in \Delta$),

45. $\sim_e(\alpha\vee_t\beta) \in \Gamma$ *implies* ($\sim_e\alpha \in \Gamma$ *and* $\sim_e\beta \in \Gamma$),

46. $\sim_e(\alpha\vee_t\beta) \in \Delta$ *implies* ($\sim_e\alpha \in \Delta$ *or* $\sim_e\beta \in \Delta$),

47. $\sim_t(\alpha\wedge_f\beta) \in \Gamma$ *implies* ($\sim_t\alpha \in \Gamma$ *and* $\sim_t\beta \in \Gamma$),

48. $\sim_t(\alpha\wedge_f\beta) \in \Delta$ *implies* ($\sim_t\alpha \in \Delta$ *or* $\sim_t\beta \in \Delta$),

49. $\sim_t(\alpha\vee_f\beta) \in \Gamma$ *implies* ($\sim_t\alpha \in \Gamma$ *or* $\sim_t\beta \in \Gamma$),

50. $\sim_t(\alpha\vee_f\beta) \in \Delta$ *implies* ($\sim_t\alpha \in \Delta$ *and* $\sim_t\beta \in \Delta$),

51. $\sim_b(\alpha\wedge_f\beta) \in \Gamma$ *implies* ($\sim_b\alpha \in \Gamma$ *or* $\sim_b\beta \in \Gamma$),

52. $\sim_b(\alpha\wedge_f\beta) \in \Delta$ *implies* ($\sim_b\alpha \in \Delta$ *and* $\sim_b\beta \in \Delta$),

53. $\sim_b(\alpha\vee_f\beta) \in \Gamma$ *implies* ($\sim_b\alpha \in \Gamma$ *and* $\sim_b\beta \in \Gamma$),

54. $\sim_b(\alpha\vee_f\beta) \in \Delta$ *implies* ($\sim_b\alpha \in \Delta$ *or* $\sim_b\beta \in \Delta$),

55. $\sim_e \forall_t x\alpha \in \Gamma$ *implies* ($\sim_e \alpha[z/x] \in \Gamma$ *for some individual variable* z),
56. $\sim_e \forall_t x\alpha \in \Delta$ *implies* ($\sim_e \alpha[y/x] \in \Delta$ *for any individual variable* y),
57. $\sim_e \exists_t x\alpha \in \Gamma$ *implies* ($\sim_e \alpha[y/x] \in \Gamma$ *for any individual variable* y),
58. $\sim_e \exists_t x\alpha \in \Delta$ *implies* ($\sim_e \alpha[z/x] \in \Delta$ *for some individual variable* z),
59. $\sim_t \forall_f x\alpha \in \Gamma$ *implies* ($\sim_t \alpha[y/x] \in \Gamma$ *for any individual variable* y),
60. $\sim_t \forall_f x\alpha \in \Delta$ *implies* ($\sim_t \alpha[z/x] \in \Delta$ *for some individual variable* z),
61. $\sim_t \exists_f x\alpha \in \Gamma$ *implies* ($\sim_t \alpha[z/x] \in \Gamma$ *for some individual variable* z),
62. $\sim_t \exists_f x\alpha \in \Delta$ *implies* ($\sim_t \alpha[y/x] \in \Delta$ *for any individual variable* y),
63. $\sim_b \forall_f x\alpha \in \Gamma$ *implies* ($\sim_b \alpha[z/x] \in \Gamma$ *for some individual variable* z),
64. $\sim_b \forall_f x\alpha \in \Delta$ *implies* ($\sim_b \alpha[y/x] \in \Delta$ *for any individual variable* y),
65. $\sim_b \exists_f x\alpha \in \Gamma$ *implies* ($\sim_b \alpha[y/x] \in \Gamma$ *for any individual variable* y),
66. $\sim_b \exists_f x\alpha \in \Delta$ *implies* ($\sim_b \alpha[z/x] \in \Delta$ *for some individual variable* z),
67. $\sim_f(\alpha \rightarrow \beta) \in \Gamma$ *implies* ($\sim_f \alpha \in \Delta$ *or* $\sim_f \beta \in \Gamma$),
68. $\sim_f(\alpha \rightarrow \beta) \in \Delta$ *implies* ($\sim_f \alpha \in \Gamma$ *and* $\sim_f \beta \in \Delta$),
69. $\sim_f \neg\alpha \in \Gamma$ *implies* $\sim_f \alpha \in \Delta$,
70. $\sim_f \neg\alpha \in \Delta$ *implies* $\sim_f \alpha \in \Gamma$,
71. $\sim_f(\alpha \wedge_t \beta) \in \Gamma$ *implies* ($\sim_f \alpha \in \Gamma$ *and* $\sim_f \beta \in \Gamma$),
72. $\sim_f(\alpha \wedge_t \beta) \in \Delta$ *implies* ($\sim_f \alpha \in \Delta$ *or* $\sim_f \beta \in \Delta$),
73. $\sim_f(\alpha \vee_t \beta) \in \Gamma$ *implies* ($\sim_f \alpha \in \Gamma$ *or* $\sim_f \beta \in \Gamma$),
74. $\sim_f(\alpha \vee_t \beta) \in \Delta$ *implies* ($\sim_f \alpha \in \Delta$ *and* $\sim_f \beta \in \Delta$),
75. $\sim_f(\alpha \wedge_f \beta) \in \Gamma$ *implies* ($\sim_f \alpha \in \Gamma$ *and* $\sim_f \beta \in \Gamma$),
76. $\sim_f(\alpha \wedge_f \beta) \in \Delta$ *implies* ($\sim_f \alpha \in \Delta$ *or* $\sim_f \beta \in \Delta$),
77. $\sim_f(\alpha \vee_f \beta) \in \Gamma$ *implies* ($\sim_f \alpha \in \Gamma$ *or* $\sim_f \beta \in \Gamma$),
78. $\sim_f(\alpha \vee_f \beta) \in \Delta$ *implies* ($\sim_f \alpha \in \Delta$ *and* $\sim_f \beta \in \Delta$),
79. $\sim_f \forall_t x\alpha \in \Gamma$ *implies* ($\sim_f \alpha[y/x] \in \Gamma$ *for any individual variable* y),
80. $\sim_f \forall_t x\alpha \in \Delta$ *implies* ($\sim_f \alpha[z/x] \in \Delta$ *for some individual variable* z),
81. $\sim_f \exists_t x\alpha \in \Gamma$ *implies* ($\sim_f \alpha[z/x] \in \Gamma$ *for some individual variable* z),
82. $\sim_f \exists_t x\alpha \in \Delta$ *implies* ($\sim_f \alpha[y/x] \in \Delta$ *for any individual variable* y),
83. $\sim_f \forall_f x\alpha \in \Gamma$ *implies* ($\sim_f \alpha[y/x] \in \Gamma$ *for any individual variable* y),
84. $\sim_f \forall_f x\alpha \in \Delta$ *implies* ($\sim_f \alpha[z/x] \in \Delta$ *for some individual variable* z),
85. $\sim_f \exists_f x\alpha \in \Gamma$ *implies* ($\sim_f \alpha[z/x] \in \Gamma$ *for some individual variable* z),
86. $\sim_f \exists_f x\alpha \in \Delta$ *implies* ($\sim_f \alpha[y/x] \in \Delta$ *for any individual variable* y).

We now generalize the notion of a sequent.

Definition 3.3.6 *An expression* $\Gamma \Rightarrow \Delta$ *is called an* infinite sequent *if* Γ *and* Δ *are infinite (countable) sets of formulas. An infinite sequent* $\Gamma \Rightarrow \Delta$ *is called provable if a finite part* $\Gamma' \Rightarrow \Delta'$ *of the sequent is provable, i.e.,* Γ' *and* Δ' *are finite subsets of* Γ *and* Δ*, respectively.*

Definition 3.3.7 *Let* $\sim_d \in \{\sim_t\sim_t, \sim_f\sim_f, \sim_b\sim_b\}$ *and* $\sim_e \in \{\sim_t, \sim_b\}$*. A* decomposition *of a sequent (or infinite sequent) S is defined as having the form* S' *or* $S'; S''$ *by*

1. $\alpha, \Gamma \Rightarrow \Delta, \alpha{\to}\beta, \beta$ *is a decomposition of* $\Gamma \Rightarrow \Delta, \alpha{\to}\beta$,
2. $\alpha{\to}\beta, \Gamma \Rightarrow \Delta, \alpha$; $\beta, \alpha{\to}\beta, \Gamma \Rightarrow \Delta$ *is a decomposition of* $\alpha{\to}\beta, \Gamma \Rightarrow \Delta$,
3. $\alpha, \Gamma \Rightarrow \Delta, \neg\alpha$ *is a decomposition of* $\Gamma \Rightarrow \Delta, \neg\alpha$,
4. $\neg\alpha, \Gamma \Rightarrow \Delta, \alpha$ *is a decomposition of* $\neg\alpha, \Gamma \Rightarrow \Delta$,
5. $\Gamma \Rightarrow \Delta, \alpha\wedge_t\beta, \alpha$; $\Gamma \Rightarrow \Delta, \alpha\wedge_t\beta, \beta$ *is a decomposition of* $\Gamma \Rightarrow \Delta, \alpha\wedge_t\beta$,
6. $\alpha, \beta, \alpha\wedge_t\beta, \Gamma \Rightarrow \Delta$ *is a decomposition of* $\alpha\wedge_t\beta, \Gamma \Rightarrow \Delta$,
7. $\Gamma \Rightarrow \Delta, \alpha\vee_t\beta, \alpha, \beta$ *is a decomposition of* $\Gamma \Rightarrow \Delta, \alpha\vee_t\beta$,
8. $\alpha, \alpha\vee_t\beta, \Gamma \Rightarrow \Delta$; $\beta, \alpha\vee_t\beta, \Gamma \Rightarrow \Delta$ *is a decomposition of* $\alpha\vee_t\beta, \Gamma \Rightarrow \Delta$,
9. $\Gamma \Rightarrow \Delta, \alpha\wedge_f\beta, \alpha, \beta$ *is a decomposition of* $\Gamma \Rightarrow \Delta, \alpha\wedge_f\beta$,
10. $\alpha, \alpha\wedge_f\beta, \Gamma \Rightarrow \Delta$; $\beta, \alpha\wedge_f\beta, \Gamma \Rightarrow \Delta$ *is a decomposition of* $\alpha\wedge_f\beta, \Gamma \Rightarrow \Delta$,
11. $\Gamma \Rightarrow \Delta, \alpha\vee_f\beta, \alpha$; $\Gamma \Rightarrow \Delta, \alpha\vee_f\beta, \beta$ *is a decomposition of* $\Gamma \Rightarrow \Delta, \alpha\vee_f\beta$,
12. $\alpha, \beta, \alpha\vee_f\beta, \Gamma \Rightarrow \Delta$ *is a decomposition of* $\alpha\vee_f\beta, \Gamma \Rightarrow \Delta$,
13. $\Gamma \Rightarrow \Delta, \forall_t x\alpha, \alpha[z/x]$ *is a decomposition of* $\Gamma \Rightarrow \Delta, \forall_t x\alpha$ *where z is a* fresh *free individual variable, i.e., z is not occurring in* $\Gamma \Rightarrow \Delta, \forall_t x\alpha$,
14. $\alpha[y_1/x], ..., \alpha[y_m/x], \forall_t x\alpha, \Gamma \Rightarrow \Delta$ *is a decomposition of* $\forall_t x\alpha, \Gamma \Rightarrow \Delta$ *where* $y_1, ..., y_m$ *are the free individual variables occurring in* $\forall_t x\alpha, \Gamma \Rightarrow \Delta$,[2]
15. $\Gamma \Rightarrow \Delta, \exists_t x\alpha, \alpha[y_1/x], ..., \alpha[y_m/x]$ *is a decomposition of* $\Gamma \Rightarrow \Delta, \exists_t x\alpha$ *where* $y_1, ..., y_m$ *are the free individual variables occurring in* $\Gamma \Rightarrow \Delta, \exists_t x\alpha$,
16. $\alpha[z/x], \exists_t x\alpha, \Gamma \Rightarrow \Delta$ *is a decomposition of* $\exists_t x\alpha, \Gamma \Rightarrow \Delta$ *where z is a fresh free individual variable,*
17. $\Gamma \Rightarrow \Delta, \forall_f x\alpha, \alpha[y_1/x], ..., \alpha[y_m/x]$ *is a decomposition of* $\Gamma \Rightarrow \Delta, \forall_f x\alpha$ *where* $y_1, ..., y_m$ *are the free individual variables occurring in* $\Gamma \Rightarrow \Delta, \forall_f x\alpha$,

[2] If $\forall_t x\alpha, \Gamma \Rightarrow \Delta$ has no free individual variable, then we adopt any free variable in $\mathcal{L}$. Such a condition is also adopted in the following corresponding items.

18. $\alpha[z/x], \forall_f x\alpha, \Gamma \Rightarrow \Delta$ *is a decomposition of* $\forall_f x\alpha, \Gamma \Rightarrow \Delta$ *where* z *is a fresh free individual variable,*

19. $\Gamma \Rightarrow \Delta, \exists_f x\alpha, \alpha[z/x]$ *is a decomposition of* $\Gamma \Rightarrow \Delta, \exists_f x\alpha$ *where* z *is a* fresh *free individual variable, i.e.,* z *is not occurring in* $\Gamma \Rightarrow \Delta, \exists_f x\alpha$,

20. $\alpha[y_1/x], ..., \alpha[y_m/x], \exists_f x\alpha, \Gamma \Rightarrow \Delta$ *is a decomposition of* $\exists_f x\alpha, \Gamma \Rightarrow \Delta$ *where* $y_1, ..., y_m$ *are the free individual variables occurring in* $\exists_f x\alpha, \Gamma \Rightarrow \Delta$,

21. $\alpha, \sim_d\alpha, \Gamma \Rightarrow \Delta$ *is a decomposition of* $\sim_d\alpha, \Gamma \Rightarrow \Delta$,

22. $\Gamma \Rightarrow \Delta, \sim_d\alpha, \alpha$ *is a decomposition of* $\Gamma \Rightarrow \Delta, \sim_d\alpha$,

23. $\sim_f\alpha, \sim_t\sim_f\sim_t\alpha, \Gamma \Rightarrow \Delta$ *is a decomposition of* $\sim_t\sim_f\sim_t\alpha, \Gamma \Rightarrow \Delta$,

24. $\Gamma \Rightarrow \Delta, \sim_t\sim_f\sim_t\alpha, \sim_f\alpha$ *is a decomposition of* $\Gamma \Rightarrow \Delta, \sim_t\sim_f\sim_t\alpha$,

25. $\sim_t\alpha, \sim_f\sim_t\sim_f\alpha, \Gamma \Rightarrow \Delta$ *is a decomposition of* $\sim_f\sim_t\sim_f\alpha, \Gamma \Rightarrow \Delta$,

26. $\Gamma \Rightarrow \Delta, \sim_f\sim_t\sim_f\alpha, \sim_t\alpha$ *is a decomposition of* $\Gamma \Rightarrow \Delta, \sim_f\sim_t\sim_f\alpha$,

27. $\neg\alpha, \sim_t\neg\sim_t\alpha, \Gamma \Rightarrow \Delta$ *is a decomposition of* $\sim_t\neg\sim_t\alpha, \Gamma \Rightarrow \Delta$,

28. $\Gamma \Rightarrow \Delta, \sim_t\neg\sim_t\alpha, \neg\alpha$ *is a decomposition of* $\Gamma \Rightarrow \Delta, \sim_t\neg\sim_t\alpha$,

29. $\neg\alpha, \sim_f\neg\sim_f\alpha, \Gamma \Rightarrow \Delta$ *is a decomposition of* $\sim_f\neg\sim_f\alpha, \Gamma \Rightarrow \Delta$,

30. $\Gamma \Rightarrow \Delta, \sim_f\neg\sim_f\alpha, \neg\alpha$ *is a decomposition of* $\Gamma \Rightarrow \Delta, \sim_f\neg\sim_f\alpha$,

31. $\sim_f\alpha, \sim_f\sim_t\sim_t\alpha, \Gamma \Rightarrow \Delta$ *is a decomposition of* $\sim_f\sim_t\sim_t\alpha, \Gamma \Rightarrow \Delta$,

32. $\Gamma \Rightarrow \Delta, \sim_f\sim_t\sim_t\alpha, \sim_f\alpha$ *is a decomposition of* $\Gamma \Rightarrow \Delta, \sim_f\sim_t\sim_t\alpha$,

33. $\sim_t\alpha, \sim_t\sim_f\sim_f\alpha, \Gamma \Rightarrow \Delta$ *is a decomposition of* $\sim_t\sim_f\sim_f\alpha, \Gamma \Rightarrow \Delta$,

34. $\Gamma \Rightarrow \Delta, \sim_t\sim_f\sim_f\alpha, \sim_t\alpha$ *is a decomposition of* $\Gamma \Rightarrow \Delta, \sim_t\sim_f\sim_f\alpha$,

35. $\neg\alpha, \neg\sim_t\sim_t\alpha, \Gamma \Rightarrow \Delta$ *is a decomposition of* $\neg\sim_t\sim_t\alpha, \Gamma \Rightarrow \Delta$,

36. $\Gamma \Rightarrow \Delta, \neg\sim_t\sim_t\alpha, \sim_f\alpha$ *is a decomposition of* $\Gamma \Rightarrow \Delta, \neg\sim_t\sim_t\alpha$,

37. $\neg\alpha, \neg\sim_f\sim_f\alpha, \Gamma \Rightarrow \Delta$ *is a decomposition of* $\neg\sim_f\sim_f\alpha, \Gamma \Rightarrow \Delta$,

38. $\Gamma \Rightarrow \Delta, \neg\sim_f\sim_f\alpha, \neg\alpha$ *is a decomposition of* $\Gamma \Rightarrow \Delta, \neg\sim_f\sim_f\alpha$,

39. $\Gamma \Rightarrow \Delta, \sim_e(\alpha\rightarrow\beta), \sim_e\neg\alpha$; $\Gamma \Rightarrow \Delta, \sim_e(\alpha\rightarrow\beta), \sim_e\beta$ *is a decomposition of* $\Gamma \Rightarrow \Delta, \sim_e(\alpha\rightarrow\beta)$,

40. $\sim_e\neg\alpha, \sim_e\beta, \sim_e(\alpha\rightarrow\beta), \Gamma \Rightarrow \Delta$ *is a decomposition of* $\sim_e(\alpha\rightarrow\beta), \Gamma \Rightarrow \Delta$,

41. $\sim_e\alpha, \Gamma \Rightarrow \Delta, \sim_e\neg\alpha$ *is a decomposition of* $\Gamma \Rightarrow \Delta, \sim_e\neg\alpha$,

42. $\sim_e\neg\alpha, \Gamma \Rightarrow \Delta, \sim_e\alpha$ *is a decomposition of* $\sim_e\neg\alpha, \Gamma \Rightarrow \Delta$,

43. $\Gamma \Rightarrow \Delta, \sim_e(\alpha\wedge_t\beta), \sim_e\alpha, \sim_e\beta$ *is a decomposition of* $\Gamma \Rightarrow \Delta, \sim_e(\alpha\wedge_t\beta)$,

44. $\sim_e\alpha, \sim_e(\alpha\wedge_t\beta), \Gamma \Rightarrow \Delta$; $\sim_e\beta, \sim_e(\alpha\wedge_t\beta), \Gamma \Rightarrow \Delta$ *is a decomposition of* $\sim_e(\alpha\wedge_t\beta), \Gamma \Rightarrow \Delta$,

45. $\Gamma \Rightarrow \Delta, \sim_e(\alpha\vee_t\beta), \sim_e\alpha$; $\Gamma \Rightarrow \Delta, \sim_e(\alpha\vee_t\beta), \sim_e\beta$ *is a decomposition of* $\Gamma \Rightarrow \Delta, \sim_e(\alpha\vee_t\beta)$,

46. $\sim_e\alpha, \sim_e\beta, \sim_e(\alpha\vee_t\beta), \Gamma \Rightarrow \Delta$ *is a decomposition of* $\sim_e(\alpha\vee_t\beta), \Gamma \Rightarrow \Delta$,

47. $\Gamma \Rightarrow \Delta, \sim_t(\alpha\wedge_f\beta), \sim_t\alpha$; $\Gamma \Rightarrow \Delta, \sim_t(\alpha\wedge_f\beta), \sim_t\beta$ *is a decomposition of* $\Gamma \Rightarrow \Delta, \sim_t(\alpha\wedge_f\beta)$,

48. $\sim_t\alpha, \sim_t\beta, \sim_t(\alpha\wedge_f\beta), \Gamma \Rightarrow \Delta$ *is a decomposition of* $\sim_t(\alpha\wedge_f\beta), \Gamma \Rightarrow \Delta$,

49. $\Gamma \Rightarrow \Delta, \sim_t(\alpha\vee_f\beta), \sim_t\alpha, \sim_t\beta$ *is a decomposition of* $\Gamma \Rightarrow \Delta, \sim_t(\alpha\vee_f\beta)$,

50. $\sim_t\alpha, \sim_t(\alpha\vee_f\beta), \Gamma \Rightarrow \Delta$; $\sim_t\beta, \sim_t(\alpha\vee_f\beta), \Gamma \Rightarrow \Delta$ *is a decomposition of* $\sim_t(\alpha\vee_f\beta), \Gamma \Rightarrow \Delta$,

51. $\Gamma \Rightarrow \Delta, \sim_b(\alpha\wedge_f\beta), \sim_b\alpha, \sim_b\beta$ *is a decomposition of* $\Gamma \Rightarrow \Delta, \sim_b(\alpha\wedge_f\beta)$,

52. $\sim_b\alpha, \sim_b(\alpha\wedge_f\beta), \Gamma \Rightarrow \Delta$; $\sim_b\beta, \sim_b(\alpha\wedge_f\beta), \Gamma \Rightarrow \Delta$ *is a decomposition of* $\sim_b(\alpha\wedge_f\beta), \Gamma \Rightarrow \Delta$,

53. $\Gamma \Rightarrow \Delta, \sim_b(\alpha\vee_f\beta), \sim_b\alpha$; $\Gamma \Rightarrow \Delta, \sim_b(\alpha\vee_f\beta), \sim_b\beta$ *is a decomposition of* $\Gamma \Rightarrow \Delta, \sim_b(\alpha\vee_f\beta)$,

54. $\sim_b\alpha, \sim_b\beta, \sim_b(\alpha\vee_f\beta), \Gamma \Rightarrow \Delta$ *is a decomposition of* $\sim_b(\alpha\vee_f\beta), \Gamma \Rightarrow \Delta$,

55. $\Gamma \Rightarrow \Delta, \sim_e\forall_t x\alpha, \sim_e\alpha[y_1/x], ..., \sim_e\alpha[y_m/x]$ *is a decomposition of* $\Gamma \Rightarrow \Delta, \sim_e\forall_t x\alpha$ *where* $y_1, ..., y_m$ *are the free individual variables occurring in* $\Gamma \Rightarrow \Delta, \sim_e\forall_t x\alpha$,

56. $\sim_e\alpha[z/x], \sim_e\forall_t x\alpha, \Gamma \Rightarrow \Delta$ *is a decomposition of* $\sim_e\forall_t x\alpha, \Gamma \Rightarrow \Delta$ *where* z *is a fresh free individual variable,*

57. $\Gamma \Rightarrow \Delta, \sim_e\exists_t x\alpha, \sim_e\alpha[z/x]$ *is a decomposition of* $\Gamma \Rightarrow \Delta, \sim_e\exists_t x\alpha$ *where* z *is a fresh free individual variable,*

58. $\sim_e\alpha[y_1/x], ..., \sim_e\alpha[y_m/x], \sim_e\exists_t x\alpha, \Gamma \Rightarrow \Delta$ *is a decomposition of* $\sim_e\exists_t x\alpha, \Gamma \Rightarrow \Delta$ *where* $y_1, ..., y_m$ *are the free individual variables occurring in* $\sim_e\exists_t x\alpha, \Gamma \Rightarrow \Delta$,

59. $\Gamma \Rightarrow \Delta, \sim_t\forall_f x\alpha, \sim_t\alpha[z/x]$ *is a decomposition of* $\Gamma \Rightarrow \Delta, \sim_t\forall_f x\alpha$ *where* z *is a fresh free individual variable,*

60. $\sim_t\alpha[y_1/x], ..., \sim_t\alpha[y_m/x], \sim_t\forall_f x\alpha, \Gamma \Rightarrow \Delta$ *is a decomposition of* $\sim_t\forall_f x\alpha, \Gamma \Rightarrow \Delta$ *where* $y_1, ..., y_m$ *are the free individual variables occurring in* $\sim_t\forall_f x\alpha, \Gamma \Rightarrow \Delta$,

61. $\Gamma \Rightarrow \Delta, \sim_t\exists_f x\alpha, \sim_t\alpha[y_1/x], ..., \sim_t\alpha[y_m/x]$ *is a decomposition of* $\Gamma \Rightarrow \Delta, \sim_t\exists_f x\alpha$ *where* $y_1, ..., y_m$ *are the free individual variables occurring in* $\Gamma \Rightarrow \Delta, \sim_t\exists_f x\alpha$,

62. $\sim_t\alpha[z/x], \sim_t\exists_f x\alpha, \Gamma \Rightarrow \Delta$ *is a decomposition of* $\sim_t\exists_f x\alpha, \Gamma \Rightarrow \Delta$ *where* z *is a fresh free individual variable,*

63. $\Gamma \Rightarrow \Delta, \sim_b\forall_f x\alpha, \sim_b\alpha[y_1/x], ..., \sim_b\alpha[y_m/x]$ *is a decomposition of* $\Gamma \Rightarrow \Delta, \sim_b\forall_f x\alpha$ *where* $y_1, ..., y_m$ *are the free individual variables occurring in* $\Gamma \Rightarrow \Delta, \sim_b\forall_f x\alpha$,

64. $\sim_b\alpha[z/x], \sim_b\forall_f x\alpha, \Gamma \Rightarrow \Delta$ *is a decomposition of* $\sim_b\forall_f x\alpha, \Gamma \Rightarrow \Delta$ *where* z *is a fresh free individual variable,*

65. $\Gamma \Rightarrow \Delta, \sim_b\exists_f x\alpha, \sim_b\alpha[z/x]$ *is a decomposition of* $\Gamma \Rightarrow \Delta, \sim_b\exists_f x\alpha$ *where* z *is a fresh free individual variable,*

66. $\sim_b\alpha[y_1/x], ..., \sim_b\alpha[y_m/x], \sim_b\exists_f x\alpha, \Gamma \Rightarrow \Delta$ *is a decomposition of* $\sim_b\exists_f x\alpha, \Gamma \Rightarrow \Delta$ *where* $y_1, ..., y_m$ *are the free individual variables occurring in* $\sim_b\exists_f x\alpha, \Gamma \Rightarrow \Delta$,

67. $\sim_f\alpha, \Gamma \Rightarrow \Delta, \sim_f(\alpha \rightarrow \beta), \sim_f\beta$ *is a decomposition of* $\Gamma \Rightarrow \Delta, \sim_f(\alpha \rightarrow \beta)$,

68. $\sim_f(\alpha \rightarrow \beta), \Gamma \Rightarrow \Delta, \sim_f\alpha$; $\sim_f\beta, \sim_f(\alpha \rightarrow \beta), \Gamma \Rightarrow \Delta$ *is a decomposition of* $\sim_f(\alpha \rightarrow \beta), \Gamma \Rightarrow \Delta$,

69. $\sim_f\alpha, \Gamma \Rightarrow \Delta, \sim_f\neg\alpha$ *is a decomposition of* $\Gamma \Rightarrow \Delta, \sim_f\neg\alpha$,

70. $\sim_f\neg\alpha, \Gamma \Rightarrow \Delta, \sim_f\alpha$ *is a decomposition of* $\sim_f\neg\alpha, \Gamma \Rightarrow \Delta$,

71. $\Gamma \Rightarrow \Delta, \sim_f(\alpha \wedge_t \beta), \sim_f\alpha$; $\Gamma \Rightarrow \Delta, \sim_f(\alpha \wedge_t \beta), \sim_f\beta$ *is a decomposition of* $\Gamma \Rightarrow \Delta, \sim_f(\alpha \wedge_t \beta)$,

72. $\sim_f\alpha, \sim_f\beta, \sim_f(\alpha \wedge_t \beta), \Gamma \Rightarrow \Delta$ *is a decomposition of* $\sim_f(\alpha \wedge_t \beta), \Gamma \Rightarrow \Delta$,

73. $\Gamma \Rightarrow \Delta, \sim_f(\alpha \vee_t \beta), \sim_f\alpha, \sim_f\beta$ *is a decomposition of* $\Gamma \Rightarrow \Delta, \sim_f(\alpha \vee_t \beta)$,

74. $\sim_f\alpha, \sim_f(\alpha \vee_t \beta), \Gamma \Rightarrow \Delta$; $\sim_f\beta, \sim_f(\alpha \vee_t \beta), \Gamma \Rightarrow \Delta$ *is a decomposition of* $\sim_f(\alpha \vee_t \beta), \Gamma \Rightarrow \Delta$,

75. $\Gamma \Rightarrow \Delta, \sim_f(\alpha \wedge_f \beta), \sim_f\alpha$; $\Gamma \Rightarrow \Delta, \sim_f(\alpha \wedge_f \beta), \sim_f\beta$ *is a decomposition of* $\Gamma \Rightarrow \Delta, \sim_f(\alpha \wedge_f \beta)$,

76. $\sim_f\alpha, \sim_f\beta, \sim_f(\alpha \wedge_f \beta), \Gamma \Rightarrow \Delta$ *is a decomposition of* $\sim_f(\alpha \wedge_f \beta), \Gamma \Rightarrow \Delta$,

77. $\Gamma \Rightarrow \Delta, \sim_f(\alpha \vee_f \beta), \sim_f\alpha, \sim_f\beta$ *is a decomposition of* $\Gamma \Rightarrow \Delta, \sim_f(\alpha \vee_f \beta)$,

78. $\sim_f\alpha, \sim_f(\alpha \vee_f \beta), \Gamma \Rightarrow \Delta$; $\sim_f\beta, \sim_f(\alpha \vee_f \beta), \Gamma \Rightarrow \Delta$ *is a decomposition of* $\sim_f(\alpha \vee_f \beta), \Gamma \Rightarrow \Delta$,

79. $\Gamma \Rightarrow \Delta, \sim_f\forall_t x\alpha, \sim_f\alpha[z/x]$ *is a decomposition of* $\Gamma \Rightarrow \Delta, \sim_f\forall_t x\alpha$ *where* z *is a* fresh *free individual variable, i.e.,* z *is not occurring in* $\Gamma \Rightarrow \Delta, \sim_f\forall_t x\alpha$,

80. $\sim_f\alpha[y_1/x], ..., \sim_f\alpha[y_m/x], \sim_f\forall_t x\alpha, \Gamma \Rightarrow \Delta$ *is a decomposition of* $\sim_f\forall_t x\alpha, \Gamma \Rightarrow \Delta$ *where* $y_1, ..., y_m$ *are the free individual variables occurring in* $\sim_f\forall_t x\alpha, \Gamma \Rightarrow \Delta$,

81. $\Gamma \Rightarrow \Delta, \sim_f\exists_t x\alpha, \sim_f\alpha[y_1/x], ..., \sim_f\alpha[y_m/x]$ *is a decomposition of* $\Gamma \Rightarrow \Delta, \sim_f\exists_t x\alpha$ *where* $y_1, ..., y_m$ *are the free individual variables occurring in* $\Gamma \Rightarrow \Delta, \sim_f\exists_t x\alpha$,

82. $\sim_f\alpha[z/x], \sim_f\exists_t x\alpha, \Gamma \Rightarrow \Delta$ *is a decomposition of* $\sim_f\exists_t x\alpha, \Gamma \Rightarrow \Delta$ *where* z *is a fresh free individual variable,*

83. $\Gamma \Rightarrow \Delta, \sim_f\forall_f x\alpha, \sim_f\alpha[z/x]$ *is a decomposition of* $\Gamma \Rightarrow \Delta, \sim_f\forall_f x\alpha$ *where* z *is a* fresh *free individual variable, i.e.,* z *is not occurring in* $\Gamma \Rightarrow \Delta, \sim_f\forall_f x\alpha$,

84. $\sim_f\alpha[y_1/x], ..., \sim_f\alpha[y_m/x], \sim_f\forall_f x\alpha, \Gamma \Rightarrow \Delta$ *is a decomposition of* $\sim_f\forall_f x\alpha, \Gamma \Rightarrow \Delta$ *where* $y_1, ..., y_m$ *are the free individual variables occurring in* $\sim_f\forall_f x\alpha, \Gamma \Rightarrow \Delta$,

85. $\Gamma \Rightarrow \Delta, \sim_f \exists_f x\alpha, \sim_f \alpha[y_1/x], ..., \sim_f \alpha[y_m/x]$ *is a decomposition of* $\Gamma \Rightarrow \Delta, \sim_f \exists_f x\alpha$ *where* $y_1, ..., y_m$ *are the free individual variables occurring in* $\Gamma \Rightarrow \Delta, \sim_f \exists_f x\alpha$,

86. $\sim_f \alpha[z/x], \sim_f \exists_f x\alpha, \Gamma \Rightarrow \Delta$ *is a decomposition of* $\sim_f \exists_f x\alpha, \Gamma \Rightarrow \Delta$ *where* z *is a fresh free individual variable.*

Definition 3.3.8 *A* decomposition tree of S *is a tree which expresses a process of some repeated decomposition of* S.

A decomposition tree corresponds to a bottom up proof search tree of F_{16} − (cut). In every decomposition of S (i.e. S' or $S'; S''$), if S is unprovable in F_{16} − (cut), then so is S' or S''.

Lemma 3.3.9 *Let* $\Gamma \Rightarrow \Delta$ *be a given unprovable sequent in* F_{16} − (cut). *There exists an unprovable, saturated (infinite) sequent* $\Gamma^\omega \Rightarrow \Delta^\omega$ *such that* $\Gamma \subseteq \Gamma^\omega$ *and* $\Delta \subseteq \Delta^\omega$.

Proof. Let $\Gamma \Rightarrow \Delta$ be an unprovable sequent in F_{16} − (cut). We construct $\Gamma^\omega \Rightarrow \Delta^\omega$ from $\Gamma \Rightarrow \Delta$ as follows.

1. We apply the decomposition instructions from Definition 3.3.7 to $\Gamma \Rightarrow \Delta$, in the following order, but without some decomposition procedures, which are not related to the formulas in $\Gamma \Rightarrow \Delta$.

$$(1) \longrightarrow (2) \longrightarrow (3) \longrightarrow \cdots \longrightarrow (84).$$

In such a decomposition process, one of the decomposed elements S' and S'' of S is an unprovable sequent.

2. We repeat the same procedure (Step 1.), infinitely often. Then, we obtain an infinite, finitely branching decomposition tree.

3. By König's Lemma, we have an infinite path on this decomposition tree as follows:

$$\Gamma_0 \Rightarrow \Delta_0 \mid \Gamma_1 \Rightarrow \Delta_1 \mid \Gamma_2 \Rightarrow \Delta_2 \mid \cdots \infty,$$

where $\Gamma_0 \Rightarrow \Delta_0$ is $\Gamma \Rightarrow \Delta$. In this sequence of the sequents on the infinite path, we have that $\Gamma_0 \subseteq \Gamma_1 \subseteq \Gamma_2 \subseteq \cdots$ and $\Delta_0 \subseteq \Delta_1 \subseteq \Delta_2 \subseteq \cdots$.

4. We put $\Gamma^\omega := \bigcup_{i=0}^{\infty} \Gamma_i$ and $\Delta^\omega := \bigcup_{i=0}^{\infty} \Delta_i$. We note that $\Gamma^\omega \cap \Delta^\omega = \varnothing$.

Then, we have that $\Gamma \subseteq \Gamma^\omega$ and $\Delta \subseteq \Delta^\omega$, and can verify that $\Gamma^\omega \Rightarrow \Delta^\omega$ is an unprovable, saturated sequent. ∎

Lemma 3.3.10 *Let* $\Gamma \Rightarrow \Delta$ *be an unprovable sequent in* F_{16}−(cut), *and* $\Gamma^\omega \Rightarrow \Delta^\omega$ *be an unprovable, saturated sequent constructed from* $\Gamma \Rightarrow \Delta$ *by Lemma 3.3.9. We define a canonical model* $\mathcal{A} := \langle U, I^n, I^t, I^f, I^b \rangle$ *as follows:*

$$U := \{z \mid z \textit{ is a free individual variable occurring in } \Gamma^\omega \Rightarrow \Delta^\omega\},$$

$$p^{I^n} := \{(z_1, ..., z_m) \mid p(z_1, ..., z_m) \in \Gamma^\omega\},$$

$$p^{I^t} := \{(z_1, ..., z_m) \mid {\sim_t} p(z_1, ..., z_m) \in \Gamma^\omega\},$$

$$p^{I^f} := \{(z_1, ..., z_m) \mid {\sim_f} p(z_1, ..., z_m) \in \Gamma^\omega\},$$

$$p^{I^b} := \{(z_1, ..., z_m) \mid {\sim_b} p(z_1, ..., z_m) \in \Gamma^\omega\}.$$

Then, for any formula α,

1. $[(\alpha \in \Gamma^\omega$ *implies* $\mathcal{A} \models^n \underline{\alpha})$ *and* $(\alpha \in \Delta^\omega$ *implies not-*$(\mathcal{A} \models^n \underline{\alpha}))]$,
2. $[({\sim_t}\alpha \in \Gamma^\omega$ *implies* $\mathcal{A} \models^t \underline{\alpha})$ *and* $({\sim_t}\alpha \in \Delta^\omega$ *implies not-*$(\mathcal{A} \models^t \underline{\alpha}))]$,
3. $[({\sim_f}\alpha \in \Gamma^\omega$ *implies* $\mathcal{A} \models^f \underline{\alpha})$ *and* $({\sim_f}\alpha \in \Delta^\omega$ *implies not-*$(\mathcal{A} \models^f \underline{\alpha}))]$,
4. $[({\sim_b}\alpha \in \Gamma^\omega$ *implies* $\mathcal{A} \models^b \underline{\alpha})$ *and* $({\sim_b}\alpha \in \Delta^\omega$ *implies not-*$(\mathcal{A} \models^b \underline{\alpha}))]$

where $\underline{\alpha}$ *is obtained form* α *by replacing every individual variable* x *occurring in* α *by the name* $\underline{x}$.

Proof. By (simultaneous) induction on the complexity of α.

• Base step: Obvious by the definitions of I^* ($* \in \{n, t, f, b\}$).

• Induction step for (1): We show some cases.

(Case $\alpha \equiv \beta \wedge_t \gamma$): First, we show that $\beta \wedge_t \gamma \in \Gamma^\omega$ implies $\mathcal{A} \models^n \underline{\beta} \wedge_t \underline{\gamma}$. Suppose $\beta \wedge_t \gamma \in \Gamma^\omega$. Then, we obtain $[\beta \in \Gamma^\omega$ and $\gamma \in \Gamma^\omega]$ by Definition 3.3.5. By the induction hypothesis for (1), we obtain $[\mathcal{A} \models^n \underline{\beta}$ and $\mathcal{A} \models^n \underline{\gamma}]$. This means $\mathcal{A} \models^n \underline{\beta} \wedge_t \underline{\gamma}$. Second, we show that $\beta \wedge_t \gamma \in \Delta^\omega$ implies not-$(\mathcal{A} \models^n \underline{\beta} \wedge_t \underline{\gamma})$. Suppose $\beta \wedge_t \gamma \in \Delta^\omega$. Then, we obtain $[\beta \in \Delta^\omega$ or $\gamma \in \Delta^\omega]$ by Definition 3.3.5. By the induction hypothesis for (1), we obtain [not-$(\mathcal{A} \models^n \underline{\beta})$ or not-$(\mathcal{A} \models^n \underline{\gamma})$]. This means not-$(\mathcal{A} \models^n \underline{\beta} \wedge_t \underline{\gamma})$.

(Case $\alpha \equiv {\sim_t}\beta$): First, we show that ${\sim_t}\beta \in \Gamma^\omega$ implies $\mathcal{A} \models^n {\sim_t}\underline{\beta}$. Suppose ${\sim_t}\beta \in \Gamma^\omega$. Then we obtain $\mathcal{A} \models^t \underline{\beta}$ by the induction hypothesis for (2). Thus, we have $\mathcal{A} \models^t \underline{\beta}$, i.e., $\mathcal{A} \models^n {\sim_t}\underline{\beta}$. Second, we show that ${\sim_t}\beta \in \Delta^\omega$ implies not-$(\mathcal{A} \models^n {\sim_t}\underline{\beta})$. Suppose ${\sim_t}\beta \in \Delta^\omega$. Then, we obtain not-$(\mathcal{A} \models^t \underline{\beta})$ by the induction hypothesis for (2). Thus, we have not-$(\mathcal{A} \models^n {\sim_t}\underline{\beta})$.

• Induction step for (2): We show some cases.

(Case $\alpha \equiv \beta \rightarrow \gamma$): First, we show that ${\sim_t}(\beta \rightarrow \gamma) \in \Gamma^\omega$ implies $\mathcal{A} \models^t \underline{\beta} \rightarrow \underline{\gamma}$. Suppose ${\sim_t}(\beta \rightarrow \gamma) \in \Gamma^\omega$. Then, we obtain $[{\sim_t}\neg\beta \in \Gamma^\omega$ and

$\sim_t\gamma \in \Gamma^\omega$] by Definition 3.3.5. By the induction hypothesis for (2), we obtain [$\mathcal{A} \models^t \neg\underline{\beta}$ and $\mathcal{A} \models^t \underline{\gamma}$], i.e., [not-($\mathcal{A} \models^t \underline{\beta}$) and $\mathcal{A} \models^t \underline{\gamma}$]. This means $\mathcal{A} \models^t \underline{\beta} \rightarrow \underline{\gamma}$. Second, we show that $\sim_t(\beta \rightarrow \gamma) \in \Delta^\omega$ implies not-($\mathcal{A} \models^- \underline{\beta} \rightarrow \underline{\gamma}$). Suppose $\sim_t(\beta \rightarrow \gamma) \in \Delta^\omega$. Then, we obtain [$\sim_t\neg\beta \in \Delta^\omega$ or $\sim_t\gamma \in \Delta^\omega$] by Definition 3.3.5. By the induction hypothesis for (2), we obtain [not-($\mathcal{A} \models^t \neg\underline{\beta}$) or not-($\mathcal{A} \models^t \underline{\gamma}$)], i.e., [$\mathcal{A} \models^t \underline{\beta}$ or not-($\mathcal{A} \models^t \underline{\gamma}$)]. This means not-($\mathcal{A} \models^t \underline{\beta} \rightarrow \underline{\gamma}$).

(Case $\alpha \equiv \sim_t\beta$): First, we show that $\sim_t\sim_t\beta \in \Gamma^\omega$ implies $\mathcal{A} \models^t \sim_t\underline{\beta}$. Suppose $\sim_t\sim_t\beta \in \Gamma^\omega$. Then, we obtain $\beta \in \Gamma^\omega$ by Definition 3.3.5. By the induction hypothesis for (1) and $\beta \in \Gamma^\omega$, we obtain $\mathcal{A} \models^n \underline{\beta}$, and hence $\mathcal{A} \models^t \sim_t\underline{\beta}$. Second, we show that $\sim_t\sim_t\beta \in \Delta^\omega$ implies not-($\mathcal{A} \models^t \sim_t\underline{\beta}$). Suppose $\sim_t\sim_t\beta \in \Delta^\omega$. Then, we obtain $\beta \in \Delta^\omega$ by Definition 3.3.5. By the induction hypothesis for (1) and $\beta \in \Delta^\omega$, we obtain not-($\mathcal{A} \models^n \underline{\beta}$) and hence not-($\mathcal{A} \models^t \sim_t\underline{\beta}$).

(Case $\alpha \equiv \forall_t x\beta$): First, we show that $\sim_t\forall_t x\beta \in \Gamma^\omega$ implies $\mathcal{A} \models^t \forall_t x\underline{\beta}$. Suppose $\sim_t\forall_t x\beta \in \Gamma^\omega$. Then, we obtain $\sim_t\beta[z/x] \in \Gamma^\omega$ for some $z \in U$, by Definition 3.3.5. By the induction hypothesis for (2), we obtain that $\mathcal{A} \models^t \underline{\beta}[\underline{z}/x]$ for some $z \in U$. This means $\mathcal{A} \models^t \forall_t x\underline{\beta}$. Second, we show that $\sim_t\forall_t x\beta \in \Delta^\omega$ implies not-($\mathcal{A} \models^t \forall_t x\underline{\beta}$). Suppose $\sim_t\forall_t x\beta \in \Delta^\omega$. Then, we obtain [$\sim_t\beta[y_i/x] \in \Delta^\omega$ for all $y_i \in U$] by Definition 3.3.5. By the induction hypothesis for (2), we obtain not-($\mathcal{A} \models^t \underline{\beta}[\underline{y_i}/x]$) for all $y_i \in U$. This means not-($\mathcal{A} \models^t \forall_t x\underline{\beta}$).

(Case $\alpha \equiv \exists_t x\beta$): First, we show that $\sim_t\exists_t x\beta \in \Gamma^\omega$ implies $\mathcal{A} \models^t \exists_t x\underline{\beta}$. Suppose $\sim_t\exists_t x\beta \in \Gamma^\omega$. Then we obtain [$\sim_t\beta[y_i/x] \in \Gamma^\omega$ for all $y_i \in U$] by Definition 3.3.5. By the induction hypothesis, we obtain that $\mathcal{A} \models^t \underline{\beta}[\underline{y_i}/x]$ for all $y_i \in U$. This means $\mathcal{A} \models^t \exists_t x\underline{\beta}$. Second, we show that $\sim_t\exists_t x\beta \in \Delta^\omega$ implies not-($\mathcal{A} \models^t \exists_t x\underline{\beta}$). Suppose $\sim_t\exists_t x\beta \in \Delta^\omega$. Then, we obtain [$\sim_t\beta[z/x] \in \Delta^\omega$ for some $z \in U$] by Definition 3.3.5. By the induction hypothesis for (2), we obtain not-($\mathcal{A} \models^t \underline{\beta}[\underline{z}/x]$) for some $z \in U$. This means not-($\mathcal{A} \models^t \exists_t x\underline{\beta}$).

- Induction step for (3): We show some cases.

(Case $\alpha \equiv \sim_t\beta$): First, we show that $\sim_f\sim_t\beta \in \Gamma^\omega$ implies $\mathcal{A} \models^f \sim_t\underline{\beta}$. Suppose $\sim_f\sim_t\beta \in \Gamma^\omega$, i.e., $\sim_b\beta \in \Gamma^\omega$. Then we obtain $\mathcal{A} \models^b \underline{\beta}$ by the induction hypothesis for (4). Thus, we have $\mathcal{A} \models^f \sim_t\underline{\beta}$. Second, we show that $\sim_f\sim_t\beta \in \Delta^\omega$ implies not-($\mathcal{A} \models^f \sim_t\underline{\beta}$). Suppose $\sim_f\sim_t\beta \in \Delta^\omega$, i.e., $\sim_b\beta \in \Delta^\omega$. Then, we obtain not-($\mathcal{A} \models^b \underline{\beta}$) by the induction hypothesis for (4). Thus, we have not-($\mathcal{A} \models^f \sim_t\underline{\beta}$).

(Case $\alpha \equiv \sim_b\beta$): First, we show that $\sim_f\sim_b\beta \in \Gamma^\omega$ implies $\mathcal{A} \models^f \sim_b\underline{\beta}$. Suppose $\sim_f\sim_b\beta \in \Gamma^\omega$. Then, we obtain $\sim_t\beta \in \Gamma^\omega$ by Definition 3.3.5. By the induction hypothesis for (2) and $\sim_t\beta \in \Gamma^\omega$, we obtain $\mathcal{A} \models^t \underline{\beta}$, and hence $\mathcal{A} \models^f \sim_f\sim_t\underline{\beta}$, i.e., $\mathcal{A} \models^f \sim_b\underline{\beta}$. Second, we show that

$\sim_f\sim_b\beta \in \Delta^\omega$ implies not-($\mathcal{A} \models^f \sim_b\underline{\beta}$). Suppose $\sim_t\sim_b\beta \in \Delta^\omega$. Then, we obtain $\sim_t\beta \in \Delta^\omega$ by Definition 3.3.5. By the induction hypothesis for (2) and $\sim_t\beta \in \Delta^\omega$, we obtain not-($\mathcal{A} \models^t \underline{\beta}$) and hence not-($\mathcal{A} \models^f \sim_b\underline{\beta}$).

- Induction step for (4): Similar to that for (2). ∎

Theorem 3.3.11 (Strong completeness for $\mathbf{F}_{16}$) *For any sequent S, if S is valid, then* F_{16} − (cut) $\vdash S$.

Proof. We prove the following: if $\Gamma \Rightarrow \Delta$ is unprovable in F_{16} − (cut), then there exists a model $\mathcal{A}$ such that $\Gamma \Rightarrow \Delta$ is not valid in $\mathcal{A}$. Suppose that $\Gamma \Rightarrow \Delta$ is not provable in F_{16} − (cut). Then, by Lemma 3.3.10, we can construct a canonical model $\mathcal{A}$ satisfying the condition (1) in this lemma. Thus, we have $\mathcal{A} \models^n \underline{\gamma}$ and not-($\mathcal{A} \models^n \underline{\delta}$) for any $\gamma \in \Gamma \subseteq \Gamma^\omega$ and any $\delta \in \Delta \subseteq \Delta^\omega$. Hence, we obtain "not-($\mathcal{A} \models^n \underline{\Gamma}_*{\rightarrow}\underline{\Delta}^*$)," and hence "not-($\mathcal{A} \models^n cl(\Gamma_*{\rightarrow}\Delta^*)$)." Therefore, $\Gamma \Rightarrow \Delta$ is not valid in $\mathcal{A}$. ∎

Theorem 3.3.12 (Cut-elimination for $\mathbf{F}_{16}$) *The rule* (cut) *is admissible in cut-free* F_{16}.

Proof. By combining Theorem 3.3.11 and Theorem 3.3.4. ∎

4. Generalized paraconsistent logics

New propositional and first-order paraconsistent logics (called L_ω and FL_ω, respectively) are introduced as Gentzen-type sequent calculi with classical and paraconsistent negations. Embedding theorems of L_ω and FL_ω into propositional (first-order, respectively) classical logic are shown, and completeness theorems with respect to simple semantics for L_ω and FL_ω are proved. The cut-elimination theorems for L_ω and FL_ω are shown using both syntactical ways via the embedding theorems and semantical ways via the completeness theorems.[1]

4.1. Introduction

New propositional and first-order paraconsistent logics (called L_ω and FL_ω, respectively) are introduced as Gentzen-type sequent calculi with classical and paraconsistent negations. The aim of introducing these logics is to give a proof-theoretic framework combining classical and paraconsistent negations. The proposed paraconsistent negations are intended to (partially) include or simulate the negation connectives of the well-known "useful" many-valued paraconsistent logics: Belnap and Dunn's four-valued logic B4 [21, 47] (presented by four-valued truth tables), first-degree entailment FDE [7] (i.e., B4 presented as a "first-degree" proof system, and referred to in Chapter 2 as E_{fde}), Nelson's paraconsistent logic N4 [6], and Shramko-Wansing's 16-valued logic [180].

The "useful" many-valued paraconsistent logics arise from the objective of generalizing truth values by many-valued truth functions based on multilattices including bilattices and trilattices [12, 179, 181]. With such aims, logics based on multilattices have been studied by many researchers (see, e.g., [12, 62, 179, 180, 181] and the references therein). These multilattice-based logics are considered to be natural extensions of B4, FDE and N4. For example, Shramko-Wansing's trilattice-based 16-valued logic $FDE^{t+\sim_f}$ [180] may be regarded as an extension of FDE and B4.

[1]This chapter contains the material of [95].

Gentzen-type proof systems for some bilattice- and trilattice-based logics have been studied by several researchers. For example, sequent calculi for the logics of logical bilattices were studied by Arieli and Avron [12], sequent calculi for some bilattice-based logics, which are natural extensions of N4, were investigated and surveyed by Gargov [62], some substructural versions of these systems were studied by Kamide [88], and a cut-free sequent calculus L_{16} that includes Shramko-Wansing's $FDE^{t+\sim_f}$ was introduced by Kamide [89]. Since $FDE^{t+\sim_f}$ has both negation and involution operators, L16 needed a slightly complicated formalization to obtain a cut-free system.

In the present chapter, a Gentzen-type framework combining classical and paraconsistent negations is presented, introducing a new negation operator, and such a new operator can simultaneously represent both the paraconsistent negation and involution operators. The uncertainty level of the truth (or falsehood) of a proposition can be represented by a nesting of the new negation operator. In a sense, the proposed paraconsistent negations (and involutions) are regarded as the boundary between negation and modality: the negations behave like modal operators, but admit a kind of de Morgan laws. The proposed logics, which have two different (i.e., classical and paraconsistent) interacting negations, can also be developed in the direction of the recent field of combination of logics.

The contents of this chapter can be summarized as follows. In Section 4.2, the propositional paraconsistent logic L_ω is discussed. First, L_ω is introduced as a Gentzen-type sequent calculus by extending a sequent calculus LK for propositional classical logic. Second, an embedding theorem of L_ω into LK is shown, and by using this theorem, the cut-elimination theorem, the decidability and the paraconsistency (w.r.t. new negation expressions) are shown for L_ω. Third, a simple semantics for L_ω is introduced, and the completeness theorem w.r.t. this semantics is proved using Maehara's decomposition method that can simultaneously derive the cut-elimination theorem.

In Section 4.3, the first-order predicate version FL_ω of L_ω is discussed. The first-order versions of the embedding and cut-elimination theorems are shown for FL_ω. An extended version of the semantics for L_ω is introduced for FL_ω, and the completeness theorem w.r.t. this semantics is proved using Schütte's method that can simultaneously derive the cut-elimination theorem.

In Section 4.4, some remarks on finite-valued versions are given. Although L_ω and FL_ω may be regarded as a kind of infinite-valued logics, some finite-valued versions are proposed by using a cyclic valuation condition.

4.2. Propositional case

4.2.1. Sequent calculus

The following list of symbols is adopted for the language of the underlying logic: (countable) propositional variables (or atomic formulas) $p_0, p_1, ...$, logical connectives $\to$ (implication), $\wedge$ (conjunction), $\vee$ (disjunction), $\neg$ (classical negation) and $\sim$ (paraconsistent negation). Greek lower-case letters $\alpha, \beta, ...$ are used to denote formulas, and Greek capital letters $\Gamma, \Delta, ...$ are used to represent finite (possibly empty) sets of formulas. We write $A \equiv B$ to indicate the syntactical identity between A and B. The symbol ω is used to represent the set of natural numbers. The symbols ω_e and ω_o are used to represent $\{i \in \omega \mid i \text{ is even}\}$ and $\{i \in \omega \mid i \text{ is odd}\}$, respectively. An expression ${\sim}^i\alpha$ for any $i \in \omega$ is used to denote $\overbrace{\sim\sim\cdots\sim}^{i}\alpha$, which is defined inductively by $({\sim}^0\alpha := \alpha)$ and $({\sim}^{n+1}\alpha := \sim{\sim}^n\alpha)$. Lower-case letters i, j and k are used to denote any natural numbers. An expression ${\sim}^i\Gamma$ is used to denote the set $\{{\sim}^i\gamma \mid \gamma \in \Gamma\}$. An expression of the form $\Gamma \Rightarrow \Delta$ is called a *sequent*. An expression $L \vdash S$ is used to denote the fact that a sequent S is provable in a sequent calculus L. A rule R of inference is said to be *admissible* in a sequent calculus L if the following condition is satisfied: for any instance

$$\frac{S_1 \quad \cdots \quad S_n}{S}$$

of R, if $L \vdash S_i$ for all i, then $L \vdash S$.

Definition 4.2.1 ($\mathbf{L}_\omega$) *The initial sequents of* L_ω *are of the form: for any atomic formula p and any $i \in \omega$,*

$${\sim}^i p \Rightarrow {\sim}^i p.$$

The structural inference rules of L_ω *are of the form:*

$$\frac{\Gamma \Rightarrow \Delta, \alpha \quad \alpha, \Sigma \Rightarrow \Pi}{\Gamma, \Sigma \Rightarrow \Delta, \Pi}\ (\text{cut})$$

$$\frac{\Gamma \Rightarrow \Delta}{\alpha, \Gamma \Rightarrow \Delta}\ (\text{w-left}) \qquad \frac{\Gamma \Rightarrow \Delta}{\Gamma \Rightarrow \Delta, \alpha}\ (\text{w-right}).$$

The even *logical inference rules of* L_ω *are of the form: for any $i \in \omega_e$,*

$$\frac{\Gamma \Rightarrow \Sigma, {\sim}^i\alpha \quad {\sim}^i\beta, \Delta \Rightarrow \Pi}{{\sim}^i(\alpha\to\beta), \Gamma, \Delta \Rightarrow \Sigma, \Pi}\ (\to\text{left}^e) \qquad \frac{{\sim}^i\alpha, \Gamma \Rightarrow \Delta, {\sim}^i\beta}{\Gamma \Rightarrow \Delta, {\sim}^i(\alpha\to\beta)}\ (\to\text{right}^e)$$

$$\frac{{\sim}^i\alpha, {\sim}^i\beta, \Gamma \Rightarrow \Delta}{{\sim}^i(\alpha \wedge \beta), \Gamma \Rightarrow \Delta}\ (\wedge\text{left}^e) \qquad \frac{\Gamma \Rightarrow \Delta, {\sim}^i\alpha \quad \Gamma \Rightarrow \Delta, {\sim}^i\beta}{\Gamma \Rightarrow \Delta, {\sim}^i(\alpha \wedge \beta)}\ (\wedge\text{right}^e)$$

$$\frac{{\sim}^i\alpha, \Gamma \Rightarrow \Delta \quad {\sim}^i\beta, \Gamma \Rightarrow \Delta}{{\sim}^i(\alpha \vee \beta), \Gamma \Rightarrow \Delta}\ (\vee\text{left}^e) \qquad \frac{\Gamma \Rightarrow \Delta, {\sim}^i\alpha, {\sim}^i\beta}{\Gamma \Rightarrow \Delta, {\sim}^i(\alpha \vee \beta)}\ (\vee\text{right}^e).$$

The odd *logical inference rules of* L_ω *are of the form: for any* $j \in \omega_o$,

$$\frac{{\sim}^{j-1}\alpha, {\sim}^j\beta, \Gamma \Rightarrow \Delta}{{\sim}^j(\alpha{\rightarrow}\beta), \Gamma \Rightarrow \Delta}\ ({\rightarrow}\text{left}^o) \qquad \frac{\Gamma \Rightarrow \Delta, {\sim}^{j-1}\alpha \quad \Gamma \Rightarrow \Delta, {\sim}^j\beta}{\Gamma \Rightarrow \Delta, {\sim}^j(\alpha{\rightarrow}\beta)}\ ({\rightarrow}\text{right}^o)$$

$$\frac{{\sim}^j\alpha, \Gamma \Rightarrow \Delta \quad {\sim}^j\beta, \Gamma \Rightarrow \Delta}{{\sim}^j(\alpha \wedge \beta), \Gamma \Rightarrow \Delta}\ (\wedge\text{left}^o) \qquad \frac{\Gamma \Rightarrow \Delta, {\sim}^j\alpha, {\sim}^j\beta}{\Gamma \Rightarrow \Delta, {\sim}^j(\alpha \wedge \beta)}\ (\wedge\text{right}^o)$$

$$\frac{{\sim}^j\alpha, {\sim}^j\beta, \Gamma \Rightarrow \Delta}{{\sim}^j(\alpha \vee \beta), \Gamma \Rightarrow \Delta}\ (\vee\text{left}^o) \qquad \frac{\Gamma \Rightarrow \Delta, {\sim}^j\alpha \quad \Gamma \Rightarrow \Delta, {\sim}^j\beta}{\Gamma \Rightarrow \Delta, {\sim}^j(\alpha \vee \beta)}\ (\vee\text{right}^o).$$

The even-odd *logical inference rules of* L_ω *are of the form: for any* $i \in \omega$,

$$\frac{\Gamma \Rightarrow \Delta, {\sim}^i\alpha}{{\sim}^i(\neg\alpha), \Gamma \Rightarrow \Delta}\ (\neg\text{left}^{eo}) \qquad \frac{{\sim}^i\alpha, \Gamma \Rightarrow \Delta}{\Gamma \Rightarrow \Delta, {\sim}^i(\neg\alpha)}\ (\neg\text{right}^{eo}).$$

The sequents of the form ${\sim}^i\alpha \Rightarrow {\sim}^i\alpha$ for any formula α and any $i \in \omega$ are provable in cut-free L_ω. This fact can be proved by induction on the complexity of α. Hence, these sequents can also be regarded as the initial sequents of L_ω.

The $\neg$-less fragment of L_ω with both $i = 0$ and $j = 1$ implies an extension of a sequent system for Nelson's 4-valued logic N4 [6] without the double-negation-elimination axiom for $\sim$. Also, the $\{\rightarrow, \neg\}$-less fragment of L_ω with both $i = 0$ and $j = 1$ is a sequent system for Belnap and Dunn's 4-valued logic B4 [21, 47] without the double-negation-elimination axiom for $\sim$. For a detailed explanation for sequent calculi for N4 and B4, see Chapter 2.

The following propositions show that the expressions ${\sim}^i$ (i: even) and ${\sim}^j$ (j: odd) are regarded as an involution-like connective and a negation-like connective, respectively.

Proposition 4.2.2 *The rule*

$$\frac{\Gamma \Rightarrow \Delta}{{\sim}^i\Gamma \Rightarrow {\sim}^i\Delta}\ (\text{regu})$$

for any positive integer $i \in \omega_e$ *is admissible in cut-free* L_ω.

Proof. By induction on the length of the proof P of $\Gamma \Rightarrow \Delta$ in cut-free L_ω. We distinguish the cases according to the last inference of P. We show only the following case.

Case ($\to$left^e): The last inference of P is of the form:

$$\frac{\Gamma_1 \Rightarrow \Delta_1, {\sim}^k\alpha \quad {\sim}^k\beta, \Gamma_2 \Rightarrow \Delta_2}{{\sim}^k(\alpha \to \beta), \Gamma_1, \Gamma_2 \Rightarrow \Delta_1, \Delta_2}\ (\to\text{left}^e)$$

where $k \in \omega_e$, $\Gamma \equiv {\sim}^k(\alpha\to\beta), \Gamma_1, \Gamma_2$ and $\Delta \equiv \Delta_1, \Delta_2$. By induction hypothesis, L_ω - (cut) $\vdash {\sim}^i\Gamma_1 \Rightarrow {\sim}^i\Delta_1, {\sim}^{i+k}\alpha$ and L_ω - (cut) $\vdash {\sim}^{i+k}\beta, {\sim}^i\Gamma_2 \Rightarrow {\sim}^i\Delta_2$. Since $i+k$ is in ω_e, we can apply ($\to$left^e), and hence obtain the required fact:

$$\frac{\begin{array}{c}\vdots \\ {\sim}^i\Gamma_1 \Rightarrow {\sim}^i\Delta_1, {\sim}^{i+k}\alpha\end{array} \quad \begin{array}{c}\vdots \\ {\sim}^{i+k}\beta, {\sim}^i\Gamma_2 \Rightarrow {\sim}^i\Delta_2\end{array}}{{\sim}^{i+k}(\alpha\to\beta), {\sim}^i\Gamma_1, {\sim}^i\Gamma_2 \Rightarrow {\sim}^i\Delta_1, {\sim}^i\Delta_2}\ (\to\text{left}^e).$$

∎

Proposition 4.2.3 *Let Γ and Δ be sets of $\{\to, \neg\}$-less formulas. Then, the rule*

$$\frac{\Gamma \Rightarrow \Delta}{{\sim}^j\Delta \Rightarrow {\sim}^j\Gamma}\ (\text{twist})$$

for any $j \in \omega_o$ is admissible in cut-free L_ω.

Proof. By induction on the length of the proof P of $\Gamma \Rightarrow \Delta$ in cut-free L_ω. We distinguish the cases according to the last inference of P. We do not need to consider the cases for the inference rules concerning $\to$ and $\neg$. We show only the following case.

Case ($\wedge$left^o): The last inference of P is of the form:

$$\frac{{\sim}^k\alpha, \Gamma_1 \Rightarrow \Delta \quad {\sim}^k\beta, \Gamma_1 \Rightarrow \Delta}{{\sim}^k(\alpha\wedge\beta), \Gamma_1 \Rightarrow \Delta}\ (\wedge\text{left}^o)$$

where $k \in \omega_o$ and $\Gamma \equiv {\sim}^k(\alpha\wedge\beta), \Gamma_1$. By induction hypothesis, we have L_ω – (cut) $\vdash {\sim}^j\Delta \Rightarrow {\sim}^j\Gamma_1, {\sim}^{j+k}\alpha$ and L_ω – (cut) $\vdash {\sim}^j\Delta \Rightarrow {\sim}^j\Gamma_1, {\sim}^{j+k}\beta$. Since $j+k$ is in ω_e, we can apply ($\wedge$right^e), and hence obtain the required fact:

$$\frac{\begin{array}{c}\vdots \\ {\sim}^j\Delta \Rightarrow {\sim}^j\Gamma_1, {\sim}^{j+k}\alpha\end{array} \quad \begin{array}{c}\vdots \\ {\sim}^j\Delta \Rightarrow {\sim}^j\Gamma_1, {\sim}^{j+k}\beta\end{array}}{{\sim}^j\Delta \Rightarrow {\sim}^j\Gamma_1, {\sim}^{j+k}(\alpha\wedge\beta)}\ (\wedge\text{right}^e).$$

∎

Proposition 4.2.4 *An expression $\alpha \Leftrightarrow \beta$ is an abbreviation for the pair of sequents $\alpha \Rightarrow \beta$ and $\beta \Rightarrow \alpha$. Then, the following sequents are provable in L_ω: for any formulas α, β, any $i \in \omega_e$, any $j \in \omega_o$ and any $k \in \omega$,*

1. $\sim^i(\alpha \circ \beta) \Leftrightarrow \sim^i\alpha \circ \sim^i\beta$ *where* $\circ \in \{\rightarrow, \wedge, \vee\}$,
2. $\sim^j(\alpha \rightarrow \beta) \Leftrightarrow \sim^{j-1}\alpha \wedge \sim^j\beta$ *(esp.,* $\sim(\alpha\rightarrow\beta) \Leftrightarrow \alpha \wedge \sim\beta$*)*,
3. $\sim^j(\alpha \wedge \beta) \Leftrightarrow \sim^j\alpha \vee \sim^j\beta$,
4. $\sim^j(\alpha \vee \beta) \Leftrightarrow \sim^j\alpha \wedge \sim^j\beta$,
5. $\sim^k(\neg\alpha) \Leftrightarrow \neg(\sim^k\alpha)$.

Proof. We show some cases.

(2):

$$\dfrac{\dfrac{\dfrac{\sim^{j-1}\alpha \Rightarrow \sim^{j-1}\alpha}{\sim^{j-1}\alpha, \sim^j\beta \Rightarrow \sim^{j-1}\alpha}\ (\text{w-left}) \qquad \dfrac{\sim^j\beta \Rightarrow \sim^j\beta}{\sim^{j-1}\alpha, \sim^j\beta \Rightarrow \sim^j\beta}\ (\text{w-left})}{\sim^{j-1}\alpha, \sim^j\beta \Rightarrow \sim^{j-1}\alpha \wedge \sim^j\beta}\ (\wedge\text{right}^e)}{\sim^j(\alpha\rightarrow\beta) \Rightarrow \sim^{j-1}\alpha \wedge \sim^j\beta}\ (\rightarrow\text{left}^o).$$

$$\dfrac{\dfrac{\dfrac{\sim^{j-1}\alpha \Rightarrow \sim^{j-1}\alpha}{\sim^{j-1}\alpha, \sim^j\beta \Rightarrow \sim^{j-1}\alpha}\ (\text{w-left}) \qquad \dfrac{\sim^j\beta \Rightarrow \sim^j\beta}{\sim^{j-1}\alpha, \sim^j\beta \Rightarrow \sim^j\beta}\ (\text{w-left})}{\sim^{j-1}\alpha, \sim^j\beta \Rightarrow \sim^j(\alpha\rightarrow\beta)}\ (\rightarrow\text{right}^o)}{\sim^{j-1}\alpha \wedge \sim^j\beta \Rightarrow \sim^j(\alpha\rightarrow\beta)}\ (\wedge\text{left}^e).$$

(3):

$$\dfrac{\dfrac{\dfrac{\sim^j\alpha \Rightarrow \sim^j\alpha}{\sim^j\alpha \Rightarrow \sim^j\alpha, \sim^j\beta}\ (\text{w-right}) \qquad \dfrac{\sim^j\beta \Rightarrow \sim^j\beta}{\sim^j\beta \Rightarrow \sim^j\alpha, \sim^j\beta}\ (\text{w-right})}{\sim^j(\alpha \wedge \beta) \Rightarrow \sim^j\alpha, \sim^j\beta}\ (\wedge\text{left}^o)}{\sim^j(\alpha \wedge \beta) \Rightarrow \sim^j\alpha \vee \sim^j\beta}\ (\vee\text{right}^e).$$

$$\dfrac{\dfrac{\dfrac{\sim^j\alpha \Rightarrow \sim^j\alpha}{\sim^j\alpha \Rightarrow \sim^j\alpha, \sim^j\beta}\ (\text{w-right})}{\sim^j\alpha \Rightarrow \sim^j(\alpha \wedge \beta)}\ (\wedge\text{right}^o) \qquad \dfrac{\dfrac{\sim^j\beta \Rightarrow \sim^j\beta}{\sim^j\beta \Rightarrow \sim^j\alpha, \sim^j\beta}\ (\text{w-right})}{\sim^j\beta \Rightarrow \sim^j(\alpha \wedge \beta)}\ (\wedge\text{right}^o)}{\sim^j\alpha \vee \sim^j\beta \Rightarrow \sim^j(\alpha \wedge \beta)}\ (\vee\text{left}^e).$$

(5):

$$\dfrac{\dfrac{\sim^k\alpha \Rightarrow \sim^k\alpha}{\Rightarrow \neg(\sim^k\alpha), \sim^k\alpha}\ (\neg\text{right}^{eo})}{\sim^k(\neg\alpha) \Rightarrow \neg(\sim^k\alpha)}\ (\neg\text{left}^{eo}) \qquad \dfrac{\dfrac{\sim^k\alpha \Rightarrow \sim^k\alpha}{\neg(\sim^k\alpha), \sim^k\alpha \Rightarrow}\ (\neg\text{left}^{eo})}{\neg(\sim^k\alpha) \Rightarrow \sim^k(\neg\alpha)}\ (\neg\text{right}^{eo}).$$

■

Definition 4.2.5 (LK) *The sequent calculus* LK *for (propositional) classical logic is obtained from* L_ω *by deleting the odd logical inference rules and replacing* i *in the initial sequents and the even and even-odd logical inference rules by* 0 *(i.e., deleting every occurrence of* $\sim$*). The modified inference rules for* LK *by replacing* i *by* 0 *are denoted by deleting the superscript "e" or "eo", e.g.,* ($\to$left) *and* ($\neg$right).

As is well-known, LK admits cut-elimination.

4.2.2. Embedding and cut-elimination

Definition 4.2.6 *Let* $\Phi := \{p, q, r, ...\}$ *be a fixed countable non-empty set of atomic formulas. Then, we define the sets* $\Phi_0 := \Phi$ *and* $\Phi_i := \{p_i \mid p \in \Phi\}$ $(1 \leq i \in \omega)$ *of atomic formulas. The language* $\mathcal{L}_{L_\omega}$ *of* L_ω *is defined using* $\Phi, \to, \wedge, \vee, \neg$ *and* $\sim$*. The language* $\mathcal{L}_{LK}$ *of* LK *is defined using* $\bigcup_{i\in\omega} \Phi_i$, $\to$, $\wedge$, $\vee$ *and* $\neg$.

A mapping f *from* $\mathcal{L}_{L_\omega}$ *to* $\mathcal{L}_{LK}$ *is defined as follows.*

1. $f({\sim}^i p) := p_i \in \Phi_i$ *for each* $p \in \Phi$ *and each* $i \in \omega$ *(esp.,* $f(p) := p \in \Phi$*),*
2. $f({\sim}^i(\neg\alpha)) := \neg f({\sim}^i\alpha)$ *for each* $i \in \omega$,
3. $f({\sim}^i(\alpha \circ \beta)) := f({\sim}^i\alpha) \circ f({\sim}^i\beta)$ $(\circ \in \{\to, \wedge, \vee\})$ *for each* $i \in \omega_e$,
4. $f({\sim}^j(\alpha{\to}\beta)) := f({\sim}^{j-1}\alpha) \wedge f({\sim}^j\beta)$ *for each* $j \in \omega_o$,
5. $f({\sim}^j(\alpha \wedge \beta)) := f({\sim}^j\alpha) \vee f({\sim}^j\beta)$ *for each* $j \in \omega_o$,
6. $f({\sim}^j(\alpha \vee \beta)) := f({\sim}^j\alpha) \wedge f({\sim}^j\beta)$ *for each* $j \in \omega_o$.

An expression $f(\Gamma)$ denotes the result of replacing every occurrence of a formula α in Γ by an occurrence of $f(\alpha)$.

Theorem 4.2.7 *Let* Γ *and* Δ *be sets of formulas in* $\mathcal{L}_{L_\omega}$*, and* f *be the mapping defined in Definition 4.2.6. Then:*

1. $L_\omega \vdash \Gamma \Rightarrow \Delta$ *iff* $\text{LK} \vdash f(\Gamma) \Rightarrow f(\Delta)$.
2. $L_\omega - (\text{cut}) \vdash \Gamma \Rightarrow \Delta$ *iff* $\text{LK} - (\text{cut}) \vdash f(\Gamma) \Rightarrow f(\Delta)$.

Proof. (2) immediately follows from (1). Thus, we only examine (1).

(Left-to-right): By induction on the length of the proof P of $\Gamma \Rightarrow \Delta$ in L_ω. We distinguish the cases according to the last inference of P. We only show the following cases.

Case ($\sim^i p \Rightarrow \sim^i p$): The last inference of P is of the form: $\sim^i p \Rightarrow \sim^i p$. In this case, we obtain $f(\sim^i p) \Rightarrow f(\sim^i p)$, i.e., $p_i \Rightarrow p_i$ ($p_i \in \Phi_i$), which is an initial sequent of LK.

Case ($\rightarrow$left^e): The last inference of P is of the form:

$$\frac{\Gamma_1 \Rightarrow \Delta_1, \sim^i\alpha \quad \sim^i\beta, \Gamma_2 \Rightarrow \Delta_2}{\sim^i(\alpha{\rightarrow}\beta), \Gamma_1, \Gamma_2 \Rightarrow \Delta_1, \Delta_2}\ (\rightarrow\text{left}^e).$$

By induction hypothesis, we have LK $\vdash f(\Gamma_1) \Rightarrow f(\Delta_1), f(\sim^i\alpha)$ and LK $\vdash f(\sim^i\beta), f(\Gamma_2) \Rightarrow f(\Delta_2)$. Then, we obtain

$$\frac{\begin{matrix}\vdots \\ f(\Gamma_1) \Rightarrow f(\Delta_1), f(\sim^i\alpha)\end{matrix} \quad \begin{matrix}\vdots \\ f(\sim^i\beta), f(\Gamma_2) \Rightarrow f(\Delta_2)\end{matrix}}{f(\sim^i\alpha){\rightarrow}f(\sim^i\beta), f(\Gamma_1), f(\Gamma_2) \Rightarrow f(\Delta_1), f(\Delta_2)}\ (\rightarrow\text{left})$$

where $f(\sim^i\alpha){\rightarrow}f(\sim^i\beta)$ coincides with $f(\sim^i(\alpha{\rightarrow}\beta))$ by the definition of f.

Case ($\rightarrow$right^o): The last inference of P is of the form:

$$\frac{\Gamma \Rightarrow \Delta', \sim^{j-1}\alpha \quad \Gamma \Rightarrow \Delta', \sim^j\beta}{\Gamma \Rightarrow \Delta', \sim^j(\alpha{\rightarrow}\beta)}\ (\rightarrow\text{right}^o).$$

By induction hypothesis, we have LK $\vdash f(\Gamma) \Rightarrow f(\Delta'), f(\sim^{j-1}\alpha)$ and LK $\vdash f(\Gamma) \Rightarrow f(\Delta'), f(\sim^j\beta)$. Then, we obtain

$$\frac{\begin{matrix}\vdots \\ f(\Gamma) \Rightarrow f(\Delta'), f(\sim^{j-1}\alpha)\end{matrix} \quad \begin{matrix}\vdots \\ f(\Gamma) \Rightarrow f(\Delta'), f(\sim^j\beta)\end{matrix}}{f(\Gamma) \Rightarrow f(\Delta'), f(\sim^{j-1}\alpha) \wedge f(\sim^j\beta)}\ (\wedge\text{right})$$

where $f(\sim^{j-1}\alpha) \wedge f(\sim^j\beta)$ coincides with $f(\sim^j(\alpha{\rightarrow}\beta))$ by the definition of f.

(Right-to-left): By induction on the length of the proof Q of $f(\Gamma) \Rightarrow f(\Delta)$ in LK. We distinguish the cases according to the last inference of Q, and show only the case ($\wedge$left).

Subcase (1): The last inference of Q is of the form:

$$\frac{f(\sim^{j-1}\alpha), f(\sim^j\beta), f(\Gamma') \Rightarrow f(\Delta)}{f(\sim^{j-1}\alpha) \wedge f(\sim^j\beta), f(\Gamma') \Rightarrow f(\Delta)}\ (\wedge\text{left})$$

where $f(\sim^{j-1}\alpha) \wedge f(\sim^j\beta)$ coincides with $f(\sim^j(\alpha{\rightarrow}\beta))$ by the definition of f. By induction hypothesis, we have $\mathrm{L}_\omega \vdash \sim^{j-1}\alpha, \sim^j\beta, \Gamma' \Rightarrow \Delta$, and hence obtain:

$$\frac{\begin{matrix}\vdots \\ \sim^{j-1}\alpha, \sim^j\beta, \Gamma' \Rightarrow \Delta\end{matrix}}{\sim^j(\alpha{\rightarrow}\beta), \Gamma' \Rightarrow \Delta}\ (\rightarrow\text{left}^o).$$

Subcase (2): The last inference of Q is of the form:

$$\frac{f(\sim^j\alpha), f(\sim^j\beta), f(\Gamma') \Rightarrow f(\Delta)}{f(\sim^j\alpha) \wedge f(\sim^j\beta), f(\Gamma') \Rightarrow f(\Delta)} \ (\wedge\text{left})$$

where $f(\sim^j\alpha) \wedge f(\sim^j\beta)$ coincides with $f(\sim^j(\alpha \vee \beta))$ by the definition of f. By induction hypothesis, we have $\mathrm{L}_\omega \vdash \sim^j\alpha, \sim^j\beta, \Gamma' \Rightarrow \Delta$, and hence obtain:

$$\frac{\begin{array}{c}\vdots\\ \sim^j\alpha, \sim^j\beta, \Gamma' \Rightarrow \Delta\end{array}}{\sim^j(\alpha \vee \beta), \Gamma' \Rightarrow \Delta} \ (\vee\text{left}^o).$$

Subcase (3): The last inference of Q is of the form:

$$\frac{f(\sim^i\alpha), f(\sim^i\beta), f(\Gamma') \Rightarrow f(\Delta)}{f(\sim^i\alpha) \wedge f(\sim^i\beta), f(\Gamma') \Rightarrow f(\Delta)} \ (\wedge\text{left})$$

where $f(\sim^i\alpha) \wedge f(\sim^i\beta)$ coincides with $f(\sim^i(\alpha \wedge \beta))$ by the definition of f. By induction hypothesis, we have $\mathrm{L}_\omega \vdash \sim^i\alpha, \sim^i\beta, \Gamma' \Rightarrow \Delta$, and hence obtain:

$$\frac{\begin{array}{c}\vdots\\ \sim^i\alpha, \sim^i\beta, \Gamma' \Rightarrow \Delta\end{array}}{\sim^i(\alpha \wedge \beta), \Gamma' \Rightarrow \Delta} \ (\wedge\text{left}^e).$$

∎

Using Theorem 4.2.7, we can obtain the following theorems.

Theorem 4.2.8 *The rule* (cut) *is admissible in cut-free* L_ω.

Proof. Suppose $\mathrm{L}_\omega \vdash \Gamma \Rightarrow \Delta$. Then, we have $\mathrm{LK} \vdash f(\Gamma) \Rightarrow f(\Delta)$ by Theorem 4.2.7 (1), and hence $\mathrm{LK} - (\mathrm{cut}) \vdash f(\Gamma) \Rightarrow f(\Delta)$ by the cut-elimination property of LK. By Theorem 4.2.7 (2), we obtain $\mathrm{L}_\omega - (\mathrm{cut}) \vdash \Gamma \Rightarrow \Delta$. ∎

Corollary 4.2.9 L_ω *is decidable.*

Proof. By decidability of LK, for each α, it is possible to decide if $f(\alpha)$ is LK-provable. Then, by Theorem 4.2.7, L_ω is decidable. ∎

Definition 4.2.10 *Let* $\sharp$ *be a unary connective. A sequent calculus* L *is called* explosive *with respect to* $\sharp$ *if each pair of formulas* α *and* β*, the sequent* $\alpha, \sharp\alpha \Rightarrow \beta$ *is provable in* L*. It is called* paraconsistent *with respect to* $\sharp$ *if it is not explosive with respect to* $\sharp$.

Proposition 4.2.11 *Let $\sharp$ be $\sim^i$ ($i \in \omega_e$) or $\sim^j$ ($j \in \omega_o$). Then, L_ω is paraconsistent with respect to $\sharp$.*

Proof. Consider a sequent $p, \sharp p \Rightarrow q$ where p and q are distinct atomic formulas. Then, the unprovability of this sequent is guaranteed by using Theorem 4.2.8. ∎

Note that L_ω is explosive with respect to $\neg$, since it includes classical logic.

4.2.3. Semantics and completeness

We observe that for any formulas $\alpha_1, ..., \alpha_m, \beta_1, ..., \beta_n$, the sequent $\alpha_1, ..., \alpha_m \Rightarrow \beta_1, ..., \beta_n$ is provable in L_ω if and only if so is $\alpha_1 \wedge \cdots \wedge \alpha_m \Rightarrow \beta_1 \vee \cdots \vee \beta_n$. Let Γ be a set $\{\alpha_1, ..., \alpha_m\}$ ($m \geq 0$). Then, Γ^* means $\alpha_1 \vee \cdots \vee \alpha_m$ if $m \geq 1$, and otherwise $\neg(p \rightarrow p)$ where p is a fixed atomic formula. Also Γ_* means $\alpha_1 \wedge \cdots \wedge \alpha_m$ if $m \geq 1$, and otherwise $p \rightarrow p$ where p is a fixed atomic formula.

Definition 4.2.12 *Valuations v_i, $i \in \omega$, are mappings from the set of all atomic formulas to the set of truth values $\{t, f\}$, such that for each i, the truth value of any propositional letter at the level i is assigned. Each v_i is extended to a mapping from the set of all formulas to $\{t, f\}$ by the following prescriptions:*

1. *for each $i \in \omega$, $v_i(\sim\alpha) = t$ iff $v_{i+1}(\alpha) = t$,*
2. *for each $i \in \omega$, $v_i(\neg\alpha) = t$ iff $v_i(\alpha) = f$,*
3. *for each $i \in \omega_e$, $v_i(\alpha \wedge \beta) = t$ iff $v_i(\alpha) = t$ and $v_i(\beta) = t$,*
4. *for each $i \in \omega_e$, $v_i(\alpha \vee \beta) = t$ iff $v_i(\alpha) = t$ or $v_i(\beta) = t$,*
5. *for each $i \in \omega_e$, $v_i(\alpha \rightarrow \beta) = t$ iff $v_i(\alpha) = f$ or $v_i(\beta) = t$,*
6. *for any $j \in \omega_o$, $v_j(\alpha \wedge \beta) = t$ iff $v_j(\alpha) = t$ or $v_j(\beta) = t$,*
7. *for each $j \in \omega_o$, $v_j(\alpha \vee \beta) = t$ iff $v_j(\alpha) = t$ and $v_j(\beta) = t$,*
8. *for each $j \in \omega_o$, $v_j(\alpha \rightarrow \beta) = t$ iff $v_{j-1}(\alpha) = t$ and $v_j(\beta) = t$.*

An assignment *is any set $V_k := \{v_r^k \mid v_r^k : \text{valuation}, r \in \omega\}$. A formula α is called* tautology *if for each assignment V_k, we have that $v_0^k(\alpha) = t$ holds for any valuations v_i^k ($i \in \omega$). A sequent of the form $\Gamma \Rightarrow \Delta$ is called tautology if so is the formula $\Gamma_* \rightarrow \Delta^*$.*

The intended meanings of $v_i(\alpha) = t$ (i: even) and $v_j(\alpha) = t$ (j: odd) may be "α is verified (or accepted) at the uncertainty level i" and "α is refuted (or falsified) at the uncertainty level j", respectively. Also, the intended meanings of $v_i(\alpha) = f$ (i: even) and $v_j(\alpha) = f$ (j: odd) may be "α is *not* verified (or accepted) at the uncertainty level i" and "α is *not* refuted (or falsified) at the uncertainty level j", respectively. The "uncertainty level" means the uncertainty of the truth (or falsehood) of a proposition. For example, the uncertainty level 0 means that it has no uncertainty. If $i < j$, then the uncertainty level j is higher than the uncertainty level i.

The following soundness theorem is obtained.

Theorem 4.2.13 *For any sequent S, if $\mathrm{L}_\omega \vdash S$, then S is a tautology.*

Proof. By induction on the length of the proof P of S in L_ω. We distinguish the cases according to the last inference of P. We show only the cases ($\wedge$right^e) and ($\to$left^o). We now fix an assignment $\{v_0, v_1, v_2, ...\}$ on propositional letters.

Case ($\wedge$right^e): The last inference of P is of the form:

$$\frac{\Gamma \Rightarrow \Delta, {\sim}^i\alpha \quad \Gamma \Rightarrow \Delta, {\sim}^i\beta}{\Gamma \Rightarrow \Delta, {\sim}^i(\alpha \wedge \beta)}\ (\wedge\text{right}^e).$$

We show that $\Gamma_*{\to}\Delta^* \vee {\sim}^i(\alpha \wedge \beta)$ is a tautology. Suppose that (1) $v_0(\Gamma_*) = t$. Then, we show that $v_0(\Delta^* \vee {\sim}^i(\alpha\wedge\beta)) = t$. If $v_0({\sim}^i(\alpha\wedge\beta)) = t$, then $v_0(\Delta^*\vee{\sim}^i(\alpha\wedge\beta)) = t$. Thus, suppose that $v_0({\sim}^i(\alpha\wedge\beta)) = v_i(\alpha\wedge\beta) = v_i(\alpha) = v_i(\beta) = f$. Then, we have (2) $v_0({\sim}^i\alpha) = v_i(\alpha) = f$ and (3) $v_0({\sim}^i\beta) = v_i(\beta) = f$. On the other hand, by induction hypothesis, we have that $\Gamma_*{\to}\Delta^* \vee {\sim}^i\alpha$ and $\Gamma_*{\to}\Delta^* \vee {\sim}^i\beta$ are tautology. Then, we have (4) $v_0(\Gamma_*{\to}\Delta^* \vee {\sim}^i\alpha) = t$ and (5) $v_0(\Gamma_*{\to}\Delta^* \vee {\sim}^i\beta) = t$. By (1–5), we obtain $v_0(\Delta^*) = t$. Therefore we have $v_0(\Delta^* \vee {\sim}^i(\alpha \wedge \beta)) = t$.

Case ($\to$left^o): The last inference of P is of the form:

$$\frac{{\sim}^{j-1}\alpha, {\sim}^j\beta, \Gamma \Rightarrow \Delta}{{\sim}^j(\alpha{\to}\beta), \Gamma \Rightarrow \Delta}\ (\wedge\text{left}^o).$$

We show that ${\sim}^j(\alpha{\to}\beta) \wedge \Gamma_*{\to}\Delta^*$ is a tautology. Suppose that $v_0({\sim}^j(\alpha{\to}\beta) \wedge \Gamma_*) = t$, i.e., (1) $v_j(\alpha{\to}\beta) = v_0(\Gamma_*) = t$. We show $v_0(\Delta^*) = t$. By (1), we have (2) $v_{j-1}(\alpha) = v_j(\beta) = t$. On the other hand, we have $v_0({\sim}^{j-1}\alpha \wedge {\sim}^j\beta \wedge \Gamma_*{\to}\Delta^*) = t$ by induction hypothesis. Thus, we have that (3) $v_{j-1}(\alpha) = v_j(\beta) = v_0(\Gamma_*) = t$ implies $v_0(\Delta^*) = t$. By (1–3), we obtain $v_0(\Delta^*) = t$.

∎

We now start to prove the completeness theorem by modifying Maehara's decomposition method for LK.

Definition 4.2.14 *Let i, j and k be any natural numbers respectively in ω_e, ω_o and ω.* A decomposition *of a sequent S is defined as of the form (S_0) or $(S_0; S_1)$ by*

(1a) $\Gamma \Rightarrow \Delta, \sim^i\alpha$; $\Gamma \Rightarrow \Delta, \sim^i\beta$ *is a decomposition of* $\Gamma \Rightarrow \Delta, \sim^i(\alpha \wedge \beta)$,

(1b) $\sim^i\alpha, \sim^i\beta, \Gamma \Rightarrow \Delta$ *is a decomposition of* $\sim^i(\alpha \wedge \beta), \Gamma \Rightarrow \Delta$,

(2a) $\Gamma \Rightarrow \Delta, \sim^i\alpha, \sim^i\beta$ *is a decomposition of* $\Gamma \Rightarrow \Delta, \sim^i(\alpha \vee \beta)$,

(2b) $\sim^i\alpha, \Gamma \Rightarrow \Delta$; $\sim^i\beta, \Gamma \Rightarrow \Delta$ *is a decomposition of* $\sim^i(\alpha \vee \beta), \Gamma \Rightarrow \Delta$,

(3a) $\sim^i\alpha, \Gamma \Rightarrow \Delta, \sim^i\beta$ *is a decomposition of* $\Gamma \Rightarrow \Delta, \sim^i(\alpha \rightarrow \beta)$,

(3b) $\Gamma \Rightarrow \Delta, \sim^i\alpha$; $\sim^i\beta, \Gamma \Rightarrow \Delta$ *is a decomposition of* $\sim^i(\alpha \rightarrow \beta), \Gamma \Rightarrow \Delta$,

(4a) $\Gamma \Rightarrow \Delta, \sim^j\alpha, \sim^j\beta$ *is a decomposition of* $\Gamma \Rightarrow \Delta, \sim^j(\alpha \wedge \beta)$,

(4b) $\sim^j\alpha, \Gamma \Rightarrow \Delta$; $\sim^j\beta, \Gamma \Rightarrow \Delta$ *is a decomposition of* $\sim^j(\alpha \wedge \beta), \Gamma \Rightarrow \Delta$,

(5a) $\Gamma \Rightarrow \Delta, \sim^j\alpha$; $\Gamma \Rightarrow \Delta, \sim^j\beta$ *is a decomposition of* $\Gamma \Rightarrow \Delta, \sim^j(\alpha \vee \beta)$,

(5b) $\sim^j\alpha, \sim^j\beta, \Gamma \Rightarrow \Delta$ *is a decomposition of* $\sim^j(\alpha \vee \beta), \Gamma \Rightarrow \Delta$,

(6a) $\Gamma \Rightarrow \Delta, \sim^{j-1}\alpha$; $\Gamma \Rightarrow \Delta, \sim^j\beta$ *is a decomposition of* $\Gamma \Rightarrow \Delta, \sim^j(\alpha \rightarrow \beta)$,

(6b) $\sim^{j-1}\alpha, \sim^j\beta, \Gamma \Rightarrow \Delta$ *is a decomposition of* $\sim^j(\alpha \rightarrow \beta), \Gamma \Rightarrow \Delta$,

(7a) $\sim^k\alpha, \Gamma \Rightarrow \Delta$ *is a decomposition of* $\Gamma \Rightarrow \Delta, \sim^k(\neg\alpha)$,

(7b) $\Gamma \Rightarrow \Delta, \sim^k\alpha$ *is a decomposition of* $\sim^k(\neg\alpha), \Gamma \Rightarrow \Delta$.

Definition 4.2.15 *A* decomposition tree of S *is a tree produced by a decomposition procedure starting from S.*[2] *A* complete decomposition tree of S *is such that the formulas occurring in the leaves have the form $\sim^i p$, with p atomic formula and $i \in \omega$.*

We note that, by construction, the decomposition procedure terminates.

Lemma 4.2.16 *Let (S_0) or $(S_0; S_1)$ be a decomposition of S. If S is a tautology, then so are S_0 and S_1.*

[2] In other words, a decomposition tree corresponds to a bottom-up proof search tree.

Proof. We fix an assignment $\{v_0, v_1, v_2, ...\}$ on propositional letters and prove that if $v_0(S) = t$ then $v_0(S_0) = t$ and $v_0(S_1) = t$. We show only the following cases.

(3b): Suppose that $\sim^i(\alpha\to\beta) \wedge \Gamma_*\to\Delta^*$ $(i \in \omega_e)$ is a tautology. First, we show that $\Gamma_*\to\Delta^* \vee \sim^i\alpha$ is a tautology. Suppose that (1) $v_0(\Gamma_*) = t$. We show $v_0(\Delta^* \vee \sim^i\alpha) = t$. If $v_0(\sim^i\alpha) = t$, then $v_0(\Delta^* \vee \sim^i\alpha) = t$. Thus, suppose that $v_0(\sim^i\alpha) = v_i(\alpha) = f$. Then, $v_i(\alpha\to\beta) = t$, and hence (2) $v_0(\sim^i(\alpha\to\beta)) = v_i(\alpha\to\beta) = t$. On the other hand, we have (3) $v_0(\sim^i(\alpha\to\beta) \wedge \Gamma_*\to\Delta^*) = t$ by the hypothesis. Thus, we obtain $v_0(\Delta^*) = t$ by (1), (2) and (3). Therefore $v_0(\Delta^* \vee \sim^i\alpha) = t$. Second, we show that $\sim^i\beta\wedge\Gamma_*\to\Delta^*$ is a tautology. Suppose that (4) $v_0(\sim^i\beta\wedge\Gamma_*) = t$. Then, (5) $v_0(\sim^i\beta) = v_i(\beta) = t$ and (6) $v_0(\Gamma_*) = t$. By (5), we have $v_i(\alpha\to\beta) = t$, and hence (7) $v_0(\sim^i(\alpha\to\beta)) = v_i(\alpha\to\beta) = t$. By (3), (6) and (7), we obtain $v_0(\Delta^*) = t$.

(6a): Suppose that $\Gamma_*\to\Delta^* \vee \sim^j(\alpha\to\beta)$ $(j \in \omega_o)$ is a tautology. First, we show that $\Gamma_*\to\Delta^* \vee \sim^{j-1}\alpha$ is a tautology. Suppose that (1) $v_0(\Gamma_*) = t$. We show $v_0(\Delta^* \vee \sim^{j-1}\alpha) = t$. By the hypothesis, we have (2) $v_0(\Gamma_*\to\Delta^* \vee \sim^j(\alpha\to\beta)) = t$, i.e., $v_0(\Gamma_*) = t$ implies $[v_0(\Delta^*) = t$ or $v_0(\sim^j(\alpha\to\beta)) = t]$ where $v_0(\sim^j(\alpha\to\beta)) = t$ iff $v_{j-1}(\alpha) = v_0(\sim^{j-1}\alpha) = t$ and $v_j(\beta) = t$. By (1) and (2), if $v_0(\Delta^*) = t$, then $v_0(\Delta^* \vee \sim^{j-1}\alpha) = t$, and if $v_0(\sim^j(\alpha\to\beta)) = t$, i.e., $v_0(\sim^{j-1}\alpha) = t$, then $v_0(\Delta^* \vee \sim^{j-1}\alpha) = t$. Second, we show that $\Gamma_*\to\Delta^* \vee \sim^j\beta$ is a tautology. Suppose (1). We show $v_0(\Delta^* \vee \sim^j\beta) = t$. By (1) and (2), if $v_0(\Delta^*) = t$, then $v_0(\Delta^* \vee \sim^j\beta) = t$, and if $v_0(\sim^j(\alpha\to\beta)) = t$, i.e., $v_0(\sim^j\beta) = t$, then $v_0(\Delta^* \vee \sim^j\beta) = t$. ∎

Lemma 4.2.17 *Suppose that each α_j or β_k in $\{\alpha_1, ..., \alpha_m, \beta_1, ..., \beta_n\}$ is a formula of the form $\sim^i p$ where p is an atomic formula. Then, the sequent $\alpha_1, ..., \alpha_m \Rightarrow \beta_1, ..., \beta_n$ is a tautology if and only if there are α_j $(j \le m)$ and β_k $(k \le n)$ such that $\alpha_j \equiv \beta_k$.*

We note that the sequents of the form $\alpha, \Gamma \Rightarrow \Delta, \alpha$ are obviously provable in cut-free L_ω.

Lemma 4.2.18 *Let S_0 (or $S_0; S_1$) be a decomposition of S. If S_0 (S_0 and S_1) is (are) provable in cut-free L_ω, then so is S.*

The following (strong) completeness theorem is then obtained.

Theorem 4.2.19 *For any sequent S, if S is a tautology, then $L_\omega -$ (cut) $\vdash S$.*

Proof. Suppose that a sequent S is a tautology. Then, all the leaves of a complete decomposition tree of S are tautologies by using Lemma 4.2.16 repeatedly. Then, these leaves are provable in cut-free L_ω by Lemma 4.2.17. By using Lemma 4.2.18 repeatedly for the complete decomposition tree of S, all the sequents in the tree are provable in cut-free L_ω. Therefore, in particular, S is provable in cut-free L_ω. ∎

An alternative proof of Theorem 4.2.8 (cut-elimination) is obtained as follows by combining Theorems 4.2.19 and 4.2.13. Suppose that $\mathrm{L}_\omega \vdash S$ for any sequent S. Then, S is a tautology by Theorem 4.2.13. We then obtain $\mathrm{L}_\omega - (\mathrm{cut}) \vdash S$ by Theorem 4.2.19.

4.3. First-order case

4.3.1. Sequent calculus, embedding and cut-elimination

The notations used in the previous section are also adopted in this section. *Formulas* of the underlying first-order (predicate) logic are constructed from (countable) predicate symbols $p, q, ...$, individual variables $x, y, ...$, individual constants $c, d, ...$, function symbols $f, g, ...$, logical connectives $\rightarrow$, $\wedge$, $\vee$, $\neg$, $\sim$, $\forall$ (universal quantifier) and $\exists$ (existential quantifier). Small letters $t, s, ...$ are used to denote terms. We adopt the notation $\alpha[t/x]$ as the formula which is obtained from a formula α by replacing all free occurrences of an individual variable x in α by a term t, but avoiding a clash of variables.

Definition 4.3.1 ($\mathbf{FL}_\omega$) FL_ω *is obtained from* L_ω *by adding the* even *and* odd *inference rules of the form: for any* $i \in \omega_e$ *and* $j \in \omega_o$,

$$\frac{\sim^i\alpha[t/x], \Gamma \Rightarrow \Delta}{\sim^i\forall x\alpha, \Gamma \Rightarrow \Delta}\ (\forall\mathrm{left}^e) \qquad \frac{\Gamma \Rightarrow \Delta, \sim^i\alpha[z/x]}{\Gamma \Rightarrow \Delta, \sim^i\forall x\alpha}\ (\forall\mathrm{right}^e)$$

$$\frac{\sim^i\alpha[z/x], \Gamma \Rightarrow \Delta}{\sim^i\exists x\alpha, \Gamma \Rightarrow \Delta}\ (\exists\mathrm{left}^e)$$

$$\frac{\Gamma \Rightarrow \Delta, \sim^i\alpha[t/x]}{\Gamma \Rightarrow \Delta, \sim^i\exists x\alpha}\ (\exists\mathrm{right}^e) \qquad \frac{\sim^j\alpha[z/x], \Gamma \Rightarrow \Delta}{\sim^j\forall x\alpha, \Gamma \Rightarrow \Delta}\ (\forall\mathrm{left}^o)$$

$$\frac{\Gamma \Rightarrow \Delta, \sim^j\alpha[t/x]}{\Gamma \Rightarrow \Delta, \sim^j\forall x\alpha}\ (\forall\mathrm{right}^o)$$

$$\frac{\sim^j\alpha[t/x], \Gamma \Rightarrow \Delta}{\sim^j\exists x\alpha, \Gamma \Rightarrow \Delta}\ (\exists\mathrm{left}^o) \qquad \frac{\Gamma \Rightarrow \Delta, \sim^j\alpha[z/x]}{\Gamma \Rightarrow \Delta, \sim^j\exists x\alpha}\ (\exists\mathrm{right}^o)$$

where z in ($\forall$right^e)*,* ($\exists$left^e)*,* ($\forall$left^o) *and* ($\exists$right^o) *is an individual variable which has the eigenvariable condition, i.e., z does not occur as a free individual variable in the lower sequent of the rule.*

We note that for any formula α and any $i \in \omega$, the sequent ${\sim}^i\alpha \Rightarrow {\sim}^i\alpha$ is provable in cut-free FL_ω.

Definition 4.3.2 (FLK) *A sequent calculus* FLK *for first-order (predicate) classical logic is obtained from* LK *by adding the ${\sim}$-free versions of the quantifier rules* ($\forall$left^e), ($\forall$right^e), ($\exists$left^e) and ($\exists$right^e)*, i.e., these rules are obtained from the even quantifier rules by replacing i by* 0*. Such modified rules are denoted by deleting the superscript "e".*

As well-known, FLK admits cut-elimination.

Definition 4.3.3 *Let $\Phi := \{p, q, r, ...\}$ be a fixed countable non-empty set of atomic formulas. Then, we define the sets $\Phi_0 := \Phi$ and $\Phi_i := \{p_i \mid p \in \Phi\}$ $(1 \leq i \in \omega)$ of atomic formulas. The language $\mathcal{L}_{\mathrm{FL}_\omega}$ of FL_ω (also denoted $\mathcal{L}$ for short) is defined by using $\Phi, \rightarrow, \wedge, \vee, \neg, \forall, \exists$ and ${\sim}$. The language $\mathcal{L}_{\mathrm{FLK}}$ of* FLK *is defined by using $\bigcup_{i\in\omega}\Phi_i$, $\rightarrow$, $\wedge$, $\vee$, $\neg$, $\forall$ and $\exists$.*

A mapping f from $\mathcal{L}_{\mathrm{FL}_\omega}$ to $\mathcal{L}_{\mathrm{FLK}}$ is defined by the conditions presented in Definition 4.2.6 and the following conditions:

1. $f({\sim}^i(Q\alpha)) := Qf({\sim}^i\alpha)$ $\;Q \in \{\forall x, \exists x\}$ *for each* $i \in \omega_e$,
2. $f({\sim}^j(\forall x\alpha)) := \exists x f({\sim}^j\alpha)$ *for each* $j \in \omega_o$,
3. $f({\sim}^j(\exists x\alpha)) := \forall x f({\sim}^j\alpha)$ *for each* $j \in \omega_o$.

Theorem 4.3.4 *Let Γ and Δ be sequences of formulas in $\mathcal{L}_{\mathrm{FL}_\omega}$, and f be the mapping defined in Definition 4.3.3. Then:*

1. $\mathrm{FL}_\omega \vdash \Gamma \Rightarrow \Delta$ *iff* $\mathrm{FLK} \vdash f(\Gamma) \Rightarrow f(\Delta)$.
2. $\mathrm{FL}_\omega - (\mathrm{cut}) \vdash \Gamma \Rightarrow \Delta$ *iff* $\mathrm{FLK} - (\mathrm{cut}) \vdash f(\Gamma) \Rightarrow f(\Delta)$.

Using Theorem 4.3.4 and the cut-elimination theorem for FLK, we can obtain the following theorem and proposition.

Theorem 4.3.5 *The rule* (cut) *is admissible in cut-free* FL_ω.

Proposition 4.3.6 *Let $\sharp$ be ${\sim}^i$ $(i \in \omega_e)$ or ${\sim}^j$ $(j \in \omega_o)$. Then, FL_ω is paraconsistent with respect to $\sharp$.*

4.3.2. Semantics and completeness

In this subsection, for the sake of simplicity, we adopt the language without individual constants and function symbols. We also use the same names $\mathcal{L}$ and FL_ω for the reduced language and the corresponding system, respectively.

Definition 4.3.7 *A structure $\mathcal{A} := \langle U, \{I^i\}_{i\in\omega}\rangle$ is called a* model *if the following conditions hold:*

1. *U is a non-empty set,*
2. *I^i $(i \in \omega)$ are mappings such that $p^{I^i} \subseteq U^n$ (i.e. p^{I^i} are n-ary relations on U) for each n-ary predicate symbol p.*

We introduce the notation $\underline{u}$ for the name of $u \in U$, and denotes $\mathcal{L}[\mathcal{A}]$ for the language obtained from $\mathcal{L}$ by adding the names of all the elements of U. A formula α is called a closed formula *if α has no free individual variable. A formula of the form $\forall x_1 \cdots \forall x_m\alpha$ is called the* universal closure *of α if the free variables of α are $x_1, ..., x_m$. We write $cl(\alpha)$ for the universal closure of α.*

Definition 4.3.8 *Let $\mathcal{A} := \langle U, \{I^i\}_{i\in\omega}\rangle$ be a model. The consequence relations $\mathcal{A} \models^i \alpha$ $(i \in \omega)$ for any closed formula α of $\mathcal{L}[\mathcal{A}]$ are defined inductively as follows:*

1. *for each n-ary atomic formula $p(\underline{x}_1, ..., \underline{x}_n)$ and each $i \in \omega$,* $\mathcal{A} \models^i p(\underline{x}_1, ..., \underline{x}_n)$ *iff* $(x_1, ..., x_n) \in p^{I^i}$,
2. *for each $i \in \omega$, $\mathcal{A} \models^i {\sim}\alpha$ iff $\mathcal{A} \models^{i+1} \alpha$,*
3. *for each $i \in \omega$, $\mathcal{A} \models^i \neg\alpha$ iff not-$(\mathcal{A} \models^i \alpha)$,*
4. *for each $i \in \omega_e$, $\mathcal{A} \models^i \alpha \wedge \beta$ iff $\mathcal{A} \models^i \alpha$ and $\mathcal{A} \models^i \beta$,*
5. *for each $i \in \omega_e$, $\mathcal{A} \models^i \alpha \vee \beta$ iff $\mathcal{A} \models^i \alpha$ or $\mathcal{A} \models^i \beta$,*
6. *for each $i \in \omega_e$, $\mathcal{A} \models^i \alpha \to \beta$ iff not-$(\mathcal{A} \models^i \alpha)$ or $\mathcal{A} \models^i \beta$,*
7. *for each $i \in \omega_e$, $\mathcal{A} \models^i \forall x\alpha$ iff $\mathcal{A} \models^i \alpha[\underline{u}/x]$ for all $u \in U$,*
8. *for each $i \in \omega_e$, $\mathcal{A} \models^i \exists x\alpha$ iff $\mathcal{A} \models^i \alpha[\underline{u}/x]$ for some $u \in U$,*
9. *for each $j \in \omega_o$, $\mathcal{A} \models^j \alpha \wedge \beta$ iff $\mathcal{A} \models^j \alpha$ or $\mathcal{A} \models^j \beta$,*
10. *for each $j \in \omega_o$, $\mathcal{A} \models^j \alpha \vee \beta$ iff $\mathcal{A} \models^j \alpha$ and $\mathcal{A} \models^j \beta$,*
11. *for each $j \in \omega_o$, $\mathcal{A} \models^j \alpha \to \beta$ iff $\mathcal{A} \models^{j-1} \alpha$ and $\mathcal{A} \models^j \beta$,*

12. for each $j \in \omega_o$, $\mathcal{A} \models^j \forall x\alpha$ *iff* $\mathcal{A} \models^j \alpha[\underline{u}/x]$ *for some* $u \in U$,

13. for each $j \in \omega_o$, $\mathcal{A} \models^j \exists x\alpha$ *iff* $\mathcal{A} \models^j \alpha[\underline{u}/x]$ *for all* $u \in U$.

The consequence relations $\mathcal{A} \models^i \alpha$ $(i \in \omega)$ *for any formula* α *of* $\mathcal{L}$ *are defined by* ($\mathcal{A} \models^i \alpha$ *iff* $\mathcal{A} \models^i cl(\alpha)$). *A formula* α *of* $\mathcal{L}$ *is called* valid *if* $\mathcal{A} \models^0 \alpha$ *holds for each model* $\mathcal{A}$. *A sequent* $\Gamma \Rightarrow \Delta$ *of* $\mathcal{L}$ *is called valid if so is the formula* $\Gamma_* \rightarrow \Delta^*$.

Theorem 4.3.9 *For any sequent* S, *if* $\mathrm{FL}_\omega \vdash S$, *then* S *is valid.*

We now start to prove the completeness theorem for FL_ω by using an extended Schütte's method (for this method, see e.g., [185]).

Definition 4.3.10 *Let* i, j *and* k *be any natural numbers respectively in* ω_e, ω_o *and* ω. *A sequent* $\Gamma \Rightarrow \Delta$ *is called* saturated *if for any formulas* α *and* β,

(s1) $\sim^i(\alpha \wedge \beta) \in \Gamma$ *implies* ($\sim^i\alpha \in \Gamma$ *and* $\sim^i\beta \in \Gamma$),

(s2) $\sim^i(\alpha \wedge \beta) \in \Delta$ *implies* ($\sim^i\alpha \in \Delta$ *or* $\sim^i\beta \in \Delta$),

(s3) $\sim^i(\alpha \vee \beta) \in \Gamma$ *implies* ($\sim^i\alpha \in \Gamma$ *or* $\sim^i\beta \in \Gamma$),

(s4) $\sim^i(\alpha \vee \beta) \in \Delta$ *implies* ($\sim^i\alpha \in \Delta$ *and* $\sim^i\beta \in \Delta$),

(s5) $\sim^i(\alpha \rightarrow \beta) \in \Gamma$ *implies* ($\sim^i\alpha \in \Delta$ *or* $\sim^i\beta \in \Gamma$),

(s6) $\sim^i(\alpha \rightarrow \beta) \in \Delta$ *implies* ($\sim^i\alpha \in \Gamma$ *and* $\sim^i\beta \in \Delta$),

(s7) $\sim^k\neg\alpha \in \Gamma$ *implies* $\sim^k\alpha \in \Delta$,

(s8) $\sim^k\neg\alpha \in \Delta$ *implies* $\sim^k\alpha \in \Gamma$,

(s9) $\sim^i\forall x\alpha \in \Gamma$ *implies* ($\sim^i\alpha[y/x] \in \Gamma$ *for any individual variable* y),

(s10) $\sim^i\forall x\alpha \in \Delta$ *implies* ($\sim^i\alpha[z/x] \in \Delta$ *for some indiv. variable* z),

(s11) $\sim^i\exists x\alpha \in \Gamma$ *implies* ($\sim^i\alpha[z/x] \in \Gamma$ *for some individual variable* z),

(s12) $\sim^i\exists x\alpha \in \Delta$ *implies* ($\sim^i\alpha[y/x] \in \Delta$ *for any individual variable* y),

(s13) $\sim^j(\alpha \wedge \beta) \in \Gamma$ *implies* ($\sim^j\alpha \in \Gamma$ *or* $\sim^j\beta \in \Gamma$),

(s14) $\sim^j(\alpha \wedge \beta) \in \Delta$ *implies* ($\sim^j\alpha \in \Delta$ *and* $\sim^j\beta \in \Delta$),

(s15) $\sim^j(\alpha \vee \beta) \in \Gamma$ *implies* ($\sim^j\alpha \in \Gamma$ *and* $\sim^j\beta \in \Gamma$),

(s16) $\sim^j(\alpha \vee \beta) \in \Delta$ *implies* ($\sim^j\alpha \in \Delta$ *or* $\sim^j\beta \in \Delta$),

(s17) $\sim^j(\alpha \rightarrow \beta) \in \Gamma$ *implies* ($\sim^{j-1}\alpha \in \Gamma$ *and* $\sim^j\beta \in \Gamma$),

(s18) $\sim^j(\alpha \rightarrow \beta) \in \Delta$ *implies* ($\sim^{j-1}\alpha \in \Delta$ *or* $\sim^j\beta \in \Delta$), (s19) $\sim^j \forall x\alpha \in \Gamma$ *implies* ($\sim^j \alpha[z/x] \in \Gamma$ *for some individual variable* z),

(s20) $\sim^j \forall x\alpha \in \Delta$ *implies* ($\sim^j \alpha[y/x] \in \Delta$ *for any individual variable* y),

(s21) $\sim^j \exists x\alpha \in \Gamma$ *implies* ($\sim^j \alpha[y/x] \in \Gamma$ *for any individual variable* y),

(s22) $\sim^j \exists x\alpha \in \Delta$ *implies* ($\sim^j \alpha[z/x] \in \Delta$ *for some indiv. variable* z).

A sequent $\Gamma \Rightarrow \Delta$ is called an *infinite sequent* if Γ and Δ are infinite (countable) sets of formulas. An infinite sequent $\Gamma \Rightarrow \Delta$ is called *provable* if two finite subsets $\Gamma' \subseteq \Gamma$ and $\Delta' \subseteq \Delta$ exist, such that $\Gamma' \Rightarrow \Delta'$ is provable in cut-free FL_ω.

Definition 4.3.11 *Let i, j and k be any natural numbers respectively in ω_e, ω_o and ω. A* decomposition *of a sequent (or infinite sequent) S is defined as of the form (S_0) or ($S_0; S_1$) by*

(1a) $\Gamma \Rightarrow \Delta, \sim^i(\alpha \wedge \beta), \sim^i\alpha$; $\Gamma \Rightarrow \Delta, \sim^i(\alpha \wedge \beta), \sim^i\beta$ *is a decomposition of* $\Gamma \Rightarrow \Delta, \sim^i(\alpha \wedge \beta)$,

(1b) $\sim^i\alpha, \sim^i\beta, \sim^i(\alpha \wedge \beta), \Gamma \Rightarrow \Delta$ *is a decomposition of* $\sim^i(\alpha \wedge \beta), \Gamma \Rightarrow \Delta$,

(2a) $\Gamma \Rightarrow \Delta, \sim^i(\alpha \vee \beta), \sim^i\alpha, \sim^i\beta$ *is a decomposition of* $\Gamma \Rightarrow \Delta, \sim^i(\alpha \vee \beta)$,

(2b) $\sim^i\alpha, \sim^i(\alpha \vee \beta), \Gamma \Rightarrow \Delta$; $\sim^i\beta, \sim^i(\alpha \vee \beta), \Gamma \Rightarrow \Delta$ *is a decomposition of* $\sim^i(\alpha \vee \beta), \Gamma \Rightarrow \Delta$,

(3a) $\sim^i\alpha, \Gamma \Rightarrow \Delta, \sim^i(\alpha \to \beta), \sim^i\beta$ *is a decomposition of* $\Gamma \Rightarrow \Delta, \sim^i(\alpha \to \beta)$,

(3b) $\sim^i(\alpha \to \beta), \Gamma \Rightarrow \Delta, \sim^i\alpha$; $\sim^i\beta, \sim^i(\alpha \to \beta), \Gamma \Rightarrow \Delta$ *is a decomposition of* $\sim^i(\alpha \to \beta), \Gamma \Rightarrow \Delta$,

(4a) $\sim^k\alpha, \Gamma \Rightarrow \Delta, \sim^k\neg\alpha$ *is a decomposition of* $\Gamma \Rightarrow \Delta, \sim^k\neg\alpha$,

(4b) $\sim^k\neg\alpha, \Gamma \Rightarrow \Delta, \sim^k\alpha$ *is a decomposition of* $\sim^k\neg\alpha, \Gamma \Rightarrow \Delta$,

(5a) $\Gamma \Rightarrow \Delta, \sim^i\forall x\alpha, \sim^i\alpha[z/x]$ *is a decomposition of* $\Gamma \Rightarrow \Delta, \sim^i\forall x\alpha$ *where z is a* fresh *free individual variable, i.e., the variable z is not occurring in* $\Gamma \Rightarrow \Delta, \sim^i\forall x\alpha$,

(5b) $\sim^i\alpha[y_1/x], ..., \sim^i\alpha[y_m/x], \sim^i\forall x\alpha, \Gamma \Rightarrow \Delta$ *is a decomposition of* $\sim^i\forall x\alpha, \Gamma \Rightarrow \Delta$ *where $y_1, ..., y_m$ are the free individual variables occurring in* $\sim^i\forall x\alpha, \Gamma \Rightarrow \Delta$,[3]

(6a) $\Gamma \Rightarrow \Delta, \sim^i\exists x\alpha, \sim^i\alpha[y_1/x], ..., \sim^i\alpha[y_m/x]$ *is a decomposition of* $\Gamma \Rightarrow \Delta, \sim^i\exists x\alpha$ *where $y_1, ..., y_m$ are the free individual variables occurring in* $\Gamma \Rightarrow \Delta, \sim^i\exists x\alpha$,

[3] Strictly speaking, if $\sim^i\forall x\alpha, \Gamma \Rightarrow \Delta$ has no free individual variable, then we adopt any free variable in $\mathcal{L}$. Such a condition is also adopted in the corresponding items below.

(6b) $\sim^i\alpha[z/x], \sim^i\exists x\alpha, \Gamma \Rightarrow \Delta$ *is a decomposition of* $\sim^i\exists x\alpha, \Gamma \Rightarrow \Delta$ *where* z *is a fresh free individual variable,*

(7a) $\Gamma \Rightarrow \Delta, \sim^j(\alpha \wedge \beta), \sim^j\alpha, \sim^j\beta$ *is a decomposition of* $\Gamma \Rightarrow \Delta, \sim^j(\alpha \wedge \beta)$,

(7b) $\sim^j\alpha, \sim^j(\alpha \wedge \beta), \Gamma \Rightarrow \Delta\ ;\ \sim^j\beta, \sim^j(\alpha \wedge \beta), \Gamma \Rightarrow \Delta$ *is a decomposition of* $\sim^j(\alpha \wedge \beta), \Gamma \Rightarrow \Delta$,

(8a) $\Gamma \Rightarrow \Delta, \sim^j(\alpha \vee \beta), \sim^j\alpha\ ;\ \Gamma \Rightarrow \Delta, \sim^j(\alpha \vee \beta), \sim^j\beta$ *is a decomposition of* $\Gamma \Rightarrow \Delta, \sim^j(\alpha \vee \beta)$,

(8b) $\sim^j\alpha, \sim^j\beta, \sim^j(\alpha \vee \beta), \Gamma \Rightarrow \Delta$ *is a decomposition of* $\sim^j(\alpha \vee \beta), \Gamma \Rightarrow \Delta$,

(9a) $\Gamma \Rightarrow \Delta, \sim^j(\alpha\rightarrow\beta), \sim^{j-1}\alpha\ ;\ \Gamma \Rightarrow \Delta, \sim^j(\alpha\rightarrow\beta), \sim^j\beta$ *is a decomposition of* $\Gamma \Rightarrow \Delta, \sim^j(\alpha\rightarrow\beta)$,

(9b) $\sim^{j-1}\alpha, \sim^j\beta, \sim^j(\alpha\rightarrow\beta), \Gamma \Rightarrow \Delta$ *is a decomposition of* $\sim^j(\alpha\rightarrow\beta), \Gamma \Rightarrow \Delta$,

(10a) $\Gamma \Rightarrow \Delta, \sim^j\forall x\alpha, \sim^j\alpha[y_1/x], ..., \sim^j\alpha[y_m/x]$ *is a decomposition of* $\Gamma \rightarrow \Delta, \sim^j\forall x\alpha$ *where* $y_1, ..., y_m$ *are the free individual variables occurring in* $\Gamma \rightarrow \Delta, \sim^j\forall x\alpha$,

(10b) $\sim^j\alpha[z/x], \sim^j\forall x\alpha, \Gamma \Rightarrow \Delta$ *is a decomposition of* $\sim^j\forall x\alpha, \Gamma \Rightarrow \Delta$ *where* z *is a fresh free individual variable,*

(11a) $\Gamma \Rightarrow \Delta, \sim^j\exists x\alpha, \sim^j\alpha[z/x]$ *is a decomposition of* $\Gamma \Rightarrow \Delta, \sim^j\exists x\alpha$ *where* z *is a* fresh *free individual variable, i.e., the variable* z *is not occurring in* $\Gamma \Rightarrow \Delta, \sim^j\exists x\alpha$,

(11b) $\sim^j\alpha[y_1/x], ..., \sim^j\alpha[y_m/x], \sim^j\exists x\alpha, \Gamma \Rightarrow \Delta$ *is a decomposition of* $\sim^j\exists x\alpha,\ \Gamma \Rightarrow \Delta$ *where* $y_1, ..., y_m$ *are the free individual variables occurring in* $\sim^j\exists x\alpha, \Gamma \Rightarrow \Delta$.

A *decomposition tree of* S is a tree obtained through repeated decomposition of S.

We note that, at each step of the decomposition procedure, if the input sequent is unprovable in cut-free FL_ω, then the output sequents must be unprovable in cut-free FL_ω. Observe moreover that the steps of the procedure where new fresh variables are introduced play a central role.

Lemma 4.3.12 *Let* $\Gamma \Rightarrow \Delta$ *be an unprovable sequent in* FL_ω – (cut). *Then, an unprovable saturated and possibly infinite sequent* $\Gamma^\omega \Rightarrow \Delta^\omega$ *exists, such that* $\Gamma \subseteq \Gamma^\omega$ *and* $\Delta \subseteq \Delta^\omega$.

Proof. Let $\Gamma \Rightarrow \Delta$ be an unprovable sequent in FL_ω − (cut). We construct $\Gamma^\omega \Rightarrow \Delta^\omega$ from $\Gamma \Rightarrow \Delta$ as follows.

(1) We apply the decomposition procedure in Definition 4.3.11 to $\Gamma \Rightarrow \Delta$, in the following order:

$$(1a) \longrightarrow (1b) \longrightarrow (2a) \longrightarrow \cdots \longrightarrow (11b)$$

where the application of any step may be empty if the corresponding formula lacks in the input sequent. By hypothesis, at least one among the output sequents must be an unprovable sequent.

(2) We repeat the same procedure (1) without halting. Then, we obtain a finitely branching decomposition tree with infinitely many nodes.

(3) By König's Lemma, we have an infinite path of this decomposition tree as follows.

$$\Gamma_0 \Rightarrow \Delta_0 \mid \Gamma_1 \Rightarrow \Delta_1 \mid \Gamma_2 \Rightarrow \Delta_2 \mid \cdots \infty,$$

where $\Gamma_0 \Rightarrow \Delta_0$ is $\Gamma \Rightarrow \Delta$. In this sequence of the sequents on the infinite path, we have that $\Gamma_0 \subseteq \Gamma_1 \subseteq \Gamma_2 \subseteq \cdots$ and $\Delta_0 \subseteq \Delta_1 \subseteq \Delta_2 \subseteq \cdots$.

(4) We put $\Gamma^\omega := \bigcup_{i=0}^{\infty} \Gamma_i$ and $\Delta^\omega := \bigcup_{i=0}^{\infty} \Delta_i$.[4]

Then, we have that $\Gamma \subseteq \Gamma^\omega$ and $\Delta \subseteq \Delta^\omega$, and can verify that $\Gamma^\omega \Rightarrow \Delta^\omega$ is an unprovable, saturated (infinite) sequent. ∎

Lemma 4.3.13 *Let $\Gamma \Rightarrow \Delta$ be an unprovable sequent in* FL_ω − (cut), *and $\Gamma^\omega \Rightarrow \Delta^\omega$ be an unprovable, saturated (infinite) sequent constructed from $\Gamma \Rightarrow \Delta$ by Lemma 4.3.12. We define a canonical model $\mathcal{A} := \langle U, \{I^i\}_{i \in \omega}\rangle$ by*

$$U := \{z \mid z \textit{ is a free individual variable occurring in } \Gamma^\omega \Rightarrow \Delta^\omega\},$$
$$p^{I^i} := \{(z_1, ..., z_m) \mid {\sim}^i p(z_1, ..., z_m) \in \Gamma^\omega\}.$$

Then, for any formula α and any $i \in \omega$,

$$[({\sim}^i\alpha \in \Gamma^\omega \textit{ implies } \mathcal{A} \models^i \underline{\alpha}) \textit{ and } ({\sim}^i\alpha \in \Delta^\omega \textit{ implies not-}(\mathcal{A} \models^i \underline{\alpha}))]$$

where $\underline{\alpha}$ is obtained form α by replacing every individual variable x occurring in α by the name $\underline{x}$.

Proof. By (simultaneous) induction on the complexity of α.[5]

- Base step: Obvious by the definitions of I^i.
- Induction step: We show some cases.

Case ($\alpha \equiv {\sim}\beta$ for any $i \in \omega$): We show only the case ${\sim}^i{\sim}\beta \in \Gamma^\omega$ implies $\mathcal{A} \models^i {\sim}\underline{\beta}$, since the other case (concerning Δ^ω) can be shown

[4] Note that $\Gamma^\omega \cap \Delta^\omega = \varnothing$.

[5] We must carry out the induction on α for all $i \in \omega$ simultaneously.

similarly. Suppose $\sim^i\sim\beta \in \Gamma^\omega$, i.e., $\sim^{i+1}\beta \in \Gamma^\omega$. By the induction hypothesis, we obtain $\mathcal{A} \models^{i+1} \underline{\beta}$, i.e., $\mathcal{A} \models^i \sim\underline{\beta}$.

Case ($\alpha \equiv \beta \rightarrow \gamma$ for any $i \in \omega_e$): First we show that for any $i \in \omega_e$, $\sim^i(\beta\rightarrow\gamma) \in \Gamma^\omega$ implies $\mathcal{A} \models^i \underline{\beta\rightarrow\gamma}$. Suppose $\sim^i(\beta\rightarrow\gamma) \in \Gamma^\omega$. Then, we obtain that $[\sim^i\beta \in \Delta^\omega$ or $\sim^i\gamma \in \Gamma^\omega]$ by Definition 4.3.10 (s5). By the induction hypothesis, we obtain that [not-($\mathcal{A} \models^i \underline{\beta}$) or $\mathcal{A} \models^i \underline{\gamma}$]. This means $\mathcal{A} \models^i \underline{\beta\rightarrow\gamma}$. Second, we show that for any $i \in \omega_e$, $\sim^i(\beta\rightarrow\gamma) \in \Delta^\omega$ implies not-($\mathcal{A} \models^i \underline{\beta\rightarrow\gamma}$). Suppose $\sim^i(\beta\rightarrow\gamma) \in \Delta^\omega$. Then, we obtain that $[\sim^i\beta \in \Gamma^\omega$ and $\sim^i\gamma \in \Delta^\omega]$ by Definition 4.3.10 (s6). By the induction hypothesis, we obtain that [$\mathcal{A} \models^i \underline{\beta}$ and not-($\mathcal{A} \models^i \underline{\gamma}$)], i.e., not-[not-($\mathcal{A} \models^i \underline{\beta}$) or $\mathcal{A} \models^i \underline{\gamma}$]. This means not-($\mathcal{A} \models^i \underline{\beta\rightarrow\gamma}$).

Case ($\alpha \equiv \beta\rightarrow\gamma$ for any $j \in \omega_o$): First we show that for any $j \in \omega_o$, $\sim^j(\beta\rightarrow\gamma) \in \Gamma^\omega$ implies $\mathcal{A} \models^j \underline{\beta\rightarrow\gamma}$. Suppose $\sim^j(\beta\rightarrow\gamma) \in \Gamma^\omega$. Then, we obtain that $[\sim^{j-1}\beta \in \Gamma^\omega$ and $\sim^j\gamma \in \Gamma^\omega]$ by Definition 4.3.10 (s17). By the induction hypothesis, we obtain that [$\mathcal{A} \models^{j-1} \underline{\beta}$ and $\mathcal{A} \models^j \underline{\gamma}$].[6] This means $\mathcal{A} \models^j \underline{\beta\rightarrow\gamma}$. Second, we show that for any $j \in \omega_o$, $\sim^j(\beta\rightarrow\gamma) \in \Delta^\omega$ implies not-($\mathcal{A} \models^j \underline{\beta\rightarrow\gamma}$). Suppose $\sim^j(\beta\rightarrow\gamma) \in \Delta^\omega$. Then, we obtain that $[\sim^{j-1}\beta \in \Delta^\omega$ or $\sim^j\gamma \in \Delta^\omega]$ by Definition 4.3.10 (s18). By the induction hypothesis, we obtain that [not-($\mathcal{A} \models^{j-1} \underline{\beta}$) or not-($\mathcal{A} \models^j \underline{\gamma}$)], i.e., not-[$\mathcal{A} \models^{j-1} \underline{\beta}$ and $\mathcal{A} \models^j \underline{\gamma}$]. This means not-($\mathcal{A} \models^j \underline{\beta\rightarrow\gamma}$).

Case ($\alpha \equiv \forall x\beta$ for any $i \in \omega_e$): We show only that $\sim^i\forall x\beta \in \Gamma^\omega$ implies $\mathcal{A} \models^i \forall x\underline{\beta}$. Suppose $\sim^i\forall x\beta \in \Gamma^\omega$. Then we obtain that $\sim^i\beta[y_i/x] \in \Gamma^\omega$ for all $y_i \in U$, by Definition 4.3.10 (s9). By the induction hypothesis, we obtain that $\mathcal{A} \models^i \underline{\beta}[\underline{y_i}/x]$ for all $y_i \in U$. This means $\mathcal{A} \models^i \forall x\underline{\beta}$.

Case ($\alpha \equiv \forall x\beta$ for any $j \in \omega_o$): We show only that $\sim^j\forall x\beta \in \Gamma^\omega$ implies $\mathcal{A} \models^j \forall x\underline{\beta}$. Suppose $\sim^j\forall x\beta \in \Gamma^\omega$. Then, we obtain that $\sim^j\beta[z/x] \in \Gamma^\omega$ for some $z \in U$, by Definition 4.3.10 (s20). By the induction hypothesis, we obtain that $\mathcal{A} \models^j \underline{\beta}[\underline{z}/x]$ for some $z \in U$. This means $\mathcal{A} \models^j \forall x\underline{\beta}$. ∎

Theorem 4.3.14 *For any sequent S, if S is valid, then* FL_ω − (cut) ⊢ S.

Proof. It is sufficient to show that if $\Gamma \Rightarrow \Delta$ is not provable in FL_ω − (cut), then there exists a model $\mathcal{A}$ such that $\Gamma \Rightarrow \Delta$ is not valid in $\mathcal{A}$. Suppose that $\Gamma \Rightarrow \Delta$ is not provable in FL_ω − (cut). Then, by Lemma 4.3.13, we can construct a canonical model $\mathcal{A}$. Thus, we have $\mathcal{A} \models^i \underline{\gamma}$ and not-($\mathcal{A} \models^i \underline{\delta}$) for any $\gamma \in \Gamma \subseteq \Gamma^\omega$ and any $\delta \in \Delta \subseteq \Delta^\omega$. Hence, we obtain "not-($\mathcal{A} \models^i \underline{\Gamma_*\rightarrow\Delta^*}$)", and hence "not-($\mathcal{A} \models^i cl(\Gamma_*\rightarrow\Delta^*)$)". Therefore, $\Gamma \Rightarrow \Delta$ is not valid in $\mathcal{A}$. ∎

[6] For the case $\sim^{j-1}\beta \in \Gamma^\omega$, we have to use the induction hypothesis w.r.t. $j-1 \in \omega_e$.

Similar to the propositional case, an alternative proof of Theorem 4.3.5 (cut-elimination) is obtained by combining Theorems 4.3.14 and 4.3.9.

4.4. Remarks

Although L_ω and FL_ω may be regarded as a kind of infinite-valued logics, some finite-valued versions of these logics can be obtained by adding some inference rules w.r.t. $\sim$. For example, a finite-valued version of L_ω can be obtained by adding the inference rules of the form: for a fixed positive integer $n \geq 2$,

$$\frac{\alpha, \Gamma \Rightarrow \Delta}{{\sim}^n\alpha, \Gamma \Rightarrow \Delta}\ ({\sim}^n\text{left}) \qquad \frac{\Gamma \Rightarrow \Delta, \alpha}{\Gamma \Rightarrow \Delta, {\sim}^n\alpha}\ ({\sim}^n\text{right}).$$

In these rules, the case $n = 2$ corresponds to the double-negation-elimination axiom ${\sim}{\sim}\alpha \leftrightarrow \alpha$. The completeness, cut-elimination and embedding results for such extended logics can be obtained by imposing some appropriate modifications.

For example, consider the logic $L_n = L_\omega + ({\sim}^n\text{left}) + ({\sim}^n\text{right})$ for a fixed positive integer $n \geq 2$. The embedding function f w.r.t. L_n, which is like an embedding function presented in Definition 4.2.6, needs the condition:

$$f({\sim}^n\alpha) := f(\alpha),$$

and the semantics for L_n needs the following *cyclic* valuation condition instead of the condition 1 of Definition 4.2.12:

1′. $v_i({\sim}\alpha) = t$ iff $v_{i+1}(\alpha) = t$ if $i < n-1$, and $v_0(\alpha) = t$ otherwise.

It is noted that the logic L_2 (i.e., the case $n = 2$) is just a 4-valued extension of B4, FDE and N4, since the cyclic valuations v_0 and v_1 respectively correspond to the well-known dual consequence relations $\models^+$ (verification) and $\models^-$ (falsification) used in N4.

Part II.

Variations of bi-intuitionistic logic

5. Paraconsistent logics extending bi-intuitionistic logic

In this chapter, a family of propositional logics with constructive implication, constructive co-implication and three kinds of negation is introduced. The logics extend bi-intuitionistic propositional logic, in which both intuitionistic negation and dual intuitionistic co-negation can be defined, by a primitive paraconsistent strong negation connective. A relational possible worlds semantics as well as sound and complete display sequent calculi for the logics under consideration are presented. Moreover, an extended inferentialist semantics is presented. This semantics generalizes the familiar Brouwer-Heyting-Kolmogorov (BHK) interpretation of intuitionistic logic in terms of canonical proofs and the slightly less well-known interpretation of N4 in terms of canonical proofs and disproofs. The logics under consideration are shown to be sound with respect to the inferentialist semantics. We will first present motivating considerations that focus on the notion of constructiveness and we shall later present motivating ideas based on considering the speech act of denial.[1]

5.1. Introduction

It is sometimes said that classical logic admits of a constructive interpretation if it is assumed that every proposition is decidable, but this does not imply that classical logic *is* constructive, and although classical logic has been called a (new) constructive logic by Girard [65], there seems to be a broad agreement among logicians that classical logic is *not* constructive. But what is a constructive logic? Sometimes the term "constructive logic" is used as a synonym for "intuitionistic logic". However, logics other than intuitionistic logic have also been said to be constructive, like, for instance, Johansson's minimal logic, Heyting-Brouwer logic, or David Nelsons's logics with strong negation. Whereas there exists *the* system of classical propositional and predicate logic, it is far from clear whether there exists exactly one system of constructive

[1] This chapter brings together material from [209] and [211].

logic. In a situation where there are no clear, agreed-upon, individually necessary and jointly sufficient conditions for the constructiveness of a logical system, it seems quite difficult or next to pointless to designate one particular logic as *the* correct constructive logic. Nevertheless, for some reasons certain logics may still be regarded as constructive logics.

5.1.1. Positive constructive propositional logics

It is well known that the implicational fragments of intuitionistic and classical logic differ, as Peirce's law $((\alpha \to \beta) \to \alpha) \to \alpha$ is classically but not intuitionistically valid, and it seems that there is a consensus among logicians that, among other things, the failure of Peirce's law indicates that

(∗) Intuitionistic implicational logic is a constructive logic.

Intuitionistic logic and classical logic (understood as consequence relations) share their conjunction-disjunction fragment, and the constructiveness of this fragment appears to be uncontroversial.[2] In their disjunction-negation fragments, however, intuitionistic and classical logic differ. In particular, intuitionistic logic enjoys a constructive feature which classical logic fails to have in its disjunction-negation fragment, the *disjunction property*: If a disjunction $(\alpha \vee \beta)$ is provable, then α is provable or β is provable. In classical logic $p \vee {\sim}p$ is provable, but neither the atomic formula p nor its classical negation ${\sim}p$ is provable. Moreover, the conjunction-negation fragment of intuitionistic logic lacks a constructive feature which Nelsons's constructive logics enjoy, namely the *constructible falsity property*: If ${\sim}(\alpha \wedge \beta)$ is provable, then ${\sim}\alpha$ is provable or ${\sim}\beta$ is provable. In intuitionistic logic ${\sim}(p \wedge {\sim}p)$ is provable, but neither the literal ${\sim}p$ nor its negation ${\sim}{\sim}p$ is provable. Still, there appears to be an agreement among logicians that

(∗∗) Positive intuitionistic propositional logic, IPL^+, is a constructive logic.

This view is supported by the observation that IPL^+ is a fragment not only of intuitionistic logic, but also of Johansson's minimal logic, Heyting-Brouwer logic, and Nelsons's logics with strong negation.

Heyting-Brouwer logic adds to intuitionistic logic a binary connective which is a natural companion to implication and which is often called

[2] Gödel [66] noticed that intuitionistic and classical propositional logic understood as sets of formulas share their conjunction-negation fragment.

co-implication. Whereas intuitionistic implication, $\rightarrow$, is the residual of conjunction in IPL$^+$ in the sense that

$$\alpha \wedge \beta \vdash \gamma \text{ iff } \alpha \vdash \beta \rightarrow \gamma, \tag{5.1}$$

co-implication (also referred to as "pseudo-difference"), $\prec$[3], is the residual of disjunction in the $\{\wedge, \vee, \prec\}$-fragment of Heyting-Brouwer logic:

$$\gamma \vdash \alpha \vee \beta \text{ iff } \gamma \prec \beta \vdash \alpha.^4 \tag{5.2}$$

Let us refer to the $\{\wedge, \vee, \prec\}$-fragment of Heyting-Brouwer logic as HB$^+$. Is HB$^+$ a constructive logic?[5] To justify $(**)$, one may point to the so-called proof (alias Brouwer-Heyting-Kolmogorov) interpretation of IPL$^+$, see, for instance [42]. According to this interpretation, a (canonical) proof of an implication $\alpha \rightarrow \beta$ is a construction that transforms any proof of α into a proof of β, and a proof of a conjunction $\alpha \wedge \beta$ is a pair (π_1, π_2) consisting of a proof π_1 of α and a proof π_2 of β. A proof of a disjunction $\alpha \vee \beta$ is a pair (i, π) such that $i = 0$ and π is a proof of α or $i = 1$ and π is a proof of β. One can then show that positive propositional intuitionistic logic is sound with respect to its proof interpretation: For every formula α provable (derivable from the empty set) in IPL$^+$, there exists a construction of α. That is, one possible criterion for the constructiveness of a logic is its correctness with respect to an interpretation in terms of canonical proofs.[6] For HB$^+$ we

[3] We use the symbol for co-implication suggested in [68]. The more familiar symbol used, for example, in [163] is $\dot{-}$. As Goré [68] explains, the left-right symmetry of the more familiar symbol hides the asymmetry of the pseudo-difference operation. In $\gamma \prec \beta$, γ is in a positive position and β in a negative position. This becomes clear, for instance, if the Boolean understanding of $\beta \rightarrow \gamma$ as $\neg\beta \vee \gamma$ is analogously applied to co-implication by reading $\gamma \prec \beta$ as $\gamma \wedge \neg\beta$. Wolter [217] uses $\varphi \breve{\rightarrow} \psi$ instead of $\psi \prec \varphi$. $\gamma \prec \beta$ may be read as "β co-implies γ" or as "γ excludes β". Co-implication has been thoroughly investigated by Cecylia Rauszer [161], [162], [163], who added co-implication (and co-negation, see below) to intuitionistic logic to obtain Heyting-Brouwer logic. See also [189], [68], [28], and the references therein.

[4] Classical implication is the residual of conjunction in classical logic. One may therefore ask whether there exists a purely co-implicational formula which stands to the result of dropping implication and intuitionistic negation from Heyting-Brouwer logic as Peirce's law stands to intuitionistic logic. This co-implicative analogue of Peirce's law is stated in Section 5.3.

[5] In Heyting-Brouwer logic H-B, intuitionistic negation and the co-negation of Heyting-Brouwer logic can be defined using $\rightarrow$ and $\prec$, see Observation 5.2.4. The addition of $\prec$ to IPL$^+$ allows one to define intuitionistic negation, and the addition of $\rightarrow$ to HB$^+$ allows one to define co-negation.

[6] It turns out that for logics with strong negation *disproofs* naturally enter the picture in addition to proofs, see [119], [196].

may consider dual proofs: *reductiones ad absurdum*. According to this interpretation, a (canonical) reductio ad absurdum of a co-implication $\beta \mathbin{-\!\!\!\prec} \alpha$ is a construction that transforms any reductio of α into a reductio of β. A reductio of a disjunction $\alpha \vee \beta$ is a pair (π_1, π_2) consisting of a reductio π_1 of α and a reductio π_2 of β. A reductio of a conjunction $\alpha \wedge \beta$ is a pair (i, π) such that $i = 0$ and π is a reductio of α or $i = 1$ and π is a reductio of β. One can then show that HB^+ is sound with respect to its dual proof interpretation: For every formula α reducible to absurdity (formula α from which the empty set can be derived) in HB^+, there exists a construction of α, see p. 155. In view of this observation, we draw the conclusion that

$(***)$ HB^+ is a constructive logic.

5.1.2. Adding strong negation

The result of adding $\mathbin{-\!\!\!\prec}$ to IPL^+ (alias the result of adding $\rightarrow$ to HB^+) is propositional Heyting-Brouwer logic H-B (also called bi-intuitionistic logic, [68] or subtractive logic [38]), cf. Chapter 2. As we will see, in this logic, intuitionistic negation and co-negation are definable. Is H-B a constructive logic? The status not only of classical negation but also of intuitionistic negation and co-negation as a constructive connective is contentious. The addition of classical negation to the $\{\wedge, \vee\}$-fragment of intuitionistic logic results in a failure of the desirable disjunction property, and so does the addition of co-negation (see Section 5.6), whereas the addition of intuitionistic negation results in a failure of the desirable constructible falsity property. Also, intuitionistic negation has been criticized, because it does not express the idea of direct falsification. An intuitionistically negated formula $\sim\alpha$ is verified at a possible world (alias state) s in an intuitionistic Kripke model iff at every state related to s by the pre-order of the model, α fails to be verified. There is no way of falsifying α at s in the sense of verifying the negation of α by considering just s. In Nelson's logics with strong negation (see, among many other sources, [6], [51], [73], [83], [92], [134], [137], [138], [141] [170], [186], [190], [191], [192], [196], [201], [205]) the situation is different. In the relational semantics of these logics, support of truth and support of falsity conditions are stated separately. A state s supports the truth of an atom p iff p is verified at s, and s supports the falsity of p iff p is falsified at s. Verification and falsification of atomic formulas may vary from model to model. Strong negation is interpreted as leading from the support of truth to the support of falsity, and vice versa: A state s supports the truth (falsity) of $\sim\alpha$ iff s supports the falsity (truth) of

α. In the relational semantics of intuitionistic logic and H-B, only verification conditions are specified for all kinds of formulas. If in addition to verification falsification is acknowledged as a semantic category in its own right, and if falsity is expressed in the object language by a unary negation operation, then the separate consideration of support of falsity conditions for all kinds of formulas leads to separate support of truth conditions for all kinds of negated formulas. This may well be interpreted as a constructive treatment of negation. The following question thus arises:

> What are the correct support of truth conditions for negated complex formulas? (Or, equivalently, what are the support of falsity conditions for complex formulas?)

In intuitionistic logic the double-negation elimination law $\sim\sim\alpha \rightarrow \alpha$ and the De Morgan law $\sim(\alpha \wedge \beta) \rightarrow (\sim\alpha \vee \sim\beta)$ fail to be valid. If one considers intuitionistic logic as the correct system of constructive logic, these failures indicate that the double negation law and the above De Morgan law are not constructively valid. But we have already seen that the constructive nature of intuitionistic negation is doubtful. If one is not prejudiced by the assumption that intuitionistic logic is the correct constructive logic, then nothing stands in the way of accepting both double negation laws and all the familiar De Morgan laws. And indeed, in Nelson's constructive logics with strong negation, all these principles are valid. The view that a situation supports the falsity of a conjunction $(\alpha \wedge \beta)$ (disjunction $(\alpha \vee \beta)$) iff it supports the falsity of α or (and) it supports the falsity of β seems to be deeply rooted in our intuitive understanding of conjunction, disjunction, truth, and falsity. Moreover, if negation as falsity is a bridge from support of truth to support of falsity, and vice versa, then there is no way around both double negation laws.

The picture is less clear when we consider the support of truth conditions of negated implications, and it gets more complicated when we at the same time consider the support of truth conditions for negated co-implications. On the classical understanding of negated implications, a formula $\sim(\alpha \rightarrow \beta)$ is true iff α is true and β is *not true*. On the intuitionistic reading, $\sim(\alpha \rightarrow \beta)$ is verified at a state s iff for every "later" state t (every possible expansion t of s), there is a state t' later than t such that α is verified at t', whereas β is not verified at t'. In Nelsons's logic, support of falsity of a formula α (support of truth of $\sim\alpha$) is always a matter determined at the state of evaluation, and a state s supports the truth of $\sim(\alpha \rightarrow \beta)$ iff s supports the truth of α and s supports the *falsity* of β. According to this view, $\sim(\alpha \rightarrow \beta)$ is equivalent to

$(\alpha \wedge {\sim}\beta)$, where ${\sim}$ expresses falsity and not the absence of truth. Since the support of truth and the support of falsity are persistent along a model's pre-order, a state s supports the truth of ${\sim}(\alpha \to \beta)$ iff every possible expansion t of s supports the truth of α and the falsity of β. If the semantics is set up such that the equivalence

$$ {\sim}(\alpha \to \beta) \leftrightarrow (\alpha \wedge {\sim}\beta) \tag{5.3} $$

is valid (and atomic formulas may not only be neither verified nor falsified at some state but also both verified and falsified at some state), we obtain Nelson's constructive four-valued propositional logic N4 in the co-implication-free language $\{{\sim}, \wedge, \vee, \to\}$.[7]

This is not the end of the story concerning the language $\{{\sim}, \wedge, \vee, \to\}$, however, because another understanding of the relation between implication and negation has been proposed already since ancient times. It turns out that a slight modification of the support of truth conditions for negated implications leads from N4 to a system of *connexive* logic in which the support of falsity of implications is *not* interpreted as falsification at the world of evaluation, see [206] for a survey and references. Connexive logics have a standard logical vocabulary but contain certain non-theorems of classical logic as theorems. Since classical propositional logic is Post-complete, any additional axiom in its language gives rise to the trivial system, so that any non-trivial system of connexive logic in this vocabulary must leave out some theorems of classical logic. Among the characteristic theorems of connexive logics are *Aristotle's Theses*:

$$ {\sim}({\sim}\alpha \to \alpha),\ {\sim}(\alpha \to {\sim}\alpha), \tag{5.4} $$

and *Boethius' Theses*

$$ (\alpha \to \beta) \to {\sim}(\alpha \to {\sim}\beta),\ (\alpha \to {\sim}\beta) \to {\sim}(\alpha \to \beta) \tag{5.5} $$

which are not theorems of classical logic. A connective $\to$ that satisfies the above theses is sometimes said to be a connexive implication.

5.1.3. Motivations of connexive logic

Since connexive logic is not a well-established area of non-classical logic, we will briefly look at motivations of it. In addition to an appeal to certain intuitions about meaning connections between the antecedent and

[7] Note that in N4 a truth constant $\top$ can be defined as $p \to p$ for some atom p, but no falsity constant $\bot$. Odintsov [138] investigates extensions of the system $\mathrm{N4}^{\bot}$ in the language $\{{\sim}, \wedge, \vee, \to, \bot\}$, which is axiomatized by adding the formulas $\bot \to \alpha$ and $\alpha \to {\sim}\bot$ to the axioms of N4, cf. Chapter 2.

the succedent of valid implications, there exist at least two motivating ideas for connexive logic. The first comes from Aristotle's syllogistic. It is well known that the syllogistic contains inferences that are not classically valid under the standard translation into predicate logic. One of the most prominent examples is the inference from "Every P is Q" to "Some Ps are Qs":

$$\forall x(P(x) \to Q(x)) \vdash \exists x(P(x) \wedge Q(x)) \tag{5.6}$$

Normally, we do not quantify over the empty set. If we assume that the interpretation of P is empty, there is hardly any reason to assume that every P is Q, but if the interpretation of P is non-empty, (5.6) is a valid inference. Inference (5.6) cannot be consistently added as a rule to a proof system for classical predicate logic, as is obvious from the following instance of (5.6):

$$\forall x((P(x) \wedge {\sim}P(x)) \to Q(x)) \vdash \exists x((P(x) \wedge {\sim}P(x)) \wedge Q(x)) \tag{5.7}$$

The premise of (5.7) is classically valid, whereas the conclusion is classically unsatisfiable. Now, in classical logic, inference (5.6) is interchangeable with

$$\forall x(P(x) \to Q(x)) \vdash \exists x{\sim}(P(x) \to {\sim}Q(x)). \tag{5.8}$$

Storrs McCall [123] pointed out that in a system of connexive logic (5.8) is a valid inference. This is especially perspicuous in the quantified connexive logic QC introduced in [204], because there

$$\sim(\alpha \to \beta) \leftrightarrow (\alpha \to {\sim}\beta) \tag{5.9}$$

is an axiom. One might therefore suggest to translate statements of the form "Some Ps are Qs" not as $\exists x(P(x) \wedge Q(x))$ but as $\exists x{\sim}(P(x) \to {\sim}Q(x))$, which in the system QC is equivalent to $\exists x(P(x) \to Q(x))$.

Another motivation comes from Categorial Grammar, see also Chapter 12. In the (various versions of) the Lambek Calculus, there are two implications, $\backslash$ and $/$, which are the residuals of a non-commutative, so-called multiplicative (intensional) conjunction ("fusion"), $\cdot$. In one version of the calculus, $\cdot$ is assumed to be associative; in another version, it is non-associative. The formulas of the Lambek Calculus stand for syntactic types, and a derivability statement (sequent) $x \vdash y$ is to be understood as "every expression of type x is also of type y". An expression e is of type $x \backslash y$ iff for every expression e' of type x, the string $e'e$ is of type y, and e is of type y/x iff for every expression e' of type x, the string ee' is of type y. A transitive verb like `loves`, for example, may be syntactically typed as $((n \backslash s)/n)$, because it combines with any name of syntactic type n from the right to an expression of type $(n \setminus s)$ that looks to the left for a name to result in an expression of type s, a sentence. It then makes sense to introduce a negation $\sim x$ to designate the class of expressions that are definitely not of type x. An expression e is of type $\sim (x \setminus y)$ iff for every expression e' of type x, the string $e'e$ is of type $\sim y$, and e is of type $\sim (y/x)$ iff for every expression e' of type x, the string ee' is of type $\sim y$. These definitions validate the sequents $\sim (x \setminus y) \vdash x \setminus \sim y$, $\sim (y/x) \vdash (\sim y/x)$, and their converses. The expression `loves Mary`, for example, is of type $\sim (n \setminus (n \setminus s))$, because in combination with any name from the left it results in an expression which is definitely not an intransitive verb, namely in a sentence. Clearly, the suggested reading of $\sim$ also justifies the double negation laws. As a result of these considerations, we obtain directional versions Boethius' Theses (as sequents) such as:[8]

$$(x \setminus y) \vdash \sim (x \setminus \sim y). \tag{5.10}$$

5.1.4. Completing the picture

Not only the equivalences (5.3) and (5.9) are serious candidates for expressing the support of truth conditions for negated implications. If we think of the classical understanding of a co-implication $(\alpha \mathbin{-\!\prec} \beta)$ as $(\alpha \wedge \sim \beta)$, the following equivalence must also be taken into account:

$$\sim (\alpha \to \beta) \leftrightarrow (\alpha \mathbin{-\!\prec} \beta), \tag{5.11}$$

and classical De Morgan duality then suggests yet another equivalence:[9]

$$\sim (\alpha \to \beta) \leftrightarrow (\sim \beta \mathbin{-\!\prec} \sim \alpha) \tag{5.12}$$

Eventually, we have to specify the support of truth conditions for constructively negated co-implications. In analogy to what we have done for negated implications, we may consider the classical (or rather Nelson-like) reading of negated co-implications, the connexive understanding of negated co-implications, the reading of negated co-implications as implications, and the understanding of negated co-implications as contraposed implications. Altogether, this range of readings will give us *sixteen* systems of constructive propositional logic. For want of a better terminology and notation, in Table 5.1 the characteristic equivalences in question are listed as equivalences $I_1 - I_4$ and $C_1 - C_4$. For convenience,

---

[8] In Categorial Grammar, the left-hand side of a sequent may not be empty, because the empty string has no syntactic type.

[9] This equivalence was pointed out to us by Greg Restall.

the constructive propositional logics in the language $\{\wedge, \vee, \rightarrow, \prec, \sim\}$ that differ from each other only with respect to validating a certain pair of these equivalences (one from the I-equivalences and one from the C-equivalences) will be referred to as systems (I_i, C_j), $i, j \in \{1, 2, 3, 4\}$.

I_1	$\sim(\alpha \rightarrow \beta) \leftrightarrow (\alpha \wedge \sim\beta)$	negated implication, classical reading
I_2	$\sim(\alpha \rightarrow \beta) \leftrightarrow (\alpha \rightarrow \sim\beta)$	negated implication, connexive reading
I_3	$\sim(\alpha \rightarrow \beta) \leftrightarrow (\alpha \prec \beta)$	negated implication as co-implication
I_4	$\sim(\alpha \rightarrow \beta) \leftrightarrow (\sim\beta \prec \sim\alpha)$	negated impl. as contraposed co-impl.
C_1	$\sim(\alpha \prec \beta) \leftrightarrow (\sim\alpha \vee \beta)$	negated co-implication, classical reading
C_2	$\sim(\alpha \prec \beta) \leftrightarrow (\sim\alpha \prec \beta)$	negated co-impl., connexive reading
C_3	$\sim(\alpha \prec \beta) \leftrightarrow (\alpha \rightarrow \beta)$	negated co-implication as implication
C_4	$\sim(\alpha \prec \beta) \leftrightarrow (\sim\beta \rightarrow \sim\alpha)$	negated co-impl. as contraposed impl.

Table 5.1.: Constructively negated implications and co-implications.

5.2. Syntax and relational semantics

We will consider a propositional language $\mathcal{L}$ defined in Backus–Naur form as follows:

$$\begin{aligned} \text{atomic formulas: } & p \in \mathit{Atom} \\ \text{formulas: } & \alpha \in \mathit{Form(Atom)} \\ \alpha ::= \ & p \mid \sim\alpha \mid (\alpha \wedge \alpha) \mid (\alpha \vee \alpha) \mid (\alpha \rightarrow \alpha) \mid (\alpha \prec \alpha). \end{aligned}$$

The language without $\prec$ is the language of intuitionistic propositional logic, IPL, of David Nelson's propositional logics with strong negation, and of connexive propositional logic (if we do *not* use distinct symbols for the "corresponding" connectives from distinct logics). In $\mathcal{L}$, where both implication $\rightarrow$ and co-implication $\prec$ are present, two distinct unary negation connectives can be defined: intuitionistic negation, which we now denote as $\neg$, and co-negation, $-$. We will focus, however, on the single *primitive* strong negation $\sim$. Equivalence, $\leftrightarrow$, is defined as usual, and co-equivalence, $\succ\!\prec$, is defined as expected, by setting $\alpha \succ\!\prec \beta := (\alpha \prec \beta) \vee (\beta \prec \alpha)$.

In this section, we will introduce the sixteen constructive logics (I_i, C_j), $i, j \in \{1, 2, 3, 4\}$, semantically. Since all these logics are interpreted in models based on (Kripke) frames, the semantic presentation admits of a transparent comparison between the logics under consideration.

Definition 5.2.1 *A frame is a pre-order* $\langle I, \leq \rangle$. *Intuitively, I is a non-empty set of information states, and $\leq$ is a reflexive transitive binary relation of possible expansion of states on I.*

Instead of $w \leq w'$, we also write $w' \geq w$.

Definition 5.2.2 *A model is a structure* $\langle I, \leq, v^+, v^- \rangle$, *where* $\langle I, \leq \rangle$ *is a frame and v^+ (v^-) is a function that maps every $p \in$ Atom to a subset of I (namely the states that support the truth (falsity) of p). It is assumed that the functions v^+ and v^- satisfy the following persistence conditions for atoms: if $w \leq w'$, then $w \in v^+(p)$ implies $w' \in v^+(p)$; if $w \leq w'$, then $w \in v^-(p)$ implies $w' \in v^-(p)$. The relations $\mathcal{M}, w \models^+ \alpha$ ("state w supports the truth of $\mathcal{L}$-formula α in model $\mathcal{M}$") and $\mathcal{M}, w \models^- \alpha$ ("state w supports the falsity of $\mathcal{L}$-formula α in model $\mathcal{M}$") are inductively defined as follows:*

$$
\begin{array}{lll}
\mathcal{M}, w \models^+ p & \textit{iff} & w \in v^+(p) \\
\mathcal{M}, w \models^- p & \textit{iff} & w \in v^-(p) \\[1ex]
\mathcal{M}, w \models^+ {\sim}\alpha & \textit{iff} & \mathcal{M}, w \models^- \alpha \\
\mathcal{M}, w \models^- {\sim}\alpha & \textit{iff} & \mathcal{M}, w \models^+ \alpha \\[1ex]
\mathcal{M}, w \models^+ (\alpha \wedge \beta) & \textit{iff} & \mathcal{M}, w \models^+ \alpha \textit{ and } \mathcal{M}, w \models^+ \beta \\
\mathcal{M}, w \models^- (\alpha \wedge \beta) & \textit{iff} & \mathcal{M}, w \models^- \alpha \textit{ or } \mathcal{M}, w \models^- \beta \\[1ex]
\mathcal{M}, w \models^+ (\alpha \vee \beta) & \textit{iff} & \mathcal{M}, w \models^+ \alpha \textit{ or } \mathcal{M}, w \models^+ \beta \\
\mathcal{M}, w \models^- (\alpha \vee \beta) & \textit{iff} & \mathcal{M}, w \models^- \alpha \textit{ and } \mathcal{M}, w \models^- \beta \\[1ex]
\mathcal{M}, w \models^+ (\alpha \to \beta) & \textit{iff} & \textit{for all } w' \geq w : \mathcal{M}, w' \not\models^+ \alpha \textit{ or } \mathcal{M}, w' \models^+ \beta \\
\mathcal{M}, w \models^+ (\alpha \prec \beta) & \textit{iff} & \textit{there exists } w' \leq w : \mathcal{M}, w' \models^+ \alpha \textit{ and} \\
 & & \mathcal{M}, w' \not\models^+ \beta
\end{array}
$$

where $\mathcal{M}, w \not\models^+ \alpha$ is the classical negation of $\mathcal{M}, w \models^+ \alpha$.

In Table 5.2, we list the support of falsity conditions corresponding to the equivalences $I_1 - I_4$ and $C_1 - C_4$ from Table 5.1. No matter which equivalences we choose, support of truth and support of falsity is persistent for arbitrary formulas.

Observation 5.2.3 (Persistence) *For every $\mathcal{L}$-formula α, model $\langle I, \leq, v^+, v^- \rangle$, and $w, w' \in I$: if $w \leq w'$, then $w \in v^+(\alpha)$ implies $w' \in v^+(\alpha)$; if $w \leq w'$, then $w \in v^-(\alpha)$ implies $w' \in v^-(\alpha)$.*

We can make the following simple but important observation concerning the expressive power of the logics we are about to define.

cI_1	$\mathcal{M}, w \models^- (\alpha \to \beta)$	iff	$\mathcal{M}, w \models^+ \alpha$ and $\mathcal{M}, w \models^- \beta$
cI_2	$\mathcal{M}, w \models^- (\alpha \to \beta)$	iff	for every $w' \geq w : \mathcal{M}, w' \not\models^+ \alpha$ or $\mathcal{M}, w' \models^- \beta$
cI_3	$\mathcal{M}, w \models^- (\alpha \to \beta)$	iff	there is $w' \leq w : \mathcal{M}, w' \models^+ \alpha$ and $\mathcal{M}, w' \not\models^+ \beta$
cI_4	$\mathcal{M}, w \models^- (\alpha \to \beta)$	iff	there is $w' \leq w : \mathcal{M}, w' \not\models^- \alpha$ and $\mathcal{M}, w' \models^- \beta$
cC_1	$\mathcal{M}, w \models^- (\alpha \prec \beta)$	iff	$\mathcal{M}, w \models^- \alpha$ or $\mathcal{M}, w \models^+ \beta$
cC_2	$\mathcal{M}, w \models^- (\alpha \prec \beta)$	iff	there is $w' \leq w : \mathcal{M}, w' \models^- \alpha$ and $\mathcal{M}, w' \not\models^+ \beta$
cC_3	$\mathcal{M}, w \models^- (\alpha \prec \beta)$	iff	for every $w' \geq w : \mathcal{M}, w' \not\models^+ \alpha$ or $\mathcal{M}, w' \models^+ \beta$
cC_4	$\mathcal{M}, w \models^- (\alpha \prec \beta)$	iff	for every $w' \geq w : \mathcal{M}, w' \models^- \alpha$ or $\mathcal{M}, w' \not\models^- \beta$

Table 5.2.: Support of falsity conditions for negated implications and co-implications.

Observation 5.2.4 *Let p be a certain atomic formula, let $\top := p \to p$, and let $\bot := p \prec p$. For every model $\mathcal{M}$ and every state w from $\mathcal{M}$, $\mathcal{M}, w \models^+ \top$ and $\mathcal{M}, w \not\models^+ \bot$. Thus, we can define the co-negation $-$ of Heyting-Brouwer logic by setting $-\alpha := \top \prec \alpha$ and intuitionistic negation $\neg$, by setting $\neg\alpha := \alpha \to \bot$.*

The support of truth clause for co-negation then is:

$$\mathcal{M}, w \models^+ -\alpha \quad \text{iff} \quad \text{there exists } w' \leq w \text{ and } \mathcal{M}, w' \not\models^+ \alpha,$$

whereas the support of truth conditions for intuitionistic negation are the familiar ones:

$$\mathcal{M}, w \models^+ \neg\alpha \quad \text{iff} \quad \text{for every } w' \geq w,\ \mathcal{M}, w' \not\models^+ \alpha.$$

Note that if $\mathcal{M} = \langle I, \leq \rangle$ is a frame, v is a function from *Atom* to subsets of I, and $\mathcal{M}, w \models \alpha$ is defined exactly as $\mathcal{M}, w \models^+ \alpha$, except that $\mathcal{M}, w \models p$ iff $w \in v(p)$, then $\langle I, \leq, v \rangle$ is a model for H-B. The logic H-B, understood as a set of formulas, is the set of all $\sim$-free $\mathcal{L}$-formulas α such that for every model $\mathcal{M} = \langle I, \leq, v \rangle$, and every $w \in I$, $\mathcal{M}, w \models \alpha$.

Definition 5.2.5 *The propositional logics (I_i, C_j) are defined as the triples $(\mathcal{L}, \models^+_{I_i,C_j}, \models^-_{I_i,C_j})$, where the entailment relations $\models^+_{I_i,C_j}, \models^-_{I_i,C_j} \subseteq \mathcal{P}(\mathcal{L}) \times \mathcal{P}(\mathcal{L})$ are defined as follows:*
$\Delta \models^+_{I_i,C_j} \Gamma$ iff for every model $\mathcal{M} = \langle I, \leq, v^+, v^- \rangle$ defined with clauses cI_i and cC_j and every $w \in I$, if $\mathcal{M}, w \models^+ \alpha$ for every $\alpha \in \Delta$, then $\mathcal{M}, w \models^+ \beta$ for some $\beta \in \Gamma$, and
$\Delta \models -_{I_i,C_j}\Gamma$ iff for every model $\mathcal{M} = \langle I, \leq, v^+, v^- \rangle$ defined with clauses cI_i and cC_j and every $w \in I$, if $\mathcal{M}, w \models^- \alpha$ for every $\alpha \in \Gamma$, then

$\mathcal{M}, w \models^- \beta$ for some $\beta \in \Delta$.
For singleton sets $\{\alpha\}$ and $\{\beta\}$, we write $\alpha \models^+_{I_i,C_j} \beta$ ($\alpha \models^-_{I_i,C_j} \beta$) instead of $\{\alpha\} \models^+_{I_i,C_j} \{\beta\}$ ($\{\alpha\} \models^-_{I_i,C_j} \{\beta\}$). If the context is clear, we shall sometimes omit the subscript $_{I_i,C_j}$.

This definition of a logic as comprising *two* entailment relations instead of just one is unusual but not at all unnatural, see, for instance, Chapter 2, [183, 214, 213]. The set of all constructively false sentences is not the complement of the set of all constructively true sentences, and we can make the following observation.

Observation 5.2.6 *If $(I_i, C_j) \neq (I_4, C_4)$, then $\models^+_{I_i,C_j} \neq \models^-_{I_i,C_j}$.*

Proof. For every logic (I_i, C_j), it holds that $(p \wedge (p \to q)) \models^+ q$. However, for no logic (I_1, C_j) and for no logic (I_3, C_j), we have $\sim q \models^+ \sim(p \wedge (p \to q))$. To see this, a one-element countermodel suffices, where the following holds for the single state w: $w \in v^-(q)$, $w \notin v^+(p)$, and $w \notin v^-(p)$. Other counterexamples work for the logics (I_2, C_j) and (I_4, C_1) - (I_4, C_3). For every logic (I_i, C_j), it holds that $p \models^+ (q \to p)$. But in a singleton model where $w \notin v^-(p)$ and $w \notin v^+(q)$, we have $w \models^+ (q \to \sim p)$ and $w \not\models^+ \sim p$, which shows for the logics (I_2, C_j) that $\sim(q \to p) \not\models^+ \sim p$. For every logic (I_i, C_j), it holds that $r \wedge (r \to (p \prec p)) \models^+ q$. Consider a singleton model with $w \models^- q$, $w \not\models^- r$, $w \not\models^- p$, and $w \not\models^+ p$. This model shows that $\sim q \not\models^+ \sim r \vee ((\sim p \vee p) \prec \sim r)$ in the case of logic (I_4, C_1) and that $\sim p \not\models^+ \sim r \vee ((\sim p \prec p) \prec \sim r)$ in the case of logic (I_4, C_2). In (I_4, C_3) we have $\sim (p \prec q) \models^+ p \to q$. A singleton model in which $w \not\models^+ p$, $w \models^- q$ and $w \not\models^- p$ shows that $\sim q \prec \sim p \not\models^+ p \prec q$. ∎

We do not require that for atomic formulas p, $v^+(p) \cap v^-(p) = \varnothing$. Therefore, the logics under consideration are *paraconsistent*. Neither is it the case that for any formula β, $\{p, \sim p\} \models^+_{I_i,C_j} \beta$ nor is it the case that $\beta \models^-_{I_i,C_j} \{p, \sim p\}$.[10]

The next observation on negation normal forms will be used in the completeness proof of Section 5.3. A formula is in *negation normal form* if it contains $\sim$ only in front of atoms. The following translations ρ_{I_i,C_j} send every formula α to a formula in negation normal form, where

[10] Co-negation is, of course, also a paraconsistent negation, see [189], [27], whereas intuitionistic negation is "paracomplete".

$p \in Atom$ and $\odot \in \{\vee, \wedge, \rightarrow, \prec\}$:

$$
\begin{array}{lcl}
\rho_{I_i,C_j}(p) & = & p \\
\rho_{I_i,C_j}(\sim p) & = & \sim p \\
\rho_{I_i,C_j}(\sim\sim \alpha) & = & \rho_{I_i,C_j}(\alpha) \\
\rho_{I_i,C_j}(\alpha \odot \beta) & = & \rho_{I_i,C_j}(\alpha) \odot \rho_{I_i,C_j}(\beta) \\
\rho_{I_i,C_j}(\sim(\alpha \vee \beta)) & = & \rho_{I_i,C_j}(\sim\alpha) \wedge \rho_{I_i,C_j}(\sim\beta) \\
\rho_{I_i,C_j}(\sim(\alpha \wedge \beta)) & = & \rho_{I_i,C_j}(\sim\alpha) \vee \rho_{I_i,C_j}(\sim\beta) \\
\rho_{I_1,C_j}(\sim(\alpha \rightarrow \beta)) & = & \rho_{I_1,C_j}(\alpha) \wedge \rho_{I_1,C_j}(\sim\beta) \\
\rho_{I_2,C_j}(\sim(\alpha \rightarrow \beta)) & = & \rho_{I_2,C_j}(\alpha) \rightarrow \rho_{I_2,C_j}(\sim\beta) \\
\rho_{I_3,C_j}(\sim(\alpha \rightarrow \beta)) & = & \rho_{I_3,C_j}(\alpha) \prec \rho_{I_3,C_j}(\beta) \\
\rho_{I_4,C_j}(\sim(\alpha \rightarrow \beta)) & = & \rho_{I_4,C_j}(\sim\beta) \prec \rho_{I_4,C_j}(\sim\alpha) \\
\rho_{I_i,C_1}(\sim(\alpha \prec \beta)) & = & \rho_{I_i,C_1}(\sim\alpha) \vee \rho_{I_i,C_1}(\beta) \\
\rho_{I_i,C_2}(\sim(\alpha \prec \beta)) & = & \rho_{I_i,C_2}(\sim\alpha) \prec \rho_{I_i,C_2}(\beta) \\
\rho_{I_i,C_3}(\sim(\alpha \prec \beta)) & = & \rho_{I_i,C_3}(\alpha) \rightarrow \rho_{I_i,C_3}(\beta) \\
\rho_{I_i,C_4}(\sim(\alpha \prec \beta)) & = & \rho_{I_i,C_4}(\sim\beta) \rightarrow \rho_{I_i,C_4}(\sim\alpha)
\end{array}
$$

Lemma 5.2.7 *For every formula α, $\rho_{I_i,C_j}(\alpha)$ is in negation normal form and $\alpha \models^+_{I_i,C_j} \rho_{I_i,C_j}(\alpha)$, $\rho_{I_i,C_j}(\alpha) \models^+_{I_i,C_j} \alpha$, $\alpha \models^-_{I_i,C_j} \rho_{I_i,C_j}(\alpha)$, $\rho_{I_i,C_j}(\alpha) \models^-_{I_i,C_j} \alpha$.*

5.3. Display calculi

Developing a proof system for logics with both intuitionistic implication *and* co-implication encounters some problems. The standard sequent calculus for intuitionistic logic is asymmetric; it uses sequents with multiple antecedents and (at most) single conclusions in order to avoid the provability of Peirce's law. If one admits symmetric sequents (with multiple antecedents and succedents) and just adds the natural and obvious sequent rules for introducing co-implication (in the style of Gentzen's sequent calculus for classical logic, *LK*), namely:

$$
\frac{\Gamma, \beta \vdash \alpha, \Delta}{\Gamma, (\beta \prec \alpha) \vdash \Delta} \qquad \frac{\Gamma \vdash \beta, \Delta \quad \Sigma, \alpha \vdash \Pi}{\Sigma, \Gamma \vdash (\beta \prec \alpha), \Delta, \Pi} \tag{5.13}
$$

one can not only prove Peirce's law, but also a sequent which contains just one co-implicative formula and is an analogue of the sequent expressing the provability of Peirce's law:[11]

$$
\dfrac{\dfrac{\alpha \vdash \alpha \qquad \dfrac{\dfrac{\alpha \vdash \alpha}{\alpha, \beta \vdash \alpha}}{\alpha, \beta \prec \alpha \vdash \varnothing}}{\alpha, \alpha \vdash \alpha \prec (\beta \prec \alpha)}}{\dfrac{\alpha \vdash \alpha \prec (\beta \prec \alpha)}{\alpha \prec (\alpha \prec (\beta \prec \alpha)) \vdash \varnothing}}
$$

The formula $\alpha \prec (\alpha \prec (\beta \prec \alpha))$ may be called Peirce's co-law.

The sequent calculus for Heyting-Brouwer logic H-B in [38] uses single-conclusion sequents but imposes a "singleton on the left" constraint on the left introduction rule for co-implication (and a "singleton on the right" constraint on the right introduction rule for implication). This sequent calculus is thus asymmetric, but it does not enjoy cut-elimination. Nor does the sequent system for H-B in [161] (see Chapter 2) allow cut-elimination. A counterexample due to T. Uustalu is presented in [28]. These problems can be overcome in display logic.[12] We will employ this very general and flexible sequent-style proof-theoretical framework and present display sequent calculi for the logics (I_i, C_j), which add strong negation $\sim$ to H-B. We may then apply a very general cut-elimination theorem stating that every "properly displayable" logic enjoys cut-elimination, a theorem due to Nuel Belnap [23], see Theorem 2.3.40.

As we have noted in earlier chapters, one fundamental idea of display calculi is to exploit the fact that certain logical operations are residuated pairs to specify rules for introducing these operations on the left and

[11] The corresponding proof of Peirce's law in the multiple-conclusion sequent calculus is:

$$
\dfrac{\dfrac{\dfrac{\dfrac{\dfrac{\alpha \vdash \alpha}{\alpha \vdash \alpha, \beta}}{\varnothing \vdash (\alpha \to \beta), \alpha} \qquad \alpha \vdash \alpha}{(\alpha \to \beta) \to \alpha \vdash \alpha, \alpha}}{(\alpha \to \beta) \to \alpha \vdash \alpha}}{\varnothing \vdash ((\alpha \to \beta) \to \alpha) \to \alpha}
$$

[12] Recall that Buisman and Goré [28] have presented a non-standard cut-free sequent calculus for Heyting-Brouwer logic. In this calculus, the sequent rule for implications in succedent position of a sequent and the rule for co-implications in antecedent position of a sequent come with side conditions on variables for families of sets of formulas. Two other, cut-free sequent calculi for Heyting-Brouwer logic are presented in [70]. The first calculus is intermediate between display calculi and standard sequent systems. From this system a variant is defined, which is amenable to automated proof-search.

the right side of the derivability sign $\vdash$, that is, in antecedent and in succedent position. Moreover, it is characteristic of display logic to associate a single structural connective $\diamond$ in the language of sequents with a pair $(\diamond_1, \diamond_2)$ of connectives from the logical object language, so that in antecedent position $\diamond$ is interpreted as $\diamond_1$ and in succedent position as $\diamond_2$. The left introduction rule for $\diamond_1$ and the right introduction rule for $\diamond_2$ may then be stated as follows:

$$\frac{\alpha \diamond \beta \vdash X}{\alpha \diamond_1 \beta \vdash X} \qquad \frac{X \vdash \alpha \diamond \beta}{X \vdash \alpha \diamond_2 \beta},$$

where X is a structure, a term in the language of sequents. The connectives $\diamond_1$ and $\diamond_2$ may be said to be Gentzen duals of each other.

As mention in Chapter 2, a cut-free sound and complete display calculus for Heyting-Brouwer logic has been presented by Goré [68]. In this section, we present a variant of Goré's system and extend it by rules for constructively negated formulas. Whereas Goré treats the pair of commutative operations $\wedge$ and $\vee$ as Gentzen duals and the non-commutative operations $\to$ and $\prec$, we here will treat the residuated pairs $(\wedge, \to)$ and $(\prec, \vee)$ as pairs of Gentzen duals.

Recall that in Gentzen's sequents, the comma may bee seen as a context sensitive structural connective to be understood as conjunction in antecedent position and as disjunction in succedent position of a sequent. In our display calculi, we will use the binary operations $\circ$ and $\bullet$ as structural connectives. In antecedent position, $\circ$ is to be interpreted as conjunction and in succedent position as implication. In antecedent position, $\bullet$ is to be read as co-implication and in succedent position as disjunction. A sequent is an expression of the shape $X \vdash Y$, where X and Y are structures. We also assume the empty structure $\mathbf{I}$, and the set of structures is defined in the obvious way as follows:

$$\begin{aligned} \text{formulas:} \quad & \alpha \in \mathit{Form(Atom)} \\ \text{structures:} \quad & X \in \mathit{Struc(Form)} \\ X ::= \quad & \alpha \mid \mathbf{I} \mid (X \circ X) \mid (X \bullet X). \end{aligned}$$

The intuitive interpretation of the structural connectives justifies certain "display postulates" (*dp*) (we omit outer brackets):

$$\frac{\dfrac{Y \vdash X \circ Z}{X \circ Y \vdash Z}}{X \vdash Y \circ Z} \qquad \frac{\dfrac{X \vdash Y \circ Z}{X \circ Y \vdash Z}}{Y \vdash X \circ Z} \qquad \frac{\dfrac{X \bullet Z \vdash Y}{X \vdash Y \bullet Z}}{X \bullet Y \vdash Z} \qquad \frac{\dfrac{X \bullet Y \vdash Z}{X \vdash Y \bullet Z}}{X \bullet Z \vdash Y}$$

Moreover, we assume certain rules ($\mathbf{I}r$) which govern the empty structure:

$$\frac{\dfrac{X \circ \mathbf{I} \vdash Y}{X \vdash Y}}{\mathbf{I} \circ X \vdash Y} \qquad \frac{\dfrac{\mathbf{I} \circ X \vdash Y}{X \vdash Y}}{X \circ \mathbf{I} \vdash Y} \qquad \frac{\dfrac{X \vdash Y \bullet \mathbf{I}}{X \vdash Y}}{X \vdash \mathbf{I} \bullet Y} \qquad \frac{\dfrac{X \vdash \mathbf{I} \bullet Y}{X \vdash Y}}{X \vdash Y \bullet \mathbf{I}}$$

certain "logical" structural rules:

$$\frac{}{p \vdash p}\,(id) \qquad \frac{}{\sim p \vdash \sim p}\,(id{\sim}) \qquad \frac{X \vdash \alpha \quad \alpha \vdash Y}{X \vdash Y}\,(cut)$$

and versions of the standard structural rules from ordinary Gentzen calculi for classical logic, monotonicity (alias thinning or weakening), exchange (alias permutation), and contraction, together with associativity (presented in Table 5.3). Note that the failure of left (right) monotonicity for $\bullet$ (respectively $\circ$) blocks the provability of Peirce's co-law (respectively Peirce's law).

$$\frac{X \vdash Y}{X \vdash Y \bullet Z}\,(rm) \qquad \frac{X \vdash Y}{X \circ Z \vdash Y}\,(lm)$$

$$\frac{X \vdash Y \bullet Z}{X \vdash Z \bullet Y}\,(re) \qquad \frac{X \circ Z \vdash Y}{Z \circ X \vdash Y}\,(le)$$

$$\frac{X \vdash Y \bullet Y}{X \vdash Y}\,(rc) \qquad \frac{X \circ X \vdash Y}{X \vdash Y}\,(lc)$$

$$\frac{X \vdash (Y \bullet Z) \bullet X'}{X \vdash Y \bullet (Z \bullet X')}\,(ra) \qquad \frac{(X \circ Y) \circ Z \vdash X'}{X \circ (Y \circ Z) \vdash X'}\,(la)$$

Table 5.3.: Structural sequent rules.

The display sequent calculi $\delta(I_i, C_j)$, $i, j \in \{1, 2, 3, 4\}$, for the constructive logics (I_i, C_j) share these rules and the introduction rules stated in Table 5.3. The particular display calculus $\delta(I_i, C_j)$ then is the proof system obtained by adding the rules rI_i and rC_j from Table 5.5.

A derivation of a sequent s from a set of sequents $\{\mathsf{s}_1, \ldots, \mathsf{s}_n\}$ in $\delta(I_i, C_j)$ is defined as a tree with root s such that every leaf is an instantiation of (id), $(id{\sim})$, or a sequent from $\{\mathsf{s}_1, \ldots, \mathsf{s}_n\}$, and every other node is obtained by an application of one of the remaining rules. A proof of a sequent s in $\delta(I_i, C_j)$ is a derivation of s from $\varnothing$. Sequents s and s' are said to be interderivable iff s is derivable from $\{\mathsf{s}'\}$ and s' is derivable from s.

Two sequents s and s' are said to be structurally equivalent if they are interderivable by means of display postulates only. It is characteristic for display calculi that any substructure of a given sequent s may be displayed as the entire antecedent or succedent of a structurally equivalent sequent s'.

If $\mathsf{s} = X \vdash Y$ is a sequent, then the displayed occurrence of X (Y) is an antecedent (succedent) part of s. If an occurrence of $(Z \circ W)$ is an antecedent part of s, then the displayed occurrences of Z and W are antecedent parts of s. If an occurrence of $(Z \bullet W)$ is an antecedent part of s, then the displayed occurrence of Z (W) is an antecedent (succedent) part of s. If an occurrence of $(Z \circ W)$ is a succedent part of s, then the displayed occurrence of Z (W) is an antecedent (succedent) part of s. If an occurrence of $(Z \bullet W)$ is a succedent part of s, then the displayed occurrences of Z and W are succedent parts of s.

$$\frac{X \vdash \alpha \quad Y \vdash \beta}{X \circ Y \vdash (\alpha \wedge \beta)} \ (\vdash \wedge) \qquad \frac{\alpha \circ \beta \vdash X}{(\alpha \wedge \beta) \vdash X} \ (\wedge \vdash)$$

$$\frac{X \vdash \alpha \bullet \beta}{X \vdash (\alpha \vee \beta)} \ (\vdash \vee) \qquad \frac{\alpha \vdash X \quad \beta \vdash Y}{(\alpha \vee \beta) \vdash X \bullet Y} \ (\vee \vdash)$$

$$\frac{X \vdash \alpha \circ \beta}{X \vdash (\alpha \to \beta)} \ (\vdash \to) \qquad \frac{X \vdash \alpha \quad \beta \vdash Y}{(\alpha \to \beta) \vdash X \circ Y} \ (\to \vdash)$$

$$\frac{X \vdash \beta \quad \alpha \vdash Y}{X \bullet Y \vdash \beta \prec\!\!\!- \alpha} \ (\vdash \prec\!\!\!-) \qquad \frac{\beta \bullet \alpha \vdash X}{\beta \prec\!\!\!- \alpha \vdash X} \ (\prec\!\!\!- \vdash)$$

$$\frac{X \vdash {\sim}\alpha \bullet {\sim}\beta}{X \vdash {\sim}(\alpha \wedge \beta)} \ (\vdash {\sim}\wedge) \qquad \frac{{\sim}\alpha \vdash X \quad {\sim}\beta \vdash Y}{{\sim}(\alpha \wedge \beta) \vdash X \bullet Y} \ ({\sim}\wedge \vdash)$$

$$\frac{X \vdash {\sim}\alpha \quad Y \vdash {\sim}\beta}{X \circ Y \vdash {\sim}(\alpha \vee \beta)} \ (\vdash {\sim}\vee) \qquad \frac{{\sim}\alpha \circ {\sim}\beta \vdash X}{{\sim}(\alpha \vee \beta) \vdash X} \ ({\sim}\vee \vdash)$$

$$\frac{X \vdash \alpha}{X \vdash {\sim}{\sim}\alpha} \ (\vdash {\sim}{\sim}) \qquad \frac{\alpha \vdash X}{{\sim}{\sim}\alpha \vdash X} \ ({\sim}{\sim} \vdash)$$

Table 5.4.: Introduction rules shared by all logics (I_i, C_j)

Theorem 5.3.1 (cf. [23]) *For every sequent* s *and every antecedent*

rI_1	$\dfrac{X \vdash \alpha \quad Y \vdash {\sim}\beta}{X \circ Y \vdash {\sim}(\alpha \to \beta)}$	$\dfrac{\alpha \circ {\sim}\beta \vdash X}{{\sim}(\alpha \to \beta) \vdash X}$
rI_2	$\dfrac{X \vdash \alpha \circ {\sim}\beta}{X \vdash {\sim}(\alpha \to \beta)}$	$\dfrac{X \vdash \alpha \quad {\sim}\beta \vdash Y}{{\sim}(\alpha \to \beta) \vdash X \circ Y}$
rI_3	$\dfrac{X \vdash \alpha \quad \beta \vdash Y}{X \bullet Y \vdash {\sim}(\alpha \to \beta)}$	$\dfrac{\alpha \bullet \beta \vdash X}{{\sim}(\alpha \to \beta) \vdash X}$
rI_4	$\dfrac{X \vdash {\sim}\beta \quad {\sim}\alpha \vdash Y}{X \bullet Y \vdash {\sim}(\alpha \to \beta)}$	$\dfrac{{\sim}\beta \bullet {\sim}\alpha \vdash X}{{\sim}(\alpha \to \beta) \vdash X}$
rC_1	$\dfrac{X \vdash {\sim}\alpha \bullet \beta}{X \vdash {\sim}(\alpha -\!\!< \beta)}$	$\dfrac{{\sim}\alpha \vdash X \quad \beta \vdash Y}{{\sim}(\alpha -\!\!< \beta) \vdash X \bullet Y}$
rC_2	$\dfrac{X \vdash {\sim}\alpha \quad \beta \vdash Y}{X \bullet Y \vdash {\sim}(\alpha -\!\!< \beta)}$	$\dfrac{{\sim}\alpha \bullet \beta \vdash X}{{\sim}(\alpha -\!\!< \beta) \vdash X}$
rC_3	$\dfrac{X \vdash \alpha \circ \beta}{X \vdash {\sim}(\alpha -\!\!< \beta)}$	$\dfrac{Y \vdash \alpha \quad \beta \vdash X}{{\sim}(\alpha -\!\!< \beta) \vdash Y \circ X}$
rC_4	$\dfrac{X \vdash {\sim}\beta \circ {\sim}\alpha}{X \vdash {\sim}(\alpha -\!\!< \beta)}$	$\dfrac{Y \vdash {\sim}\beta \quad {\sim}\alpha \vdash X}{{\sim}(\alpha -\!\!< \beta) \vdash Y \circ X}$

Table 5.5.: Sequent rules for negated implications and co-implications.

(succedent) part X of s, there exists a sequent s' structurally equivalent to s such that X is the entire antecedent (succedent) of s'.

Observation 5.3.2 *For every $\mathcal{L}$-formula α and every display calculus $\delta(I_i, C_j)$, $\alpha \vdash \alpha$ is provable (and hence $\mathbf{I} \vdash \alpha \to \alpha$ and $\alpha -\!\!< \alpha \vdash \mathbf{I}$ are provable).*

Proof. The proof is by induction on the number of occurrences of connectives in α. We here display two cases for $\delta(I_4, C_3)$:

$$\dfrac{\dfrac{{\sim}\beta \vdash {\sim}\beta \quad {\sim}\alpha \vdash {\sim}\alpha}{{\sim}\beta \bullet {\sim}\alpha \vdash {\sim}(\alpha \to \beta)}}{{\sim}(\alpha \to \beta) \vdash {\sim}(\alpha \to \beta)}$$

$$\dfrac{\dfrac{\alpha \vdash \alpha \quad \beta \vdash \beta}{{\sim}(\alpha -\!\!< \beta) \vdash \alpha \circ \beta}}{{\sim}(\alpha -\!\!< \beta) \vdash {\sim}(\alpha -\!\!< \beta)}$$

The remaining cases are equally simple. ∎

One can define translations τ_1 and τ_2 from structures into formulas such that these translations reflect the intuitive, context-sensitive interpretation of the structural connectives: τ_1 translates structures which are antecedent parts of a sequent, whereas τ_2 translates structures which are succedent parts of a sequent.

Definition 5.3.3 *The translations τ_1 and τ_2 from structures into formulas are inductively defined as follows, where α is a formula and p is a certain atom:*

$$
\begin{array}{rclrcl}
\tau_1(\alpha) & = & \alpha & \tau_2(\alpha) & = & \alpha \\
\tau_1(\mathbf{I}) & = & p \to p & \tau_2(\mathbf{I}) & = & p \prec p \\
\tau_1(X \circ Y) & = & \tau_1(X) \wedge \tau_1(Y) & \tau_2(X \circ Y) & = & \tau_1(X) \to \tau_2(Y) \\
\tau_1(X \bullet Y) & = & \tau_1(X) \prec \tau_2(Y) & \tau_2(X \bullet Y) & = & \tau_2(X) \vee \tau_2(Y)
\end{array}
$$

Theorem 5.3.4 (Soundness) *(1) If the sequent $X \vdash Y$ is provable in $\delta(I_i, C_j)$, then $\tau_1(X) \models^+_{I_i,C_j} \tau_2(Y)$. (2) If the sequent $X \vdash Y$ is provable in $\delta(I_i, C_j)$, then $\sim\tau_2(Y) \models^-_{I_i,C_j} \sim\tau_1(X)$.*

Proof. (1) can be proved by induction on derivations in the display calculi
$\delta(I_i, C_j)$. We present here just two cases and omit some subscripts. (a) rC_2 right-hand side of $\vdash$. Suppose (*) $\tau_1(X) \models^+ \tau_2(\sim\alpha)$ and $\tau_1(\beta) \models^+ \tau_2(Y)$. To show: $\tau_1(X \bullet Y) \models^+ \tau_2(\sim(\alpha \prec \beta))$. $\tau_1(X \bullet Y) = \tau_1(X) \prec \tau_2(Y)$. Let $w \models^+ \tau_1(X) \prec \tau_2(Y)$. Then $\exists w'$ with $w' \leq w$, $w' \models^+ \tau_1(X)$, and $w' \not\models^+ \tau_2(Y)$. By (*), $w' \models^+ \tau_2(\sim\alpha)$ (i.e., $w' \models^- \alpha$) and $w' \not\models^+ \tau_1(\beta)$. Thus, $w \models^+ \tau_2(\sim(\alpha \prec \beta))$. (b) rC_4 left-hand side of $\vdash$. Suppose (*) $\tau_1(Y) \models^+ \tau_2(\sim\beta)$ and $\tau_1(\sim\alpha) \models^+ \tau_2(X)$. To show: $\tau_1(\sim(\alpha \prec \beta)) \models^+ \tau_2(Y \circ X)$. $\tau_2(Y \circ X) = \tau_1(Y) \to \tau_2(X)$. Let $w \models^+ \tau_1(\sim(\alpha \prec \beta))$. Then, by cC_4, $\forall w' \geq w$: $w' \models^- \alpha$ or $w' \not\models^- \beta$. By (*), $\forall w' \geq w$: $w' \models^+ \tau_2(X)$ or $w' \not\models^+ \tau_1(Y)$. Thus, $w \models^+ \tau_1(Y) \to \tau_2(X)$. (2) follows from (1), the definition of $\models^-_{I_i,C_j}$ and the fact that $w \models^+ \sim\alpha$ iff $w \models^- \alpha$. (Indeed, the succedents of the two claims are equivalent.) ∎

In order to prove completeness, we will apply some lemmata. We add to the language $\mathcal{L}$ for every atomic formula p a new atom p^* to obtain the language $\mathcal{L}^*$. If α is an $\mathcal{L}$-formula, $(\alpha)^*$ is the result of replacing every strongly negated atom $\sim p$ in α by p^*.

Lemma 5.3.5 *For every $\mathcal{L}$-formula α, if $\varnothing \models^+_{I_i,C_j} \alpha$, then $(\rho_{I_i,C_j}(\alpha))^*$ is valid in* H-B.

Proof. Let $\varnothing \models^{+}_{I_i,C_j} \alpha$. By Lemma 5.2.7, this is the case iff $\varnothing \models^{+}_{I_i,C_j} \rho_{I_i,C_j}(\alpha)$. If $(\rho_{I_i,C_j}(\alpha))^{*}$ is not valid in H-B, then there is a model $\mathcal{M} = \langle I, \leq, v\rangle$ and $w \in I$ with $\mathcal{M}, w \not\models (\rho_{I_i,C_j}(\alpha))^{*}$. Define the structure $\mathcal{M}' = \langle I', \leq', v^{+}, v^{-}\rangle$ by setting $I' := I$, $\leq' := \leq$, $v^{+} := v$ and $w \in v^{-}(p)$ iff $w \in v(p^{*})$, for every atomic $\mathcal{L}$-formula p. Clearly, $\mathcal{M}'$ is a model. By induction on $\mathcal{L}$-formulas α, one can show that $\mathcal{M}, w \not\models (\rho_{I_i,C_j}(\alpha))^{*}$ iff $\mathcal{M}', w \not\models^{+} \rho_{I_i,C_j}(\alpha)$, which contradicts $\varnothing \models^{+}_{I_i,C_j} \rho_{I_i,C_j}(\alpha)$. ∎

Lemma 5.3.6 *For every $\sim$-free $\mathcal{L}$-formula α, if α is provable in* H-B, *then $\mathbf{I} \vdash \alpha$ is provable in $\delta(I_i, C_j)$ without using any sequent rules for strongly negated formulas.*

Proof. It is enough to show that the axiom schemata for H-B stated in [161, p. 24] and [163, p. 18] are provable in $\delta(I_i, C_j)$ and that modus ponens and the rule

$$\frac{\alpha}{\neg - \alpha}$$

preserve provability in $\delta(I_i, C_j)$ without making appeal to sequent rules for strongly negated formulas. For the latter and for Axiom (A_{11}) from [163], for example, see:

$$\begin{array}{c}
\mathbf{I} \vdash \alpha \\ \hline
\mathbf{I} \vdash \alpha \bullet (p \mathbin{-\!\!<} p) \\ \hline
\mathbf{I} \circ (p \to p) \vdash \alpha \bullet (p \mathbin{-\!\!<} p) \\ \hline
(p \to p) \vdash \alpha \bullet (p \mathbin{-\!\!<} p) \\ \hline
((p \to p) \bullet \alpha) \vdash (p \mathbin{-\!\!<} p) \\ \hline
((p \to p) \mathbin{-\!\!<} \alpha) \vdash (p \mathbin{-\!\!<} p) \\ \hline
\mathbf{I} \circ ((p \to p) \mathbin{-\!\!<} \alpha) \vdash (p \mathbin{-\!\!<} p) \\ \hline
\mathbf{I} \vdash ((p \to p) \mathbin{-\!\!<} \alpha) \circ (p \mathbin{-\!\!<} p) \\ \hline
\mathbf{I} \vdash ((p \to p) \mathbin{-\!\!<} \alpha) \to (p \mathbin{-\!\!<} p)
\end{array}
\qquad
\begin{array}{c}
\alpha \vdash \alpha \quad \beta \vdash \beta \\ \hline
\alpha \bullet \beta \vdash (\alpha \mathbin{-\!\!<} \beta) \\ \hline
\alpha \vdash \beta \bullet (\alpha \mathbin{-\!\!<} \beta) \\ \hline
\alpha \circ \mathbf{I} \vdash \beta \bullet (\alpha \mathbin{-\!\!<} \beta) \\ \hline
\alpha \circ \mathbf{I} \vdash \beta \vee (\alpha \mathbin{-\!\!<} \beta) \\ \hline
\mathbf{I} \vdash \alpha \circ (\beta \vee (\alpha \mathbin{-\!\!<} \beta)) \\ \hline
\mathbf{I} \vdash \alpha \to (\beta \vee (\alpha \mathbin{-\!\!<} \beta))
\end{array}$$

∎

Lemma 5.3.7 *For every $\mathcal{L}$-formula α, the sequents $\alpha \vdash \rho_{I_i,C_j}(\alpha)$ and $\rho_{I_i,C_j}(\alpha) \vdash \alpha$ are provable in $\delta(I_i, C_j)$.*

Lemma 5.3.8 *Every sequent $X \vdash \tau_1(X)$ and $\tau_2(X) \vdash X$ is provable in $\delta(I_i, C_j)$, for all $i, j \in \{1, 2, 3, 4\}$.*

Proof. By simultaneous induction on X. For instance, we have:

$$\frac{X \vdash \tau_1(X) \qquad \tau_2(Y) \vdash Y}{X \bullet Y \vdash \tau_1(X) \mathbin{-\!\!<} \tau_2(Y)}$$

∎

Theorem 5.3.9 (Completeness)
(1) If $\rho_{I_i,C_j}(\tau_1(X)) \models^{+}_{I_i,C_j} \rho_{I_i,C_j}(\tau_2(Y))$*, then the sequent* $X \vdash Y$ *is provable in* $\delta(I_i, C_j)$*. (2) If* $\rho_{I_i,C_j}({\sim}\tau_2(Y)) \models^{-}_{I_i,C_j} \rho_{I_i,C_j}({\sim}\tau_1(X))$*, then the sequent* $X \vdash Y$ *is provable in* $\delta(I_i, C_j)$*.*

Proof. (1) Suppose $\rho_{I_i,C_j}(\tau_1(X)) \models^{+}_{I_i,C_j} \rho_{I_i,C_j}(\tau_2(Y))$. Then,

$$\varnothing \models^{+}_{I_i,C_j} \rho_{I_i,C_j}(\tau_1(X)) \to \rho_{I_i,C_j}(\tau_2(Y)).$$

Using Lemma 5.3.5, we obtain that $(\rho_{I_i,C_j}(\tau_1(X)))^* \to (\rho_{I_i,C_j}(\tau_2(Y)))^*$ is valid in H-B. By completeness of Rauszer's axiomatization of H-B, it follows that

$$(\rho_{I_i,C_j}(\tau_1(X)))^* \to (\rho_{I_i,C_j}(\tau_2(Y)))^*$$

is provable in this axiom system. By Lemma 5.3.6, we obtain a proof of the sequent $\mathbf{I} \vdash (\rho_{I_i,C_j}(\tau_1(X)))^* \to (\rho_{I_i,C_j}(\tau_2(Y)))^*$. By applying (*cut*) to this sequent and the provable sequent

$$(\rho_{I_i,C_j}(\tau_1(X)))^* \to (\rho_{I_i,C_j}(\tau_2(Y)))^* \vdash (\rho_{I_i,C_j}(\tau_1(X)))^* \circ (\rho_{I_i,C_j}(\tau_2(Y)))^*,$$

we may see that $(\rho_{I_i,C_j}(\tau_1(X)))^* \vdash (\rho_{I_i,C_j}(\tau_2(Y)))^*$ is provable in $\delta(I_i, C_j)$. Since
$(\rho_{I_i,C_j}(\tau_1(X)))^* \vdash (\rho_{I_i,C_j}(\tau_2(Y)))^*$ is provable without any appeal to sequent rules for strongly negated formulas, the sequent $\rho_{I_i,C_j}(\tau_1(X)) \vdash \rho_{I_i,C_j}(\tau_2(Y))$ is provable in $\delta(I_i, C_j)$, and then, by Lemma 5.3.7, $\tau_1(X) \vdash \tau_2(Y)$ is provable in $\delta(I_i, C_j)$. Finally, by Lemma 5.3.8, $X \vdash Y$ is provable in $\delta(I_i, C_j)$. (2): Obvious. ∎

As explained already in Chapter 2, Belnap [23] presents a very general cut-elimination theorem covering all "properly displayable" logics, which are logics satisfying a number of conditions (C1) – (C8). Recall that condition (C8) is the requirement of eliminability of principal cuts, i.e., applications of (*cut*) in which the two premise sequents have been obtained by introducing the main connective of the cut-formula α. The display calculi $\delta(I_i, C_j)$ do not satisfy condition (C1), which says that each formula which is a constituent of some premise of a sequent rule is a subformula of the conclusion sequent. We may note, however, that $X \vdash Y$ is provable in $\delta(I_i, C_j)$ iff $(\rho_{Ii,C_j}(\tau_1(X)))^* \vdash (\rho_{Ii,C_j}(\tau_2(Y)))^*$ is provable in $\delta(I_i, C_j)$ without any appeal to rules involving $\sim$. Let $\delta(I_i, C_j)^+$ denote the result of dropping all sequent rules exhibiting $\sim$ from $\delta(I_i, C_j)$.

Theorem 5.3.10 *If* $X \vdash Y$ *is provable in system* $\delta(I_i, C_j)$*, then*

$$(\rho_{Ii,C_j}(\tau_1(X)))^* \vdash (\rho_{Ii,C_j}(\tau_2(Y)))^*$$

is provable in $\delta(I_i, C_j)^+$ *without any applications of (cut).*

Proof. The system $\delta(I_i, C_j)^+$ satisfies Belnap's conditions (C1)–(C8). The principal cut-elimination step for $-\!\!\!<$ is:

$$\dfrac{\dfrac{X \vdash \beta \quad \alpha \vdash Y}{X \bullet Y \vdash \beta -\!\!\!< \alpha} \qquad \dfrac{\beta \bullet \alpha \vdash Z}{\beta -\!\!\!< \alpha \vdash Z}}{X \bullet Y \vdash Z}$$

is replaced by

$$\dfrac{\dfrac{\dfrac{\dfrac{\dfrac{X \vdash \beta \quad \dfrac{\beta \bullet \alpha \vdash Z}{\beta \vdash \alpha \bullet Z}}{X \vdash \alpha \bullet Z}}{X \bullet Z \vdash \alpha} \qquad \alpha \vdash Y}{X \bullet Z \vdash Y}}{X \vdash Z \bullet Y}}{X \bullet Y \vdash Z}$$

∎

We noted above that intuitionistic logic enjoys the disjunction property but does not enjoy the constructible falsity property with respect to intuitionistic negation. In Heyting-Brouwer logic, the disjunction property fails. If we take co-negation as primitive, the disjunction property already fails in the $\{-, \wedge, \vee, -\!\!\!<\}$-fragment of H-B (alias dual intuitionistic logic), since for every atom p, $p \vee -p$ is valid, but obviously neither p nor $-p$ is valid. However,

Observation 5.3.11 *If* $-(\alpha \wedge \beta)$ *is valid in* H-B*, then so are* $-\alpha$ *or* $-\beta$*.*

Proof. By "gluing" of models. Suppose there are models $\mathcal{M}_1$ and $\mathcal{M}_2$ and states w_1, w_2 with $\mathcal{M}_1, w_1 \not\models^+ -\alpha$ and $\mathcal{M}_2, w_2 \not\models^+ -\beta$. We add a new state w such that the truth of no atom is supported at w and consider the relation $\leq'$, which is the reflexive, transitive closure of $\leq \cup \{\langle w, w_1\rangle, \langle w, w_2\rangle\}$. The resulting structure is a model, and at w it supports the truth of the valid $-(\alpha \wedge \beta)$, which contradicts the fact that $\mathcal{M}_1, w_1 \not\models^+ -\alpha$ and $\mathcal{M}_2, w_2 \not\models^+ -\beta$. ∎

So far, we have motivated and defined the sixteen logics (I_i, C_j), $i, j \in \{1, 2, 3, 4\}$,[13] which comprise both intuitionistic implication and co-

[13] In this chapter, we do not consider logics which combine co-implication and strong negation, but in which implication is absent. Among these logics, we can find a "dual" of N4, see also [84].

implication. These logics enrich the combination of the constructive logics IPL$^+$ and HB$^+$ by a strong negation operation $\sim$, which may be regarded as a constructive negation. Its conservative addition to IPL$^+$ in the systems of Nelson does not lead to a violation of the disjunction property and gives rise to the constructible falsity property. Moreover, we have presented strongly sound and complete display sequent calculi for the logics (I_i, C_j).

The logics (I_i, C_j) may be viewed as constructive logics, if one is not disturbed by the fact that these logics fail to enjoy the constructible falsity property for the definable intuitionistic negation and the disjunction property for the definable co-negation. The constructiveness of the logics (I_i, C_j) would have to be further justified by showing them correct with respect to an interpretation in terms of canonical proofs, dual proofs, *disproofs*, and *dual disproofs*, where a disproof (dual disproof) of α is a derivation of $\sim\alpha$ from the empty set (derivation of the empty set from $\sim\alpha$). In a first step we consider HB$^+$.

We refer to the result of dropping the sequent rules for $\rightarrow$ from $(I_i, C_j)^+$ as δHB$^+$. δHB$^+$ is a display sequent calculus for HB$^+$ in the language $\{\wedge, \vee, \prec\}$. If $X \vdash Y$ is provable in δHB$^+$, then it follows from Theorem 5.3.4 that $\tau_1(X)$ entails $\tau_2(Y)$ in HB$^+$; the converse follows by Theorem 5.3.9. Since the structural connective $\circ$ is interpreted as implication in succedent position, the proof of Theorem 5.3.12 refers to *both* proofs and dual proofs. In particular, we must say what is a canonical reductio (dual proof) of an implication $(\alpha \rightarrow \beta)$, namely a pair (π_1, π_2), where π_1 is a proof of α and π_2 is a reductio of β.[14] Moreover, we require that for no formula α, there exists both a proof and a reductio.

Theorem 5.3.12 *If* $\alpha \vdash \mathbf{I}$ *is provable in* δHB$^+$, *then there exists a construction* π *which is a reductio ad absurdum of* α.

Proof. We prove a more general claim by induction on proofs in δHB$^+$, namely: If $X \vdash Y$ is provable in δHB$^+$, then there exists a construction π such that $\pi(\pi')$ is a reductio ad absurdum of $\tau_1(X)$ whenever π' is a reductio ad absurdum of $\tau_2(Y)$. Note that any reductio of $\tau_2(\mathbf{I}) = (p \prec p)$ is the identity function.
The cases of the rules $(\prec\ \vdash)$, $(\wedge \vdash)$, and $(\vdash \vee)$ are trivial.
$(\vdash\ \prec)$: Suppose $\pi_1({\pi_1}')$ is a reductio of $\tau_1(X)$ whenever ${\pi_1}'$ is a reductio of β, and $\pi_2({\pi_2}')$ is a reductio of α whenever ${\pi_2}'$ is a reductio of $\tau_2(Y)$. We define a construction π^* such that $\pi^*(\pi^{*\prime})$ is a reductio of $\tau_1(X) \prec \tau_2(Y)$ whenever $\pi^{*\prime}$ is a reductio of $\beta \prec \alpha$. Let $\pi^{*\prime}$ be a reductio of $\beta \prec \alpha$. Then for every reductio θ of α, $\pi^{*\prime}(\theta)$ is a reductio of β.

[14] A proof of $(\alpha \prec \beta)$ then is a reductio of $(\alpha \rightarrow \beta)$.

Therefore, $\pi^{*\prime}(\pi_2)$ is a reductio of $\beta \prec \tau_2(Y)$, and $\pi^* := \pi_1(\pi^{*\prime}(\pi_2))$ is a reductio of $\tau_1(X) \prec \tau_2(Y)$.
($\vdash \wedge$): Suppose $\pi_1(\pi_1{}')$ is a reductio of $\tau_1(X)$ whenever $\pi_1{}'$ is a reductio of α, and $\pi_2(\pi_2{}')$ is a reductio of $\tau_1(Y)$ whenever $\pi_2{}'$ is a reductio of β. We define a construction π^* such that $\pi^*(\pi^{*\prime})$ is a reductio of $\tau_1(X) \wedge \tau_1(Y)$ whenever $\pi^{*\prime}$ is a reductio of $\alpha \wedge \beta$. Let $\pi^{*\prime}$ be a reductio of $\alpha \wedge \beta$. Then $\pi^{*\prime}$ is a pair (i, π), such that $i = 0$ and π is reductio of α or $i = 1$ and π is a reductio of β. Clearly, $\pi^* = (0, \pi_1(\pi))$ or $\pi^* = (1, \pi_2(\pi))$ is a reductio of $\tau_1(X) \wedge \tau_1(Y)$.
The display postulate: $X \bullet Z \vdash Y \ / X \vdash Y \bullet Z$: Suppose $\pi(\pi')$ is a reductio of $\tau_1(X \bullet Z)$ $(= \tau_1(X) \prec \tau_2(Z))$ whenever π' is a reductio of $\tau_2(Y)$. We define a construction π^* such that $\pi^*(\pi^{*\prime})$ is a reductio of $\tau_1(X)$ whenever $\pi^{*\prime}$ is a reductio of $\tau_2(Y \bullet Z)$ $(= \tau_2(Y) \vee \tau_2(Z))$. Thus, let $\pi^{*\prime} = (\pi_1, \pi_2)$, where π_1 is a reductio of $\tau_2(Y)$ and π_2 is a reductio of $\tau_2(Z)$. Then $\pi(\pi_1)$ is a reductio of $\tau_1(X) \prec \tau_2(Z)$, and $\pi(\pi_1)(\pi_2)$ is a reductio of $\tau_1(X)$.
The display postulate: $X \circ Y \vdash Z \ / X \vdash Y \circ Z$: Suppose $\pi(\pi')$ is a reductio of $\tau_1(X) \wedge \tau_1(Y)$ whenever π' is a reductio of $\tau_2(Z)$. That is, $\pi(\pi')$ is a pair (i, π'') such that $i = 0$ and π'' is a reductio of $\tau_1(X)$ or $i = 1$ and π'' is a reductio of $\tau_1(Y)$. Suppose π^* is a pair (π_1, π_2), where π_1 is a proof of $\tau_1(Y)$ and π_2 is a reductio of $\tau_2(Z)$. Then $\pi(\pi_2)$ is a pair $(0, \pi'')$ and π'' is a reductio of $\tau_1(X)$. That $\pi(\pi_2)$ is a pair $(1, \pi'')$ where π'' is a reductio of $\tau_1(Y)$ is impossible, because π_1 is a proof of $\tau_1(Y)$.
The structural rule (lm): $X \vdash Y \ / X \circ Z \vdash Y$. Suppose $\pi(\pi')$ is a reductio of $\tau_1(X)$ whenever π' is a reductio of $\tau_2(Y)$. Let π^* be a reductio of $\tau_2(Y)$. Then $(0, \pi(\pi^*))$ is a reductio of $\tau_1(X) \wedge \tau_1(Z)$.
The remaining cases are left to the reader. ∎

5.3.1. Inferential status and speech acts

We now consider the problem of defining an inferentialist semantics in terms of proofs, disproofs, and their duals for the logics (I_i, C_j). Whereas the Brouwer-Heyting-Kolmogorov interpretation of intuitionistic logic uses just the notion of proof as primitive, and López-Escobar's inferentialist interpretation of Nelson's logics with strong negation utilizes only the notions of proof and disproof as primitive, the inferentialist interpretation of bi-intuitionistic logic with strong negation employs the four notions of proofs, disproofs, dual proofs, and dual disproofs as primitive concepts. This semantics will be motivated by considerations on the speech act of denial.

It seems to be an accepted view that assertion and denial are particularly important speech acts in the context of a use-based, inferentialist account of linguistic meaning. In particular, the idea is that the rules of use that determine the meaning of linguistic expressions provide a basis for warranted assertions and denials. In order to make an assertion, it is enough to seriously utter a sentence, for example the sentence "Mary is beautiful". In order to obtain an absolutely clear case of denying that Mary is beautiful, the contrary predicate "is ugly" may be used, i.e., the sentence "Mary is ugly" may be seriously uttered. Instead of replacing the adjective "beautiful" by another, contrary item from the lexicon, namely the adjective "ugly", one may employ a suitable unary negation connective, $\sim$ "it is definitely false that". It is definitely false that Mary is beautiful if and only if Mary is ugly. The "only if" may not be clear. If it is definitely false that Mary is beautiful, then Mary not just fails to be beautiful, but she is ugly. It is then denied that Mary is beautiful by seriously uttering the negated sentence "$\sim$ Mary is beautiful". This move is supported by the existence of more systematically and regularly connected pairs of contrary predicates in the lexicon: "sane" versus "insane", "believes" versus "disbelieves", "desirable" versus "undesirable", etc. The prefixes "in", "dis" and "un" suggest the introduction of the negation connective $\sim$, so that a denial of a sentence α may be represented as an assertion of $\sim\alpha$.

Negation can be iterated. Is a denial of $\sim\alpha$ an assertion of α? Can denying be iterated? Can asserting be iterated? It seems plausible to assume that a speaker may assert that Mary is beautiful not only by seriously uttering the sentence "Mary is beautiful", but also by seriously uttering the sentence "I assert that Mary is beautiful". Similarly, a speaker may deny that Mary is beautiful not just by seriously uttering the sentences "Mary is ugly", but also by seriously uttering the sentence "I deny that Mary is beautiful". Clearly, first- and other-person asserting-that-ascriptions and denying-that-ascriptions *may* be iterated. A sentence such as "I deny that I deny that Mary is beautiful" is perfectly grammatical, though perhaps difficult to parse. Seriously uttering this sentence amounts to performing the same speech act as uttering the perhaps more idiomatic sentence "I deny that Mary is ugly". A clear case of denying that Mary is ugly is seriously uttering the sentence "Mary is beautiful". A denial of $\sim\alpha$ thus seems to be an assertion of α, and recall that an assertion of $\sim\alpha$ was introduced as a denial of α.

The notions of assertion and denial stand in a close relation to the notions of proof and disproof, respectively. If I assert that α, then I commit myself to be ready to prove α, and if I deny that α, then I commit myself to be ready to disprove α. Assertion and denial are basic speech acts which are insensitive to the complexity and composition of the asserted or denied sentence α. No matter how complex α may be and no matter how α is composed, in order to assert or deny α, it is enough to seriously utter the sentence α or its strong negation $\sim\alpha$. A (canonical) proof or a (canonical) disproof of a sentence α, however, *is* sensitive to the complexity and composition of α. A canonical proof of a conjunction $(\alpha \wedge \beta)$, for example, requires a proof of α and a proof of β, whereas a canonical disproof of $(\alpha \vee \beta)$ requires a disproof of α and a disproof of β.

If we only look at proofs and disproofs of elementary sentences representable by atomic formulas of a propositional or first-order language, then proofs and disproofs are often basic acts. We can take up an example provided by A. Grzegorczyk [72]. Suppose that l is a yellow lemon. We may prove that l is yellow just by drawing visual attention to l, and we may disprove that l is red again just by drawing visual attention to l. The falsification of the proposition that l is red is as direct as the verification of the proposition that l is yellow. Neither would we attempt to disprove that l is red by leading the assumption that l is red to an absurdity, nor would we attempt to prove that l is yellow by leading the assumption that l fails to be yellow to an absurdity. We would, under normal circumstances at least, just point to the colour of l. It might be objected that the provability of an elementary sentence such as "l is yellow" requires a theory and that in verifying by eye that l is yellow, we do not just see that l is yellow but *infer* that l is yellow from a theory based on our visual experience. But then in falsifying by eye that l is red, we still do not seem to lead the assumption that l is red to an absurdity. If in the verification case we directly *infer* from (a theory based on) our visual experience that l is yellow, then in the falsification case it seems that we directly infer from (a theory based on) our visual experience that l is definitely not red. Therefore, *if* disproving by eye that l is red is conceived of as an inference of the proposition that l is definitely not red, then this "definitely not" is not a so-called negation as inconsistency. In other words "l is definitely not red" is not to be understood as "l is red implies absurdity", cf. [60, 201].

What is absurdity? A sentence expresses absurdity, the absurd proposition, if in every model, the sentence fails to be true. If we consider possible worlds models, a sentence expresses absurdity, if the sentence fails to be true at every possible world in every model. Possible worlds are often conceived of as classical models satisfying the principle of bivalence. But they may also be conceived of as information states that may or may not support the truth or the falsity of propositions. If absence of truth is distinguished from falsity, so that the principle of bivalence is violated, a sentence may express absurdity without being false in every model or false at every state in every model. A sentence thus expresses absurdity if it is never true, and, *in general*, a reduction to absurdity is a reduction to non-truth. Of course, we may then also consider reductions to non-falsity. A sentence expresses non-falsity, if it is never false.

If an act of assertion commits a speaker to be ready to prove the asserted proposition, and an act of denial commits a speaker to be ready to disprove the denied proposition, one may wonder what kind of action is such that it commits a speaker to be ready to reduce the assumption that a certain proposition α is true to absurdity (or, more generally, to non-truth) and what kind of action is such that it commits a speaker to be ready to reduce the assumption that α is definitely false to non-truth. It seems that the *first* kind of commitment comes with asserting that nothing supports the truth of α. If I assert that nothing supports the truth of the sentence "Person *b* stabbed person *c*", I am committed to be ready to show that any piece of information (in particular any piece of information that seems to establish the truth of the assumption that *b* stabbed *c*) fails to establish the truth of the assumption that *b* stabbed *c*. If there is a witness who claims to have seen that *b* stabbed *c*, for example, I may point out that the witness used to be extremely unreliable on previous occasions. Proceeding in this way, I may try to show that there is no conclusive evidence in favour of "*b* stabbed *c*".

What makes it difficult, perhaps, to see the difference between disproofs and reductions to absurdity is that one might hold that every direct falsification of α also reduces the assumption that α to absurdity. If I present a group of very reliable witnesses who confirm that *b* was not at the crime scene, this may be viewed as a direct falsification of "*b* stabbed *c*", in addition leading the assumption that *b* stabbed *c* to absurdity. But, firstly, this does not show that there is no difference between disproving and reducing to absurdity, and, secondly, note that information may be contradictory. Someone else might present another group of highly reliable witnesses who claim that they saw that *b* stabbed *c*, so that the available testimony both supports the truth and supports the falsity of "*b* stabbed *c*". Thus, it is not at all clear that disprovability always implies reducibility to non-truth. Indeed, the implication may fail.

The *second* kind of commitment appears to come with asserting that no information supports the falsity of assumption α. If I assert that no information supports the falsity of "*b* stabbed *c*", I am committed to be ready to show that the assumption that *b* definitely did not stab *c* leads to absurdity. Again, I might try to point to certain facts that are incompatible with the assumption under consideration, although they do not prove that *b* stabbed *c*. I might, for example, point out that *b*'s fingerprints can be found on the dagger that has been removed from *c*'s corpse.

	inferential status	related speech act
$\varnothing \vdash \alpha$	α is provable	to assert that α
$\varnothing \vdash {\sim}\alpha$	α is disprovable	to deny that α
$\alpha \vdash \varnothing$	α is reducible to non-truth	to assert that no information supports the truth of α
${\sim}\alpha \vdash \varnothing$	α is reducible to non-falsity	to assert that no information supports the falsity of α

Table 5.6.: Speech acts and the inferential status of propositions.

The view that the denial of a sentence s can be profitably analyzed as the assertion of a suitable negation of s is contentious. According to Greg Restall [169]

> [d]enial is not to be analysed as the assertion of a negation,

whereas Bryson Brown [26, p. 646] explains that he has

> a modest proposal: negation is *denial* in the object language.

Graham Priest [155, p. 105] concedes that the uttering of a negated sentence sometimes may be interpreted as a denial but holds that "asserting a negation (in the Fregean sense) is not necessarily a denial." Priest regards rejection as the linguistic expression of denial and takes rejecting something to be putting a bar on accepting it. "When justified, it is so because there is evidence against the claim: positive grounds for keeping it out of one's beliefs" [155, p. 103]. This exclusion from belief is stronger than agnosticism (absence of belief) but, as it seems, weaker than disbelief. Timothy Williamson [216, p. 10] explains that "we can regard assertion as the verbal counterpart of judgement and judgement as the occurrent form of belief". The association of assertions with proofs and denials with disproofs takes the negative judgement of denial as the occurrent form of disbelief and not as the occurrent form of refusal of belief.

In this chapter, we discuss logics in which it is important to distinguish between provability, disprovability, and their duals. The term "duality" has several meanings even in mathematics. In one usage the concept of duality is related to order reversal. In this sense, the dual of provability is reducibility to non-truth. The dual of disprovability is reducibility to non-falsity. The picture summarized in Table 5.6 emerges.

5.3.2. Inferential relations and logical operations

If α is provable, then it is warranted to assert that α, if α is disprovable, then it is warranted to deny that α, if α is reducible to non-truth, then it is warranted to assert that no information supports the truth of α, and if α is reducible to non-falsity, then it is warranted to assert that no information supports the falsity of α.

The above considerations on the inferential status of a sentence α can be
generalized to proofs from a finite set of sentences assumed to be true $\alpha_1, \ldots, \alpha_n$ and reductions from a finite set of sentences $\alpha_1, \ldots, \alpha_n$ assumed not to be true. If the expression "assumptions" is reserved for sentences assumed to be true, there seems to be a semantic gap in English and other natural languages, as there is no idiomatic term for sentences assumed not to be true. Let us agree to call sentences assumed not to be true *counterassumptions*. Sentences assumed to be false may be called *rejections* (or repudiations), so that sentences assumed not to be false might be called *counterrejections* (counterrepudiations). Table 5.7 lists eight different kinds of inferential relations.

	inferential relation
$\alpha_1, \ldots, \alpha_n \vdash \alpha$	α is provable from assumptions $\alpha_1, \ldots, \alpha_n$
$\alpha_1, \ldots, \alpha_n \vdash {\sim}\alpha$	α is disprovable from assumptions $\alpha_1, \ldots, \alpha_n$
$\alpha \vdash \alpha_1, \ldots, \alpha_n$	α is reducible to absurdity from counterassumptions $\alpha_1, \ldots, \alpha_n$
${\sim}\alpha \vdash \alpha_1, \ldots, \alpha_n$	α is reducible to non-falsity from counterassumptions $\alpha_1, \ldots, \alpha_n$
${\sim}\alpha_1, \ldots, {\sim}\alpha_n \vdash \alpha$	α is provable from rejections $\alpha_1, \ldots, \alpha_n$
${\sim}\alpha_1, \ldots, {\sim}\alpha_n \vdash {\sim}\alpha$	α is disprovable from rejections $\alpha_1, \ldots, \alpha_n$
$\alpha \vdash {\sim}\alpha_1, \ldots, {\sim}\alpha_n$	α is reducible to absurdity from counterrejections $\alpha_1, \ldots, \alpha_n$
${\sim}\alpha \vdash {\sim}\alpha_1, \ldots, {\sim}\alpha_n$	α is reducible to non-falsity from counterrejections $\alpha_1, \ldots, \alpha_n$

Table 5.7.: Inferential relations.

If we want to reduce the inferential relation between the sentences $\alpha_1, \ldots, \alpha_n$ and the sentence α to the inferential status of a single formula, we may use suitable connectives: conjunction $\wedge$, disjunction $\vee$, implication $\rightarrow$, and the less well-known co-implication $\prec$, see Table 5.8.

inferential relation	inferential status
$\alpha_1, \ldots, \alpha_n \vdash \alpha$	$\varnothing \vdash (\alpha_1 \wedge \ldots \wedge \alpha_n) \to \alpha$
$\alpha_1, \ldots, \alpha_n \vdash \mathord{\sim}\alpha$	$\varnothing \vdash (\alpha_1 \wedge \ldots \wedge \alpha_n) \to \mathord{\sim}\alpha$
$\alpha \vdash \alpha_1, \ldots, \alpha_n$	$\alpha \mathbin{-\!\!\prec} (\alpha_1 \vee \ldots \vee \alpha_n) \vdash \varnothing$
$\mathord{\sim}\alpha \vdash \alpha_1, \ldots, \alpha_n$	$\mathord{\sim}\alpha \mathbin{-\!\!\prec} (\alpha_1 \vee \ldots \vee \alpha_n) \vdash \varnothing$
$\mathord{\sim}\alpha_1, \ldots, \mathord{\sim}\alpha_n \vdash \alpha$	$\varnothing \vdash (\mathord{\sim}\alpha_1 \wedge \ldots \wedge \mathord{\sim}\alpha_n) \to \alpha$
$\mathord{\sim}\alpha_1, \ldots, \mathord{\sim}\alpha_n \vdash \mathord{\sim}\alpha$	$\varnothing \vdash (\mathord{\sim}\alpha_1 \wedge \ldots \wedge \mathord{\sim}\alpha_n) \to \mathord{\sim}\alpha$
$\alpha \vdash \mathord{\sim}\alpha_1, \ldots, \mathord{\sim}\alpha_n$	$\alpha \mathbin{-\!\!\prec} (\mathord{\sim}\alpha_1 \vee \ldots \vee \mathord{\sim}\alpha_n) \vdash \varnothing$
$\mathord{\sim}\alpha \vdash \mathord{\sim}\alpha_1, \ldots, \mathord{\sim}\alpha_n$	$\mathord{\sim}\alpha \mathbin{-\!\!\prec} (\mathord{\sim}\alpha_1 \vee \ldots \vee \mathord{\sim}\alpha_n) \vdash \varnothing$

Table 5.8.: From inferential relations to inferential status.

We thereby arrive at the following vocabulary: $\{\wedge, \vee, \to, -\!\!\prec, \sim\}$. Whereas $\wedge, \vee, \to$, and $-\!\!\prec$ may be seen to emerge from the reduction of inferential relations to inferential status stated in Table 5.8, $\sim$ reflects the distinction between provability and disprovability. Conjunction $\wedge$ combines formulas in antecedent position, i.e., on the left of $\vdash$, and disjunction combines formulas in succedent position, i.e., on the right of $\vdash$. Implication is a vehicle for registering formulas that appear in antecedent position in succedent position, and co-implication is a vehicle for registering formulas that appear in succedent position in antecedent position. We read $\alpha -\!\!\prec \beta$ as "β co-implies α" or as "α excludes β". Whereas implication is the residuum of conjunction, co-implication is the residuum of disjunction:

$$(\alpha \wedge \beta) \vdash \gamma \ \text{ iff } \ \alpha \vdash (\beta \to \gamma) \ \text{ iff } \ \beta \vdash (\alpha \to \gamma),$$
$$\gamma \vdash (\alpha \vee \beta) \ \text{ iff } \ (\gamma -\!\!\prec \alpha) \vdash \beta \ \text{ iff } \ (\gamma -\!\!\prec \beta) \vdash \alpha.$$

The strong negation $\sim$ is a *primitive* negation. Other kinds of negation connectives are *definable* in the presence of $\to$ and $-\!\!\prec$. Let p be a certain propositional letter. Then we define non-falsity as follows: $\top := (p \to p)$, and non-truth in this way: $\bot := (p -\!\!\prec p)$. We can then introduce two negation connectives:

$$-\alpha := (\top -\!\!\prec \alpha) \ \text{(co-negation), and}$$
$$\neg\alpha := (\alpha \to \bot) \ \text{(intuitionistic negation).}$$

Other defined connectives of H-B are equivalence, $\leftrightarrow$, and co-equivalence, $\succ\!\!-\!\!\prec$, which are defined as follows:

$$\alpha \equiv \beta := (\alpha \to \beta) \wedge (\beta \to \alpha); \qquad \alpha \succ\!\!-\!\!\prec \beta := (\alpha -\!\!\prec \beta) \vee (\beta -\!\!\prec \alpha).$$

The connectives $\wedge, \vee, \rightarrow$, and $\prec$ are the primitive connectives of bi-intuitionistic logic BiInt, also known as Heyting-Brouwer logic H-B or as subtractive logic, cf. Chapter 2 and [28, 38, 68, 69, 70, 84, 161, 162, 163, ?, 217]. Extensions of H-B by strong negation $\sim$ have been introduced and investigated in [209], see also [101]. Logics with strong negation and intuitionistic implication have been introduced by David Nelson in the late 1940s and subsequently have been investigated by many researchers, see Chapter 2 and, for example, [6, 51, 73, 83, 92, 111, 134, 137, 138, 141, 186, 190, 191, 192, 196, 201, 205, 207].

5.4. Inferentialist (proof-theoretic) interpretation

The plan now is to interpret the connectives of $\mathcal{L}$ in the style of the Brouwer-Heyting-Kolmogorov (BHK) interpretation of the intuitionistic connectives in terms of canonical proofs, see, for example, [42, p. 154]. It is well-known that David Nelson's constructive logics with strong negation admit of a sound interpretation in terms of both proofs and disproofs, see [119, 196]. We will supplement the BHK interpretation by interpretations in terms of canonical disproofs, canonical reductions to absurdity (alias non-truth), and canonical reductions to non-falsity. That is, we will define the notions of canonical proofs, disproofs, dual proofs, and dual disproofs of complex $\mathcal{L}$-formulas by simultaneous induction. To make sure that the interpretation is correct for the logics (I_i, C_j), we will make the following assumptions:

1. for no $\mathcal{L}$-formula α there exists both a proof and a dual proof of α;
2. for no $\mathcal{L}$-formula α there exists both a disproof and a dual disproof of α;
3. every $\mathcal{L}$-formula α either has a proof or dual proof;
4. every $\mathcal{L}$-formula α either has a disproof or dual disproof.

Note that we do not need clauses for the constants $\bot$ and $\top$ and the negation operations $\neg$ and $-$, because in $\mathcal{L}$ these connectives are definable. We also assume that the conditions under which an entity is a canonical proof, disproof, dual proof, or dual disproof of an atomic sentence depend on the appropriate and relevant social practice and are not a matter of logic.

5.4.1. Canonical proofs

We first consider the inductive definition of the notion of a canonical proof of a compound $\mathcal{L}$-formula.

- A canonical proof of a strongly negated formula $\sim\alpha$ is a canonical disproof of α.
- A canonical proof of a conjunction $(\alpha \wedge \beta)$ is a pair (π_1, π_2) consisting of a canonical proof π_1 of α and a canonical proof π_2 of β.
- A canonical proof of a disjunction $(\alpha \vee \beta)$ is a pair (i, π) such that $i = 0$ and π is a canonical proof of α or $i = 1$ and π is a canonical proof of β.
- A canonical proof of an implication $(\alpha \rightarrow \beta)$ is a construction that transforms any canonical proof of α into a canonical proof of β.
- A canonical proof of a co-implication $(\alpha \prec \beta)$ is a pair (π_1, π_2), where π_1 is a canonical proof of α and π_2 is a canonical dual proof of β. (This pair is a canonical dual proof of $(\alpha \rightarrow \beta)$).

5.4.2. Canonical disproofs

- A canonical disproof of a strongly negated formula $\sim\alpha$ is a canonical proof of α.
- A canonical disproof of a conjunction $(\alpha \wedge \beta)$ is a pair (i, π) such that $i = 0$ and π is a canonical disproof of α or $i = 1$ and π is a canonical disproof of β.
- A canonical disproof of a disjunction $(\alpha \vee \beta)$ is a pair (π_1, π_2) consisting of a canonical disproof π_1 of α and a canonical disproof π_2 of β.
- A canonical disproof of an implication $(\alpha \rightarrow \beta)$ in

  (I_1C_j) is a pair (π_1, π_2) consisting of a canonical proof π_1 of α and a canonical disproof π_2 of β.

  (I_2C_j) is a construction that transforms any canonical proof of α into a canonical disproof of β.

  (I_3C_j) is a pair (π_1, π_2), where π_1 is a canonical proof of α and π_2 is a canonical dual proof of β. (This pair is a canonical dual proof of $(\alpha \rightarrow \beta)$).

(I_4C_j) is a pair (π_1, π_2), where π_1 is a canonical disproof of β and π_2 is a canonical dual disproof of α.

- A canonical disproof of a co-implication $(\alpha \prec \beta)$ in

  (I_iC_1) is a pair (i, π) such that $i = 0$ and π is a canonical disproof of α or $i = 1$ and π is a canonical proof of β.

  (I_iC_2) is a pair (π_1, π_2), where π_1 is a canonical disproof of α and π_2 is a canonical dual proof of β. (This pair is a canonical dual proof of $({\sim}\alpha \to \beta)$).

  (I_iC_3) is a construction that transforms any canonical proof of α into a canonical proof of β.

  (I_iC_4) is a construction that transforms any canonical disproof of β into a canonical disproof of α.

5.4.3. Canonical reductions to non-truth (canonical dual proofs)

- A canonical reduction to non-truth of a strongly negated formula ${\sim}\alpha$ is canonical dual disproof of α.

- A canonical reduction to non-truth of a conjunction $(\alpha \wedge \beta)$ is a pair (i, π) such that $i = 0$ and π is a canonical dual proof of α or $i = 1$ and π is a canonical dual proof of β.

- A canonical reduction to non-truth of a disjunction $(\alpha \vee \beta)$ is a pair (π_1, π_2) consisting of a dual proof π_1 of α and a dual proof π_2 of β.

- A canonical reduction to non-truth of an implication $(\alpha \to \beta)$ is a pair (π_1, π_2), where π_1 is a canonical proof of α and π_2 is a canonical dual proof of β. (This pair is a canonical proof of $(\alpha \prec \beta)$).

- A canonical reduction to non-truth of a co-implication $(\alpha \prec \beta)$ is a construction that transforms any dual proof of β into a dual proof of α.

5.4.4. Canonical reductions to non-falsity (canonical dual disproofs)

- A canonical reduction to non-falsity of a strongly negated formula ${\sim}\alpha$ is a canonical dual proof of α.

- A canonical reduction to non-falsity of a conjunction $(\alpha \wedge \beta)$ is a pair (π_1, π_2) consisting of a dual disproof π_1 of α and a dual disproof π_2 of β.

- A canonical reduction to non-falsity of a disjunction $(\alpha \vee \beta)$ is a pair (i, π) such that $i = 0$ and π is a canonical dual disproof of α or $i = 1$ and π is a canonical dual disproof of β.

- A canonical reduction to non-falsity of an implication $(\alpha \rightarrow \beta)$ in

  (I_1C_j) is a pair (i, π) such that $i = 0$ and π is a canonical dual proof of α or $i = 1$ and π is a canonical dual disproof of β.

  (I_2C_j) is a pair (π_1, π_2), where π_1 is a canonical proof of α and π_2 is a canonical dual disproof of β.

  (I_3C_j) is a construction that transforms any canonical dual proof of β into a canonical dual proof of α. (This pair is a canonical dual proof of $(\alpha \prec \beta)$).

  (I_4C_j) is a construction that transforms any canonical dual disproof of α into a canonical dual disproof of β.

- A canonical reduction to non-falsity of a co-implication $(\alpha \prec \beta)$ in

  (I_iC_1) is a pair (π_1, π_2), where π_1 is a canonical dual disproof of α and π_2 is a canonical dual proof of β.

  (I_iC_2) is a construction that transforms any canonical dual proof of β into a canonical dual disproof of α. (This construction is a canonical dual proof of $({\sim}\alpha \prec \beta)$).

  (I_iC_3) is a pair (π_1, π_2), where π_1 is a canonical proof of α and π_2 is a canonical dual proof of β. (This pair is a canonical dual proof of $(\alpha \rightarrow \beta)$).

  (I_iC_4) is a pair (π_1, π_2), where π_1 is a canonical disproof of β and π_2 is a canonical dual disproof of α.

In order to show by induction on the construction of inferences that the logics (I_i, C_j) are sound with respect to the above BHK-style interpretation in terms of proofs, disproofs, and their duals, we need proof systems for the semantically defined logics (I_i, C_j). We consider the display calculi defined in [209].

5.5. Correctness of the logics (I_i, C_j) w.r.t. the inferentialist semantics

We show that if a sequent is provable in $\delta(I_i, C_j)$, then there exists a certain construction made up from proofs, disproofs, and their duals that transforms any proof of the antecedent of the sequent into a proof of its succedent.[15]

Theorem 5.5.1 *Let $i, j \in \{1, 2, 3, 4\}$. If $X \vdash Y$ is provable in $\delta(I_i, C_j)$, then*

1. *there exists a construction π such that $\pi(\pi')$ is a canonical proof of $\tau_2(Y)$ whenever π' is a canonical proof of $\tau_1(X)$.*

2. *there exists a construction π such that $\pi(\pi')$ is a canonical dual proof of $\tau_1(X)$ whenever π' is a canonical dual proof of $\tau_2(Y)$.*

Proof. By simultaneous induction on derivations in $\delta(I_i, C_j)$.
(i): We first consider the display postulates. The first display postulates for $\circ$ are:

$$\cfrac{\cfrac{Y \vdash X \circ Z}{X \circ Y \vdash Z}}{X \vdash Y \circ Z}$$

Suppose, by the induction hypothesis for (i), that there exists a construction π that transforms any canonical proof of $\tau_1(Y)$ into a canonical proof of $\tau_2(X \circ Z)$ $(= \tau_1(X) \to \tau_2(Z))$, i.e., into a construction that transforms any canonical proof of $\tau_1(X)$ into a canonical proof of $\tau_2(Z)$. Let π' be any canonical proof of $\tau_1(X \circ Y)$ $(= \tau_1(X) \wedge \tau_1(Y))$. The proof π' is a pair (π'_1, π'_2), where π'_1 is a canonical proof of $\tau_1(X)$ and π'_2 is canonical proof of $\tau_1(Y)$. Then $(\pi(\pi'_2))(\pi'_1)$ is a proof[16] of $\tau_2(Z)$.
Suppose next, by the induction hypothesis for (i), that there is a construction π that transforms any proof of $\tau_1(X \circ Y)$ $(= \tau_1(X) \wedge \tau_1(Y))$ into a proof of $\tau_2(Z)$. Let π' be any proof of $\tau_1(X)$. Then $\pi^*(\pi') = \pi((\pi', \;\;))$ is a construction that transforms any proof of $\tau_1(Y)$ into a proof of $\tau_2(Z)$.
The second pair of display postulates for $\circ$ is dealt with similarly.
The first pair of display postulates for $\bullet$ is:

$$\cfrac{\cfrac{X \bullet Z \vdash Y}{X \vdash Y \bullet Z}}{X \bullet Y \vdash Z}$$

[15] This result may give rise to a four-sorted typed λ-calculus.
[16] In the sequel we will often omit the expression "canonical".

Suppose, by the induction hypothesis for (i), that there exists a construction π that transforms any proof of $\tau_1(X \bullet Z)$, $(= \tau_1(X) \prec \tau_2(Z))$, i.e., any pair (π_1, π_2), where π_1 is a proof of $\tau_1(X)$ and π_2 is a dual proof of $\tau_2(Z)$, into a proof of $\tau_2(Y)$. Let π' be any proof of $\tau_1(X)$. There either is a dual proof of $\tau_2(Z)$ or there is not. If there is such a dual proof, let π'' be a fixed dual proof of $\tau_2(Z)$. Then $(0, \pi((\pi', \pi'')))$ is a proof of $(\tau_2(Y) \vee \tau_2(Z))$. If there does not exist any dual proof of $\tau_2(Z)$, then there exists a proof of $\tau_2(Z)$. Let π''' be such a proof. Then $(1, \pi''')$ is a proof of $(\tau_2(Y) \vee \tau_2(Z))$.
Suppose now, by the induction hypothesis for (i), that there is a construction π that transforms any proof of $\tau_1(X)$ into a proof of $\tau_2(Y \bullet Z)$ $(= \tau_2(Y) \vee \tau_2(Z))$. Let π' be any proof of $\tau_1(X \bullet Y)$ $(= \tau_1(X) \prec \tau_2(Y))$, i.e., any pair (π_1, π_2), where π_1 is a proof of $\tau_1(X)$ and π_2 is a dual proof of $\tau_2(Y)$. Since $\tau_2(Y)$ has no proof, $\pi(\pi_1)$ is a proof of $\tau_2(Z)$.
The second pair of display postulates for $\bullet$ is dealt with similarly.
The case of the logical structural rules is simple; the axiomatic sequents are dealt with by the identity function and (*cut*) by functional application. The case of the other structural sequent rules from Table 5.3 is quite obvious.
We present here, by way of example, just the cases of three introduction rules.
$(\vdash \prec)$:

$$\frac{X \vdash \beta \quad \alpha \vdash Y}{X \bullet Y \vdash \beta \prec \alpha}$$

Suppose, by the induction hypothesis for (i), that there is a construction π that transforms any proof of $\tau_1(X)$ into a proof of $\tau_2(\beta)$ and, by the induction hypothesis for (ii), that there is a construction π' that transforms any dual proof of $\tau_2(Y)$ into a dual proof of $\tau_1(\alpha)$. Let π^* be a proof of $\tau_1(X \bullet Y)$ $(= \tau_1(X) \prec \tau_2(Y))$. Then π^* is a pair (π_1, π_2), where π_1 is a proof of $\tau_1(X)$ and π_2 is a dual proof of $\tau_2(Y)$. Therefore, the pair $(\pi(\pi_1), \pi'(\pi_2))$ is proof of $(\beta \prec \alpha)$.
rI_4, first rule:

$$\frac{X \vdash \sim \beta \qquad \sim \alpha \vdash Y}{X \bullet Y \vdash \sim (\alpha \rightarrow \beta)}$$

Suppose, by the induction hypothesis for (i), that π is a construction that transforms any proof $\tau_1(X)$ into a proof of $\sim \beta$, and, by the induction hypothesis for (ii), that π' is a construction that transforms any dual proof of $\tau_2(Y)$ into a dual proof of $\sim \alpha$, i.e., into a dual disproof of α. Let π^* be a proof of $\tau_1(X \bullet Y)$ $(= \tau_1(x) \prec \tau_2(Y))$. Then π^* is a pair (π_1, π_2), where π_1 is a proof of $\tau_1(X)$ and π_2 is a dual proof of $\tau_2(Y)$. A proof of $\sim (\alpha \rightarrow \beta)$ in (I_4, C_j) is a disproof of $(\alpha \rightarrow \beta)$ in (I_4, C_j), which is a pair (π'_1, π'_2), where π'_1 is a proof of $\sim \beta$, and π'_2 is a dual proof of $\sim \alpha$. Note that $(\pi(\pi_1), \pi'(\pi_2))$ is such a pair.

rC_1, second rule:

$$\frac{\sim \alpha \vdash X \quad \beta \vdash Y}{\sim (\alpha \!-\!\!\prec \beta) \vdash X \bullet Y}$$

Suppose, by the induction hypothesis for (i), that π' is a construction that transforms any proof of β into a proof of $\tau_2(Y)$ and that π'' is a construction that transforms any proof $\sim \alpha$ into a proof of $\tau_2(X)$. Let π^* be a proof of $\sim (\alpha \!-\!\!\prec \beta)$ in (I_i, C_1), i.e., a disproof of $(\alpha \!-\!\!\prec \beta)$ in (I_i, C_1). Then π^* is a pair (i, π) such that $i = 0$ and π is a disproof of α or $i = 1$ and π is a proof of β. But then either $(0, \pi''(\pi))$ or $(1, \pi'(\pi))$ is a proof of $\tau_2(X \bullet Y)$.

(ii): We present here just the case of the second pair of display postulates for $\circ$:

$$\frac{X \vdash Y \circ Z}{\frac{X \circ Y \vdash Z}{Y \vdash X \circ Z}}$$

Suppose, by the induction hypothesis for (ii), that π is a construction that transforms any dual proof of $\tau_1(Y) \to \tau_2(Z)$ (i.e., any pair (π_1, π_2), where π_1 is a proof of $\tau_1(Y)$ and π_2 is dual proof of $\tau_2(Z)$) into a dual proof of $\tau_1(X)$. Either there is a proof of $\tau_1(Y)$ or not. If there is a proof of $\tau_1(Y)$, let π'' be such a proof. Then $(0, \pi(\pi'', \pi'))$ is a dual proof of $\tau_1(X \circ Y)$. If there is no proof of $\tau_1(Y)$, then there is a dual proof of $\tau_1(Y)$. Let π''' be such a dual proof. Then $(1, \pi''')$ is a dual proof of $\tau_1(X \circ Y)$.

Now suppose that, by the induction hypothesis for (ii), there exists a construction π that transforms any dual proof of $\tau_2(Z)$ into a dual proof of $\tau_1(x) \wedge \tau_1(Y)$. Let π' be any dual proof of $\tau_1(X) \to \tau_2(Z)$, i.e., a pair (π_1, π_2), where π_1 is a proof of $\tau_1(X)$ and π_2 is a dual proof of $\tau_2(Z)$. Then there is no dual proof of $\tau_1(X)$ and $(0, \pi(\pi_2))$ is a dual proof of $\tau_1(Y)$. ∎

Corollary 5.5.2 *Let $i, j \in \{1, 2, 3, 4\}$. If $X \vdash Y$ is provable in $\delta(I_i, C_j)$, then*

1. *there exists a construction π such that $\pi(\pi')$ is a canonical disproof of $\tau_2(Y)$ whenever π' is a canonical disproof of $\tau_1(X)$.*

2. *there exists a construction π such that $\pi(\pi')$ is a canonical dual disproof of $\tau_1(X)$ whenever π' is a canonical dual disproof of $\tau_2(Y)$.*

Proof. Every canonical disproof of α is a canonical proof of $\sim \alpha$ and every canonical dual disproof of α is a canonical dual proof of $\sim \alpha$. ∎

The following claims are derivable from Theorem 5.5.1 and Corollary 5.5.2.

Theorem 5.5.3 *Let $i, j \in \{1, 2, 3, 4\}$.*

1. *If $\mathbf{I} \vdash \alpha$ is provable in $\delta(I_i, C_j)$, then there exists a construction π which is a proof of α.*

2. *If $\alpha \vdash \mathbf{I}$ is provable in $\delta(I_i, C_j)$, then there exists a construction π which is a dual proof of α.*

3. *If $\mathbf{I} \vdash \sim\alpha$ is provable in $\delta(I_i, C_j)$, then there exists a construction π which is a disproof of α.*

4. *If $\sim\alpha \vdash \mathbf{I}$ is provable in $\delta(I_i, C_j)$, then there exists a construction π which is a dual disproof of α.*

Proof. Note that any canonical proof of $\tau_1(\mathbf{I}) = (p \rightarrow p)$ and any canonical dual proof of $\tau_2(\mathbf{I}) = (p \prec p)$ is the identity function. ∎

Example 5.5.4 *The sequent $\mathbf{I} \vdash q \vee -q$ is provable in the logics (I_i, C_j), and it can easily be seen that there exists a construction that is a (canonical) proof of $q \vee ((p \rightarrow p) \prec q)$. A proof of $q \vee ((p \rightarrow p) \prec q)$ is a pair (i, π), where $i = 0$ and π is a proof of q, or $i = 1$ and π is a proof of $((p \rightarrow p) \prec q)$. Now, π is a proof of $((p \rightarrow p) \prec q)$ iff π is a pair (π_1, π_2), where π_1 is a proof of $(p \rightarrow p)$ and π_2 is a dual proof of q. Since the identity function is a proof of $(p \rightarrow p)$ and since every $\mathcal{L}$-formula either has a proof or a dual proof, there exists a proof of $q \vee ((p \rightarrow p) \prec q)$.*

Example 5.5.5 *There exists a construction that is a proof of $\sim (p \rightarrow q) \rightarrow (p \prec q)$ in the logics (I_3, C_j). A proof of $\sim (p \rightarrow q) \rightarrow (p \prec q)$ in (I_3, C_j) is a construction that transforms any proof of $\sim (p \rightarrow q)$ into a proof of $(p \prec q)$. A proof of $\sim (p \rightarrow q)$ in (I_3, C_j) is a disproof of $(p \rightarrow q)$, which is a pair (π_1, π_2), where π_1 is a proof of p and π_2 is a dual proof of q. But this pair is a proof of $(p \prec q)$ in (I_3, C_j), so that the identity function is a proof of $\sim (p \rightarrow q) \rightarrow (p \prec q)$ in (I_3, C_j).*

5.6. A logic stronger than bi-intuitionistic logic

We have considered sixteen extensions of propositional Brouwer-Heyting logic by strong negation. Each of these logics (I_i, C_j) ($i, j \in \{1, 2, 3, 4\}$) turned out to be correct with respect to an extended BHK-style inferentialist interpretation. The interpretation makes use of four primitive notions, namely the notions of proof, disproof, dual proof, and dual disproof. This correctness result supports the view that the logics (I_i, C_j) are indeed constructive propositional logics. The findings of this chapter can be summarized as in Table 5.9.[17]

(propositional) logic	*soundness with respect to an interpretation in terms of*
intuitionistic logic	proofs
Nelson's logics	proofs and disproofs
dual intuitionistic logic	dual proofs
bi-intuitionistic logic	proofs and dual proofs
bi-intuitionistic logic extended by strong negation	proof, disproofs, and their duals

Table 5.9.: Summary

However, the interpretation in terms of canonical proofs and canonical dual proofs is not complete with respect to Heyting-Brouwer logic. The extended inferential semantics validates the implication

$$(\mathbf{bc}) \quad (-\alpha \wedge -\beta) \rightarrow -(\alpha \vee \beta),$$

which is easily refutable in a Kripke model for H-B that violates "backward convergence": for all u, v, and w it holds that if $u \leq w$ and $v \leq w$, then there exists t with $t \leq u$ and $t \leq v$.

It is well-known that IPL is characterized not only by the class of all intuitionistic Kripke frames but also by the class of all rooted frames and the class of all (rooted) trees, see, for example [36, p. 33]. In other words, the language of IPL cannot distinguish between pre-orders, rooted frames, and trees. Trees can be defined in various ways, we here follow the presentation in [36].

Definition 5.6.1 *Let $\mathcal{F} = \langle I, \leq \rangle$ be a frame. A frame $\mathcal{F}' = \langle I', \leq' \rangle$ is a subframe of $\mathcal{F}$ iff $I' \subseteq I$ and $\leq'$ is the restriction of $\leq$ to I'. The frame $\mathcal{F}'$ is a generated subframe of $\mathcal{F}$ iff I' is an upwards closed subset of I, which means that if $w \in I'$ and $w \leq' v$, then $v \in I'$. If $\mathcal{F}'$ is generated by a singleton $\{w\}$, then $\mathcal{F}'$ is said to be rooted (in w).*

[17] In intuitionistic logic, $\bot$ is primitive and has no proof; in dual intuitionistic logic, $\top$ is primitive and has no dual proof.

Definition 5.6.2 *If $\mathcal{F} = \langle I, \leq \rangle$ is a frame, then $\mathcal{F}$ is called a tree iff (i) $\mathcal{F}$ is rooted and (ii) for every $w \in I$, the set $\{v \in I \mid v \leq w\}$ is finite and linearly ordered by $\leq$.*

Every rooted tree satisfies backward convergence, and (**bc**) corresponds with backward convergence. Suppose that $\mathcal{F} = \langle I, \leq \rangle$ satisfies backward convergence and that $s \not\models^+$ **bc** for some $s \in I$. Then for some formulas α and β and some $w \in I$ with $s \leq w$, $w \models^+ -\alpha$ and $w \models^+ -\beta$ but (*) $w \not\models^+ -(\alpha \vee \beta)$. Hence there exit states u, v with $u \leq w$, $v \leq w$, $u \not\models^+ \alpha$ and $v \not\models^+ \beta$. Since $\mathcal{F}$ satisfies backward convergence, there exists $t \in I$ with $t \leq u$ and $t \leq v$. By persistence, $t \not\models^+ \alpha$ and $t \not\models^+ \beta$. By (*), however, $t \models^+ \alpha$ or $t \models^+ \beta$. If $\mathcal{F}$ does not satisfy backward convergence, then there are states u, v, w in I with $u \leq w$, $v \leq w$ and $u \neq v$. If p is an atomic formula that is verified at every s such that $t \leq s$ and nowhere else, and q is an atomic formula that is verified at every r such that $u \leq r$ nowhere else, then obviously $w \not\models$ **bc**.

If the extended BHK-interpretation in terms of proofs and dual proofs is regarded as a natural semantics, it might be argued that the logic of rooted trees in the language $\mathcal{L}_{\text{H}-\text{B}}$ is the correct bi-intuitionistic logic.

6. Symmetric and dual paraconsistent logics

Two new systems of first-order paraconsistent logic with De Morgan-type negations and co-implication, called symmetric paraconsistent logic (SPL) and dual paraconsistent logic (DPL), are introduced as Gentzen-type sequent calculi. The logic SPL is symmetric in the sense that the rule of contraposition is admissible in cut-free SPL. By using this symmetry property, a simpler cut-free sequent calculus for SPL is obtained. The logic DPL is not symmetric, but it has the duality principle known from classical logic. Simple semantics for SPL and DPL are introduced, and the completeness theorems with respect to these semantics are proved. The cut-elimination theorems for SPL and DPL are proved in two ways: One is a syntactical way which is based on the embedding theorems of SPL and DPL into Gentzen's LK, and the other is a semantical way which is based on the completeness theorems.[1]

6.1. Introduction

6.1.1. De Morgan-type negations: Symmetry versus duality

Paraconsistent logics are logics which have the desirable property of paraconsistency (see, e.g., [41, 154], and the references therein). *Paraconsistency* is, roughly speaking, a property of negations (or negation-like operators), and is known to be useful for representing inconsistency-tolerant reasoning more appropriately. Examples of paraconsistent negations (i.e., negations enjoying paraconsistency) are De Morgan-type negations, such as strong negation [134], negations based on four-valued logic [21, 47] and negations based on bilattice logics [12].

The De Morgan-type negations have the common characteristic axioms of the De Morgan laws: ${\sim}(\alpha \wedge \beta) \leftrightarrow {\sim}\alpha \vee {\sim}\beta$ and ${\sim}(\alpha \vee \beta) \leftrightarrow {\sim}\alpha \wedge {\sim}\beta$. These axioms imply the fact that the rule of contraposition

$$\frac{\Gamma \Rightarrow \Delta}{{\sim}\Delta \Rightarrow {\sim}\Gamma}\ (\text{cont})$$

[1] This chapter makes use of [101].

(where $\sim\Delta$ stands for $\{\sim\alpha \mid \alpha \in \Delta\}$) is admissible in a sequent calculus for the $\rightarrow$-free fragment of the logic in question. These $\rightarrow$-free fragments are symmetric with respect to $\sim$ in this sense. By virtue of this symmetry property, a simpler sequent calculus or axiomatization can be obtained for these logics. The De Morgan laws without the axioms for $\rightarrow$ also imply the *duality principle*. The duality principle for the $\rightarrow$-free fragment of Gentzen's sequent calculus LK for classical logic is well-known: If $\rightarrow$-free LK $\vdash \alpha \Rightarrow \beta$, then $\rightarrow$-free LK $\vdash \tilde{\beta} \Rightarrow \tilde{\alpha}$, where $\tilde{\alpha}$ and $\tilde{\beta}$ are, respectively, obtained from α and β by replacing every occurrence of $\wedge, \vee, \forall$ and $\exists$ by those of $\vee, \wedge, \exists$ and $\forall$, respectively. This principle also holds for a sequent calculus for Belnap and Dunn's four-valued logic.

On the other hand, an extended logic (or sequent calculus) with $\rightarrow$ which has the symmetry property or duality principle has not been studied yet. The reason may be that the De Morgan-like laws with respect to $\rightarrow$ and the De Morgan dual counterpart connective of $\rightarrow$ have not been considered. In this chapter, such extended logics are proposed by using a co-implication connective $\prec$ as the De Morgan dual counterpart of $\rightarrow$.

6.1.2. Combination with co-implication

The *co-implication* connective $\prec$ has been studied by many researchers in the context of *Heyting-Brouwer logic* (*H-B logic* for short) or equivalently *bi-intuitionistic logic*, which is, roughly speaking, an extension of (positive) intuitionistic logic with $\prec$ (see, e.g., [28, 38, 120, 161] and the references therein). The connective $\prec$ is known as a subtraction (or pseudo-difference) operator, since $\prec$ has the informal interpretation $\alpha \prec \beta := \alpha \wedge \neg\beta$ where $\neg$ is classical negation. The notion of subtraction is considered to be very important in the area of computer science, in particular in database theory and software development. Indeed, subtraction is used as a basic operation of databases, and the difference of two or more program codes (or data) is frequently checked in computer systems. Thus, combining co-implication with paraconsistent negations is also an interesting issue for computer science.

Combining co-implication with paraconsistent negations has been studied in [209] in an intuitionistic setting, although the motivation is different from the present chapter's one. In [209], two extended H-B logics with certain packages of axioms concerning $\prec$, $\rightarrow$ and $\sim$ are investigated, see also Chapter 5. One is

A1: $\sim(\alpha\rightarrow\beta) \leftrightarrow \sim\beta \prec \sim\alpha$ (negated implication as contraposed co-implication)

A2: $\sim(\alpha\prec\beta) \leftrightarrow \sim\beta\rightarrow\sim\alpha$ (negated co-implication as contraposed implication)

and another is

A3: $\sim(\alpha\rightarrow\beta) \leftrightarrow \alpha\prec\beta$ (negated implication as co-implication)

A4: $\sim(\alpha\prec\beta) \leftrightarrow \alpha\rightarrow\beta$ (negated co-implication as implication).

It is shown in [209] that some intuitionistic cut-free display calculi for the logics with these axioms are complete with respect to a Kripke semantics, see also Chapter 12. However, standard cut-free Gentzen-type sequent calculi, first-order versions and classical logic based versions for logics with {A1, A2} or {A3, A4} have not been studied yet. For the first set of axioms, constructing a standard cut-free Gentzen-type sequent calculus for H-B logic is known as difficult [28].

6.1.3. Our approach

The aim of this chapter is to obtain cut-free Gentzen-type systems and simple semantics for classical first-order (symmetric and dual) paraconsistent logics with both De Morgan-type negations and co-implication. The result of this chapter may thus regarded as a supplement to the work presented in [209] Chapter 5, but now based on classical first-order logic rather than propositional (positive) intuitionistic logic. Based upon *classical logic*, a very simple framework with the natural properties of symmetry and duality can be obtained.

The results of this chapter can be summarized as follows. Two new first-order paraconsistent logics with De Morgan-type negations and co-implication, called *symmetric paraconsistent logic* (SPL) and *dual paraconsistent logic* (DPL), are introduced as Gentzen-type sequent calculi by extending LK. The logic SPL has the inference rules which correspond to the laws A1 and A2. SPL is symmetric, i.e., the rule (cont) is admissible in cut-free SPL. By using (cont), a simpler cut-free system SPL$^-$ can also be obtained for SPL. The logic DPL has the inference rules which correspond to the laws A3 and A4. DPL is not symmetric, but it has the duality principle. These proposed logics are regarded as natural variants of the H-B logic, and as extensions of the existing paraconsistent logics with De Morgan-type negations, i.e., the $\{\wedge, \vee, \sim\}$-fragments of Nelson's paraconsistent logics, Belnap and Dunn's four-valued logic and Arieli and Avron's bilattice logics. Simple semantics for SPL and DPL are introduced, and the completeness theorems with respect to these semantics are proved using Schütte's method [185]. This method simultaneously provides a semantical proof of the cut-elimination theorems for SPL and DPL. The cut-elimination theorems for SPL and DPL are also proved by using the embedding theorems of SPL and DPL into LK.

6.2. Symmetric paraconsistent logic

Prior to a detailed discussion, the language $\mathcal{L}$ used in this chapter is introduced below. *Formulas* are constructed from predicate symbols $p, q, ...$, (countably many) individual variables $x, y, ...$, individual constants $c, d, ...$, function symbols $f, g, ...$, $\rightarrow$ (implication), $\prec$ (co-implication or subtraction), $\wedge$ (conjunction), $\vee$ (disjunction), $\neg$ (classical negation), $\sim$ (paraconsistent negation), $\forall$ (universal quantifier) and $\exists$ (existential quantifier). Small letters $t, s, ...$ are used to denote terms, Greek small letters $\alpha, \beta, ...$ are used to denote formulas, and Greek capital letters $\Gamma, \Delta, ...$ are used to represent finite (possibly empty) sets of formulas. An expression $\alpha[t/x]$ means the formula which is obtained from a formula α by replacing all free occurrences of the individual variable x in α by the term t, but avoiding a clash of variables. A *sequent* is an expression of the form $\Gamma \Rightarrow \Delta$. The symbol $\equiv$ is used to denote the equality of sets of symbols. An expression $L \vdash \Gamma \Rightarrow \Delta$ means that the sequent $\Gamma \Rightarrow \Delta$ is provable in a sequent calculus L, and L in this expression will occasionally be omitted.

Firstly, we define the sequent calculus LK for classical logic, and secondly we define SPL by extending LK with $\sim$.

Definition 6.2.1 (LK) *The initial sequents of* LK *are of the form:*

$$\alpha \Rightarrow \alpha.$$

The structural inference rules of LK *are of the form:*

$$\frac{\Gamma \Rightarrow \Delta, \alpha \quad \alpha, \Sigma \Rightarrow \Pi}{\Gamma, \Sigma \Rightarrow \Delta, \Pi} \text{ (cut)}$$

$$\frac{\Gamma \Rightarrow \Delta}{\alpha, \Gamma \Rightarrow \Delta} \text{ (we-left)} \qquad \frac{\Gamma \Rightarrow \Delta}{\Gamma \Rightarrow \Delta, \alpha} \text{ (we-right).}$$

The logical inference rules of LK *are of the form:*

$$\frac{\Gamma \Rightarrow \Delta, \alpha \quad \beta, \Sigma \Rightarrow \Pi}{\alpha{\rightarrow}\beta, \Gamma, \Sigma \Rightarrow \Delta, \Pi} \text{ (}{\rightarrow}\text{left)} \qquad \frac{\alpha, \Gamma \Rightarrow \Delta, \beta}{\Gamma \Rightarrow \Delta, \alpha{\rightarrow}\beta} \text{ (}{\rightarrow}\text{right)}$$

$$\frac{\alpha, \beta, \Gamma \Rightarrow \Delta}{\alpha \wedge \beta, \Gamma \Rightarrow \Delta} \text{ (}\wedge\text{left)} \qquad \frac{\Gamma \Rightarrow \Delta, \alpha \quad \Gamma \Rightarrow \Delta, \beta}{\Gamma \Rightarrow \Delta, \alpha \wedge \beta} \text{ (}\wedge\text{right)}$$

$$\frac{\alpha, \Gamma \Rightarrow \Delta \quad \beta, \Gamma \Rightarrow \Delta}{\alpha \vee \beta, \Gamma \Rightarrow \Delta} \ (\vee\text{left}) \qquad \frac{\Gamma \Rightarrow \Delta, \alpha, \beta}{\Gamma \Rightarrow \Delta, \alpha \vee \beta} \ (\vee\text{right})$$

$$\frac{\Gamma \Rightarrow \Delta, \alpha}{\neg\alpha, \Gamma \Rightarrow \Delta} \ (\neg\text{left}) \qquad \frac{\alpha, \Gamma \Rightarrow \Delta}{\Gamma \Rightarrow \Delta, \neg\alpha} \ (\neg\text{right})$$

$$\frac{\alpha[t/x], \Gamma \Rightarrow \Delta}{\forall x\alpha, \Gamma \Rightarrow \Delta} \ (\forall\text{left}) \qquad \frac{\Gamma \Rightarrow \Delta, \alpha[z/x]}{\Gamma \Rightarrow \Delta, \forall x\alpha} \ (\forall\text{right})$$

$$\frac{\alpha[z/x], \Gamma \Rightarrow \Delta}{\exists x\alpha, \Gamma \Rightarrow \Delta} \ (\exists\text{left}) \qquad \frac{\Gamma \Rightarrow \Delta, \alpha[t/x]}{\Gamma \Rightarrow \Delta, \exists x\alpha} \ (\exists\text{right})$$

where t in (∀left) *and* (∃right) *is a term, and z in* (∀right) *and* (∃left) *is an individual variable which has the eigenvariable condition, i.e., z does not occur as a free individual variable in the lower sequent of the rule.*

Definition 6.2.2 (SPL) *A sequent calculus* SPL *for symmetric paraconsistent logic is obtained from* LK *by adding the logical inference rules of the form:*

$$\frac{\alpha, \Gamma \Rightarrow \Delta, \beta}{\alpha \prec \beta, \Gamma \Rightarrow \Delta} \ (\prec\text{left}) \qquad \frac{\Gamma \Rightarrow \Delta, \alpha \quad \beta, \Sigma \Rightarrow \Pi}{\Gamma, \Sigma \Rightarrow \Delta, \Pi, \alpha \prec \beta} \ (\prec\text{right})$$

$$\frac{\alpha, \Gamma \Rightarrow \Delta}{\sim\sim\alpha, \Gamma \Rightarrow \Delta} \ (\sim\text{left}) \qquad \frac{\Gamma \Rightarrow \Delta, \alpha}{\Gamma \Rightarrow \Delta, \sim\sim\alpha} \ (\sim\text{right})$$

$$\frac{\sim\beta, \Gamma \Rightarrow \Delta, \sim\alpha}{\sim(\alpha \rightarrow \beta), \Gamma \Rightarrow \Delta} \ (\sim\rightarrow\text{left}) \qquad \frac{\Gamma \Rightarrow \Delta, \sim\beta \quad \sim\alpha, \Sigma \Rightarrow \Pi}{\Gamma, \Sigma \Rightarrow \Delta, \Pi, \sim(\alpha \rightarrow \beta)} \ (\sim\rightarrow\text{right})$$

$$\frac{\Gamma \Rightarrow \Delta, \sim\beta \quad \sim\alpha, \Sigma \Rightarrow \Pi}{\sim(\alpha \prec \beta), \Gamma, \Sigma \Rightarrow \Delta, \Pi} \ (\sim\prec\text{left}) \qquad \frac{\sim\beta, \Gamma \Rightarrow \Delta, \sim\alpha}{\Gamma \Rightarrow \Delta, \sim(\alpha \prec \beta)} \ (\sim\prec\text{right})$$

$$\frac{\sim\alpha, \Gamma \Rightarrow \Delta \quad \sim\beta, \Gamma \Rightarrow \Delta}{\sim(\alpha \wedge \beta), \Gamma \Rightarrow \Delta} \ (\sim\wedge\text{left}) \qquad \frac{\Gamma \Rightarrow \Delta, \sim\alpha, \sim\beta}{\Gamma \Rightarrow \Delta, \sim(\alpha \wedge \beta)} \ (\sim\wedge\text{right})$$

$$\frac{\sim\alpha, \sim\beta, \Gamma \Rightarrow \Delta}{\sim(\alpha \vee \beta), \Gamma \Rightarrow \Delta} \ (\sim\vee\text{left}) \qquad \frac{\Gamma \Rightarrow \Delta, \sim\alpha \quad \Gamma \Rightarrow \Delta, \sim\beta}{\Gamma \Rightarrow \Delta, \sim(\alpha \vee \beta)} \ (\sim\vee\text{right})$$

$$\frac{\Gamma \Rightarrow \Delta, \sim\alpha}{\sim\neg\alpha, \Gamma \Rightarrow \Delta} \ (\sim\neg\text{left}) \qquad \frac{\sim\alpha, \Gamma \Rightarrow \Delta}{\Gamma \Rightarrow \Delta, \sim\neg\alpha} \ (\sim\neg\text{right})$$

$$\frac{\sim\alpha[z/x], \Gamma \Rightarrow \Delta}{\sim\forall x\alpha, \Gamma \Rightarrow \Delta} \ (\sim\forall\text{left}) \qquad \frac{\Gamma \Rightarrow \Delta, \sim\alpha[t/x]}{\Gamma \Rightarrow \Delta, \sim\forall x\alpha} \ (\sim\forall\text{right})$$

$$\frac{\sim\alpha[t/x], \Gamma \Rightarrow \Delta}{\sim\exists x\alpha, \Gamma \Rightarrow \Delta} \ (\sim\exists\text{left}) \qquad \frac{\Gamma \Rightarrow \Delta, \sim\alpha[z/x]}{\Gamma \Rightarrow \Delta, \sim\exists x\alpha} \ (\sim\exists\text{right})$$

where t and z in the quantifier rules are an arbitrary term and an individual variable with the eigenvariable condition, respectively.

Note that LK + {($\prec$ left), ($\prec$ right)} is equivalent to Crolard's sequent calculus SLK for (classical) subtractive logic [38]. It is also remarked that the connective $\prec$, which is characterized by the inference rules ($\prec$ left) and ($\prec$ right), is definable in LK, i.e., SLK and LK are theorem-equivalent. Moreover, the $\{\wedge, \vee, \sim\}$-fragment of SPL is a common fragment of Belnap and Dunn's four-valued logic [21, 47] and Arieli and Avron's bilattice logics [12].

An expression $\alpha \Leftrightarrow \beta$ means the sequents $\alpha \Rightarrow \beta$ and $\beta \Rightarrow \alpha$.

Proposition 6.2.3 *The following sequents are provable in* SPL*:*

1. $\sim\sim\alpha \Leftrightarrow \alpha$,
2. $\sim\neg\alpha \Leftrightarrow \neg\sim\alpha$,
3. $\sim(\alpha\rightarrow\beta) \Leftrightarrow (\sim\beta\prec\sim\alpha)$ *(negated implication as contraposed co-implication),*
4. $\sim(\alpha\prec\beta) \Leftrightarrow (\sim\beta\rightarrow\sim\alpha)$ *(negated co-implication as contraposed implication),*
5. $\sim(\alpha \wedge \beta) \Leftrightarrow \sim\alpha \vee \sim\beta$,
6. $\sim(\alpha \vee \beta) \Leftrightarrow \sim\alpha \wedge \sim\beta$,
7. $\sim(\forall x\alpha) \Leftrightarrow \exists x(\sim\alpha)$,
8. $\sim(\exists x\alpha) \Leftrightarrow \forall x(\sim\alpha)$.

In the laws addressed in Proposition 6.2.3, $\prec$, $\vee$ and $\exists$ are regarded as the De Morgan duals (w.r.t. $\sim$) of $\rightarrow$, $\wedge$ and $\forall$, respectively. As mentioned in [209], the laws 3 and 4 in Proposition 6.2.3 were suggested by Greg Restall.

Proposition 6.2.4 *The rules*

$$\frac{\sim\sim\alpha, \Gamma \Rightarrow \Delta}{\alpha, \Gamma \Rightarrow \Delta}\ (\sim\text{left}^{-1}) \qquad \frac{\Gamma \Rightarrow \Delta, \sim\sim\alpha}{\Gamma \Rightarrow \Delta, \alpha}\ (\sim\text{right}^{-1})$$

are admissible in cut-free SPL.

Proof. Straightforward. ∎

We then obtain the characteristic property of SPL as follows.

Theorem 6.2.5 (Admissibility of contraposition) *The rule of contraposition*

$$\frac{\Gamma \Rightarrow \Delta}{{\sim}\Delta \Rightarrow {\sim}\Gamma}\ (\text{cont})$$

is admissible in cut-free SPL.

Proof. By induction on the proof P of $\Gamma \Rightarrow \Delta$ in cut-free SPL. We distinguish the cases according to the last inference of P. We show only the following case.

Case (${\sim}{-\!\!\prec}$ left): The last inference of P is of the form:

$$\frac{\Gamma \Rightarrow \Delta, {\sim}\beta \quad {\sim}\alpha, \Sigma \Rightarrow \Pi}{{\sim}(\alpha {-\!\!\prec} \beta), \Gamma, \Sigma \Rightarrow \Delta, \Pi}\ ({\sim}{-\!\!\prec}\ \text{left}).$$

By the hypothesis of induction, we have SPL$-$(cut) $\vdash {\sim}\Pi \Rightarrow {\sim}\Sigma, {\sim}{\sim}\alpha$ and SPL$-$(cut) $\vdash {\sim}{\sim}\beta, {\sim}\Delta \Rightarrow {\sim}\Gamma$. We then obtain the required fact:

$$\dfrac{\dfrac{\dfrac{\vdots}{{\sim}\Pi \Rightarrow {\sim}\Sigma, {\sim}{\sim}\alpha}}{{\sim}\Pi \Rightarrow {\sim}\Sigma, \alpha}\ ({\sim}\text{right}^{-1}) \quad \dfrac{\dfrac{\vdots}{{\sim}{\sim}\beta, {\sim}\Delta \Rightarrow {\sim}\Gamma}}{\beta, {\sim}\Delta \Rightarrow {\sim}\Gamma}\ ({\sim}\text{left}^{-1})}{\dfrac{{\sim}\Pi, {\sim}\Delta \Rightarrow {\sim}\Sigma, {\sim}\Gamma, \alpha {-\!\!\prec} \beta}{{\sim}\Pi, {\sim}\Delta \Rightarrow {\sim}\Sigma, {\sim}\Gamma, {\sim}{\sim}(\alpha {-\!\!\prec} \beta)}\ ({\sim}\text{right})}\ ({-\!\!\prec}\ \text{right})$$

where (${\sim}$right^{-1}) and (${\sim}$left^{-1}) are admissible in cut-free SPL by Proposition 6.2.4. ∎

Note that the rule of contraposition with respect to Nelson's strong negation is not admissible in the standard cut-free sequent calculi for Nelson's logics. (For a system with contraposable strong negation see [135] and for the standard cut-free sequent calculi for Nelson's logics see [196])

Definition 6.2.6 (SPL$^-$) *The system* SPL$^-$ *is defined as* LK + $\{$(${-\!\!\prec}$ left), (${-\!\!\prec}$ right), (${\sim}$left), (${\sim}$right), (cont)$\}$.

Theorem 6.2.7 (Cut-free equivalence between SPL$^-$ and SPL)

The systems SPL $-$ (cut) *and* SPL$^-$ $-$ (cut) *are theorem-equivalent, i.e., for any sequent S,* SPL $-$ (cut) $\vdash S$ *iff* SPL$^-$ $-$ (cut) $\vdash S$.

Proof. ($\Longrightarrow$): By using the rule (cont). ($\Longleftarrow$): By Theorem 6.2.5. ∎

6.3. Embedding and cut-elimination

In order to prove the cut-elimination theorem for SPL, we give an embedding f of SPL into LK, which is a modified extension of the embedding of Nelson's logic N3 into (positive) intuitionistic logic. For the embedding of Nelson's logic, see [73, 164, 193].

Definition 6.3.1 *We fix a countable set AT of atomic formulas, and define the set $AT' := \{p' \mid p \in AT\}$ of atomic formulas. The set FO_{SPL} of formulas of* SPL *is obtained from the language $\mathcal{L}$ by using AT. The set FO_{LK} of formulas of* LK *is obtained from FO_{SPL} by adding AT' and deleting the formulas with $\sim$ or $-\!\!\!<$.*

A mapping f from FO_{SPL} to FO_{LK} is defined inductively as follows:

1. $f(p) := p$ *and* $f({\sim}p) := p' \in AT'$ *for any* $p \in AT$,
2. $f(\alpha \circ \beta) := f(\alpha) \circ f(\beta)$ *where* $\circ \in \{\wedge, \vee, \rightarrow\}$,
3. $f(\alpha -\!\!\!< \beta) := f(\alpha) \wedge \neg f(\beta)$,
4. $f(\circ\alpha) := \circ f(\alpha)$ *where* $\circ \in \{\neg, \forall x, \exists x\}$,
5. $f({\sim}(\alpha \wedge \beta)) := f({\sim}\alpha) \vee f({\sim}\beta)$,
6. $f({\sim}(\alpha \vee \beta)) := f({\sim}\alpha) \wedge f({\sim}\beta)$,
7. $f({\sim}(\alpha \rightarrow \beta)) := f({\sim}\beta) -\!\!\!< f({\sim}\alpha)$ *(i.e.,* $f({\sim}\beta) \wedge \neg f({\sim}\alpha)$*)*,
8. $f({\sim}(\alpha -\!\!\!< \beta)) := f({\sim}\beta) \rightarrow f({\sim}\alpha)$,
9. $f({\sim}(\forall x\alpha)) := \exists x f({\sim}\alpha)$,
10. $f({\sim}(\exists x\alpha)) := \forall x f({\sim}\alpha)$,
11. $f({\sim}(\neg\alpha)) := \neg f({\sim}\alpha)$,
12. $f({\sim}{\sim}\alpha) := f(\alpha)$.

Let Γ be a set of formulas in FO_{SPL}. Then, an expression $f(\Gamma)$ means the result of replacing every occurrence of a formula α in Γ by an occurrence of $f(\alpha)$.

Theorem 6.3.2 (Embedding of SPL into LK) *Let Γ and Δ be sets of formulas in FO_{SPL}, and f be the mapping defined in Definition 6.3.1.*

(1) SPL $\vdash \Gamma \Rightarrow \Delta$ *iff* LK $\vdash f(\Gamma) \Rightarrow f(\Delta)$.

(2) LK $-$ (cut) $\vdash f(\Gamma) \Rightarrow f(\Delta)$ *iff* SPL $-$ (cut) $\vdash \Gamma \Rightarrow \Delta$.

Proof. We show only (1), since (2) can be obtained by observing the proof of (1). We show only the direction ($\Longrightarrow$) by induction on the proof P of $\Gamma \Rightarrow \Delta$ in SPL. We distinguish the cases according to the last inference of P. We show some cases.

Case ($-\!\!<$ right): The last inference of P is of the form:

$$\dfrac{\Gamma \Rightarrow \Delta, \alpha \quad \beta, \Sigma \Rightarrow \Pi}{\Gamma, \Sigma \Rightarrow \Delta, \Pi, \alpha -\!\!< \beta}\ (-\!\!<\text{right}).$$

By the hypothesis of induction, we have $\mathrm{LK} \vdash f(\Gamma) \Rightarrow f(\Delta), f(\alpha)$ and $\mathrm{LK} \vdash f(\beta), f(\Sigma) \Rightarrow f(\Pi)$. Then, we obtain the required fact as follows.

$$\dfrac{\dfrac{\overset{\vdots}{f(\Gamma) \Rightarrow f(\Delta), f(\alpha)}}{f(\Gamma), f(\Sigma) \Rightarrow f(\Delta), f(\Pi), f(\alpha)}\ \vdots\,(\text{we}-\text{left/right}) \quad \dfrac{\dfrac{\overset{\vdots}{f(\beta), f(\Sigma) \Rightarrow f(\Pi)}}{f(\Sigma) \Rightarrow f(\Pi), \neg f(\beta)}\,(\neg\text{right})}{f(\Gamma), f(\Sigma) \Rightarrow f(\Delta), f(\Pi), \neg f(\beta)}\ \vdots\,(\text{we}-\text{left/right})}{f(\Gamma), f(\Sigma) \Rightarrow f(\Delta), f(\Pi), f(\alpha) \wedge \neg f(\beta)}\ (\wedge\text{right})$$

where $f(\alpha) \wedge \neg f(\beta) = f(\alpha -\!\!< \beta)$.

Case ($\sim\!-\!\!<$ left): The last inference of P is of the form:

$$\dfrac{\Gamma \Rightarrow \Delta, {\sim}\beta \quad {\sim}\alpha, \Sigma \Rightarrow \Pi}{{\sim}(\alpha -\!\!< \beta), \Gamma, \Sigma \Rightarrow \Delta, \Pi}\ ({\sim}-\!\!<\text{left}).$$

By the hypothesis of induction, we have $\mathrm{LK} \vdash f(\Gamma) \Rightarrow f(\Delta), f({\sim}\beta)$ and $\mathrm{LK} \vdash f({\sim}\alpha), f(\Sigma) \Rightarrow f(\Pi)$. Then, we obtain the required fact:

$$\dfrac{\overset{\vdots}{f(\Gamma) \Rightarrow f(\Delta), f({\sim}\beta)} \quad \overset{\vdots}{f({\sim}\alpha), f(\Sigma) \Rightarrow f(\Pi)}}{f({\sim}\beta) \to f({\sim}\alpha), f(\Gamma), f(\Sigma) \Rightarrow f(\Delta), f(\Pi)}\ (\to\text{left})$$

where $f({\sim}\beta) \to f({\sim}\alpha) = f({\sim}(\alpha -\!\!< \beta))$.

Case (${\sim}\!\to$right): The last inference of P is of the form:

$$\dfrac{\Gamma \Rightarrow \Delta, {\sim}\beta \quad {\sim}\alpha, \Sigma \Rightarrow \Pi}{\Gamma, \Sigma \Rightarrow \Delta, \Pi, {\sim}(\alpha \to \beta)}\ ({\sim}\!\to\text{right}).$$

By the hypothesis of induction, we have $\mathrm{LK} \vdash f(\Gamma) \Rightarrow f(\Delta), f({\sim}\beta)$ and $\mathrm{LK} \vdash f({\sim}\alpha), f(\Sigma) \Rightarrow f(\Pi)$. Then, we obtain the required fact as follows.

$$
\dfrac{
\begin{array}{c}\vdots\\ f(\Gamma)\Rightarrow f(\Delta),f(\sim\beta)\\ \vdots\ (\text{we}-\text{left/right})\\ f(\Gamma),f(\Sigma)\Rightarrow f(\Delta),f(\Pi),f(\sim\beta)\end{array}
\quad
\begin{array}{c}\vdots\\ \dfrac{f(\sim\alpha),f(\Sigma)\Rightarrow f(\Pi)}{f(\Sigma)\Rightarrow f(\Pi),\neg f(\sim\alpha)}\,(\neg\text{right})\\ \vdots\ (\text{we}-\text{left/right})\\ f(\Gamma),f(\Sigma)\Rightarrow f(\Delta),f(\Pi),\neg f(\sim\alpha)\end{array}
}{f(\Gamma),f(\Sigma)\Rightarrow f(\Delta),f(\Pi),f(\sim\beta)\wedge\neg f(\sim\alpha)}\,(\wedge\text{right})
$$

where $f(\sim\beta) \wedge \neg f(\sim\alpha) = f(\sim\beta) \mathbin{-\!\!<} f(\sim\alpha) = f(\sim(\alpha\to\beta))$. ∎

Theorem 6.3.3 (Cut-elimination for SPL) *The rule* (cut) *is admissible in cut-free* SPL.

Proof. Suppose that SPL $\vdash \Gamma \Rightarrow \Delta$. Then we have LK $\vdash f(\Gamma) \Rightarrow f(\Delta)$ by Theorem 6.3.2 (1). We obtain LK$-$(cut) $\vdash f(\Gamma) \Rightarrow f(\Delta)$ by the well-known cut-elimination theorem for LK. By Theorem 6.3.2 (2), we obtain the required fact SPL$-$(cut) $\vdash \Gamma \Rightarrow \Delta$. ∎

Theorem 6.3.4 (Cut-elimination for SPL$^-$) *The rule* (cut) *is admissible in cut-free* SPL$^-$.

Proof. By Theorems 6.2.7 and 6.3.3. ∎

Using Theorems 6.3.3 and 6.3.4, the paraconsistency of SPL and SPL$^-$ w.r.t. $\sim$ is shown.

Definition 6.3.5 *Let $\sharp$ be a unary connective. A sequent calculus L is called* explosive *with respect to $\sharp$ if for any formulas α and β, the sequents of the form $\alpha, \sharp\alpha \Rightarrow \beta$ are provable in L. It is called* paraconsistent *with respect to $\sharp$ if it is not explosive with respect to $\sharp$.*

Theorem 6.3.6 (Paraconsistency of SPL and SPL$^-$) *Let L be* SPL *or* SPL$^-$. *L is paraconsistent with respect to $\sim$.*

Proof. Let p and q be distinct atomic formulas. Then, the sequent $p, \sim p \Rightarrow q$ is not provable in L. The unprovability of this sequent is guaranteed by Theorems 6.3.3 and 6.3.4. ∎

Note that SPL and SPL$^-$ are explosive with respect to $\neg$.

Since (first-order) LK is known as undecidable, the extensions SPL and SPL$^-$ of LK are also undecidable. On the other hand, the *monadic fragment* of LK, the fragment in which all predicate symbols are one-place and there are no function symbols, is known to be decidable. This fact implies the following theorem.

Theorem 6.3.7 (Decidability of the monadic fragments) *The monadic fragments of* SPL *and* SPL$^-$ *are decidable.*

Proof. By (a slightly modified version of) Theorem 6.3.2, the provability relation of the fragments can be transformed into that of the monadic fragment of LK. Since the monadic fragment of LK is decidable, the monadic fragments of SPL and SPL$^-$ are also decidable. ∎

Similarly, we can also obtain the following theorem.

Theorem 6.3.8 (Decidability of the propositional fragments) *The propositional fragments of* SPL *and* SPL$^-$ *are decidable.*

6.4. Dual paraconsistent logic

Definition 6.4.1 (DPL) *A sequent calculus* DPL *for dual paraconsistent logic is obtained from* SPL *by replacing the inference rules* ($\sim\!\rightarrow$left), ($\sim\!\rightarrow$right), ($\sim\!\prec$left) *and* ($\sim\!\prec$right) *by the inference rules of the form:*

$$\frac{\alpha,\Gamma\Rightarrow\Delta,\beta}{\sim(\alpha\rightarrow\beta),\Gamma\Rightarrow\Delta}\ (\sim\!\rightarrow\text{left}^d)\qquad \frac{\Gamma\Rightarrow\Delta,\alpha\quad \beta,\Sigma\Rightarrow\Pi}{\Gamma,\Sigma\Rightarrow\Delta,\Pi,\sim(\alpha\rightarrow\beta)}\ (\sim\!\rightarrow\text{right}^d)$$

$$\frac{\Gamma\Rightarrow\Delta,\alpha\quad \beta,\Sigma\Rightarrow\Pi}{\sim(\alpha\prec\beta),\Gamma,\Sigma\Rightarrow\Delta,\Pi}\ (\sim\!\prec\text{left}^d)\qquad \frac{\alpha,\Gamma\Rightarrow\Delta,\beta}{\Gamma\Rightarrow\Delta,\sim(\alpha\prec\beta)}\ (\sim\!\prec\text{right}^d).$$

Proposition 6.4.2 *The sequents 1, 2, 5–8 in Proposition 6.2.3 and the following sequents are provable in* DPL*:*

3′. $\sim(\alpha\rightarrow\beta)\Leftrightarrow(\alpha\prec\beta)$ *(negated implication as co-implication).*

4′. $\sim(\alpha\prec\beta)\Leftrightarrow(\alpha\rightarrow\beta)$ *(negated co-implication as implication).*

Proposition 6.2.4 holds for DPL, but Theorem 6.2.5 does not hold for DPL.

Definition 6.4.3 *The set FO_{DPL} of formulas of* DPL *is the same as FO_{SPL} in Definition 6.3.1. The set FO_{LK} of formulas of* LK *is defined in Definition 6.3.1*

A mapping f from FO_{DPL} to FO_{LK} is obtained from the mapping defined in Definition 6.3.1 by replacing the conditions 11 and 12 by the following conditions:

11′. $f({\sim}(\alpha{\rightarrow}\beta)) := f(\alpha){-\!\!\prec} f(\beta)$ *(i.e.,* $f(\alpha)\wedge\neg f(\beta)$*),*

12′. $f({\sim}(\alpha{-\!\!\prec}\beta)) := f(\alpha){\rightarrow} f(\beta)$.

Theorem 6.4.4 (Embedding of DPL into LK) *Let Γ and Δ be sets of formulas in FO_{DPL}, and f be the mapping defined in Definition 6.4.3.*

1. DPL $\vdash \Gamma \Rightarrow \Delta$ *iff* LK $\vdash f(\Gamma) \Rightarrow f(\Delta)$.
2. LK $-$ (cut) $\vdash f(\Gamma) \Rightarrow f(\Delta)$ *iff* DPL $-$ (cut) $\vdash \Gamma \Rightarrow \Delta$.

Using Theorem 6.4.4, we can derive the following theorem.

Theorem 6.4.5 (Cut-elimination for DPL) *The rule* (cut) *is admissible in cut-free* DPL.

We then present the characteristic property of DPL as follows.

Theorem 6.4.6 (Duality for DPL) *Suppose that $\tilde{\alpha}$ and $\tilde{\beta}$ are the formulas obtained from formulas α and β,respectively, by replacing every occurrence of $\wedge$, $\vee$, $\forall$, $\exists$, $\rightarrow$ and ${-\!\!\prec}$ by $\vee$, $\wedge$, $\exists$, $\forall$, ${-\!\!\prec}$ and $\rightarrow$, respectively.*

1. *if* DPL $-$ (cut) $\vdash \alpha \Rightarrow \beta$, *then* DPL $-$ (cut) $\vdash \tilde{\beta} \Rightarrow \tilde{\alpha}$.
2. *if* DPL $-$ (cut) $\vdash \alpha \Leftrightarrow \beta$, *then* DPL $-$ (cut) $\vdash \tilde{\beta} \Leftrightarrow \tilde{\alpha}$.

Proof. We show only (1), since (2) is derived from (1). By the hypothesis, we have a cut-free proof P of $\alpha \Rightarrow \beta$ in DPL$-$(cut). We replace all the sequents of P by the converse sequents (i.e., the succedent and antecedent are exchanged), and replace all the occurrences of $\wedge, \vee, \rightarrow$ and ${-\!\!\prec}$ by those of $\vee$, $\wedge$, ${-\!\!\prec}$ and $\rightarrow$, respectively. We then obtain a proof of $\tilde{\beta} \Rightarrow \tilde{\alpha}$ in DPL$-$(cut). ∎

Theorem 6.4.7 (Paraconsistency of DPL) DPL *is paraconsistent with respect to* $\sim$.

Theorem 6.4.8 (Decidability of the monadic and propositional fragments) *The monadic and propositional fragments of* DPL *are both decidable.*

6.5. Semantics and completeness

In this section, the semantics and completeness for SPL and DPL are discussed. For the sake of simplicity of the discussion, the language without individual constants and function symbols is adopted in this section. The same names $\mathcal{L}$, SPL and DPL are used for the reduced language and the corresponding subsystems, respectively. Let Γ be a set $\{\alpha_1, ..., \alpha_m\}$ $(m \geq 0)$ of formulas, and p be a fixed atomic formula. Then Γ^* is defined as $\alpha_1 \vee \cdots \vee \alpha_m$ if $m \geq 1$, and $\neg(p{\rightarrow}p)$ if $m = 0$. Also Γ_* is defined as $\alpha_1 \wedge \cdots \wedge \alpha_m$ if $m \geq 1$, and $p{\rightarrow}p$ if $m = 0$.

First, the semantics and completeness for SPL are discussed. The semantics and completeness for DPL can also be obtained similarly, and hence the completeness proof for DPL is omitted here.

Definition 6.5.1 $\mathcal{A} := \langle U, I^+, I^- \rangle$ *is called a* model *if the following conditions hold:*

1. *U is a non-empty set,*
2. *I^+ and I^- are mappings such that $p^{I^+}, p^{I^-} \subseteq U^n$ (i.e., p^{I^+} and p^{I^-} are n-ary relations on U) for an n-ary predicate symbol p.*

We introduce the notation $\underline{u}$ for the name of $u \in U$, and write $\mathcal{L}[\mathcal{A}]$ for the language obtained from $\mathcal{L}$ by adding the names of all the elements of U. A formula α is called a closed formula *if α has no free individual variable. A formula of the form $\forall x_1 \cdots \forall x_m \alpha$ is called the* universal closure *of α if the free variables of α are $x_1, ..., x_m$. We write $cl(\alpha)$ for the universal closure of α.*

Definition 6.5.2 *Let $\mathcal{A} := \langle U, I^+, I^- \rangle$ be a model. The satisfaction relations $\mathcal{A} \models^+ \alpha$ and $\mathcal{A} \models^- \alpha$ for any closed formula α of $\mathcal{L}[\mathcal{A}]$ are defined inductively as follows:*

1. *$[\mathcal{A} \models^+ p(\underline{u}_1, ..., \underline{u}_n)$ iff $(u_1, ..., u_n) \in p^{I^+}]$ and $[\mathcal{A} \models^- p(\underline{u}_1, ..., \underline{u}_n)$ iff $(u_1, ..., u_n) \in p^{I^-}]$ for any n-ary atomic formula $p(\underline{u}_1, ..., \underline{u}_n)$,*
2. *$\mathcal{A} \models^+ \alpha \wedge \beta$ iff $\mathcal{A} \models^+ \alpha$ and $\mathcal{A} \models^+ \beta$,*
3. *$\mathcal{A} \models^+ \alpha \vee \beta$ iff $\mathcal{A} \models^+ \alpha$ or $\mathcal{A} \models^+ \beta$,*
4. *$\mathcal{A} \models^+ \alpha{\rightarrow}\beta$ iff not-$(\mathcal{A} \models^+ \alpha)$ or $\mathcal{A} \models^+ \beta$,*
5. *$\mathcal{A} \models^+ \alpha{\prec}\beta$ iff $\mathcal{A} \models^+ \alpha$ and not-$(\mathcal{A} \models^+ \beta)$,*
6. *$\mathcal{A} \models^+ \neg\alpha$ iff not-$(\mathcal{A} \models^+ \alpha)$,*

7. $\mathcal{A} \models^{+} {\sim}\alpha$ *iff* $\mathcal{A} \models^{-} \alpha$,

8. $\mathcal{A} \models^{+} \forall x\alpha$ *iff* $\mathcal{A} \models^{+} \alpha[\underline{u}/x]$ *for all* $u \in U$,

9. $\mathcal{A} \models^{+} \exists x\alpha$ *iff* $\mathcal{A} \models^{+} \alpha[\underline{u}/x]$ *for some* $u \in U$,

10. $\mathcal{A} \models^{-} \alpha \wedge \beta$ *iff* $\mathcal{A} \models^{-} \alpha$ *or* $\mathcal{A} \models^{-} \beta$,

11. $\mathcal{A} \models^{-} \alpha \vee \beta$ *iff* $\mathcal{A} \models^{-} \alpha$ *and* $\mathcal{A} \models^{-} \beta$,

12. $\mathcal{A} \models^{-} \alpha{\rightarrow}\beta$ *iff not-*$(\mathcal{A} \models^{-} \alpha)$ *and* $\mathcal{A} \models^{-} \beta$,

13. $\mathcal{A} \models^{-} \alpha{\prec}\beta$ *iff* $\mathcal{A} \models^{-} \alpha$ *or not-*$(\mathcal{A} \models^{-} \beta)$,

14. $\mathcal{A} \models^{-} \neg\alpha$ *iff not-*$(\mathcal{A} \models^{-} \alpha)$,

15. $\mathcal{A} \models^{-} {\sim}\alpha$ *iff* $\mathcal{A} \models^{+} \alpha$,

16. $\mathcal{A} \models^{-} \forall x\alpha$ *iff* $\mathcal{A} \models^{-} \alpha[\underline{u}/x]$ *for some* $u \in U$,

17. $\mathcal{A} \models^{-} \exists x\alpha$ *iff* $\mathcal{A} \models^{-} \alpha[\underline{u}/x]$ *for all* $u \in U$.

The satisfaction relations $\mathcal{A} \models^{+} \alpha$ *and* $\mathcal{A} \models^{-} \alpha$ *for any formula* α *of* $\mathcal{L}$ *are defined by* $(\mathcal{A} \models^{+} \alpha$ *iff* $\mathcal{A} \models^{+} cl(\alpha))$ *and* $(\mathcal{A} \models^{-} \alpha$ *iff* $\mathcal{A} \models^{-} cl(\alpha))$. *A formula* α *of* $\mathcal{L}$ *is called* valid *if* $\mathcal{A} \models^{+} \alpha$ *holds for any model* $\mathcal{A}$. *A sequent* $\Gamma \Rightarrow \Delta$ *of* $\mathcal{L}$ *is called valid if so is the formula* $\Gamma_{*}{\rightarrow}\Delta^{*}$.

The intended meanings of the satisfaction relations $\models^{+}$ and $\models^{-}$ are *verification* (or *provability*, or *support of truth*) and *falsification* (or *refutability*, or *support of falsity*), respectively.

Theorem 6.5.3 (Soundness of SPL) *For any sequent* S, *if* SPL $\vdash$ S, *then* S *is valid.*

Proof. By induction on the proof P of S. We distinguish the cases according to the last inference of P. We show only the following case.

(Case (${\sim}\exists$right)): The last inference of P is of the form:

$$\frac{\Gamma \Rightarrow \Delta, {\sim}\alpha[z/x]}{\Gamma \Rightarrow \Delta, {\sim}\exists x\alpha}\ ({\sim}\exists\text{right}).$$

We show that "$\Gamma \Rightarrow \Delta, {\sim}\alpha[z/x]$ is valid" implies "$\Gamma \Rightarrow \Delta, {\sim}\exists x\alpha$ is valid". By the hypothesis, (i): $\forall z_1 \cdots \forall z_n \forall z(\Gamma_{*}{\rightarrow}(\Delta^{*} \vee ({\sim}\alpha[z/x])))$ (where z_1, ..., z_n are the free individual variables occurring in $\Gamma \Rightarrow \Delta, {\sim}\exists x\alpha$) is valid. We show that $\mathcal{A} \models^{+} \forall z_1 \cdots \forall z_n(\Gamma_{*}{\rightarrow}(\Delta^{*} \vee ({\sim}\exists x\alpha)))$ for any model $\mathcal{A} := \langle U, I^{+}, I^{-}\rangle$, i.e., we show that for any $u_1, ..., u_n \in U$, $\mathcal{A} \models^{+} \underline{\Gamma}_{*}{\rightarrow}(\underline{\Delta}^{*} \vee ({\sim}\exists x\underline{\alpha}))$, where $\underline{\Gamma}_{*}$, $\underline{\Delta}^{*}$ and $\underline{\alpha}$ are respectively obtained from Γ_*, Δ^* and α by replacing $z_1, ..., z_n$ by $\underline{u}_1, ..., \underline{u}_n$.[2] By (i), we have $\mathcal{A} \models^+ (\underline{\Gamma}_* \to (\underline{\Delta}^* \vee (\sim\underline{\alpha}[z/x])))[\underline{w}/z]$ for any $w \in U$. By the eigenvariable condition, z is not occurring freely in $\underline{\Gamma}_*, \underline{\Delta}^*$ and $\underline{\alpha}$. Thus, $\underline{\Gamma}_*[\underline{w}/z]$ and $\underline{\Delta}^*[\underline{w}/z]$ are equivalent to $\underline{\Gamma}_*$and $\underline{\Delta}^*$ respectively, and $\underline{\alpha}[z/x][\underline{w}/z]$ is equivalent to $\underline{\alpha}[\underline{w}/z][\underline{w}/x]$, i.e., $\underline{\alpha}[\underline{w}/x]$. Therefore, for any $w \in U$, we have that (a): $\mathcal{A} \models^+ \underline{\Gamma}_* \to (\underline{\Delta}^* \vee \sim\underline{\alpha}[\underline{w}/x])$. Suppose that (b): $[\mathcal{A} \models^+ \underline{\Gamma}_*$ and not $(\mathcal{A} \models^+ \underline{\Delta}^*)$]. Then, by (a), we have that for any $w \in U$, $\mathcal{A} \models^+ \sim\underline{\alpha}[\underline{w}/x]$, i.e., $\mathcal{A} \models^- \underline{\alpha}[\underline{w}/x]$. Therefore, we obtain (c): $\mathcal{A} \models^- \exists x\underline{\alpha}$, and hence $\mathcal{A} \models^+ \sim\exists x\underline{\alpha}$. This means that (b) implies (c), i.e., $\mathcal{A} \models^+ \underline{\Gamma}_*$ implies ($\mathcal{A} \models^+ \underline{\Delta}^*$ or $\mathcal{A} \models^+ \sim\exists x\underline{\alpha}$). Therefore, we have the required fact that $\mathcal{A} \models^+ \underline{\Gamma}_* \to (\underline{\Delta}^* \vee (\sim\exists x\underline{\alpha}))$ for any $u_1, ...u_n \in U$. ∎

Now, we start to prove the completeness theorem.

Definition 6.5.4 *A sequent* $\Gamma \Rightarrow \Delta$ *is called* saturated *if for any formulas* α *and* β,

(s1) $\alpha \wedge \beta \in \Gamma$ *implies* ($\alpha \in \Gamma$ *and* $\beta \in \Gamma$),

(s2) $\alpha \wedge \beta \in \Delta$ *implies* ($\alpha \in \Delta$ *or* $\beta \in \Delta$),

(s3) $\alpha \vee \beta \in \Gamma$ *implies* ($\alpha \in \Gamma$ *or* $\beta \in \Gamma$),

(s4) $\alpha \vee \beta \in \Delta$ *implies* ($\alpha \in \Delta$ *and* $\beta \in \Delta$),

(s5) $\alpha \to \beta \in \Gamma$ *implies* ($\alpha \in \Delta$ *or* $\beta \in \Gamma$),

(s6) $\alpha \to \beta \in \Delta$ *implies* ($\alpha \in \Gamma$ *and* $\beta \in \Delta$),

(s7) $\alpha \prec \beta \in \Gamma$ *implies* ($\alpha \in \Gamma$ *and* $\beta \in \Delta$),

(s8) $\alpha \prec \beta \in \Delta$ *implies* ($\alpha \in \Delta$ *or* $\beta \in \Gamma$),

(s9) $\neg\alpha \in \Gamma$ *implies* $\alpha \in \Delta$,

(s10) $\neg\alpha \in \Delta$ *implies* $\alpha \in \Gamma$,

(s11) $\forall x\alpha \in \Gamma$ *implies* ($\alpha[y/x] \in \Gamma$ *for any individual variable* y),

(s12) $\forall x\alpha \in \Delta$ *implies* ($\alpha[z/x] \in \Delta$ *for some individual variable* z),

(s13) $\exists x\alpha \in \Gamma$ *implies* ($\alpha[z/x] \in \Gamma$ *for some individual variable* z),

(s14) $\exists x\alpha \in \Delta$ *implies* ($\alpha[y/x] \in \Delta$ *for any individual variable* y),

(s15) $\sim\sim\alpha \in \Gamma$ *implies* $\alpha \in \Gamma$,

(s16) $\sim\sim\alpha \in \Delta$ *implies* $\alpha \in \Delta$,

[2] We note that $(\sim\exists x\alpha)[\underline{w}/z_1, ..., \underline{u}_n/z_n]$ (the simultaneous substitution) is equivalent to $\sim\exists x(\alpha[\underline{u}_1/z_1, ..., \underline{u}_n/z_n])$, i.e., $\sim\exists x\underline{\alpha}$.

(s17) $\sim(\alpha \wedge \beta) \in \Gamma$ *implies* ($\sim\alpha \in \Gamma$ *or* $\sim\beta \in \Gamma$),

(s18) $\sim(\alpha \wedge \beta) \in \Delta$ *implies* ($\sim\alpha \in \Delta$ *and* $\sim\beta \in \Delta$),

(s19) $\sim(\alpha \vee \beta) \in \Gamma$ *implies* ($\sim\alpha \in \Gamma$ *and* $\sim\beta \in \Gamma$),

(s20) $\sim(\alpha \vee \beta) \in \Delta$ *implies* ($\sim\alpha \in \Delta$ *or* $\sim\beta \in \Delta$),

(s21) $\sim(\alpha \rightarrow \beta) \in \Gamma$ *implies* ($\sim\alpha \in \Delta$ *and* $\sim\beta \in \Gamma$),

(s22) $\sim(\alpha \rightarrow \beta) \in \Delta$ *implies* ($\sim\alpha \in \Gamma$ *or* $\sim\beta \in \Delta$),

(s23) $\sim(\alpha \multimap\!\!< \beta) \in \Gamma$ *implies* ($\sim\alpha \in \Gamma$ *or* $\sim\beta \in \Delta$),

(s24) $\sim(\alpha \multimap\!\!< \beta) \in \Delta$ *implies* ($\sim\alpha \in \Delta$ *and* $\sim\beta \in \Gamma$),

(s25) $\sim\neg\alpha \in \Gamma$ *implies* $\sim\alpha \in \Delta$,

(s26) $\sim\neg\alpha \in \Delta$ *implies* $\sim\alpha \in \Gamma$,

(s27) $\sim\forall x\alpha \in \Gamma$ *implies* ($\sim\alpha[z/x] \in \Gamma$ *for some individual variable* z),

(s28) $\sim\forall x\alpha \in \Delta$ *implies* ($\sim\alpha[y/x] \in \Delta$ *for any individual variable* y),

(s29) $\sim\exists x\alpha \in \Gamma$ *implies* ($\sim\alpha[y/x] \in \Gamma$ *for any individual variable* y),

(s30) $\sim\exists x\alpha \in \Delta$ *implies* ($\sim\alpha[z/x] \in \Delta$ *for some individual variable* z).

Definition 6.5.5 *An expression* $\Gamma \Rightarrow \Delta$ *is called an* infinite sequent *if* Γ *or* Δ *are infinite (countable) sets of formulas. An infinite sequent* $\Gamma \Rightarrow \Delta$ *is called provable if a sequent* $\Gamma' \Rightarrow \Delta'$ *is provable, where* Γ' *and* Δ' *are finite subsets of* Γ *and* Δ *respectively.*

Definition 6.5.6 *A* decomposition *of a sequent (or infinite sequent)* S *is*
defined as being of the form S' *or* $S'; S''$ *by*

(1a) $\Gamma \Rightarrow \Delta, \alpha \wedge \beta, \alpha$; $\Gamma \Rightarrow \Delta, \alpha \wedge \beta, \beta$ *is a decomposition of* $\Gamma \Rightarrow \Delta, \alpha \wedge \beta$,

(1b) $\alpha, \beta, \alpha \wedge \beta, \Gamma \Rightarrow \Delta$ *is a decomposition of* $\alpha \wedge \beta, \Gamma \Rightarrow \Delta$,

(2a) $\Gamma \Rightarrow \Delta, \alpha \vee \beta, \alpha, \beta$ *is a decomposition of* $\Gamma \Rightarrow \Delta, \alpha \vee \beta$,

(2b) $\alpha, \alpha \vee \beta, \Gamma \Rightarrow \Delta$; $\beta, \alpha \vee \beta, \Gamma \Rightarrow \Delta$ *is a decomposition of* $\alpha \vee \beta, \Gamma \Rightarrow \Delta$,

(3a) $\alpha, \Gamma \Rightarrow \Delta, \alpha \rightarrow \beta, \beta$ *is a decomposition of* $\Gamma \Rightarrow \Delta, \alpha \rightarrow \beta$,

(3b) $\alpha \rightarrow \beta, \Gamma \Rightarrow \Delta, \alpha$; $\beta, \alpha \rightarrow \beta, \Gamma \Rightarrow \Delta$ *is a decomposition of* $\alpha \rightarrow \beta, \Gamma \Rightarrow \Delta$,

(4a) $\Gamma \Rightarrow \Delta, \alpha \multimap \beta, \alpha$; $\beta, \Gamma \Rightarrow \Delta, \alpha \multimap \beta$ *is a decomposition of* $\Gamma \Rightarrow \Delta, \alpha \multimap \beta$,

(4b) $\alpha, \alpha \multimap \beta, \Gamma \Rightarrow \Delta, \beta$ *is a decomposition of* $\alpha \multimap \beta, \Gamma \Rightarrow \Delta$,

(5a) $\alpha, \Gamma \Rightarrow \Delta, \neg\alpha$ *is a decomposition of* $\Gamma \Rightarrow \Delta, \neg\alpha$,

(5b) $\neg\alpha, \Gamma \Rightarrow \Delta, \alpha$ *is a decomposition of* $\neg\alpha, \Gamma \Rightarrow \Delta$,

(6a) $\Gamma \Rightarrow \Delta, \forall x\alpha, \alpha[z/x]$ *is a decomposition of* $\Gamma \Rightarrow \Delta, \forall x\alpha$ *where* z *is a* fresh *free individual variable, i.e.,* z *is not occurring in* $\Gamma \Rightarrow \Delta, \forall x\alpha$,

(6b) $\alpha[y_1/x], ..., \alpha[y_m/x], \forall x\alpha, \Gamma \Rightarrow \Delta$ *is a decomposition of* $\forall x\alpha, \Gamma \Rightarrow \Delta$ *where* $y_1, ..., y_m$ *are the free individual variables occurring in* $\forall x\alpha, \Gamma \Rightarrow \Delta$,[3]

(7a) $\Gamma \Rightarrow \Delta, \exists x\alpha, \alpha[y_1/x], ..., \alpha[y_m/x]$ *is a decomposition of* $\Gamma \Rightarrow \Delta, \exists x\alpha$ *where* $y_1, ..., y_m$ *are the free individual variables occurring in* $\Gamma \Rightarrow \Delta, \exists x\alpha$,

(7b) $\alpha[z/x], \exists x\alpha, \Gamma \Rightarrow \Delta$ *is a decomposition of* $\exists x\alpha, \Gamma \Rightarrow \Delta$ *where* z *is a fresh free individual variable,*

(8a) $\alpha, {\sim}{\sim}\alpha, \Gamma \Rightarrow \Delta$ *is a decomposition of* ${\sim}{\sim}\alpha, \Gamma \Rightarrow \Delta$,

(8b) $\Gamma \Rightarrow \Delta, {\sim}{\sim}\alpha, \alpha$ *is a decomposition of* $\Gamma \Rightarrow \Delta, {\sim}{\sim}\alpha$,

(9a) $\Gamma \Rightarrow \Delta, {\sim}(\alpha \wedge \beta), {\sim}\alpha, {\sim}\beta$ *is a decomposition of* $\Gamma \Rightarrow \Delta, {\sim}(\alpha \wedge \beta)$,

(9b) ${\sim}\alpha, {\sim}(\alpha \wedge \beta), \Gamma \Rightarrow \Delta$; ${\sim}\beta, {\sim}(\alpha \wedge \beta), \Gamma \Rightarrow \Delta$ *is a decomposition of* ${\sim}(\alpha \wedge \beta), \Gamma \Rightarrow \Delta$,

(10a) $\Gamma \Rightarrow \Delta, {\sim}(\alpha \vee \beta), {\sim}\alpha$; $\Gamma \Rightarrow \Delta, {\sim}(\alpha \vee \beta), {\sim}\beta$ *is a decomposition of* $\Gamma \Rightarrow \Delta, {\sim}(\alpha \vee \beta)$,

(10b) ${\sim}\alpha, {\sim}\beta, {\sim}(\alpha \vee \beta), \Gamma \Rightarrow \Delta$ *is a decomposition of* ${\sim}(\alpha \vee \beta), \Gamma \Rightarrow \Delta$,

(11a) $\Gamma \Rightarrow \Delta, {\sim}(\alpha \to \beta), {\sim}\beta$; ${\sim}\alpha, \Gamma \Rightarrow \Delta, {\sim}(\alpha \to \beta)$ *is a decomposition of* $\Gamma \Rightarrow \Delta, {\sim}(\alpha \to \beta)$,

(11b) ${\sim}\beta, {\sim}(\alpha \to \beta), \Gamma \Rightarrow \Delta, {\sim}\alpha$ *is a decomposition of* ${\sim}(\alpha \to \beta), \Gamma \Rightarrow \Delta$,

(12a) ${\sim}\beta, \Gamma \Rightarrow \Delta, {\sim}(\alpha \multimap \beta), {\sim}\alpha$ *is a decomposition of* $\Gamma \Rightarrow \Delta, {\sim}(\alpha \multimap \beta)$,

(12b) ${\sim}(\alpha \multimap \beta), \Gamma \Rightarrow \Delta, {\sim}\beta$; ${\sim}\alpha, {\sim}(\alpha \multimap \beta), \Gamma \Rightarrow \Delta$ *is a decomposition of* ${\sim}(\alpha \multimap \beta), \Gamma \Rightarrow \Delta$,

(13a) ${\sim}\alpha, \Gamma \Rightarrow \Delta, {\sim}\neg\alpha$ *is a decomposition of* $\Gamma \Rightarrow \Delta, {\sim}\neg\alpha$,

(13b) ${\sim}\neg\alpha, \Gamma \Rightarrow \Delta, {\sim}\alpha$ *is a decomposition of* ${\sim}\neg\alpha, \Gamma \Rightarrow \Delta$,

[3] If $\forall x\alpha, \Gamma \Rightarrow \Delta$ has no free individual variable, then we replace x in α by any variable from $\mathcal{L}$. Such a condition is also adopted in (7a), (14a) and (15b).

(14a) $\Gamma \Rightarrow \Delta, {\sim}\forall x\alpha, {\sim}\alpha[y_1/x], ..., {\sim}\alpha[y_m/x]$ *is a decomposition of* $\Gamma \Rightarrow \Delta, {\sim}\forall x\alpha$ *where* $y_1, ..., y_m$ *are the free individual variables occurring in* $\Gamma \Rightarrow \Delta, {\sim}\forall x\alpha$,

(14b) ${\sim}\alpha[z/x], {\sim}\forall x\alpha, \Gamma \Rightarrow \Delta$ *is a decomposition of* ${\sim}\forall x\alpha, \Gamma \Rightarrow \Delta$ *where* z *is a fresh free individual variable,*

(15a) $\Gamma \Rightarrow \Delta, {\sim}\exists x\alpha, {\sim}\alpha[z/x]$ *is a decomposition of* $\Gamma \Rightarrow \Delta, {\sim}\exists x\alpha$ *where* z *is a fresh free individual variable,*

(15b) ${\sim}\alpha[y_1/x], ..., {\sim}\alpha[y_m/x], {\sim}\exists x\alpha, \Gamma \Rightarrow \Delta$ *is a decomposition of* ${\sim}\exists x\alpha, \Gamma \Rightarrow \Delta$ *where* $y_1, ..., y_m$ *are the free individual variables occurring in* ${\sim}\exists x\alpha, \Gamma \Rightarrow \Delta$.

Definition 6.5.7 *A* decomposition tree of S *is a tree which is the result of some repeated decomposition of* S.

In other words, a decomposition tree corresponds to a bottom up proof search tree of SPL−(cut). In every decomposition of S (i.e., S' or $S'; S''$), if S is unprovable in SPL−(cut), then so is S' or S''.

Lemma 6.5.8 *Let* $\Gamma \Rightarrow \Delta$ *be a given unprovable sequent in* SPL−(cut). *There exists an unprovable, saturated (infinite) sequent* $\Gamma^\omega \Rightarrow \Delta^\omega$ *such that* $\Gamma \subseteq \Gamma^\omega$ *and* $\Delta \subseteq \Delta^\omega$.

Proof. Let $\Gamma \Rightarrow \Delta$ be an unprovable sequent in SPL−(cut). We construct $\Gamma^\omega \Rightarrow \Delta^\omega$ from $\Gamma \Rightarrow \Delta$ as follows.

(1) We apply the decomposition procedure from Definition 6.5.6 to $\Gamma \Rightarrow \Delta$, in the following order, skipping the decomposition procedures which are not applicable to the formulas in $\Gamma \Rightarrow \Delta$.

$$\text{(1a)} \longrightarrow \text{(1b)} \longrightarrow \text{(2a)} \longrightarrow \cdots \longrightarrow \text{(15b)}.$$

In such a decomposition process, one of the decomposed elements S' and S'' of S is an unprovable sequent.

(2) We repeat the same procedure (1), infinitely often. Then, we obtain an infinite decomposition tree with finitely many branches.

(3) By König's Lemma, we have an infinite path of this decomposition tree as follows.

$$\Gamma_0 \Rightarrow \Delta_0 \mid \Gamma_1 \Rightarrow \Delta_1 \mid \Gamma_2 \Rightarrow \Delta_2 \mid \cdots \infty,$$

where $\Gamma_0 \Rightarrow \Delta_0$ is $\Gamma \Rightarrow \Delta$. In this sequence of sequents on the infinite path, we have that $\Gamma_0 \subseteq \Gamma_1 \subseteq \Gamma_2 \subseteq \cdots$ and $\Delta_0 \subseteq \Delta_1 \subseteq \Delta_2 \subseteq \cdots$.

(4) We put $\Gamma^\omega := \bigcup_{i=0}^{\infty} \Gamma_i$ and $\Delta^\omega := \bigcup_{i=0}^{\infty} \Delta_i$.[4]

Then, we have that $\Gamma \subseteq \Gamma^\omega$ and $\Delta \subseteq \Delta^\omega$, and can verify that $\Gamma^\omega \Rightarrow \Delta^\omega$ is an unprovable, saturated sequent. ∎

[4] We note that $\Gamma^\omega \cap \Delta^\omega = \varnothing$.

Lemma 6.5.9 *Let* $\Gamma \Rightarrow \Delta$ *be an unprovable sequent in* SPL−(cut), *and* $\Gamma^{\omega} \Rightarrow \Delta^{\omega}$ *be an unprovable, saturated sequent constructed from* $\Gamma \Rightarrow \Delta$ *by Lemma 6.5.8. We define a canonical model* $\mathcal{A} := \langle U, I^{+}, I^{-}\rangle$ *as follows:*

$$U := \{z \mid z \textit{ is a free individual variable occurring in } \Gamma^{\omega} \Rightarrow \Delta^{\omega}\},$$
$$p^{I^{+}} := \{(z_1, ..., z_m) \mid p(z_1, ..., z_m) \in \Gamma^{\omega}\},$$
$$p^{I^{-}} := \{(z_1, ..., z_m) \mid {\sim}p(z_1, ..., z_m) \in \Gamma^{\omega}\}.$$

Then, for any formula α,

1. $[(\alpha \in \Gamma^{\omega}$ *implies* $\mathcal{A} \models^{+} \underline{\alpha})$ *and* $(\alpha \in \Delta^{\omega}$ *implies not-*$(\mathcal{A} \models^{+} \underline{\alpha}))]$,
2. $[({\sim}\alpha \in \Gamma^{\omega}$ *implies* $\mathcal{A} \models^{-} \underline{\alpha})$ *and* $({\sim}\alpha \in \Delta^{\omega}$ *implies not-*$(\mathcal{A} \models^{-} \underline{\alpha}))]$

where $\underline{\alpha}$ *is obtained from* α *by replacing every individual variable* x *occurring in* α *by the name* $\underline{x}$.

Proof. By (simultaneous) induction on the complexity of α.

• Base step: Obvious by the definitions of I^{+} and I^{-}.

• Induction step for (1): We show some cases.

(Case $\alpha \equiv \beta{-\!\!\!<}\gamma$): First, we show that $\beta{-\!\!\!<}\gamma \in \Gamma^{\omega}$ implies $\mathcal{A} \models^{+} \underline{\beta}{-\!\!\!<}\underline{\gamma}$. Suppose $\beta{-\!\!\!<}\gamma \in \Gamma^{\omega}$. Then, we obtain $[\beta \in \Gamma^{\omega}$ and $\gamma \in \Delta^{\omega}]$ by Definition 6.5.4 (s7). By the induction hypothesis for (1), we obtain $[\mathcal{A} \models^{+} \underline{\beta}$ and not-$(\mathcal{A} \models^{+} \underline{\gamma})]$. This means $\mathcal{A} \models^{+} \underline{\beta}{-\!\!\!<}\underline{\gamma}$. Second, we show that $\beta{-\!\!\!<}\gamma \in \Delta^{\omega}$ implies not-$(\mathcal{A} \models^{+} \underline{\beta}{-\!\!\!<}\underline{\gamma})$. Suppose $\beta{-\!\!\!<}\gamma \in \Delta^{\omega}$. Then, we obtain $[\beta \in \Delta^{\omega}$ or $\gamma \in \Gamma^{\omega}]$ by Definition 6.5.4 (s8). By the induction hypothesis for (1), we obtain [not-$(\mathcal{A} \models^{+} \underline{\beta})$ or $\mathcal{A} \models^{+} \underline{\gamma}]$. This means not-$(\mathcal{A} \models^{+} \underline{\beta}{-\!\!\!<}\underline{\gamma})$.

(Case $\alpha \equiv {\sim}\beta$): First, we show that ${\sim}\beta \in \Gamma^{\omega}$ implies $\mathcal{A} \models^{+} {\sim}\underline{\beta}$. Suppose ${\sim}\beta \in \Gamma^{\omega}$. Then we obtain $\mathcal{A} \models^{-} \underline{\beta}$ by the induction hypothesis for (2). Thus, we have $\mathcal{A} \models^{+} {\sim}\underline{\beta}$. Second, we show that ${\sim}\beta \in \Delta^{\omega}$ implies not-$(\mathcal{A} \models^{+} {\sim}\underline{\beta})$. Suppose ${\sim}\beta \in \Delta^{\omega}$. Then, we obtain not-$(\mathcal{A} \models^{-} \underline{\beta})$ by the induction hypothesis for (2). Thus, we have not-$(\mathcal{A} \models^{+} {\sim}\underline{\beta})$.

• Induction step for (2): We show some cases.

(Case $\alpha \equiv \beta{\rightarrow}\gamma$): First, we show that ${\sim}(\beta{\rightarrow}\gamma) \in \Gamma^{\omega}$ implies $\mathcal{A} \models^{-} \underline{\beta}{\rightarrow}\underline{\gamma}$. Suppose ${\sim}(\beta{\rightarrow}\gamma) \in \Gamma^{\omega}$. Then, we obtain $[{\sim}\beta \in \Delta^{\omega}$ and ${\sim}\gamma \in \Gamma^{\omega}]$ by Definition 6.5.4 (s21). By the induction hypothesis for (2), we obtain [not-$(\mathcal{A} \models^{-} \underline{\beta})$ and $\mathcal{A} \models^{-} \underline{\gamma}]$. This means $\mathcal{A} \models^{-} \underline{\beta}{\rightarrow}\underline{\gamma}$. Second, we show that ${\sim}(\beta{\rightarrow}\gamma) \in \Delta^{\omega}$ implies not-$(\mathcal{A} \models^{-} \underline{\beta}{\rightarrow}\underline{\gamma})$. Suppose ${\sim}(\beta{\rightarrow}\gamma) \in \Delta^{\omega}$. Then, we obtain $[{\sim}\beta \in \Gamma^{\omega}$ or ${\sim}\gamma \in \Delta^{\omega}]$ by Definition 6.5.4 (s22). By the induction hypothesis for (2), we obtain [$\mathcal{A} \models^- \underline{\beta}$ or not-($\mathcal{A} \models^- \underline{\gamma}$)]. This means not-($\mathcal{A} \models^- \underline{\beta}{\to}\underline{\gamma}$).

(Case $\alpha \equiv \beta{-\!\!<}\gamma$): First, we show that ${\sim}(\beta{-\!\!<}\gamma) \in \Gamma^\omega$ implies $\mathcal{A} \models^- \underline{\beta}{-\!\!<}\underline{\gamma}$. Suppose ${\sim}(\beta{-\!\!<}\gamma) \in \Gamma^\omega$. Then, we obtain [${\sim}\beta \in \Gamma^\omega$ or ${\sim}\gamma \in \Delta^\omega$] by Definition 6.5.4 (s23). By the induction hypothesis for (2), we obtain [$\mathcal{A} \models^- \underline{\beta}$ or not-($\mathcal{A} \models^- \underline{\gamma}$)]. This means $\mathcal{A} \models^- \underline{\beta}{-\!\!<}\underline{\gamma}$. Second, we show that ${\sim}(\beta{-\!\!<}\gamma) \in \Delta^\omega$ implies not-($\mathcal{A} \models^- \underline{\beta}{-\!\!<}\underline{\gamma}$). Suppose ${\sim}(\beta{-\!\!<}\gamma) \in \Delta^\omega$. Then, we obtain [${\sim}\beta \in \Delta^\omega$ and ${\sim}\gamma \in \Gamma^\omega$] by Definition 6.5.4 (s24). By the induction hypothesis for (2), we obtain [not-($\mathcal{A} \models^- \underline{\beta}$) and $\mathcal{A} \models^- \underline{\gamma}$]. This means not-($\mathcal{A} \models^- \underline{\beta}{-\!\!<}\underline{\gamma}$).

(Case $\alpha \equiv {\sim}\beta$): First, we show that ${\sim}{\sim}\beta \in \Gamma^\omega$ implies $\mathcal{A} \models^- {\sim}\underline{\beta}$. Suppose ${\sim}{\sim}\beta \in \Gamma^\omega$. Then, we obtain $\beta \in \Gamma^\omega$ by Definition 6.5.4 (s15). By the induction hypothesis for (1) and $\beta \in \Gamma^\omega$, we obtain $\mathcal{A} \models^+ \underline{\beta}$, and hence $\mathcal{A} \models^- {\sim}\underline{\beta}$. Second, we show that ${\sim}{\sim}\beta \in \Delta^\omega$ implies not-($\mathcal{A} \models^- {\sim}\underline{\beta}$). Suppose ${\sim}{\sim}\beta \in \Delta^\omega$. Then, we obtain $\beta \in \Delta^\omega$ by Definition 6.5.4 (s16). By the induction hypothesis for (1) and $\beta \in \Delta^\omega$, we obtain not-($\mathcal{A} \models^+ \underline{\beta}$) and hence not-($\mathcal{A} \models^- {\sim}\underline{\beta}$).

(Case $\alpha \equiv \forall x\beta$): First, we show that ${\sim}\forall x\beta \in \Gamma^\omega$ implies $\mathcal{A} \models^- \forall x\underline{\beta}$. Suppose ${\sim}\forall x\beta \in \Gamma^\omega$. Then, we obtain ${\sim}\beta[z/x] \in \Gamma^\omega$ for some $z \in U$, by Definition 6.5.4 (s27). By the induction hypothesis for (2), we obtain that $\mathcal{A} \models^- \underline{\beta}[\underline{z}/x]$ for some $z \in U$. This means $\mathcal{A} \models^- \forall x\underline{\beta}$. Second, we show that ${\sim}\forall x\beta \in \Delta^\omega$ implies not-($\mathcal{A} \models^- \forall x\underline{\beta}$). Suppose ${\sim}\forall x\beta \in \Delta^\omega$. Then, we obtain [${\sim}\beta[y_i/x] \in \Delta^\omega$ for all $y_i \in U$] by Definition 6.5.4 (s28). By the induction hypothesis for (2), we obtain not-($\mathcal{A} \models^- \underline{\beta}[\underline{y_i}/x]$) for all $y_i \in U$. This means not-($\mathcal{A} \models^- \forall x\underline{\beta}$).

(Case $\alpha \equiv \exists x\beta$): First, we show that ${\sim}\exists x\beta \in \Gamma^\omega$ implies $\mathcal{A} \models^- \exists x\underline{\beta}$. Suppose ${\sim}\exists x\beta \in \Gamma^\omega$. Then we obtain [${\sim}\beta[y_i/x] \in \Gamma^\omega$ for all $y_i \in U$] by Definition 6.5.4 (s29). By the induction hypothesis, we obtain that $\mathcal{A} \models^- \underline{\beta}[\underline{y_i}/x]$ for all $y_i \in U$. This means $\mathcal{A} \models^- \exists x\underline{\beta}$. Second, we show that ${\sim}\exists x\beta \in \Delta^\omega$ implies not-($\mathcal{A} \models^- \exists x\underline{\beta}$). Suppose ${\sim}\exists x\beta \in \Delta^\omega$. Then, we obtain [${\sim}\beta[z/x] \in \Delta^\omega$ for some $z \in U$] by Definition 6.5.4 (s30). By the induction hypothesis for (2), we obtain not-($\mathcal{A} \models^- \underline{\beta}[\underline{z}/x]$) for some $z \in U$. This means not-($\mathcal{A} \models^- \exists x\underline{\beta}$). ∎

Theorem 6.5.10 (Strong completeness for SPL) *For any sequent S, if S is valid, then* SPL−(cut) $\vdash S$.

Proof. We prove the following: if $\Gamma \Rightarrow \Delta$ is unprovable in SPL−(cut), then there exists a model $\mathcal{A}$ such that $\Gamma \Rightarrow \Delta$ is not valid in $\mathcal{A}$. Suppose that $\Gamma \Rightarrow \Delta$ is not provable in SPL−(cut). Then, by Lemma 6.5.9, we can construct a canonical model $\mathcal{A}$ with the condition (1) in this lemma. Thus, we have $\mathcal{A} \models^+ \underline{\gamma}$ and not ($\mathcal{A} \models^+ \underline{\delta}$) for any $\gamma \in \Gamma \subseteq \Gamma^\omega$ and any $\delta \in \Delta \subseteq \Delta^\omega$. Hence, we obtain "not-($\mathcal{A} \models^+ \underline{\Gamma}_* \to \underline{\Delta}^*$)", and hence "not-($\mathcal{A} \models^+ cl(\Gamma_* \to \Delta^*)$)". Therefore, $\Gamma \Rightarrow \Delta$ is not valid in $\mathcal{A}$. ∎

Combining Theorem 6.5.10 and Theorem 6.5.3, we can obtain an alternative (semantical) proof of the cut-elimination theorem for SPL.

Definition 6.5.11 *The semantics of* DPL *is obtained from that of* SPL *by replacing the conditions 12 and 13 in Definition 6.5.2 by*

12′. $\mathcal{A} \models^- \alpha \to \beta$ *iff* $\mathcal{A} \models^+ \alpha$ *and not-*($\mathcal{A} \models^+ \beta$),

13′. $\mathcal{A} \models^- \alpha \prec \beta$ *iff not-*($\mathcal{A} \models^+ \alpha$) *or* $\mathcal{A} \models^+ \beta$.

The definition of the validity of formulas and sequents in DPL *is analogous to the definition of these notions in* SPL. *In the case of* DLP *we use the term "*d-validity*", in order to distinguish validity in* DPL *from validity in* SPL.

Theorem 6.5.12 (Soundness of DPL) *For any sequent S, if* DPL $\vdash$ *S, then S is d-valid.*

Theorem 6.5.13 (Strong completeness for DPL) *For any sequent S, if S is d-valid, then* DPL$-$(cut) $\vdash$ *S.*

Part III.

Paraconsistent temporal logics

7. Paraconsistent intuitionistic temporal logics

It is known that linear-time temporal logic (LTL), which is an extension of classical logic, is useful for expressing temporal reasoning as investigated in computer science. In this chapter, two constructive and bounded versions of LTL, which are extensions of intuitionistic logic or Nelson's paraconsistent logic, are introduced as Gentzen-type sequent calculi. These logics, $\mathrm{IB}[l]$ and $\mathrm{PB}[l]$, are intended to provide a useful theoretical basis for representing not only temporal (linear-time), but also constructive, and paraconsistent (inconsistency-tolerant) reasoning. The time domain of the proposed logics is bounded by a fixed positive integer. Despite the restriction on the time domain, the logics can derive almost all the typical temporal axioms of LTL. As a merit of bounding time, faithful embeddings into intuitionistic logic and Nelson's paraconsistent logic are shown for $\mathrm{IB}[l]$ and $\mathrm{PB}[l]$, respectively. Completeness (with respect to Kripke semantics), cut-elimination, normalization (with respect to natural deduction), and decidability theorems for the newly defined logics are proved as the main results of this chapter. Moreover, we present sound and complete display calculi for $\mathrm{IB}[l]$ and $\mathrm{PB}[l]$.

In [124] it has been emphasized that intuitionistic linear-time logic (ILTL) admits an elegant characterization of safety and liveness properties. The system ILTL, however, has been presented only in an algebraic setting. The present chapter is the first semantical *and* proof-theoretical study of bounded constructive linear-time temporal logics containing either intuitionistic or strong negation.[1]

7.1. Introduction

7.1.1. Constructive linear-time temporal logics

It is known that *linear-time temporal logic* (LTL) is very useful for verifying and specifying concurrent systems [37]. Gentzen-type sequent calculi for LTL and its neighbours have been introduced by many researchers. For example, a sequent calculus LT_ω for LTL, which is precisely a system

[1]This chapter presents the material of [100].

for *Kröger's infinitary temporal logic* [112], was introduced by Kawai [108], who proved cut-elimination and completeness theorems for this calculus. An alternative proof of the cut-elimination theorem for LT_ω was given by introducing an embedding of LT_ω into a sequent calculus for *infinitary logic*, see [94].

In the present chapter, two constructive (or intuitionistic) and bounded versions of LT_ω, which have embeddings into intuitionistic logic and Nelson's paraconsistent logic rather than infinitary logic, are studied. The first one, which is an extension of intuitionistic logic, is called *intuitionistic bounded linear-time temporal logic* (denoted as $\mathrm{IB}[l]$), and the second one, which is an extension of *Nelson's paraconsistent logic* N4 [6, 134, 196], is called *paraconsistent bounded linear-time temporal logic* (denoted as $\mathrm{PB}[l]$). Completeness (w.r.t. Kripke semantics), embedding, cut-elimination, normalization (w.r.t. natural deduction) and decidability theorems for $\mathrm{IB}[l]$ and $\mathrm{PB}[l]$ are proved as the main results of this chapter. The logics $\mathrm{IB}[l]$ and $\mathrm{PB}[l]$ are intended to give a useful theoretical basis for adequately representing not only temporal (linear-time), but also constructive and paraconsistent (inconsistency-tolerant) reasoning.

Whereas the Hilbert-style axiom scheme for the temporal operators G (globally) and X (next): $\mathrm{G}\alpha \leftrightarrow (\alpha \wedge \mathrm{X}\alpha \wedge \mathrm{X}^2\alpha \wedge \cdots\ \infty)$, where $\mathrm{X}^i\alpha$ means $\overbrace{\mathrm{XX}\cdots\mathrm{X}}^{i}\alpha$, is characteristic of LT_ω, the axiom scheme: $\mathrm{G}\alpha \leftrightarrow (\alpha \wedge \mathrm{X}\alpha \wedge \mathrm{X}^2\alpha \wedge \cdots \wedge \mathrm{X}^l\alpha)$, which may be regarded as a finite approximation of the original scheme, is characteristic of the logics $\mathrm{IB}[l]$ and $\mathrm{PB}[l]$. Then the following very informal correspondences are useful to understand these logics: $\mathrm{G}\alpha$ in LT_ω corresponds to the infinite conjunction $\bigwedge_{j=0}^{\infty} \mathrm{X}^j\alpha$ in infinitary logic (extended by X^i), and $\mathrm{G}\alpha$ in $\mathrm{IB}[l]$ and $\mathrm{PB}[l]$ corresponds to the finite conjunction $\bigwedge_{j=0}^{l} \mathrm{X}^j\alpha$ in intuitionistic or Nelson's paraconsistent logic (extended by X^i).

7.1.2. Why do we bound the time domain?

Although the standard LTL has an infinite (unbounded) time domain, namely the set ω of all natural numbers, the logics $\mathrm{IB}[l]$ and $\mathrm{PB}[l]$ have a *bounded time domain* which is restricted by a fixed positive integer l, i.e., the set $\omega_l := \{x \in \omega \mid x \leq l\}$. Despite the restriction on the time domain, $\mathrm{IB}[l]$ and $\mathrm{PB}[l]$ can derive almost all the typical temporal axioms of LTL, such as a *time induction axiom*. As mentioned before, $\mathrm{IB}[l]$ and $\mathrm{PB}[l]$ allow us to obtain simple embeddings into intuitionistic logic and Nelson's paraconsistent logic, respectively. Using the embedding results, cut-elimination and decidability theorems for these logics can be derived. Moreover, a completeness theorem (w.r.t. Kripke semantics) and a normalization theorem (w.r.t. natural deduction) can be obtained. A comparable theoretical merit may not be obtained for an unbounded and intuitionistic version of LTL, because the unbounded time domain requires some infinite inference rules. Such infinite inference rules are neither familiar to nor welcomed by researchers who study automated reasoning, since these rules cannot be implemented efficiently. Indeed, the replacement of such infinite rules of certain proof systems by finitary rules is known as an important issue.

To restrict the time domain in LTL is not a new idea. Such an idea was discussed, for instance, in [24, 35, 76]. By using and introducing a bounded time domain and the notion of bounded validity, *bounded tableaux calculi* (with temporal constraints) for propositional and first-order LTLs were studied by Cerrito, Mayer and Prand [35]. It is also known that to restrict the time domain is a technique that may be applied to obtain a decidable or efficient fragment of LTL [76]. Restricting the time domain implies not only some purely theoretical merits as mentioned above, but also some practical merits for describing temporal databases [35] and for implementing an efficient model checking algorithm, called *bounded model checking* [24]. Such practical merits are important due to the fact that there are problems in computer science and artificial intelligence where only a finite fragment of the time sequence is of interest [35]. We hope that IB[l] and PB[l] provide a good proof-theoretical basis for such practical applications as well as a good tool for automated reasoning with (bounded) linear-time formalisms.

7.1.3. Why do we use constructive and paraconsistent logics?

In (extensions of) standard classical propositional logic, the law of excluded middle $\alpha \vee \neg\alpha$ is valid. This means that the information represented by classical logic is *complete information*: every formula α is either true or not true in a model. Representing only complete information is plausible in classical mathematics, which is a discipline handling eternal truth and falsehood. The statements of classical mathematics do not change their truth value in the course of time, and the classical mathematician may assume every situation to support either the truth or the falsity of such a statement. The assumption of complete information is, however, inadequate when it comes to representing the information available to real world agents. We wish to explore the consequences of *incomplete* information about computer and information systems, and then it is desirable to avail of a logic which is *paracomplete* in the sense of not validating the law of excluded middle [140], [215].

For representing the development of incomplete information over time, it turned out that constructive logics are useful as base logics for temporal reasoning. Indeed, *constructive (intuitionistic) modal and temporal logics* have been studied by several researchers, the *constructive concurrent dynamic logic* of Wijesekera and Nerode [215] being just one example of such logics. Particularly relevant for the present concerns is the intuitionistic linear-time temporal logic (ILTL) introduced in [124], which is a system that can be used to express properties relating finite and infinite behaviours. In [124], a logical characterization of safety and liveness properties is given: For every formula α, α is (expresses) an intuitionistic safety (or liveness) property iff $(\mathrm{F}\bot \to \alpha) \to \alpha$ (or $\mathrm{F}\bot \to \alpha$, resp.) is valid in ILTL. Moreover, the following decomposition theorem holds: For every formula α, $\alpha \leftrightarrow ((\mathrm{F}\bot \to \alpha) \wedge (\mathrm{F}\bot \vee \alpha))$ is valid in ILTL. The system ILTL, however, is presented only in an algebraic setting. The present chapter is the first proof-theoretical *and* model-theoretical study of bounded constructive linear-time temporal logics containing either intuitionistic or strong negation.[2]

We wish to handle *inconsistent* as well as incomplete information, since some real systems such as software systems need to ensure inconsistency-tolerance. *Paraconsistent model checking* based on *many-valued temporal logics*, for instance, which was suggested by Easterbrook and Chechik [53], is intended to represent inconsistent information for requirements elicitation in software engineering. Whereas incomplete information calls for *paracomplete* logics, handling inconsistent information within a logic requires *paraconsistent logics* such as Nelson's N4, Dunn and Belnap's four-valued logic, da Costa's C systems, or annotated logics. The present chapter's approach is based on N4, since N4 is known as a very useful paraconsistent logic in philosophical logic, computer science, and AI (see, e.g.,[140, 142, 148, 194, 196, 197]) and because N4 is based on positive intuitionistic logic. A systematic and historical survey of paraconsistent logic can be found in [154, 157].

The idea of combining time with paraconsistency is not a new idea. In order to express inconsistent states in temporal reasoning, *annotated temporal logics* $\Delta^*\tau$, which are combinations of annotated logics and LTL, were proposed by Abe and Akama [1]. The motivation for using

[2] In [45], a Curry-Howard isomorphism for intuitionistic linear-time temporal logic in the language based on the next-time operator X and intuitionistic implication is established. Note, however, that the author extends this positive constructive logic by *classical* negation and that he uses a natural deduction system with time-annotated derivability relations inspired by [125]. Natural deduction proof systems and typed λ-calculi for bounded intuitionistic linear-time temporal logics are surveyed in [93]. See also [18].

PB[l] in the present chapter is basically the same as the motivation given in [1]. Whereas Abe and Akama's approach is only semantical, the present approach is both semantical and proof-theoretical. Moreover, a general theory of combining logics has been developed, for example, in [34].

7.2. Intuitionistic bounded linear-time temporal logic

7.2.1. Sequent calculus

Formulas of IB[l] are constructed from (countably many) propositional variables, $\bot$ (the falsity constant), $\to$ (implication), $\wedge$ (conjunction), $\vee$ (disjunction), G (globally), F (eventually) and X (next). Lower-case letters $p, q, ...$ are used to denote propositional variables, Greek lower-case letters $\alpha, \beta, ...$ are used to denote formulas, and Greek capital letters $\Gamma, \Delta, ...$ are used to represent finite (possibly empty) sequences of formulas. For any $\sharp \in \{\mathrm{G}, \mathrm{F}, \mathrm{X}\}$, the expression $\sharp\Gamma$ is used to denote the sequence $\langle \sharp\gamma \mid \gamma \in \Gamma \rangle$. The symbol $\equiv$ is used to denote the equality of sequences of symbols. The symbol ω or N is used to represent the set of natural numbers. Let l be a fixed positive integer. The symbol ω_l or N_l is used to represent the set $\{i \in \omega \mid i \leq l\}$. The expression $\mathrm{X}^i\alpha$ for any $i \in \omega$ is inductively defined by ($\mathrm{X}^0\alpha \equiv \alpha$) and ($\mathrm{X}^{n+1}\alpha \equiv \mathrm{X}^n\mathrm{X}\alpha$). Lower-case letters i, j and k are used to denote any natural numbers. An expression of the form $\Gamma \Rightarrow \Delta$ where Δ is empty or a single formula is called a *sequent* (for IB[l]). An expression $L \vdash S$ is used to denote the fact that a sequent S is provable in a sequent calculus L.

Definition 7.2.1 (IB[l]) *Let l be a fixed positive integer. In the following definition, Δ represents the empty sequence or a single formula.*

The initial sequents of IB[l] *are of the following form, where p is any propositional variable:*

$$\mathrm{X}^i p \Rightarrow \mathrm{X}^i p \qquad\qquad \mathrm{X}^i \bot \Rightarrow .$$

The structural rules of IB[l] *are of the form:*

$$\frac{\Gamma \Rightarrow \alpha \quad \alpha, \Sigma \Rightarrow \Delta}{\Gamma, \Sigma \Rightarrow \Delta}\ (\text{cut})$$

$$\frac{\Gamma \Rightarrow \Delta}{\alpha, \Gamma \Rightarrow \Delta}\ (\text{we-left}) \qquad \frac{\Gamma \Rightarrow}{\Gamma \Rightarrow \alpha}\ (\text{we-right})$$

$$\frac{\alpha, \alpha, \Gamma \Rightarrow \Delta}{\alpha, \Gamma \Rightarrow \Delta}\ (\text{co}) \qquad \frac{\Gamma, \alpha, \beta, \Sigma \Rightarrow \Delta}{\Gamma, \beta, \alpha, \Sigma \Rightarrow \Delta}\ (\text{ex}).$$

The logical inference rules of IB[l] *are of the following form, for any* $k \in \omega_l$ *and any positive integer* m*:*

$$\frac{\Gamma \Rightarrow \mathrm{X}^i\alpha \quad \mathrm{X}^i\beta, \Sigma \Rightarrow \Delta}{\mathrm{X}^i(\alpha \rightarrow \beta), \Gamma, \Sigma \Rightarrow \Delta}\ (\rightarrow\text{left}) \qquad \frac{\mathrm{X}^i\alpha, \Gamma \Rightarrow \mathrm{X}^i\beta}{\Gamma \Rightarrow \mathrm{X}^i(\alpha \rightarrow \beta)}\ (\rightarrow\text{right})$$

$$\frac{\mathrm{X}^i\alpha, \Gamma \Rightarrow \Delta}{\mathrm{X}^i(\alpha \wedge \beta), \Gamma \Rightarrow \Delta}\ (\wedge\text{left1}) \qquad \frac{\mathrm{X}^i\beta, \Gamma \Rightarrow \Delta}{\mathrm{X}^i(\alpha \wedge \beta), \Gamma \Rightarrow \Delta}\ (\wedge\text{left2})$$

$$\frac{\Gamma \Rightarrow \mathrm{X}^i\alpha \quad \Gamma \Rightarrow \mathrm{X}^i\beta}{\Gamma \Rightarrow \mathrm{X}^i(\alpha \wedge \beta)}\ (\wedge\text{right}) \qquad \frac{\mathrm{X}^i\alpha, \Gamma \Rightarrow \Delta \quad \mathrm{X}^i\beta, \Gamma \Rightarrow \Delta}{\mathrm{X}^i(\alpha \vee \beta), \Gamma \Rightarrow \Delta}\ (\vee\text{left})$$

$$\frac{\Gamma \Rightarrow \mathrm{X}^i\alpha}{\Gamma \Rightarrow \mathrm{X}^i(\alpha \vee \beta)}\ (\vee\text{right1}) \qquad \frac{\Gamma \Rightarrow \mathrm{X}^i\beta}{\Gamma \Rightarrow \mathrm{X}^i(\alpha \vee \beta)}\ (\vee\text{right2})$$

$$\frac{\mathrm{X}^l\alpha, \Gamma \Rightarrow \Delta}{\mathrm{X}^{l+m}\alpha, \Gamma \Rightarrow \Delta}\ (\text{Xleft}) \qquad \frac{\Gamma \Rightarrow \mathrm{X}^l\alpha}{\Gamma \Rightarrow \mathrm{X}^{l+m}\alpha}\ (\text{Xright})$$

$$\frac{\mathrm{X}^{i+k}\alpha, \Gamma \Rightarrow \Delta}{\mathrm{X}^i\mathrm{G}\alpha, \Gamma \Rightarrow \Delta}\ (\text{Gleft}) \qquad \frac{\{\ \Gamma \Rightarrow \mathrm{X}^{i+j}\alpha\ \}_{j \in \omega_l}}{\Gamma \Rightarrow \mathrm{X}^i\mathrm{G}\alpha}\ (\text{Gright})$$

$$\frac{\{\ \mathrm{X}^{i+j}\alpha, \Gamma \Rightarrow \Delta\ \}_{j \in \omega_l}}{\mathrm{X}^i\mathrm{F}\alpha, \Gamma \Rightarrow \Delta}\ (\text{Fleft}) \qquad \frac{\Gamma \Rightarrow \mathrm{X}^{i+k}\alpha}{\Gamma \Rightarrow \mathrm{X}^i\mathrm{F}\alpha}\ (\text{Fright}).$$

Note that for any formula α, the sequent $\mathrm{X}^i\alpha \Rightarrow \mathrm{X}^i\alpha$ is provable in IB[l]. This can be shown by induction on α. Thus, the sequents of the form $\mathrm{X}^i\alpha \Rightarrow \mathrm{X}^i\alpha$ can also be regarded as initial sequents.

It is remarked that IB[l] is just a logic parameterized by a fixed concrete positive integer l. Thus, before any detailed discussion, we have to fix IB[l] as a concrete logic such as IB[5]. Indeed, for example, IB[2] is different from IB[1]: $p \wedge \mathrm{X}p \Rightarrow \mathrm{G}p$ is provable in IB[1], but it is not provable in IB[2]. A proof of $p \wedge \mathrm{X}p \Rightarrow \mathrm{G}p$ in IB[1] is presented below:

$$\dfrac{\dfrac{\dfrac{\dfrac{\dfrac{p \Rightarrow p}{\mathrm{X}p, p \Rightarrow p}\ (\text{we-left}) \quad \dfrac{\dfrac{\mathrm{X}p \Rightarrow \mathrm{X}p}{p, \mathrm{X}p \Rightarrow \mathrm{X}p}\ (\text{we-left})}{\mathrm{X}p, p \Rightarrow \mathrm{X}p}\ (\text{ex})}{\mathrm{X}p, p \Rightarrow \mathrm{G}p}\ (\text{Gright})}{p \wedge \mathrm{X}p, p \Rightarrow \mathrm{G}p}\ (\wedge\text{left2})}{\dfrac{p, p \wedge \mathrm{X}p \Rightarrow \mathrm{G}p}{p \wedge \mathrm{X}p, p \wedge \mathrm{X}p \Rightarrow \mathrm{G}p}\ (\wedge\text{left1})}\ (\text{ex})}{p \wedge \mathrm{X}p \Rightarrow \mathrm{G}p}\ (\text{co}).$$

Note that (Gright) and (Fleft) have $l + 1$ (i.e., a finite number of) premises. In (Gleft) and (Fright), the number k is bounded by l. Then

IB[l] has the Hilbert-style axiom schemes $\mathrm{G}\alpha \leftrightarrow (\alpha \wedge \mathrm{X}\alpha \wedge \mathrm{X}^2\alpha \wedge \cdots \wedge \mathrm{X}^l\alpha)$ and $\mathrm{F}\alpha \leftrightarrow (\alpha \vee \mathrm{X}\alpha \vee \mathrm{X}^2\alpha \vee \cdots \vee \mathrm{X}^l\alpha)$. By (Xleft) and (Xright), the nest of the outermost occurrences of X in a formula can be bounded by l. Indeed, (Xleft) and (Xright) correspond to the Hilbert-style axiom scheme $\mathrm{X}^{l+m}\alpha \leftrightarrow \mathrm{X}^l\alpha$.

We may regard IB[l] as an intuitionistic and bounded version of Kawai's sequent calculus LT_ω for LTL [108]. LT_ω has no l-bounded rules {(Xleft) (Xright)}, and uses ω instead of ω_l.

Proposition 7.2.2 *Let* Δ *be the empty sequence or a single formula. The rule of the form:*

$$\frac{\Gamma \Rightarrow \Delta}{\mathrm{X}\Gamma \Rightarrow \mathrm{X}\Delta}\ (\mathrm{Xregu})$$

is admissible in cut-free IB[l].

Proof. By induction on proofs P of $\Gamma \Rightarrow \Delta$ in cut-free IB[l]. We distinguish the cases according to the last inference of P. We show some cases.

Case ($\mathrm{X}^i\bot \Rightarrow$). The last inference of P is of the form: $\mathrm{X}^i\bot \Rightarrow$. In this case, we have IB[l] $\vdash \mathrm{X}\mathrm{X}^i\bot \Rightarrow$.

Case (Gleft). The last inference of P is of the form:

$$\frac{\mathrm{X}^{i+k}\alpha, \Sigma \Rightarrow \Delta}{\mathrm{X}^i\mathrm{G}\alpha, \Sigma \Rightarrow \Delta}\ (\mathrm{Gleft}).$$

By induction hypothesis, we obtain:

$$\frac{\begin{matrix}\vdots\\ \mathrm{X}\mathrm{X}^{i+k}\alpha, \mathrm{X}\Sigma \Rightarrow \mathrm{X}\Delta\end{matrix}}{\mathrm{X}\mathrm{X}^i\mathrm{G}\alpha, \mathrm{X}\Sigma \Rightarrow \mathrm{X}\Delta}\ (\mathrm{Gleft}).$$

Case ($\rightarrow$left). The last inference of P is of the form:

$$\frac{\Pi \Rightarrow \mathrm{X}^i\alpha \quad \mathrm{X}^i\beta, \Sigma \Rightarrow \Delta}{\mathrm{X}^i(\alpha \rightarrow \beta), \Pi, \Sigma \Rightarrow \Delta}\ (\rightarrow\mathrm{left}).$$

By induction hypothesis, we obtain:

$$\frac{\begin{matrix}\vdots & & \vdots\\ \mathrm{X}\Pi \Rightarrow \mathrm{X}\mathrm{X}^i\alpha & & \mathrm{X}\mathrm{X}^i\beta, \mathrm{X}\Sigma \Rightarrow \mathrm{X}\Delta\end{matrix}}{\mathrm{X}\mathrm{X}^i(\alpha \rightarrow \beta), \mathrm{X}\Pi, \mathrm{X}\Sigma \Rightarrow \mathrm{X}\Delta}\ (\rightarrow\mathrm{left}).$$

∎

Note that the rule (Xregu) is more expressive than the following standard inference rules for the normal modal logic K and KD, respectively:

$$\frac{\Gamma \Rightarrow \alpha}{\mathrm{X}\Gamma \Rightarrow \mathrm{X}\alpha} \qquad \frac{\Gamma \Rightarrow \gamma}{\mathrm{X}\Gamma \Rightarrow \mathrm{X}\gamma}$$

where γ can be empty.

Proposition 7.2.3 *An expression $\alpha \Leftrightarrow \beta$ means the sequents $\alpha \Rightarrow \beta$ and $\beta \Rightarrow \alpha$. The following sequents are provable in* IB[l], *for any formulas α, β and any $i \in \omega$:*

1. $\mathrm{X}^i\bot \Leftrightarrow \bot$,
2. $\mathrm{X}^i(\alpha \circ \beta) \Leftrightarrow \mathrm{X}^i\alpha \circ \mathrm{X}^i\beta$ *where* $\circ \in \{\rightarrow, \wedge, \vee\}$,
3. $\mathrm{X}^i\mathrm{G}\alpha \Leftrightarrow \mathrm{G}\mathrm{X}^i\alpha$,
4. $\mathrm{G}\alpha \Rightarrow \mathrm{X}\alpha$,
5. $\mathrm{G}\alpha \Rightarrow \mathrm{X}\mathrm{G}\alpha$,
6. $\mathrm{G}\alpha \Rightarrow \mathrm{G}\mathrm{G}\alpha$,
7. $\alpha, \mathrm{G}(\alpha \rightarrow \mathrm{X}\alpha) \Rightarrow \mathrm{G}\alpha$ *(time induction)*,
8. $\mathrm{G}\alpha \Leftrightarrow \alpha \wedge \mathrm{X}\alpha \wedge \mathrm{X}^2\alpha \wedge \cdots \wedge \mathrm{X}^l\alpha$,
9. $\mathrm{F}\alpha \Leftrightarrow \alpha \vee \mathrm{X}\alpha \vee \mathrm{X}^2\alpha \vee \cdots \vee \mathrm{X}^l\alpha$,
10. $\mathrm{X}^{l+i}\alpha \Leftrightarrow \mathrm{X}^{l+i}\mathrm{G}\alpha$,
11. $\mathrm{X}^{l+i}\alpha \Leftrightarrow \mathrm{X}^{l+i}\mathrm{F}\alpha$,
12. $\mathrm{X}^{l+i}\mathrm{G}\alpha \Leftrightarrow \mathrm{X}^{l+i}\mathrm{F}\alpha$.

Proof. We show some cases.

(5); the final step is an application of (Gright).

$$\dfrac{\dfrac{\vdots \atop \mathrm{X}\alpha \Rightarrow \mathrm{X}\alpha}{\mathrm{G}\alpha \Rightarrow \mathrm{X}\alpha}\,(\mathrm{Gleft}) \quad \dfrac{\vdots \atop \mathrm{X}^2\alpha \Rightarrow \mathrm{X}^2\alpha}{\mathrm{G}\alpha \Rightarrow \mathrm{X}^2\alpha}\,(\mathrm{Gleft}) \quad \cdots \quad \dfrac{\vdots \atop \mathrm{X}^n\alpha \Rightarrow \mathrm{X}^n\alpha}{\mathrm{G}\alpha \Rightarrow \mathrm{X}^n\alpha}\,(\mathrm{Gleft}) \quad \cdots}{\mathrm{G}\alpha \Rightarrow \mathrm{X}\mathrm{G}\alpha}\,.$$

(6).

$$\dfrac{\mathrm{G}\alpha \Rightarrow \mathrm{G}\alpha \quad \overset{\vdots}{\mathrm{G}\alpha \Rightarrow \mathrm{XG}\alpha} \quad \overset{\vdots}{\mathrm{G}\alpha \Rightarrow \mathrm{X}^2\mathrm{G}\alpha} \quad \cdots \quad \overset{\vdots}{\mathrm{G}\alpha \Rightarrow \mathrm{X}^n\mathrm{G}\alpha} \quad \cdots}{\mathrm{G}\alpha \Rightarrow \mathrm{GG}\alpha} \text{ (Gright).}$$

where $\vdash \mathrm{G}\alpha \Rightarrow \mathrm{X}^j\mathrm{G}\alpha$ for any $j \in \omega_l$ can be shown in a similar way as in (5).

(7). In the following proofs, the applications of (ex) are omitted.

$$\dfrac{\overset{\vdots}{\{\alpha, \mathrm{G}(\alpha \rightarrow \mathrm{X}\alpha) \Rightarrow \mathrm{X}^k\alpha\}_{k \in \omega_l}}}{\alpha, \mathrm{G}(\alpha \rightarrow \mathrm{X}\alpha) \Rightarrow \mathrm{G}\alpha} \text{ (Gright)}$$

where $\vdash \alpha, \mathrm{G}(\alpha \rightarrow \mathrm{X}\alpha) \Rightarrow \mathrm{X}^k\alpha$ for any $k \in \omega_l$ is shown by mathematical induction on k as follows: the base step is obvious, and the induction step can be shown by

$$\dfrac{\dfrac{\dfrac{\overset{\vdots \text{ ind.hyp.}}{\alpha, \mathrm{G}(\alpha \rightarrow \mathrm{X}\alpha) \Rightarrow \mathrm{X}^k\alpha} \quad \mathrm{X}^{k+1}\alpha \Rightarrow \mathrm{X}^{k+1}\alpha}{\alpha, \mathrm{G}(\alpha \rightarrow \mathrm{X}\alpha), \mathrm{X}^k(\alpha \rightarrow \mathrm{X}\alpha) \Rightarrow \mathrm{X}^{k+1}\alpha} \text{ }(\rightarrow\text{left})}{\alpha, \mathrm{G}(\alpha \rightarrow \mathrm{X}\alpha), \mathrm{G}(\alpha \rightarrow \mathrm{X}\alpha) \Rightarrow \mathrm{X}^{k+1}\alpha} \text{ (Gleft)}}{\alpha, \mathrm{G}(\alpha \rightarrow \mathrm{X}\alpha) \Rightarrow \mathrm{X}^{k+1}\alpha} \text{ (co).}$$

(10).

$$\dfrac{\dfrac{\dfrac{\dfrac{\overset{\vdots}{\mathrm{X}^l\alpha \Rightarrow \mathrm{X}^l\alpha}}{\{\ \mathrm{X}^l\alpha \Rightarrow \mathrm{X}^{l+j}\alpha\ \}_{j \in \omega_l}} \text{ (Xright)}}{\mathrm{X}^l\alpha \Rightarrow \mathrm{X}^l\mathrm{G}\alpha} \text{ (Gright)}}{\mathrm{X}^l\alpha \Rightarrow \mathrm{X}^{l+i}\mathrm{G}\alpha} \text{ (Xright)}}{\mathrm{X}^{l+i}\alpha \Rightarrow \mathrm{X}^{l+i}\mathrm{G}\alpha} \text{ (Xleft)} \qquad \dfrac{\overset{\vdots}{\mathrm{X}^{l+i}\alpha \Rightarrow \mathrm{X}^{l+i}\alpha}}{\mathrm{X}^{l+i}\mathrm{G}\alpha \Rightarrow \mathrm{X}^{l+i}\alpha} \text{ (Gleft).}$$

∎

Definition 7.2.4 (LJ) *A sequent calculus for propositional intuitionistic logic, which in this chapter we refer to as* LJ, *is obtained from* IB$[l]$ *by deleting* (Xleft), (Xright), (Gleft), (Gright), (Fleft), (Fright), *and replacing* X^i *by* X^0. *The modified inference rules for* LJ *by replacing* i *by* 0 *are denoted by using "LJ" as a superscript, e.g.,* ($\rightarrow$left^{LJ}).

Expressions like $\bigwedge\{\alpha_i \mid i \in \omega_l\}$ and $\bigvee\{\alpha_i \mid i \in \omega_l\}$ where $\{\alpha_i \mid i \in \omega_l\}$ is a multiset mean $\alpha_0 \wedge \alpha_1 \wedge \cdots \wedge \alpha_l$ and $\alpha_0 \vee \alpha_1 \vee \cdots \vee \alpha_l$, respectively. For example, $\bigwedge\{\alpha, \alpha, \beta\}$ means $\alpha \wedge \alpha \wedge \beta$. The following definition of the embedding function f is regarded as a finite analogue of the definition of the embedding function of LT_ω into infinitary logic [94].

Definition 7.2.5 *We fix a countable non-empty set Φ of propositional variables and define the sets $\Phi_i := \{p_i \mid p \in \Phi\}$ $(1 \leq i \in \omega)$ and $\Phi_0 := \Phi$ of propositional variables. The language $\mathcal{L}_{\mathrm{IB}[l]}$ of $\mathrm{IB}[l]$ is defined by using Φ, $\bot, \rightarrow, \wedge, \vee$, X, G and F. The language $\mathcal{L}_{\mathrm{LJ}}$ of LJ is defined by using $\bigcup_{i \in \omega} \Phi_i$, $\bot, \rightarrow, \wedge$ and $\vee$.*

A mapping f from $\mathcal{L}_{\mathrm{IB}[l]}$ to $\mathcal{L}_{\mathrm{LJ}}$ is defined by the following clause, for any $i \in \omega$ and any positive integer m:

1. $f(\mathrm{X}^i \bot) := \bot$,
2. $f(\mathrm{X}^i p) := p_i \in \Phi_i$ *for any* $p \in \Phi$ *(especially,* $f(p) := p \in \Phi_0$*),*
3. $f(\mathrm{X}^i(\alpha \circ \beta)) := f(\mathrm{X}^i\alpha) \circ f(\mathrm{X}^i\beta)$ *where* $\circ \in \{\rightarrow, \wedge, \vee\}$,
4. $f(\mathrm{X}^{l+m}\alpha) := f(\mathrm{X}^l\alpha)$,
5. $f(\mathrm{X}^i\mathrm{G}\alpha) := \bigwedge\{f(\mathrm{X}^{i+j}\alpha) \mid j \in \omega_l\}$,
6. $f(\mathrm{X}^i\mathrm{F}\alpha) := \bigvee\{f(\mathrm{X}^{i+j}\alpha) \mid j \in \omega_l\}$.

The expression $f(\Gamma)$ denotes the result of replacing every occurrence of a formula α in Γ by an occurrence of $f(\alpha)$.

Strictly speaking, the embedding function f strongly depends on the time bound l, i.e., f should be denoted as f_l. Indeed, $f_3(\mathrm{G}p)$ and $f_5(\mathrm{G}p)$ are different. But, for the sake of brevity, we will just use f in the following.

Theorem 7.2.6 (Embedding) *Let Γ be a sequence of formulas in $\mathcal{L}_{\mathrm{IB}[l]}$, Δ be the empty sequence or a formula in $\mathcal{L}_{\mathrm{IB}[l]}$, and f be the mapping defined in Definition 7.2.5.*

1. $\mathrm{IB}[l] \vdash \Gamma \Rightarrow \Delta$ *iff* $\mathrm{LJ} \vdash f(\Gamma) \Rightarrow f(\Delta)$.
2. $\mathrm{IB}[l] - (\mathrm{cut}) \vdash \Gamma \Rightarrow \Delta$ *iff* $\mathrm{LJ} - (\mathrm{cut}) \vdash f(\Gamma) \Rightarrow f(\Delta)$.

Proof. Since (2) follows from (1), we show only (1).

($\Longrightarrow$) : By induction on proofs P of $\Gamma \Rightarrow \Delta$ in $\mathrm{IB}[l]$. We distinguish the cases according to the last inference of P and show some cases.

Case ($\mathrm{X}^i p \Rightarrow \mathrm{X}^i p$). The last inference of P is of the form: $\mathrm{X}^i p \Rightarrow \mathrm{X}^i p$. In this case, we obtain $f(\mathrm{X}^i p) \Rightarrow f(\mathrm{X}^i p)$, i.e., $p_i \Rightarrow p_i$ ($p_i \in \Phi_i$). This is an initial sequent of LJ.

Case ($\to$left). The last inference of P is of the form:

$$\frac{\Gamma \Rightarrow \mathrm{X}^i\alpha \quad \mathrm{X}^i\beta, \Sigma \Rightarrow \Delta}{\mathrm{X}^i(\alpha \to \beta), \Gamma, \Sigma \Rightarrow \Delta} \ (\to\text{left}).$$

By induction hypothesis, we have LJ $\vdash f(\Gamma) \Rightarrow f(\mathrm{X}^i\alpha)$ and LJ $\vdash f(\mathrm{X}^i\beta), f(\Sigma) \Rightarrow f(\Delta)$. Then we obtain

$$\frac{\begin{array}{cc}\vdots & \vdots \\ f(\Gamma) \Rightarrow f(\mathrm{X}^i\alpha) & f(\mathrm{X}^i\beta), f(\Sigma) \Rightarrow f(\Delta)\end{array}}{f(\mathrm{X}^i\alpha) \to f(\mathrm{X}^i\beta), f(\Gamma), f(\Sigma) \Rightarrow f(\Delta)} \ (\to\text{left}^{LJ})$$

where $f(\mathrm{X}^i\alpha) \to f(\mathrm{X}^i\beta) = f(\mathrm{X}^i(\alpha \to \beta))$ by the definition of f.

Case (Xleft). The last inference of P is of the form:

$$\frac{\mathrm{X}^l\alpha, \Gamma \Rightarrow \Delta}{\mathrm{X}^{l+m}\alpha, \Gamma \Rightarrow \Delta} \ (\text{Xleft}).$$

By induction hypothesis, we have LJ $\vdash f(\mathrm{X}^l\alpha), f(\Gamma) \Rightarrow f(\Delta)$, and $f(\mathrm{X}^l\alpha) = f(\mathrm{X}^{l+m}\alpha)$ by the definition of f. Thus, we obtain LJ $\vdash f(\mathrm{X}^{l+m}\alpha), f(\Gamma) \Rightarrow f(\Delta)$.

Case (Gleft). The last inference of P is of the form:

$$\frac{\mathrm{X}^{i+k}\alpha, \Gamma \Rightarrow \Delta}{\mathrm{X}^i\mathrm{G}\alpha, \Gamma \Rightarrow \Delta} \ (\text{Gleft}).$$

By induction hypothesis, we have LJ $\vdash f(\mathrm{X}^{i+k}\alpha), f(\Gamma) \Rightarrow f(\Delta)$, and hence we obtain

$$\begin{array}{c}\vdots \\ f(\mathrm{X}^{i+k}\alpha), f(\Gamma) \Rightarrow f(\Delta) \\ \vdots \ (\wedge\text{left}^{LJ}) \\ \bigwedge\{f(\mathrm{X}^{i+j}\alpha) \mid j \in \omega_l\}, f(\Gamma) \Rightarrow f(\Delta)\end{array}$$

where $\bigwedge\{f(\mathrm{X}^{i+j}\alpha) \mid j \in \omega_l\} = f(\mathrm{X}^i\mathrm{G}\alpha)$ by the definition of f, and $f(\mathrm{X}^{i+k}\alpha)$ is in the multiset $\{f(\mathrm{X}^{i+j}\alpha) \mid j \in \omega_l\}$. It is remarked that the case $i > l$ is also included in this proof. In such a case, $f(\mathrm{X}^{i+k}\alpha)$ and $\bigwedge\{f(\mathrm{X}^{i+j}\alpha) \mid j \in \omega_l\}$ mean $f(\mathrm{X}^l\alpha)$ and $\overbrace{f(\mathrm{X}^l\alpha) \wedge f(\mathrm{X}^l\alpha) \wedge \cdots \wedge f(\mathrm{X}^l\alpha)}^{l}$, respectively.

Case (Gright). The last inference of P is of the form:

$$\frac{\{\ \Gamma \Rightarrow \mathrm{X}^{i+j}\alpha\ \}_{j\in\omega_l}}{\Gamma \Rightarrow \mathrm{X}^i\mathrm{G}\alpha}\ (\text{Gright}).$$

By induction hypothesis, we have LJ $\vdash f(\Gamma) \Rightarrow f(\mathrm{X}^{i+j}\alpha)$ for all $j \in \omega_l$. Let Φ be the multiset $\{f(\mathrm{X}^{i+j}\alpha) \mid j \in \omega_l\}$. We obtain

$$\begin{array}{c} \vdots \\ \{\ f(\Gamma) \Rightarrow f(\mathrm{X}^{i+j}\alpha)\ \}_{f(\mathrm{X}^{i+j}\alpha)\in\Phi} \\ \vdots\ (\wedge\text{right}^{LJ}) \\ f(\Gamma) \Rightarrow \bigwedge\Phi \end{array}$$

where $\bigwedge\Phi = f(\mathrm{X}^i\mathrm{G}\alpha)$ by the definition of f.

($\Longleftarrow$) : By induction on proofs Q of $f(\Gamma) \Rightarrow f(\Delta)$ in LJ. We distinguish the cases according to the last inference of Q, and show only the following case.

Case ($\wedge$right^{LJ}). The last inference of Q is of the form:

$$\frac{f(\Gamma) \Rightarrow f(\mathrm{X}^i\alpha) \quad f(\Gamma) \Rightarrow f(\mathrm{X}^i\beta)}{f(\Gamma) \Rightarrow f(\mathrm{X}^i(\alpha \wedge \beta))}\ (\wedge\text{right}^{LJ})$$

where $f(\mathrm{X}^i(\alpha\wedge\beta)) = f(\mathrm{X}^i\alpha) \wedge f(\mathrm{X}^i\beta)$ by the definition of f. By induction hypothesis, we have IB[l] $\vdash \Gamma \Rightarrow \mathrm{X}^i\alpha$ and IB[l] $\vdash \Gamma \Rightarrow \mathrm{X}^i\beta$. Then we obtain

$$\frac{\begin{array}{c}\vdots \\ \Gamma \Rightarrow \mathrm{X}^i\alpha\end{array} \quad \begin{array}{c}\vdots \\ \Gamma \Rightarrow \mathrm{X}^i\beta\end{array}}{\Gamma \Rightarrow \mathrm{X}^i(\alpha \wedge \beta)}\ (\wedge\text{right}).$$

∎

Using this theorem, we can prove the following.

Theorem 7.2.7 (Cut-elimination) *The rule* (cut) *is admissible in cut-free* IB[l].

Proof. Suppose IB[l] $\vdash \Gamma \Rightarrow \Delta$. Then we have LJ $\vdash f(\Gamma) \Rightarrow f(\Delta)$ by Theorem 7.2.6 (1), and hence LJ $-$ (cut) $\vdash f(\Gamma) \Rightarrow f(\Delta)$ by the well-known cut-elimination theorem for LJ. By Theorem 7.2.6 (2), we obtain IB[l] $-$ (cut) $\vdash \Gamma \Rightarrow \Delta$. ∎

Although in this chapter the cut-elimination theorem for IB[l] is proved via an embedding theorem, a direct syntactical cut-elimination proof for IB[l] may be obtained using the standard way of Gentzen.

Theorem 7.2.8 (Decidability) IB[l] *is decidable.*

Proof. By Theorem 7.2.6, provability in IB[l] can be reduced to provability in LJ. Since LJ is decidable, IB[l] is also decidable. ∎

7.2.2. Kripke semantics

The symbols $\geq$ and $\leq$ are used to represent the linear order on ω.

Definition 7.2.9 *Let l be a fixed positive integer. A* Kripke frame *is a structure*
$\langle M, N, N_l, R\rangle$ *satisfying the following conditions.*

1. *M is a nonempty set.*

2. *N is the set of natural numbers and $N_l := \{i \in N \mid i \leq l\}$.*

3. *R is a reflexive and transitive binary relation on M.*

The set M can be understood as a set of information states, and the set N can be understood as a set of time points.

Definition 7.2.10 *A* valuation $\models$ *on a Kripke frame $\langle M, N, N_l, R\rangle$ is a mapping from the set Ψ of all propositional variables to the power set $2^{M\times N}$ of the direct product $M \times N$ such that for any $p \in \Psi$, any $i \in N$, and any $x, y \in M$, if $(x, i) \in \models (p)$ and xRy, then $(y, i) \in \models (p)$. We will write $(x, i) \models p$ for $(x, i) \in \models (p)$. Each valuation $\models$ is extended to a mapping from the set Φ of all formulas to $2^{M\times N}$ by the following clauses:*

1. *$(x, i) \models \bot$ does not hold,*

2. *$(x, i) \models \alpha \rightarrow \beta$ iff $\forall y \in M$ [xRy and $(y, i) \models \alpha$ imply $(y, i) \models \beta$],*

3. *$(x, i) \models \alpha \wedge \beta$ iff $(x, i) \models \alpha$ and $(x, i) \models \beta$,*

4. *$(x, i) \models \alpha \vee \beta$ iff $(x, i) \models \alpha$ or $(x, i) \models \beta$,*

5. *$(x, i) \models \mathrm{X}\alpha$ iff $(x, i + 1) \models \alpha$,*

6. *$(x, i) \models \mathrm{X}^l\alpha$ iff $(x, l) \models \alpha$,*

7. *$(x, i) \models \mathrm{G}\alpha$ iff $\forall j \in N_l$ [$i \leq j$ implies $(x, j) \models \alpha$] if $i < l$, and otherwise $(x, l) \models \alpha$,*

8. $(x, i) \models \mathrm{F}\alpha$ *iff* $\exists j \in N_l$ $[i \leq j$ *and* $(x, j) \models \alpha]$ *if* $i < l$, *and otherwise* $(x, l) \models \alpha$.

The conditions 5 and 6 in Definition 7.2.10 are intended to express that for any positive integer m, $(x, l+m) \models \alpha$ iff $(x, 0) \models \mathrm{X}^{l+m}\alpha$ iff $(x, 0) \models \mathrm{X}^{l}\alpha$ iff $(x, l) \models \alpha$. The statement $(x, i) \models \alpha$ can be read as "α is true at the information state x and the time i."

Proposition 7.2.11 *Let* $\models$ *be a valuation on a Kripke frame* $\langle M, N, N_l, R \rangle$. *For any formula* α, *any* $i \in N$, *and any* $x, y \in M$, *if* $(x, i) \models \alpha$ *and* xRy, *then* $(y, i) \models \alpha$.

Proof. By induction on the complexity of α. ∎

In the following discussion, Proposition 7.2.11 will often be used implicitly.

Note that the time-heredity condition: $\forall i, j \in N\ \forall x \in M\ [(x, i) \models \alpha$ and $i \leq j$ imply $(x, j) \models \alpha]$ is not assumed in this semantics.

An expression $\Gamma^{\wedge}$ means $\gamma_1 \wedge \gamma_2 \wedge \cdots \wedge \gamma_n$ if $\Gamma \equiv \langle \gamma_1, \gamma_2, ..., \gamma_n \rangle$ $(0 \leq n)$. Let Δ be the empty sequence or a sequence consisting of a single formula. An expression Δ^* means α or $\perp$ if $\Delta \equiv \langle \alpha \rangle$ or $\varnothing$, respectively. An expression $(\Gamma \Rightarrow \Delta)^*$ means $\Gamma^{\wedge} \rightarrow \Delta^*$ if Γ is not empty, and means Δ^* otherwise.

Definition 7.2.12 *A* Kripke model *is a structure* $\langle M, N, N_l, R, \models \rangle$ *such that* (1) $\langle M, N, N_l, R \rangle$ *is a Kripke frame, and* (2) $\models$ *is a valuation on* $\langle M, N, N_l, R \rangle$.

A formula α *is* true *in a Kripke model* $\langle M, N, N_l, R, \models \rangle$ *if* $(x, 0) \models \alpha$ *for any* $x \in M$, *and* valid *in a Kripke frame* $\langle M, N, N_l, R \rangle$ *if it is true for any valuation* $\models$ *on the Kripke frame.*

A sequent $\Gamma \Rightarrow \Delta$ *is true in a Kripke model* $\langle M, N, N_l, R, \models \rangle$ *if the formula* $(\Gamma \Rightarrow \Delta)^*$ *is true in the Kripke model, and valid in a Kripke frame* $\langle M, N, N_l, R \rangle$ *if it is true for any valuation* $\models$ *on the Kripke frame.*

Theorem 7.2.13 (Soundness) *Let* C *be the class of all Kripke frames,* $L := \{\Gamma \Rightarrow \Delta \mid \mathrm{IB}[l] \vdash \Gamma \Rightarrow \Delta\}$ *and* $L(C) := \{\Gamma \Rightarrow \Delta \mid \Gamma \Rightarrow \Delta$ *is valid in all frames of* $C\}$. *Then* $L \subseteq L(C)$.

Proof. It is sufficient to show that for any sequent $\Gamma \Rightarrow \Delta$, if $\mathrm{IB}[l] \vdash \Gamma \Rightarrow \Delta$, then $\Gamma \Rightarrow \Delta$ is valid in $\langle M, N, N_l, R \rangle \in C$, i.e., for any valuation $\models$ on $\langle M, N, N_l, R \rangle$ and any $x \in M$, $(x, 0) \models \Gamma \Rightarrow \Delta$. This is proved by induction on proofs P of $\Gamma \Rightarrow \Delta$ in IB[l]. To show this, we distinguish the cases according to the last inference of P. Since the proof is straightforward, we show only the following cases.

Case (Gright): The last inference of P is of the form:

$$\frac{\{\Gamma \Rightarrow \mathrm{X}^{i+j}\alpha\}_{j\in\omega_l}}{\Gamma \Rightarrow \mathrm{X}^{i}\mathrm{G}\alpha}\ \text{(Gright)}.$$

In the following, we consider only the case $\Gamma \neq \varnothing$. Let $\models$ be a valuation on $\langle M, N, N_l, R\rangle$. By the induction hypothesis, we have $\forall j \in N_l\ \forall x \in M$ $[(x,0) \models \Gamma \Rightarrow \mathrm{X}^{i+j}\alpha]$. First, we show the case for $i < l$ as follows. $\forall j \in N_l\ \forall x \in M\ [(x,0) \models \Gamma \Rightarrow \mathrm{X}^{i+j}\alpha]$ iff $\forall j \in N_l\ \forall x \in M\ [(x,0) \models \Gamma^{\wedge} \rightarrow \mathrm{X}^{i+j}\alpha]$ iff $\forall j \in N_l\ \forall x \in M\ \forall y \in M\ [xRy$ and $(y,0) \models \Gamma^{\wedge}$ imply $(y, i+j) \models \alpha]$ iff $\forall x, y \in M\ [xRy$ and $(y,0) \models \Gamma^{\wedge}$ imply $\forall j \in N_l$ $[(y, i+j) \models \alpha]]$ iff $\forall x, y \in M\ [xRy$ and $(y,0) \models \Gamma^{\wedge}$ imply $(y,i) \models \mathrm{G}\alpha]$ iff $\forall x \in M\ [(x,0) \models \Gamma \Rightarrow \mathrm{X}^{i}\mathrm{G}\alpha]$. Second, we consider the case for $i \geq l$ as follows. This case can be shown similarly. The difference is that the part "$\forall j \in N_l\ [(y, i+j) \models \alpha]$" can be replaced by the following: $\forall j \in N_l$ $[(y, i+j) \models \alpha]$ iff $(y,l) \models \alpha$ iff $(y,l) \models \mathrm{G}\alpha$ iff $(y,i) \models \mathrm{G}\alpha$. Thus, we obtain the required fact.

Case (Xright): The last inference of P is of the form:

$$\frac{\Gamma \Rightarrow \mathrm{X}^{l}\alpha}{\Gamma \Rightarrow \mathrm{X}^{l+m}\alpha}\ \text{(Xright)}.$$

In the following, we consider only the case $\Gamma \neq \varnothing$. Let $\models$ be a valuation on $\langle M, N, N_l, R\rangle$. By the induction hypothesis, we have $\forall x \in M$ $[(x,0) \models \mathrm{X}^{l}\alpha]$ where $(x,0) \models \mathrm{X}^{l}\alpha$ iff $(x,l) \models \alpha$ iff $(x,m) \models \mathrm{X}^{l}\alpha$ iff $(x,0) \models \mathrm{X}^{l+m}\alpha$. We thus obtain the required fact that $\forall x \in M$ $[(x,0) \models \mathrm{X}^{l+m}\alpha]$. ∎

Prior to a more detailed presentation of the completeness proof, we prove a Lindenbaum Lemma.

Definition 7.2.14 *Let x and y be sets of formulas. The pair (x, y) is* consistent *iff for any $\alpha_1, ..., \alpha_m \in x$ and any $\beta_1, ..., \beta_n \in y$ with $(m, n \geq 0)$, the sequent $\alpha_1, ..., \alpha_m \Rightarrow \beta_1 \vee \cdots \vee \beta_n$ is not provable in* IB[l]. *The pair (x, y) is* maximal consistent *iff it is consistent and for every formula α, $\alpha \in x$ or $\alpha \in y$.*[3]

The following lemma can be proved using (cut).

[3] For example, the pair $(\{p\}, \{q\})$ where p and q are distinct propositional variables is consistent, and the pair $(\{p\}, \{p\})$ is inconsistent.

Lemma 7.2.15 *Let x and y be sets of formulas. If the pair (x, y) is consistent, then there is a maximal consistent pair (x', y') such that $x \subseteq x'$ and $y \subseteq y'$.*

Proof. Let $\gamma_1, \gamma_2, ...$ be an enumeration of all formulas of IB$[l]$. Define a sequence of pairs (x_n, y_n) $(n = 0, 1, ...)$ inductively by $(x_0, y_0) := (x, y)$, and $(x_{m+1}, y_{m+1}) := (x_m, y_m \cup \{\gamma_{m+1}\})$ if $(x_m, y_m \cup \{\gamma_{m+1}\})$ is consistent, and $(x_{m+1}, y_{m+1}) := (x_m \cup \{\gamma_{m+1}\}, y_m)$ otherwise. We can obtain the fact that if (x_m, y_m) is consistent, then so is (x_{m+1}, y_{m+1}). To verify this, suppose (x_{m+1}, y_{m+1}) is not consistent. Then there are formulas $\alpha_1, ..., \alpha_i, \alpha'_1, ..., \alpha'_j \in x_m$ and $\beta_1, ..., \beta_k, \beta'_1, ..., \beta'_l \in y_m$ such that IB$[l] \vdash \alpha_1, ..., \alpha_i \Rightarrow \beta_1 \vee \cdots \vee \beta_k \vee \gamma_{m+1}$ and IB$[l] \vdash \alpha'_1, ..., \alpha'_j, \gamma_{m+1} \Rightarrow \beta'_1 \vee \cdots \vee \beta'_l$. By using (cut) and some other rules, we can obtain IB$[l]$ $\vdash \alpha_1, ..., \alpha_i, \alpha'_1, ..., \alpha'_j \Rightarrow \beta_1 \vee \cdots \vee \beta_k \vee \beta'_1 \vee \cdots \vee \beta'_l$. This contradicts the consistency of (x_m, y_m). Hence, a pair (x_k, y_k) produced by the construction is consistent for any k. We thus obtain a maximal consistent pair $(\bigcup_{n=0}^{\infty} x_n, \bigcup_{n=0}^{\infty} y_n)$. ∎

We now start to prove the completeness theorem for IB$[l]$.

The expression $\{\Gamma\}$ means the set of all formulas occurring in Γ. Suppose that $\Gamma \Rightarrow \Delta$ is not provable in IB$[l]$. Then the pair $(\{\Gamma\}, \{\Delta\})$ is consistent. By Lemma 7.2.15, there is a maximal consistent pair (u, v) such that $\{\Gamma\} \subseteq u$ and $\{\Delta\} \subseteq v$. Note that if $\Delta \equiv \langle\alpha\rangle$, then $\alpha \notin u$ by the consistency of (u, v).

Definition 7.2.16 *Let M_L be the set of all maximal consistent pairs. The binary relation R_L on M_L is defined by $(x, w)R_L(y, z)$ iff $x \subseteq y$. The valuation $\models_L (p)$ for any propositional variable p is defined by $\{((x, w), i) \in M_L \times N \mid \mathrm{X}^i p \in x\}$.*

Lemma 7.2.17 *The structure $\langle M_L, N, N_l, R_L, \models_L\rangle$ is a Kripke model such that for any formula α, any $i \in N$, and any $(x, w) \in M_L$, $\mathrm{X}^i\alpha \in x$ iff $((x, w), i) \models_L \alpha$.*

Proof. It can be shown that (1) M_L is a nonempty set, because $(u, v) \in M_L$ by the discussion above, (2) R_L is a reflexive and transitive relation on M_L, and (3) for any propositional variable p and any $(x, w), (y, z) \in M_L$, if $(x, w)R_L(y, z)$ and $((x, w), i) \models_L (p)$, then $((y, z), i) \models_L (p)$. Thus, the structure $\langle M_L, N, N_l, R_L, \models_L\rangle$ is a Kripke model.

It remains to be shown that in this model, for any formula α, any $i \in N$, and any $(x, w) \in M_L$, $\mathrm{X}^i\alpha \in x$ iff $((x, w), i) \models_L \alpha$. This is shown by induction on the complexity of α.

Base step. By Definition 7.2.16.

Induction step.

• Case $\alpha \equiv \bot$. By the consistency of (x, w), $X^i\bot \in x$ does not hold.

• Case $\alpha \equiv \gamma \to \delta$. Suppose $X^i(\gamma \to \delta) \in x$. We will show $((x, w), i) \models_L \gamma \to \delta$, i.e., $\forall (y, z) \in M_L$ $[(x, w)R_L(y, z)$ and $((y, z), i) \models_L \gamma$ imply $((y, z), i) \models_L \delta]$. Suppose $(x, w)R_L(y, z)$ and $((y, z), i) \models_L \gamma$. Then we have (*): $X^i(\gamma \to \delta) \in y$ by the definition of R_L, and obtain (**): $X^i\gamma \in y$ by the induction hypothesis. Since (*), (**) and $\text{IB}[l] \vdash X^i(\gamma \to \delta), X^i\gamma \Rightarrow X^i\delta$, the fact that $X^i\delta \in z$ contradicts the consistency of (y, z), and hence $X^i\delta \notin z$. By the maximality of (y, z), we obtain $X^i\delta \in y$. By the induction hypothesis, we obtain the required fact $((y, z), i) \models_L \delta$. Conversely, suppose $X^i(\gamma \to \delta) \notin x$. Then $X^i(\gamma \to \delta) \in w$ by the maximality of (x, w). Then the pair $(x \cup \{X^i\gamma\}, \{X^i\delta\})$ is consistent for the following reason. If it is not consistent, then $\text{IB}[l] \vdash \Gamma, X^i\gamma \Rightarrow X^i\delta$ for some Γ consisting of formulas in x, and hence $\text{IB}[l] \vdash \Gamma \Rightarrow X^i(\gamma \to \delta)$. This fact contradicts the consistency of (x, w). By Lemma 7.2.15, there is a maximal consistent pair (y, z) such that $x \cup \{X^i\gamma\} \subseteq y$ and $\{X^i\delta\} \subseteq z$ (thus, we have $X^i\delta \notin y$ by the consistency of (y, z)). As a consequence, we have $(x, w)R_L(y, z)$, $((y, z), i) \models_L \gamma$ and not $[((y, z), i) \models_L \delta]$ by the induction hypothesis. Therefore $((x, w), i) \models_L \gamma \to \delta$ does not hold.

• Case $\alpha \equiv \gamma \wedge \delta$. Suppose $X^i(\gamma \wedge \delta) \in x$. Since $\text{IB}[l] \vdash X^i(\gamma \wedge \delta) \Rightarrow X^i\gamma$, the fact that $X^i\gamma \in w$ contradicts the consistency of (x, w), and hence $X^i\gamma \in x$. Similarly, we obtain $X^i\delta \in x$. By the induction hypothesis, we obtain $((x, w), i) \models_L \gamma$ and $((x, w), i) \models_L \delta$, and hence $((x, w), i) \models_L \gamma \wedge \delta$. Conversely, suppose $((x, w), i) \models_L \gamma \wedge \delta$, i.e., $((x, w), i) \models_L \gamma$ and $((x, w), i) \models_L \delta$. Then we obtain $X^i\gamma \in x$ and $X^i\delta \in x$ by the induction hypothesis. Since $\text{IB}[l] \vdash X^i\gamma, X^i\delta \Rightarrow X^i(\gamma \wedge \delta)$, the fact that $X^i(\gamma \wedge \delta) \in w$ contradicts the consistency of (x, w), and hence $X^i(\gamma \wedge \delta) \notin w$. By the maximality of (x, w), we obtain $X^i(\gamma \wedge \delta) \in x$.

• Case $\alpha \equiv \gamma \vee \delta$. Suppose $X^i(\gamma \vee \delta) \in x$. Since $\text{IB}[l] \vdash X^i(\gamma \vee \delta) \Rightarrow X^i\gamma \vee X^i\delta$, the fact that $X^i\gamma, X^i\delta \in w$ contradicts the consistency of (x, w), and hence $X^i\gamma \notin w$ or $X^i\delta \notin w$. Thus, we obtain $X^i\gamma \in x$ or $X^i\delta \in x$ by the maximality of (x, w). By the induction hypothesis, we obtain $((x, w), i) \models_L \gamma$ or $((x, w), i) \models_L \delta$, and hence $((x, w), i) \models_L \gamma \vee \delta$. Conversely, suppose $((x, w), i) \models_L \gamma \vee \delta$, i.e., $((x, w), i) \models_L \gamma$ or $((x, w), i) \models_L \delta$. By the induction hypothesis, we obtain $X^i\gamma \in x$ or $X^i\delta \in x$. Since $\text{IB}[l] \vdash X^i\gamma \Rightarrow X^i(\gamma \vee \delta)$ and $\text{IB}[l] \vdash X^i\delta \Rightarrow X^i(\gamma \vee \delta)$, the fact that $X^i(\gamma \vee \delta) \in w$ contradicts the consistency of (x, w), and hence $X^i(\gamma \vee \delta) \notin w$. By the maximality of (x, w), we obtain $X^i(\gamma \vee \delta) \in x$.

• Case $\alpha \equiv X\beta$. $X^iX\beta \in x$ iff $X^{i+1}\beta \in x$ iff $((x, w), i + 1) \models_L \beta$ (by the induction hypothesis) iff $((x, w), i) \models_L X\beta$.

• Case $\alpha \equiv X^l\beta$. $X^iX^l\beta \in x$ iff $X^{l+i}\beta \in x$ iff $((x, w), l + i) \models_L \beta$ (by the induction hypothesis) iff $((x, w), i) \models_L X^l\beta$.

- Case $\alpha \equiv \mathrm{G}\beta$.

(Subcase $i < l$): Suppose $\mathrm{X}^i\mathrm{G}\beta \in x$ with $i < l$. Since $\mathrm{IB}[l] \vdash \mathrm{X}^i\mathrm{G}\beta \Rightarrow \mathrm{X}^{i+k}\beta$ for any $k \in N_l$, the fact that $\mathrm{X}^{i+k}\beta \in w$ contradicts the consistency of (x, w), and hence $\mathrm{X}^{i+k}\beta \notin w$. Thus, by the maximality of (x, w), we obtain $\mathrm{X}^{i+k}\beta \in x$ for any $k \in N_l$, i.e., $\forall j \in N_l$ [$i \leq j$ implies $\mathrm{X}^j\beta \in x$]. By the induction hypothesis, we obtain $\forall j \in N_l$ [$i \leq j$ implies $((x, w), j) \models_L \beta$], i.e., $((x, w), i) \models_L \mathrm{G}\beta$. Conversely, suppose $((x, w), i) \models_L \mathrm{G}\beta$, i.e., $\forall j \in N_l$ [$i \leq j$ implies $((w, w), j) \models_L \beta$]. By the induction hypothesis, we obtain $\forall j \in N_l$ [$i \leq j$ implies $\mathrm{X}^i\beta \in x$], i.e., $\forall k \in N_l$ [$\mathrm{X}^{i+k}\beta \in x$]. Since $\mathrm{IB}[l] \vdash \mathrm{X}^i\beta, \mathrm{X}^{i+1}\beta, ..., \mathrm{X}^{i+l}\beta \Rightarrow \mathrm{X}^i\mathrm{G}\beta$, the fact that $\mathrm{X}^i\mathrm{G}\beta \in w$ contradicts the consistency of (x, w), and hence $\mathrm{X}^i\mathrm{G}\beta \notin w$. Thus, we obtain $\mathrm{X}^i\mathrm{G}\beta \in x$ by the maximality of (x, w).

(Subcase $i \geq l$): Suppose $\mathrm{X}^i\mathrm{G}\beta \in x$ with $i \geq l$. Since $\mathrm{IB}[l] \vdash \mathrm{X}^i\mathrm{G}\beta \Rightarrow \mathrm{X}^i\beta$ ($i \geq l$), the fact that $\mathrm{X}^i\beta \in w$ contradicts the consistency of (x, w), and hence $\mathrm{X}^i\beta \notin w$. Thus, by the maximality of (x, w), we obtain $\mathrm{X}^i\beta \in x$. By the induction hypothesis, we obtain $((x, w), i) \models_L \beta$, and then $((x, w), i) \models_L \beta$ iff $((x, w), i) \models_L \mathrm{G}\beta$. Conversely, suppose $((x, w), i) \models_L \mathrm{G}\beta$, i.e., $((w, w), i) \models_L \beta$. By the induction hypothesis, we obtain $\mathrm{X}^i\beta \in x$. Since $\mathrm{IB}[l] \vdash \mathrm{X}^i\beta \Rightarrow \mathrm{X}^i\mathrm{G}\beta$ ($i \geq l$), the fact that $\mathrm{X}^i\mathrm{G}\beta \in w$ contradicts the consistency of (x, w), and hence $\mathrm{X}^i\mathrm{G}\beta \notin w$. Thus, we obtain $\mathrm{X}^i\mathrm{G}\beta \in x$ by the maximality of (x, w).

- Case $\alpha \equiv \mathrm{F}\beta$.

(Subcase $i < l$): Suppose $\mathrm{X}^i\mathrm{F}\beta \in x$ with $i < l$. Since $\mathrm{IB}[l] \vdash \mathrm{X}^i\mathrm{F}\beta \Rightarrow \mathrm{X}^i\beta \vee \mathrm{X}^{i+1}\beta \vee \cdots \vee \mathrm{X}^{i+l}\beta$, the fact that $\forall l \in N_l$ [$\mathrm{X}^{i+l}\beta \in w$] contradicts the consistency of (x, w), and hence $\exists k \in N_l$ [$\mathrm{X}^{i+k}\beta \notin w$]. By the maximality of (x, w), we obtain $\exists k \in N_l$ [$\mathrm{X}^{i+k}\beta \in x$], i.e., $\exists j \in N_l$ [$i \leq j$ and $\mathrm{X}^j\beta \in x$]. By the induction hypothesis, we obtain $\exists j \in N_l$ [$i \leq j$ and $((x, w), j) \models_L \beta$], i.e., $((x, w), i) \models_L \mathrm{F}\beta$. Conversely, suppose $((x, w), i) \models_L \mathrm{F}\beta$, i.e., $\exists j \in N_l$ [$i \leq j$ and $((x, w), j) \models_L \beta$]. By the induction hypothesis, we obtain $\exists j \in N_l$ [$i \leq j$ and $\mathrm{X}^j\beta \in x$], i.e., $\exists k \in N_l$ [$\mathrm{X}^{i+k}\beta \in x$]. Since $\mathrm{IB}[l] \vdash \mathrm{X}^{i+k}\beta \Rightarrow \mathrm{X}^i\mathrm{F}\beta$ (for any $k \in N_l$), the fact that $\mathrm{X}^i\mathrm{F}\beta \in w$ contradicts the consistency of (x, w), and hence $\mathrm{X}^i\mathrm{F}\beta \notin w$. By the maximality of (x, w), we obtain $\mathrm{X}^i\mathrm{F}\beta \in x$.

(Subcase $i \geq l$): Suppose $\mathrm{X}^i\mathrm{F}\beta \in x$ with $i \geq l$. Since $\mathrm{IB}[l] \vdash \mathrm{X}^i\mathrm{F}\beta \Rightarrow \mathrm{X}^i\beta$ ($i \geq l$), the fact that $\mathrm{X}^i\beta \in w$ contradicts the consistency of (x, w), and hence $\mathrm{X}^i\beta \notin w$. By the maximality of (x, w), we obtain $\mathrm{X}^i\beta \in x$. By the induction hypothesis, we obtain $((x, w), i) \models_L \beta$, and then $((x, w), i) \models_L \beta$ iff $((x, w), i) \models_L \mathrm{F}\beta$. Conversely, suppose $((x, w), i) \models_L \mathrm{F}\beta$, i.e., $((x, w), i) \models_L \beta$. By the induction hypothesis, we obtain $\mathrm{X}^i\beta \in x$. Since $\mathrm{IB}[l] \vdash \mathrm{X}^i\beta \Rightarrow \mathrm{X}^i\mathrm{F}\beta$ ($i \geq l$), the fact that $\mathrm{X}^i\mathrm{F}\beta \in w$ contradicts the consistency of (x, w), and hence $\mathrm{X}^i\mathrm{F}\beta \notin w$.

By the maximality of (x, w), we obtain $\mathrm{X}^i\mathrm{F}\beta \in x$. ∎

Theorem 7.2.18 (Completeness) *Let C be the class of all Kripke frames, $L := \{\Gamma \Rightarrow \Delta \mid \mathrm{IB}[l] \vdash \Gamma \Rightarrow \Delta\}$ and $L(C) := \{\Gamma \Rightarrow \Delta \mid \Gamma \Rightarrow \Delta$ is valid in all frames of $C\}$. Then $L = L(C)$.*

Proof. In order to prove this theorem, by Theorem 7.2.13, it is sufficient to show that $L(C) \subseteq L$, i.e., for any sequent $\Gamma \Rightarrow \Delta$, if $\Gamma \Rightarrow \Delta$ is valid in an arbitrary frame in C, then it is provable in IB[l]. To show this, we show that if $\Gamma \Rightarrow \Delta$ is not provable in IB[l], then there is a frame $F = \langle M_L, N, N_l, R_L \rangle \in C$ such that $\Gamma \Rightarrow \Delta$ is not valid in F, i.e., there is a Kripke model $\langle M_L, N, N_l, R_L, \models_L \rangle$ such that $\Gamma \Rightarrow \Delta$ is not true in it.

Then, our goal is to show that $((u, v), 0) \models_L \Gamma \Rightarrow \Delta$ does not hold in the constructed model. Here we consider only the case $\Gamma \neq \varnothing$. We show that $((u, v), 0) \models_L \Gamma^\wedge \rightarrow \Delta^*$ does not hold, i.e., $\exists (x, z) \in M_L$ $[[(u, v)R_L(x, z)$ and $((x, z), 0) \models_L \Gamma^\wedge]$ and $[((x, z), 0) \models_L \Delta^*$ does not hold $]]$. Taking (u, v) for (x, z) and 0 for i, we can verify that there is $(u, v) \in M_L$ such that $[(u, v)R_L(u, v)$ and $((u, v), 0) \models_L \Gamma^\wedge]$ and $[((u, v), 0) \models_L \Delta^*$ does not hold$]$. The first argument is obvious because of the reflexivity of R_L and the fact that $\{\Gamma\} \subseteq u$. The second argument is shown below. The case $\Delta \equiv \varnothing$ is obvious because $((u, v), 0) \models_L \bot$ does not hold. The case $\Delta \equiv \langle \alpha \rangle$ can be proved by using Lemma 7.2.17 and the fact that $\alpha \notin u$, because we have the fact that $\alpha \notin u$ iff $[((u, v), 0) \models_L \alpha$ does not hold$]$ by Lemma 7.2.17. ∎

7.3. Paraconsistent bounded linear-time temporal logic

7.3.1. Sequent calculus

The language of PB[l] is obtained from that of IB[l] by adding a strong negation connective $\sim$. The notations and conventions used for PB[l] are almost the same as those for IB[l]. An expression of the form $\Gamma \Rightarrow \gamma$ where γ is a single formula is called a *sequent* (for PB[l]).

Definition 7.3.1 PB[l] *is obtained from* IB[l] *by restricting each succedent of sequents to a single formula (i.e., Δ used in* IB[l] *is just a single formula γ), deleting the initial sequents of the form $\mathrm{X}^i\bot \Rightarrow$ and the structural rule* (we-right), *adding initial sequents $\mathrm{X}^i{\sim}p \Rightarrow \mathrm{X}^i{\sim}p$, and* *adding (for any $k \in \omega_l$) the logical inference rules of the form:*

$$\frac{\mathrm{X}^i\alpha, \Gamma \Rightarrow \gamma}{\mathrm{X}^i{\sim}{\sim}\alpha, \Gamma \Rightarrow \gamma}\ ({\sim}{\sim}\text{left}) \qquad \frac{\Gamma \Rightarrow \mathrm{X}^i\alpha}{\Gamma \Rightarrow \mathrm{X}^i{\sim}{\sim}\alpha}\ ({\sim}{\sim}\text{right})$$

$$\frac{\mathrm{X}^i\alpha, \mathrm{X}^i{\sim}\beta, \Gamma \Rightarrow \gamma}{\mathrm{X}^i{\sim}(\alpha{\rightarrow}\beta), \Gamma \Rightarrow \gamma}\ ({\sim}{\rightarrow}\text{left}) \qquad \frac{\Gamma \Rightarrow \mathrm{X}^i\alpha \quad \Gamma \Rightarrow \mathrm{X}^i{\sim}\beta}{\Gamma \Rightarrow \mathrm{X}^i{\sim}(\alpha{\rightarrow}\beta)}\ ({\sim}{\rightarrow}\text{right})$$

$$\frac{\mathrm{X}^i{\sim}\alpha, \Gamma \Rightarrow \gamma \quad \mathrm{X}^i{\sim}\beta, \Gamma \Rightarrow \gamma}{\mathrm{X}^i{\sim}(\alpha \wedge \beta), \Gamma \Rightarrow \gamma}\ ({\sim}\wedge\text{left})$$

$$\frac{\Gamma \Rightarrow \mathrm{X}^i{\sim}\alpha}{\Gamma \Rightarrow \mathrm{X}^i{\sim}(\alpha \wedge \beta)}\ ({\sim}\wedge\text{right1}) \qquad \frac{\Gamma \Rightarrow \mathrm{X}^i{\sim}\beta}{\Gamma \Rightarrow \mathrm{X}^i{\sim}(\alpha \wedge \beta)}\ ({\sim}\wedge\text{right2})$$

$$\frac{\mathrm{X}^i{\sim}\alpha, \Gamma \Rightarrow \gamma}{\mathrm{X}^i{\sim}(\alpha \vee \beta), \Gamma \Rightarrow \gamma}\ ({\sim}\vee\text{left1}) \qquad \frac{\mathrm{X}^i{\sim}\beta, \Gamma \Rightarrow \gamma}{\mathrm{X}^i{\sim}(\alpha \vee \beta), \Gamma \Rightarrow \gamma}\ ({\sim}\vee\text{left2})$$

$$\frac{\Gamma \Rightarrow \mathrm{X}^i{\sim}\alpha \quad \Gamma \Rightarrow \mathrm{X}^i{\sim}\beta}{\Gamma \Rightarrow \mathrm{X}^i{\sim}(\alpha \vee \beta)}\ ({\sim}\vee\text{right})$$

$$\frac{\{\ \mathrm{X}^{i+j}{\sim}\alpha, \Gamma \Rightarrow \gamma\ \}_{j \in \omega_l}}{\mathrm{X}^i{\sim}\mathrm{G}\alpha, \Gamma \Rightarrow \gamma}\ ({\sim}\text{Gleft}) \qquad \frac{\Gamma \Rightarrow \mathrm{X}^{i+k}{\sim}\alpha}{\Gamma \Rightarrow \mathrm{X}^i{\sim}\mathrm{G}\alpha}\ ({\sim}\text{Gright})$$

$$\frac{\mathrm{X}^{i+k}{\sim}\alpha, \Gamma \Rightarrow \gamma}{\mathrm{X}^i{\sim}\mathrm{F}\alpha, \Gamma \Rightarrow \gamma}\ ({\sim}\text{Fleft}) \qquad \frac{\{\ \Gamma \Rightarrow \mathrm{X}^{i+j}{\sim}\alpha\ \}_{j \in \omega_l}}{\Gamma \Rightarrow \mathrm{X}^i{\sim}\mathrm{F}\alpha}\ ({\sim}\text{Fright})$$

$$\frac{\mathrm{X}^i{\sim}\alpha, \Gamma \Rightarrow \gamma}{{\sim}\mathrm{X}^i\alpha, \Gamma \Rightarrow \gamma}\ ({\sim}\text{Xleft}) \qquad \frac{\Gamma \Rightarrow \mathrm{X}^i{\sim}\alpha}{\Gamma \Rightarrow {\sim}\mathrm{X}^i\alpha}\ ({\sim}\text{Xright}).$$

We use the same names for the modified single-succedent inference rules.

Note that the rules (∼Xleft) and (∼Xright) imply $\mathrm{PB}[l] \vdash {\sim}\mathrm{X}^i\alpha \Leftrightarrow \mathrm{X}^i{\sim}\alpha$ for any formula α. Also remark that the following sequents are provable in cut-free PB[l]: for any formulas α and β,

1. ${\sim}{\sim}\alpha \leftrightarrow \alpha$,

2. ${\sim}(\alpha \wedge \beta) \leftrightarrow {\sim}\alpha \vee {\sim}\beta$,

3. ${\sim}(\alpha \vee \beta) \leftrightarrow {\sim}\alpha \wedge {\sim}\beta$,

4. ${\sim}(\alpha{\rightarrow}\beta) \leftrightarrow \alpha \wedge {\sim}\beta$,

6. ${\sim}\mathrm{G}\alpha \leftrightarrow \mathrm{F}{\sim}\alpha$,

7. ${\sim}\mathrm{F}\alpha \leftrightarrow \mathrm{G}{\sim}\alpha$,

8. ${\sim}\mathrm{X}\alpha \leftrightarrow \mathrm{X}{\sim}\alpha$.

Definition 7.3.2 (LN4) *A sequent calculus* LN4 *for Nelson's paraconsistent logic* N4 *is obtained from* PB[l] *by deleting the inference rules* (Xleft), (Xright), ($\sim$Xleft), ($\sim$Xright), (Gleft), (Gright), (Fleft), (Fright), ($\sim$Gleft), ($\sim$Gright), ($\sim$Fleft), ($\sim$Fright), *and replacing* X^i *by* X^0.

For more information on sequent calculi for N4, see, for instance, [148, 196].

Definition 7.3.3 *We fix a countable non-empty set* Φ *of propositional variables and define the sets* $\Phi_i := \{p_i \mid p \in \Phi\}$ $(1 \leq i \in \omega)$ *and* $\Phi_0 := \Phi$ *of propositional variables. The language* $\mathcal{L}_{\mathrm{PB}[l]}$ *of* PB[l] *is defined by using* $\Phi, \to, \wedge, \vee, \sim, \mathrm{X}, \mathrm{G}$ *and* F. *The language* $\mathcal{L}_{\mathrm{LN4}}$ *of* LN4 *is defined by using* $\bigcup_{i\in\omega} \Phi_i, \to, \wedge, \vee$ *and* $\sim$.

A mapping f *from* $\mathcal{L}_{\mathrm{PB}[l]}$ *to* $\mathcal{L}_{\mathrm{LN4}}$ *is obtained from Definition 7.2.5 by replacing Condition 1 by the following condition, for any* $i \in \omega$:

1'. $f(\mathrm{X}^i{\sim}\alpha) = f({\sim}\mathrm{X}^i\alpha) = {\sim}f(\mathrm{X}^i\alpha)$.

Theorem 7.3.4 (Embedding) *Let* Γ *be a sequence of formulas in* $\mathcal{L}_{\mathrm{PB}[l]}$, γ *be a formula in* $\mathcal{L}_{\mathrm{PB}[l]}$, *and* f *be the mapping defined in Definition 7.3.3.*

1. $\mathrm{PB}[l] \vdash \Gamma \Rightarrow \gamma$ *iff* $\mathrm{LN4} \vdash f(\Gamma) \Rightarrow f(\gamma)$.

2. $\mathrm{PB}[l] - (\mathrm{cut}) \vdash \Gamma \Rightarrow \gamma$ *iff* $\mathrm{LN4} - (\mathrm{cut}) \vdash f(\Gamma) \Rightarrow f(\gamma)$.

Proof. Similar to the proof of Theorem 7.2.6. We show only the direction ($\Longrightarrow$) of (1) by induction on proofs P of $\Gamma \Rightarrow \Delta$ in PB[l]. We distinguish the cases according to the last inference of P and show some cases.

Case ($\sim$Xright). The last inference of P is of the form:

$$\frac{\Gamma \Rightarrow \mathrm{X}^i{\sim}\alpha}{\Gamma \Rightarrow {\sim}\mathrm{X}^i\alpha}\ ({\sim}\mathrm{Xright}).$$

By induction hypothesis, we have $\mathrm{LN4} \vdash f(\Gamma) \Rightarrow f(\mathrm{X}^i{\sim}\alpha)$, and hence obtain the required fact $\mathrm{LN4} \vdash f(\Gamma) \Rightarrow f({\sim}\mathrm{X}^i\alpha)$, since $f(\mathrm{X}^i{\sim}\alpha) = f({\sim}\mathrm{X}^i\alpha)$ by the definition of f.

Case ($\sim\wedge$right1). The last inference of P is of the form:

$$\frac{\Gamma \Rightarrow \mathrm{X}^i{\sim}\alpha}{\Gamma \Rightarrow \mathrm{X}^i{\sim}(\alpha \wedge \beta)}\ ({\sim}\wedge\mathrm{right1}).$$

By induction hypothesis, we have LN4 $\vdash f(\Gamma) \Rightarrow f(\mathrm{X}^i{\sim}\alpha)$ where $f(\mathrm{X}^i{\sim}\alpha) = {\sim}f(\mathrm{X}^i\alpha)$ by the definition of f. Then we obtain LN4 $\vdash f(\Gamma) \Rightarrow f(\mathrm{X}^i{\sim}(\alpha \wedge \beta))$ by:

$$\dfrac{\begin{array}{c}\vdots\\ f(\Gamma) \Rightarrow {\sim}f(\mathrm{X}^i\alpha)\end{array}}{f(\Gamma) \Rightarrow {\sim}(f(\mathrm{X}^i\alpha) \wedge f(\mathrm{X}^i\beta))}\ ({\sim}\wedge\,\mathrm{right1}^{LN4})$$

where ${\sim}(f(\mathrm{X}^i\alpha) \wedge f(\mathrm{X}^i\beta)) = {\sim}(f(\mathrm{X}^i(\alpha \wedge \beta))) = f(\mathrm{X}^i{\sim}(\alpha \wedge \beta))$ by the definition of f. ∎

Theorem 7.3.5 (Cut-elimination) *The rule* (cut) *is admissible in cut-free* PB[l].

Proof. Similar to the proof of Theorem 7.2.7. ∎

Theorem 7.3.6 (Decidability) PB[l] *is decidable.*

Proof. Similar to the proof of Theorem 7.2.8. ∎

Definition 7.3.7 *Let $\sharp$ be a negation connective. A sequent calculus L is called* explosive *with respect to $\sharp$ iff for any formulas α and β, the sequent $\alpha, \sharp\alpha \Rightarrow \beta$ is provable in L. It is called* paraconsistent *with respect to $\sharp$ iff it is not explosive with respect to $\sharp$.*

Proposition 7.3.8 (Paraconsistency) PB[l] *is paraconsistent with respect to ${\sim}$.*

Proof. Consider a sequent $p, {\sim}p \Rightarrow q$ where p and q are distinct propositional variables. Then the unprovability of this sequent is guaranteed by using Theorem 7.3.5, i.e., we cannot construct a *cut-free* proof of $p, {\sim}p \Rightarrow q$. ∎

7.3.2. Kripke semantics

The same kind of Kripke frames which is used for IB[l] is also used for PB[l]. It is known that the Kripke semantics for logics with strong negation uses two kinds of valuations $\models^+$ (representing *verification*) and $\models^-$ (representing *refutation*). For information on this type of semantics see, for example, [170, 186, 196]. Kripke models for PB[l] also use such valuations.

Definition 7.3.9 Valuations $\models^+$ *and* $\models^-$ *on a Kripke frame* $\langle M, N, N_l, R\rangle$ *are mappings from the set Ψ of all propositional variables to the power set $2^{M\times N}$ such that for any $p \in \Psi$, any $i \in N$, and any $x, y \in M$,*

1. *if* $(x,i) \in \models^+ (p)$ *and* xRy, *then* $(y,i) \in \models^+ (p)$,
2. *if* $(x,i) \in \models^- (p)$ *and* xRy, *then* $(y,i) \in \models^- (p)$.

We will write $(x,i) \models^* p$ *for* $(x,i) \in \models^* (p)$ *where* $* \in \{+,-\}$. *The valuations* $\models^+$ *and* $\models^-$ *are extended to mappings from the set* Φ *of all formulas to* $2^{M\times N}$ *by the following clauses:*

1. $(x,i) \models^+ \alpha{\rightarrow}\beta$ *iff* $\forall y \in M$ [xRy *and* $(y,i) \models^+ \alpha$ *imply* $(y,i) \models^+ \beta$],
2. $(x,i) \models^+ \alpha \wedge \beta$ *iff* $(x,i) \models^+ \alpha$ *and* $(x,i) \models^+ \beta$,
3. $(x,i) \models^+ \alpha \vee \beta$ *iff* $(x,i) \models^+ \alpha$ *or* $(x,i) \models^+ \beta$,
4. $(x,i) \models^+ \mathrm{X}\alpha$ *iff* $(x,i+1) \models^+ \alpha$,
5. $(x,i) \models^+ \mathrm{X}^l\alpha$ *iff* $(x,l) \models^+ \alpha$,
6. $(x,i) \models^+ \mathrm{G}\alpha$ *iff* $\forall j \in N_l$ [$i \leq j$ *implies* $(x,j) \models^+ \alpha$] *if* $i < l$, *and otherwise* $(x,l) \models^+ \alpha$,
7. $(x,i) \models^+ \mathrm{F}\alpha$ *iff* $\exists j \in N_l$ [$i \leq j$ *and* $(x,j) \models^+ \alpha$] *if* $i < l$, *and otherwise* $(x,l) \models^+ \alpha$.
8. $(x,i) \models^+ {\sim}\alpha$ *iff* $(x,i) \models^- \alpha$,
9. $(x,i) \models^- \alpha{\rightarrow}\beta$ *iff* $(x,i) \models^+ \alpha$ *and* $(x,i) \models^- \beta$,
10. $(x,i) \models^- \alpha \wedge \beta$ *iff* $(x,i) \models^- \alpha$ *or* $(x,i) \models^- \beta$,
11. $(x,i) \models^- \alpha \vee \beta$ *iff* $(x,i) \models^- \alpha$ *and* $(x,i) \models^- \beta$,
12. $(x,i) \models^- \mathrm{X}\alpha$ *iff* $(x,i+1) \models^- \alpha$,
13. $(x,i) \models^- \mathrm{X}^l\alpha$ *iff* $(x,l) \models^- \alpha$,
14. $(x,i) \models^- \mathrm{G}\alpha$ *iff* $\exists j \in N_l$ [$i \leq j$ *and* $(x,j) \models^- \alpha$] *if* $i < l$, *and otherwise* $(x,l) \models^- \alpha$,
15. $(x,i) \models^- \mathrm{F}\alpha$ *iff* $\forall j \in N_l$ [$i \leq j$ *implies* $(x,j) \models^- \alpha$] *if* $i < l$, *and otherwise* $(x,l) \models^- \alpha$,
16. $(x,i) \models^- {\sim}\alpha$ *iff* $(x,i) \models^+ \alpha$.

Proposition 7.3.10 *Let* $\models^+$ *and* $\models^-$ *be valuations on a Kripke frame* $\langle M, N, N_l, R\rangle$. *For any formula* α, *any* $i \in N$, *and any* $x, y \in M$, (1) *if* $(x,i) \models^+ \alpha$ *and* xRy, *then* $(y,i) \models^+ \alpha$, *and* (2) *if* $(x,i) \models^- \alpha$ *and* xRy, *then* $(y,i) \models^- \alpha$.

Definition 7.3.11 *A* paraconsistent Kripke model *is a structure* $\langle M, N, N_l, R, \models^+, \models^-\rangle$ *such that* (1) $\langle M, N, N_l, R\rangle$ *is a Kripke frame, and* (2) $\models^+$ *and* $\models^-$ *are valuations on* $\langle M, N, N_l, R\rangle$.

A formula α *is* true *in a paraconsistent Kripke model* $\langle M, N, N_l, R, \models^+, \models^-\rangle$ *if* $(x, 0) \models^+ \alpha$ *for any* $x \in M$, *and* p-valid *in a Kripke frame* $\langle M, N, N_l, R\rangle$ *if it is true for any valuations* $\models^+$ *and* $\models^-$ *on the Kripke frame.*

A sequent $\Gamma \Rightarrow \gamma$ *is true in a paraconsistent Kripke model* $\langle M, N, N_l, R, \models^+, \models^-\rangle$ *if the formula* $(\Gamma \Rightarrow \gamma)^*$ *is true in the paraconsistent Kripke model, and p-valid in a Kripke frame* $\langle M, N, N_l, R\rangle$ *if it is true for any valuations* $\models^+$ *and* $\models^-$ *on the Kripke frame.*

We sketch the proof of the following completeness theorem for PB[l].

Theorem 7.3.12 (Completeness) *Let* C *be the class of all Kripke frames,* $L := \{\Gamma \Rightarrow \gamma \mid \mathrm{PB}[l] \vdash \ \Gamma \Rightarrow \gamma\}$ *and* $L(C) := \{\Gamma \Rightarrow \gamma \mid \Gamma \Rightarrow \gamma$ *is p-valid in all frames of* $C\}$. *Then* $L = L(C)$.

In order to prove $L(C) \subseteq L$, almost the same arguments as those for IB[l] will be employed, i.e., the notions of consistent and maximal consistent pairs, and modifications of Lemma 7.2.15 and Lemma 7.2.17 are used. In the following, only some particularly different points will be explained.

The canonical model $\langle M_L, N, N_l, R_L, \models^+_L, \models^-_L\rangle$ for PB[l] is defined as follows.

Definition 7.3.13 *Let* M_L *be the set of all maximal consistent pairs. The binary relation* R_L *on* M_L *is defined in the same manner as in Definition 7.2.16. Valuations* $\models^+_L (p)$ *and* $\models^-_L (p)$ *for any propositional variable* p *are defined by* $\{((x, w), i) \in M_L \times N \mid \mathrm{X}^i p \in x\}$ *and* $\{((x, w), i) \in M_L \times N \mid \mathrm{X}^i{\sim}p \in x\}$, *respectively.*

Using this definition, the PB[l] version of Lemma 7.2.17 can be formalized and proved as follows.

Lemma 7.3.14 *The structure* $\langle M_L, N, N_l, R_L, \models^+_L, \models^-_L\rangle$ *is a paraconsistent Kripke model such that for any formula* α, *any* $i \in N$, *and any* $(x, w) \in M_L$, (1) $\mathrm{X}^i\alpha \in x$ *iff* $((x, w), i) \models^+_L \alpha$, *and* (2) $\mathrm{X}^i{\sim}\alpha \in x$ *iff* $((x, w), i) \models^-_L \alpha$.

Proof. Since the structure $\langle M_L, N, N_l, R_L, \models^+_L, \models^-_L\rangle$ is a paraconsistent Kripke model, it is shown that in this model, for any formula α, any $i \in N$, and any $(x, w) \in M_L$, (1) $\mathrm{X}^i\alpha \in x$ iff $((x, w), i) \models^+_L \alpha$, and

(2) $X^i{\sim}\alpha \in x$ iff $((x,w),i) \models^-_L \alpha$. This is shown by (simultaneous) induction on the complexity of α. We show only the following critical cases.

• Case $\alpha \equiv {\sim}\beta$. First we show (1). $X^i{\sim}\beta \in x$ iff $((x,w),i) \models^-_L \beta$ (by the induction hypothesis for (2)) iff $((x,w),i) \models^+_L {\sim}\beta$. Next, we show (2). Suppose $X^i{\sim}{\sim}\beta \in x$. Since $\mathrm{PB}[l] \vdash X^i{\sim}{\sim}\beta \Rightarrow X^i\beta$, the fact that $X^i\beta \in w$ contradicts the consistency of (x,w), and hence $X^i\beta \notin w$. By the maximality of (x,w), we obtain $X^i\beta \in x$. By the induction hypothesis for (1), we obtain $((x,w),i) \models^+_L \beta$, and hence $((x,w),i) \models^-_L {\sim}\beta$.

• Case $\alpha \equiv X\beta$. First, we show (1). $X^i(X\beta) \in x$ iff $X^{i+1}\beta \in x$ iff $((x,w),i{+}1) \models^+_L \beta$ (by the induction hypothesis for (1)) iff $((x,w),i) \models^+_L X\beta$. Second, we show (2). Suppose $X^i{\sim}(X\beta) \in x$. Since $\mathrm{PB}[l] \vdash X^i{\sim}(X\beta) \Rightarrow X^{i+1}{\sim}\beta$ by using (${\sim}$Xleft) and (${\sim}$Xright), the fact that $X^{i+1}{\sim}\beta \in w$ contradicts the consistency of (x,w), and hence $X^{i+1}{\sim}\beta \notin w$. By the maximality of (x,w), we obtain $X^{i+1}{\sim}\beta \in x$. By the induction hypothesis for (2), we obtain $((x,w),i+1) \models^-_L \beta$, and hence $((x,w),i) \models^-_L X\beta$. ∎

7.4. Natural deduction

This section assumes basic knowledge of Gentzen-type natural deduction systems (for detailed information, see e.g., [153, 42]). First, we introduce a natural deduction system NPB[l] for PB[l] and show the normalization theorem for NPB[l]. Second, we discuss a natural deduction system NIB[l] for IB[l] and the normalization theorem for NIB[l]. The systems NIB[l] and NPB[l] are defined as (modified) extensions of the natural deduction system NJ for intuitionistic logic and a natural deduction system for Nelson's N4, respectively. A survey of natural deduction systems for N4 is presented in [90]. The treatment of linear time in NIB[l] and NPB[l] is adopted from [93]. In [93], the strong normalization theorem for a typed λ-calculus for the $\{\rightarrow, \wedge, \mathrm{X}, \mathrm{G}\}$-fragment of IB[$l$] is shown, but the (strong) normalization theorem for the full system is not discussed. There are a lot of natural deduction systems and typed-λ-calculi for LTL and its neighbours, and a survey of such systems is also given in [18, 45, 93]. Note that our systems somewhat resemble Baratella's and Masini's system PNJ for an intuitionistic LTL which is called a *logic of positions* [18].

Definition 7.4.1 (NPB[l]) *The inference rules of* NPB[l] *are of the following form, for any* $k \in \omega_l := \{i \in \omega \mid i \leq l\}$ *and any positive integer* *m:*

$$\frac{\begin{array}{c}[\mathrm{X}^i\alpha]\\ \vdots \\ \mathrm{X}^i\beta\end{array}}{\mathrm{X}^i(\alpha\rightarrow\beta)}\ (\rightarrow I) \qquad \frac{\mathrm{X}^i(\alpha\rightarrow\beta) \quad \mathrm{X}^i\alpha}{\mathrm{X}^i\beta}\ (\rightarrow E)$$

$$\frac{\mathrm{X}^i\alpha_1 \quad \mathrm{X}^i\alpha_2}{\mathrm{X}^i(\alpha_1\wedge\alpha_2)}\ (\wedge I) \qquad \frac{\mathrm{X}^i(\alpha_1\wedge\alpha_2)}{\mathrm{X}^i\alpha_1}\ (\wedge E1) \qquad \frac{\mathrm{X}^i(\alpha_1\wedge\alpha_2)}{\mathrm{X}^i\alpha_2}\ (\wedge E2)$$

$$\frac{\mathrm{X}^i\alpha_1}{\mathrm{X}^i(\alpha_1\vee\alpha_2)}\ (\vee I1) \qquad \frac{\mathrm{X}^i\alpha_2}{\mathrm{X}^i(\alpha_1\vee\alpha_2)}\ (\vee I2)$$

$$\frac{\mathrm{X}^i(\alpha_1\vee\alpha_2) \quad \begin{array}{c}[\mathrm{X}^i\alpha_1]\\ \vdots \\ \beta\end{array} \quad \begin{array}{c}[\mathrm{X}^i\alpha_2]\\ \vdots \\ \beta\end{array}}{\beta}\ (\vee E)$$

$$\frac{\mathrm{X}^l\alpha}{\mathrm{X}^{l+m}\alpha}\ (\mathrm{X}I) \qquad \frac{\mathrm{X}^{l+m}\alpha}{\mathrm{X}^l\alpha}\ (\mathrm{X}E)$$

$$\frac{\{\ \mathrm{X}^{i+j}\alpha\ \}_{j\in\omega_l}}{\mathrm{X}^i\mathrm{G}\alpha}\ (\mathrm{G}I) \qquad \frac{\mathrm{X}^i\mathrm{G}\alpha}{\mathrm{X}^{i+k}\alpha}\ (\mathrm{G}E)$$

$$\frac{\mathrm{X}^{i+k}\alpha}{\mathrm{X}^i\mathrm{F}\alpha}\ (\mathrm{F}I) \qquad \frac{\mathrm{X}^i\mathrm{F}\alpha \quad \begin{array}{c}[\mathrm{X}^i\alpha]\\ \vdots \\ \beta\end{array} \quad \begin{array}{c}[\mathrm{X}^{i+1}\alpha]\\ \vdots \\ \beta\end{array} \quad \cdots \quad \begin{array}{c}[\mathrm{X}^{i+l}\alpha]\\ \vdots \\ \beta\end{array}}{\beta}\ (\mathrm{F}E)$$

$$\frac{\mathrm{X}^i\alpha}{\mathrm{X}^i{\sim}{\sim}\alpha}\ ({\sim}I) \qquad \frac{\mathrm{X}^i{\sim}{\sim}\alpha}{\mathrm{X}^i\alpha}\ ({\sim}E)$$

$$\frac{\mathrm{X}^i(\alpha\wedge{\sim}\beta)}{\mathrm{X}^i{\sim}(\alpha\rightarrow\beta)}\ ({\sim}{\rightarrow}I) \qquad \frac{\mathrm{X}^i{\sim}(\alpha\rightarrow\beta)}{\mathrm{X}^i(\alpha\wedge{\sim}\beta)}\ ({\sim}{\rightarrow}E)$$

$$\frac{\mathrm{X}^i({\sim}\alpha\vee{\sim}\beta)}{\mathrm{X}^i{\sim}(\alpha\wedge\beta)}\ ({\sim}\wedge I) \qquad \frac{\mathrm{X}^i{\sim}(\alpha\wedge\beta)}{\mathrm{X}^i({\sim}\alpha\vee{\sim}\beta)}\ ({\sim}\wedge E)$$

$$\frac{\mathrm{X}^i({\sim}\alpha\wedge{\sim}\beta)}{\mathrm{X}^i{\sim}(\alpha\vee\beta)}\ ({\sim}\vee I) \qquad \frac{\mathrm{X}^i{\sim}(\alpha\vee\beta)}{\mathrm{X}^i({\sim}\alpha\wedge{\sim}\beta)}\ ({\sim}\vee E)$$

$$\frac{\mathrm{X}^i\mathrm{F}{\sim}\alpha}{\mathrm{X}^i{\sim}\mathrm{G}\alpha}\ ({\sim}\mathrm{G}I) \qquad \frac{\mathrm{X}^i{\sim}\mathrm{G}\alpha}{\mathrm{X}^i\mathrm{F}{\sim}\alpha}\ ({\sim}\mathrm{G}E)$$

$$\frac{\mathrm{X}^i\mathrm{G}{\sim}\alpha}{\mathrm{X}^i{\sim}\mathrm{F}\alpha}\ ({\sim}\mathrm{F}I) \qquad \frac{\mathrm{X}^i{\sim}\mathrm{F}\alpha}{\mathrm{X}^i\mathrm{G}{\sim}\alpha}\ ({\sim}\mathrm{F}E)$$

$$\frac{\mathrm{X}^i{\sim}\alpha}{{\sim}\mathrm{X}^i\alpha}\ ({\sim}\mathrm{X}I) \qquad \frac{{\sim}\mathrm{X}^i\alpha}{\mathrm{X}^i{\sim}\alpha}\ ({\sim}\mathrm{X}E).$$

For the sake of simplicity, the $(l+1)$-premises rule (FE) is sometimes denoted as:

$$\dfrac{\mathrm{X}^i\mathrm{F}\alpha \qquad \begin{matrix}\{[\mathrm{X}^{i+j}\alpha]\}_{j\in\omega_l}\\ \vdots \\ \beta\end{matrix}}{\beta}\ (\mathrm{F}E).$$

The inference rules $(\to I)$, $(\wedge I)$, $(\vee I1)$, $(\vee I2)$, $(\mathrm{X}I)$, $(\mathrm{G}I)$, $(\mathrm{F}I)$, $(\sim I)$, $(\sim\to I)$, $(\sim\wedge I)$, $(\sim\vee I)$, $(\sim\mathrm{G}I)$, $(\sim\mathrm{F}I)$, and $(\sim\mathrm{X}I)$ are called *introduction rules*, and the inference rules $(\to E)$, $(\wedge E1)$, $(\wedge E2)$, $(\vee E)$, $(\mathrm{X}E)$, $(\mathrm{G}E)$, $(\mathrm{F}E)$, $(\sim E)$, $(\sim\to E)$, $(\sim\wedge E)$, $(\sim\vee E)$, $(\sim\mathrm{G}E)$, $(\sim\mathrm{F}E)$, and $(\sim\mathrm{X}E)$ are called *elimination rules*. The usual terminology of *major and minor premises* of inference rules is used. The notions of *proof* (of NPB$[l]$), *(open and discharged) assumptions* of a proof, and *end-formula* of a proof are defined as usual. A formula α is said to be *provable* in NPB$[l]$ iff there exists a proof of NPB$[l]$ with no open assumption whose end-formula is α. This terminology and these standard notions are from the well-known text books [153, 42]. For example, the major and minor premises of $(\mathrm{F}E)$ are $\mathrm{X}^i\mathrm{F}\alpha$ and β, respectively, and the discharged assumptions of $(\mathrm{F}E)$ are the square bracketed assumptions $[\mathrm{X}^i\alpha], [\mathrm{X}^{i+1}\alpha], ..., [\mathrm{X}^{i+l}\alpha]$.

Note that NPB$[l]$ includes the Gentzen-type natural deduction system GN for positive intuitionistic logic. Taking 0 for i in X^i, the rules of NPB$[l]$ comprise the usual inference rules for GN. As a result, all the provable formulas (without temporal operators) in GN can be proved in NPB$[l]$. Thus, NPB$[l]$ is an extension and generalization of GN.

We give an example proof in NPB$[l]$ below, where we prove the temporal induction axiom for the case $l = 1$.

$$\dfrac{\dfrac{\dfrac{[\alpha\wedge\mathrm{G}(\alpha\to\mathrm{X}\alpha)]}{\alpha}\,(\wedge E1) \qquad \dfrac{\dfrac{[\alpha\wedge\mathrm{G}(\alpha\to\mathrm{X}\alpha)]}{\alpha}\,(\wedge E1) \qquad \dfrac{\dfrac{[\alpha\wedge\mathrm{G}(\alpha\to\mathrm{X}\alpha)]}{\mathrm{G}(\alpha\to\mathrm{X}\alpha)}\,(\wedge E2)}{\alpha\to\mathrm{X}\alpha}\,(\mathrm{G}E)}{\mathrm{X}\alpha}\,(\to I)}{\mathrm{G}\alpha}\,(\mathrm{G}I)}{(\alpha\wedge\mathrm{G}(\alpha\to\mathrm{X}\alpha))\to\mathrm{G}\alpha}\,(\to I).$$

Definition 7.4.2 *Let α be a formula occurring in a proof D in* NPB$[l]$. *Then α is called a* maximum formula *in D iff α satisfies the following conditions:* (1) *α is the conclusion of an introduction rule,* $(\vee E)$ *or* $(\mathrm{F}E)$, *and* (2) *α is the major premise of an elimination rule. A proof is said to be* normal *iff it contains no maximum formula.*

In order to define a reduction relation $\rhd$ on the set of proofs, we assume the usual definition of substitution of proofs (for assumptions). The set of proofs is closed under substitution.

Definition 7.4.3 *Let γ be a maximum formula in a proof which is the conclusion of an inference rule R. The reduction relation $\rhd$ at γ is defined as follows.*

1. R is $(\to I)$, and γ is $\mathrm{X}^i(\alpha\to\beta)$:

$$\dfrac{\dfrac{\begin{array}{c}[\mathrm{X}^i\alpha]\\ \vdots\, D\\ \mathrm{X}^i\beta\end{array}}{\mathrm{X}^i(\alpha\to\beta)}\,R \qquad \begin{array}{c}\vdots\, E\\ \mathrm{X}^i\alpha\end{array}}{\mathrm{X}^i\beta} \qquad \rhd \qquad \begin{array}{c}\vdots\, E\\ \mathrm{X}^i\alpha\\ \vdots\, D\\ \mathrm{X}^i\beta\end{array}$$

2. R is $(\wedge I)$, and γ is $\mathrm{X}^i(\alpha_1\wedge\alpha_2)$:

$$\dfrac{\dfrac{\begin{array}{c}\vdots\, D_1\\ \mathrm{X}^i\alpha_1\end{array} \quad \begin{array}{c}\vdots\, D_2\\ \mathrm{X}^i\alpha_2\end{array}}{\mathrm{X}^i(\alpha_1\wedge\alpha_2)}\,R}{\mathrm{X}^i\alpha_n} \qquad \rhd \qquad \begin{array}{c}\vdots\, D_n\\ \mathrm{X}^i\alpha_n\end{array}$$

where n is 1 or 2.

3. R is $(\vee I1)$ or $(\vee I2)$, and γ is $\mathrm{X}^i(\alpha_1\vee\alpha_2)$:

$$\dfrac{\dfrac{\begin{array}{c}\vdots\, D\\ \mathrm{X}^i\alpha_n\end{array}}{\mathrm{X}^i(\alpha_1\vee\alpha_2)}\,R \qquad \begin{array}{c}[\mathrm{X}^i\alpha_1]\\ \vdots\, E_1\\ \mathrm{X}^i\beta\end{array} \qquad \begin{array}{c}[\mathrm{X}^i\alpha_2]\\ \vdots\, E_2\\ \mathrm{X}^i\beta\end{array}}{\mathrm{X}^i\beta} \qquad \rhd \qquad \begin{array}{c}\vdots\, D\\ \mathrm{X}^i\alpha_n\\ \vdots\, E_n\\ \mathrm{X}^i\beta\end{array}$$

where n is 1 or 2.

4. R is $(\vee E)$:

$$\dfrac{\dfrac{\begin{array}{c}\vdots\, D_1\\ \mathrm{X}^i(\alpha_1\vee\alpha_2)\end{array} \qquad \begin{array}{c}[\mathrm{X}^i\alpha_1]\\ \vdots\, D_2\\ \mathrm{X}^i\gamma\end{array} \qquad \begin{array}{c}[\mathrm{X}^i\alpha_2]\\ \vdots\, D_3\\ \mathrm{X}^i\gamma\end{array}}{\mathrm{X}^i\gamma}\,R \qquad E_1 \quad \cdots \quad E_l}{\mathrm{X}^i\beta}\,R'$$

$$\rhd \quad \dfrac{\begin{array}{c}\vdots\, D_1\\ \mathrm{X}^i(\alpha_1\vee\alpha_2)\end{array} \qquad \dfrac{\begin{array}{c}[\mathrm{X}^i\alpha_1]\\ \vdots\, D_2\\ \mathrm{X}^i\gamma\end{array} \quad E_1 \quad \cdots \quad E_l}{\mathrm{X}^i\beta}\,R' \qquad \dfrac{\begin{array}{c}[\mathrm{X}^i\alpha_2]\\ \vdots\, D_3\\ \mathrm{X}^i\gamma\end{array} \quad E_1 \quad \cdots \quad E_l}{\mathrm{X}^i\beta}\,R'}{\mathrm{X}^i\beta}\,R$$

where R' is an arbitrary inference rule, and both E_1, ..., E_l are proofs of the minor premises of R' if they exist.

5. R is (XI)*, and γ is* $\mathrm{X}^{l+m}\alpha$*:*

$$\dfrac{\dfrac{\begin{matrix}\vdots\ D\\ \mathrm{X}^{l}\alpha\end{matrix}}{\mathrm{X}^{l+m}\alpha}\,R}{\mathrm{X}^{l}\alpha} \qquad \rhd \qquad \begin{matrix}\vdots\ D\\ \mathrm{X}^{l}\alpha\end{matrix}$$

6. R is (GI)*, and γ is* $\mathrm{X}^{i}\mathrm{G}\alpha$*:*

$$\dfrac{\dfrac{\begin{matrix}\vdots\ D_j\\ \{\mathrm{X}^{i+j}\alpha\}_{j\in\omega_l}\end{matrix}}{\mathrm{X}^{i}\mathrm{G}\alpha}\,R}{\mathrm{X}^{i+k}\alpha} \qquad \rhd \qquad \begin{matrix}\vdots\ D_k\\ \mathrm{X}^{i+k}\alpha\end{matrix}$$

7. R is (FI)*, and γ is* $\mathrm{X}^{i}\mathrm{F}\alpha$*:*

$$\dfrac{\dfrac{\begin{matrix}\vdots\ D_k\\ \mathrm{X}^{i+k}\alpha\end{matrix}}{\mathrm{X}^{i}\mathrm{F}\alpha}\,R \qquad \begin{matrix}\{[\mathrm{X}^{i+j}\alpha]\}_{j\in\omega_l}\\ \vdots\ E_j\\ \mathrm{X}^{i}\beta\end{matrix}}{\mathrm{X}^{i}\beta} \qquad \rhd \qquad \begin{matrix}\vdots\ D_k\\ \mathrm{X}^{i+k}\alpha\\ \vdots\ E_k\\ \mathrm{X}^{i}\beta\end{matrix}$$

8. R is (FE)*:*

$$\dfrac{\dfrac{\begin{matrix}\vdots\ D\\ \mathrm{X}^{i}\mathrm{F}\alpha\end{matrix} \qquad \begin{matrix}\{[\mathrm{X}^{i+j}\alpha]\}_{j\in\omega_l}\\ \vdots\ D_j\\ \mathrm{X}^{i}\gamma\end{matrix}}{\mathrm{X}^{i}\gamma}\,R \qquad E_1 \quad \cdots \quad E_l}{\mathrm{X}^{i}\beta}\,R'$$

$$\rhd \qquad \dfrac{\begin{matrix}\vdots\ D\\ \mathrm{X}^{i}\mathrm{F}\alpha\end{matrix} \qquad \dfrac{\begin{matrix}\{[\mathrm{X}^{i+j}\alpha]\}_{j\in\omega_l}\\ \vdots\ D_j\\ \mathrm{X}^{i}\gamma\end{matrix} \qquad E_1 \quad \cdots \quad E_l}{\mathrm{X}^{i}\beta}\,R'}{\mathrm{X}^{i}\beta}\,R$$

where R' is an arbitrary inference rule, and both E_1, ..., E_l are proofs of the minor premises of R' if they exist.

9. R is ($\sim I$)*, and γ is* $\mathrm{X}^{i}{\sim}{\sim}\alpha$*:*

$$\dfrac{\dfrac{\begin{matrix}\vdots\ D\\ \mathrm{X}^{i}\alpha\end{matrix}}{\mathrm{X}^{i}{\sim}{\sim}\alpha}\,R}{\mathrm{X}^{i}\alpha} \qquad \rhd \qquad \begin{matrix}\vdots\ D\\ \mathrm{X}^{i}\alpha\end{matrix}$$

10. R is $(\sim\!\to\! I)$*, and* γ *is* $X^i\!\sim\!(\alpha\to\beta)$*:*

$$\dfrac{\dfrac{\dfrac{\vdots\, D}{X^i(\alpha\wedge\sim\!\beta)}}{X^i\!\sim\!(\alpha\to\beta)}\,R}{X^i(\alpha\wedge\sim\!\beta)} \quad \rhd \quad \begin{array}{c}\vdots\, D\\ X^i(\alpha\wedge\sim\!\beta)\end{array}$$

11. R is $(\sim\wedge I)$*, and* γ *is* $X^i\!\sim\!(\alpha\wedge\beta)$*:*

$$\dfrac{\dfrac{\dfrac{\vdots\, D}{X^i(\sim\!\alpha\vee\sim\!\beta)}}{X^i\!\sim\!(\alpha\wedge\beta)}\,R}{X^i(\sim\!\alpha\vee\sim\!\beta)} \quad \rhd \quad \begin{array}{c}\vdots\, D\\ X^i(\sim\!\alpha\vee\sim\!\beta)\end{array}$$

12. R is $(\sim\vee I)$*, and* γ *is* $X^i\!\sim\!(\alpha\vee\beta)$*:*

$$\dfrac{\dfrac{\dfrac{\vdots\, D}{X^i(\sim\!\alpha\wedge\sim\!\beta)}}{X^i\!\sim\!(\alpha\vee\beta)}\,R}{X^i(\sim\!\alpha\wedge\sim\!\beta)} \quad \rhd \quad \begin{array}{c}\vdots\, D\\ X^i(\sim\!\alpha\wedge\sim\!\beta)\end{array}$$

13. R is $(\sim\! GI)$*, and* γ *is* $X^i\!\sim\! G\alpha$*:*

$$\dfrac{\dfrac{\dfrac{\vdots\, D}{X^iF\!\sim\!\alpha}}{X^i\!\sim\! G\alpha}\,R}{X^iF\!\sim\!\alpha} \quad \rhd \quad \begin{array}{c}\vdots\, D\\ X^iF\!\sim\!\alpha\end{array}$$

14. R is $(\sim\! FI)$*, and* γ *is* $X^i\!\sim\! F\alpha$*:*

$$\dfrac{\dfrac{\dfrac{\vdots\, D}{X^iG\!\sim\!\alpha}}{X^i\!\sim\! F\alpha}\,R}{X^iG\!\sim\!\alpha} \quad \rhd \quad \begin{array}{c}\vdots\, D\\ X^iG\!\sim\!\alpha\end{array}$$

15. R is $(\sim\! XI)$*, and* γ *is* $\sim\! X^i\alpha$*:*

$$\dfrac{\dfrac{\dfrac{\vdots\, D}{X^i\!\sim\!\alpha}}{\sim\! X^i\alpha}\,R}{X^i\!\sim\!\alpha} \quad \rhd \quad \begin{array}{c}\vdots\, D\\ X^i\!\sim\!\alpha\end{array}$$

16. Let D, D', E, F, D_j ($j \in \omega_l$) be proofs. If $D \rhd D'$, then

$$\frac{D}{\alpha}\ (I) \quad \rhd \quad \frac{D'}{\alpha}\ (I)$$

$$\frac{D \quad E}{\alpha}\ (R) \quad \rhd \quad \frac{D' \quad E}{\alpha}\ (R)$$

$$\frac{E \quad D}{\alpha}\ (R) \quad \rhd \quad \frac{E \quad D'}{\alpha}\ (R)$$

$$\frac{D \quad E \quad F}{\alpha}\ (\vee E) \quad \rhd \quad \frac{D' \quad E \quad F}{\alpha}\ (\vee E)$$

$$\frac{E \quad D \quad F}{\alpha}\ (\vee E) \quad \rhd \quad \frac{E \quad D' \quad F}{\alpha}\ (\vee E)$$

$$\frac{E \quad F \quad D}{\alpha}\ (\vee E) \quad \rhd \quad \frac{E \quad F \quad D'}{\alpha}\ (\vee E)$$

$$\frac{D_0 \cdots D \cdots D_l}{\alpha}\ (\mathrm{G}I) \quad \rhd \quad \frac{D_0 \cdots D' \cdots D_l}{\alpha}\ (\mathrm{G}I)$$

$$\frac{D_0 \cdots D \cdots D_{l+1}}{\alpha}\ (\mathrm{F}E) \quad \rhd \quad \frac{D_0 \cdots D' \cdots D_{l+1}}{\alpha}\ (\mathrm{F}E)$$

where $I \in \{\to I, \wedge E1, \wedge E2, \vee I1, \vee I2, \mathrm{X}I, \mathrm{X}E, \mathrm{G}E, \mathrm{F}I, {\sim}I, {\sim}E, {\sim}{\to}I, {\sim}{\to}E, {\sim}\wedge I, {\sim}\wedge E, {\sim}\vee I, {\sim}\vee E, {\sim}\mathrm{G}I, {\sim}\mathrm{G}E, {\sim}\mathrm{F}I, {\sim}\mathrm{F}E, {\sim}\mathrm{X}I, {\sim}\mathrm{X}E\}$, $R \in \{\to E, \wedge I\}$ *and* $D_0, ..., D_{l+1}$ *can be* D *or* D'.

Definition 7.4.4 *If D' is obtained from D by the reduction at γ, then such a fact is denoted by $D \rhd D'$. A sequence $D_0, D_1, ...$ of proofs is called a* reduction sequence *iff it satisfies the following conditions:* (1) $D_i \rhd D_{i+1}$ *for all* $i \geq 0$ *and* (2) *the last proof in the sequence is normal iff the sequence is finite. A proof D is called* normalizable *iff there is a finite reduction sequence starting from D.*

Let P be a proof. The expression oa(P) denotes the set of open assumptions of P, and the expression end(P) denotes the end-formula of P.

From the following theorem, we can obtain the fact that a formula α is provable in NPB[l] if and only if the sequent $\Rightarrow \alpha$ is provable in PB[l].

Theorem 7.4.5 (Equivalence between NPB[l] and PB[l]) *We have the following.*

1. *If P is a proof in* NPB[l] *such that* oa(P) $= \Gamma$ *and* end(P) $= \{\beta\}$, *then the sequent* $\Gamma \Rightarrow \beta$ *is provable in* PB[l].

2. *If a sequent* $\Gamma \Rightarrow \beta$ *is provable in* PB[l] − (cut), *then there is a proof Q in* NPB[l] *which satisfies the following conditions:* (a) oa(Q) $= \Gamma$, (b) end(Q) $= \{\beta\}$, *and* (c) *Q is normal.*

Proof. First, we show (1) by induction on a proof P of PB[l] such that $\mathrm{oa}(P) = \Gamma$ and $\mathrm{end}(P) = \{\beta\}$. We distinguish the cases according to the last inference of P. We show some cases.

Case (GI): P is of the form:

$$\dfrac{\begin{array}{c}\Gamma\\ \vdots\ P'\\ \{\mathrm{X}^{i+j}\alpha\}_{j\in\omega_l}\end{array}}{\mathrm{X}^i\mathrm{G}\alpha}\ (\mathrm{G}I)$$

where $\mathrm{oa}(P) = \Gamma$ and $\mathrm{end}(P) = \{\mathrm{X}^i\mathrm{G}\alpha\}$. By the hypothesis of induction, the sequents $\Gamma_j \Rightarrow \mathrm{X}^{i+j}\alpha$ for any $j \in \omega_l$ where Γ_j is a subset of Γ are provable in PB[l]. Then the sequent $\Gamma \Rightarrow \mathrm{X}^i\mathrm{G}\alpha$ is provable in PB[l] by using (we-left) and (Gright).

Case (FE): P is of the form:

$$\dfrac{\begin{array}{c}\Gamma_{l+1}\\ \vdots\\ \mathrm{X}^i\mathrm{F}\alpha\end{array}\quad \begin{array}{c}\Gamma_0[\mathrm{X}^i\alpha]\\ \vdots\\ \gamma\end{array}\quad \begin{array}{c}\Gamma_1[\mathrm{X}^{i+1}\alpha]\\ \vdots\\ \gamma\end{array}\quad \cdots \quad \begin{array}{c}\Gamma_l[\mathrm{X}^{i+l}\alpha]\\ \vdots\\ \gamma\end{array}}{\gamma}\ (\mathrm{F}E)$$

where $\mathrm{oa}(P) = \Gamma = \bigcup_{j\in\omega_{l+1}}\Gamma_j$ and $\mathrm{end}(P) = \{\gamma\}$. By the hypothesis of induction, the following sequents are provable in PB[l]: $(\Gamma_{l+1} \Rightarrow \mathrm{X}^i\mathrm{F}\alpha)$, $(\mathrm{X}^i\alpha, \Gamma_0 \Rightarrow \gamma)$, $(\mathrm{X}^{i+1}\alpha, \Gamma_1 \Rightarrow \gamma)$, ..., $(\mathrm{X}^{i+l}\alpha, \Gamma_l \Rightarrow \gamma)$. Then we obtain the required fact:

$$\begin{array}{c}\dfrac{\begin{array}{c}\Gamma_{l+1} \Rightarrow \mathrm{X}^i\mathrm{F}\alpha\\ \vdots\ \text{(we-left)}\\ \Gamma \Rightarrow \mathrm{X}^i\mathrm{F}\alpha\end{array}\qquad \dfrac{\begin{array}{c}\mathrm{X}^i\alpha, \Gamma_0 \Rightarrow \gamma\\ \vdots\ \text{(we-left)}\\ \mathrm{X}^i\alpha, \Gamma \Rightarrow \gamma\end{array}\quad \begin{array}{c}\mathrm{X}^{i+1}\alpha, \Gamma_1 \Rightarrow \gamma\\ \vdots\ \text{(we-left)}\\ \mathrm{X}^{i+1}\alpha, \Gamma \Rightarrow \gamma\end{array}\quad \cdots\quad \begin{array}{c}\mathrm{X}^{i+l}\alpha, \Gamma_l \Rightarrow \gamma\\ \vdots\ \text{(we-left)}\\ \mathrm{X}^{i+l}\alpha, \Gamma \Rightarrow \gamma\end{array}}{\mathrm{X}^i\mathrm{F}\alpha, \Gamma \Rightarrow \gamma}\ \text{(Fleft)}}{\Gamma, \Gamma \Rightarrow \gamma}\ \text{(cut)}\\ \vdots\ \text{(co)}\\ \Gamma \Rightarrow \gamma\end{array}$$

Case ($\sim\!\to\! I$): P is of the form:

$$\dfrac{\begin{array}{c}\Gamma\\ \vdots\ P'\\ \mathrm{X}^i(\alpha \wedge {\sim}\gamma)\end{array}}{\mathrm{X}^i{\sim}(\alpha{\to}\gamma)}\ ({\sim}{\to}I)$$

where $\mathrm{oa}(P) = \Gamma$ and $\mathrm{end}(P) = \{\mathrm{X}^i{\sim}(\alpha{\to}\gamma)\}$. By the hypothesis of induction, the sequent $\Gamma \Rightarrow \mathrm{X}^i(\alpha \wedge {\sim}\gamma)$ is provable in PB[l]. Then the sequent $\Gamma \Rightarrow X^i{\sim}(\alpha{\to}\gamma)$ is provable in PB[l] as follows.

$$\dfrac{\vdots \qquad\qquad \dfrac{\dfrac{\dfrac{X^i\alpha \Rightarrow X^i\alpha}{X^i\alpha, X^i{\sim}\gamma \Rightarrow X^i\alpha}\,(\text{we-left}) \quad \dfrac{X^i{\sim}\gamma \Rightarrow X^i{\sim}\gamma}{X^i\alpha, X^i{\sim}\gamma \Rightarrow X^i{\sim}\gamma}\,(\text{we-left})}{X^i\alpha, X^i{\sim}\gamma \Rightarrow X^i{\sim}(\alpha{\to}\gamma)}\,({\sim}{\to}\text{right})}{\vdots\ (\wedge\text{left}1{,}2), (\text{co})}}{\Gamma \Rightarrow X^i(\alpha \wedge {\sim}\gamma) \qquad X^i(\alpha \wedge {\sim}\gamma) \Rightarrow X^i{\sim}(\alpha{\to}\gamma)}$$

$$\dfrac{\Gamma \Rightarrow X^i(\alpha \wedge {\sim}\gamma) \qquad X^i(\alpha \wedge {\sim}\gamma) \Rightarrow X^i{\sim}(\alpha{\to}\gamma)}{\Gamma \Rightarrow X^i{\sim}(\alpha{\to}\gamma)}\,(\text{cut}).$$

Second, we prove (2) by induction on a cut-free proof P of $\Gamma \Rightarrow \beta$ in PB[l] − (cut). We distinguish the cases according to the inference of P. We show some cases.

Case (Gleft): P is of the form:

$$\dfrac{X^{i+k}\alpha, \Gamma \Rightarrow \gamma}{X^i G\alpha, \Gamma \Rightarrow \gamma}\,(\text{Gleft}).$$

By the hypothesis of induction, there is a normal proof Q' in NPB[l] of the form:

$$\begin{array}{c} \Gamma, X^{i+k}\alpha \\ \vdots\ Q' \\ \gamma \end{array}$$

where $\text{oa}(Q') = \Gamma \cup \{X^{i+k}\alpha\}$ and $\text{end}(Q') = \{\gamma\}$. Then we obtain a normal proof Q as follows.

$$\begin{array}{c} \overline{\Gamma}\ \dfrac{X^i G\alpha}{X^{i+k}\alpha}\,(GE) \\ \vdots\ Q' \\ \gamma \end{array}$$

where $\text{oa}(Q) = \Gamma \cup \{X^i G\alpha\}$ and $\text{end}(Q) = \{\gamma\}$.

Case (Fleft): P is of the form:

$$\dfrac{\{X^{i+j}\alpha, \Gamma \Rightarrow \gamma\}_{j \in \omega_l}}{X^i F\alpha, \Gamma \Rightarrow \gamma}\,(\text{Fleft}).$$

By the hypothesis of induction, there are normal proofs $\{Q_j\}_{j\in\omega_l}$ in NPB[l] of the form: for any $j \in \omega_l$,

$$\begin{array}{c} \Gamma, X^{i+j}\alpha \\ \vdots\ Q_j \\ \gamma \end{array}$$

where $\mathrm{oa}(Q_j) = \Gamma \cup \{\mathrm{X}^{i+j}\alpha\}$ and $\mathrm{end}(Q_j) = \{\gamma\}$. Then we obtain a normal proof Q as follows:

$$\dfrac{\mathrm{X}^i\mathrm{F}\alpha \quad \begin{array}{c}\Gamma[\mathrm{X}^i\alpha] \\ \vdots\, Q_0 \\ \gamma\end{array} \quad \cdots \quad \begin{array}{c}\Gamma[\mathrm{X}^{i+l}\alpha] \\ \vdots\, Q_l \\ \gamma\end{array}}{\gamma}\ (\mathrm{F}E)$$

where $\mathrm{oa}(Q) = \Gamma \cup \{\mathrm{X}^i\mathrm{F}\alpha\}$ and $\mathrm{end}(Q) = \{\gamma\}$.

Case ($\sim\!\rightarrow$right): P is of the form:

$$\dfrac{\begin{array}{c}\vdots\, P_1 \\ \Gamma \Rightarrow \mathrm{X}^i\alpha\end{array} \quad \begin{array}{c}\vdots\, P_2 \\ \Gamma \Rightarrow \mathrm{X}^i{\sim}\gamma\end{array}}{\Gamma \Rightarrow \mathrm{X}^i{\sim}(\alpha\rightarrow\gamma)}\ ({\sim}\!\rightarrow\!\mathrm{right}).$$

By the hypothesis of induction, there are normal proofs Q_1 and Q_2 in NPB$[l]$ of the form:

$$\begin{array}{c}\Gamma \\ \vdots\, Q_1 \\ \mathrm{X}^i\alpha,\end{array} \qquad \begin{array}{c}\Gamma \\ \vdots\, Q_2 \\ \mathrm{X}^i{\sim}\gamma\end{array}$$

where $\mathrm{oa}(Q_1) = \mathrm{oa}(Q_2) = \Gamma$, $\mathrm{end}(Q_1) = \{\mathrm{X}^i\alpha\}$ and $\mathrm{end}(Q_2) = \{\mathrm{X}^i{\sim}\gamma\}$. Then we obtain a normal proof Q as follows.

$$\dfrac{\dfrac{\begin{array}{c}\Gamma \\ \vdots\, Q_1 \\ \mathrm{X}^i\alpha\end{array} \quad \begin{array}{c}\Gamma \\ \vdots\, Q_2 \\ \mathrm{X}^i{\sim}\gamma\end{array}}{\mathrm{X}^i(\alpha \wedge {\sim}\gamma)}\ (\wedge I)}{\mathrm{X}^i{\sim}(\alpha\rightarrow\gamma)}\ ({\sim}\!\rightarrow\! I)$$

where $\mathrm{oa}(Q) = \Gamma$ and $\mathrm{end}(Q) = \{\mathrm{X}^i{\sim}(\alpha\rightarrow\gamma)\}$. ∎

Theorem 7.4.6 (Normalization for NPB$[l]$) *All proofs in* NPB$[l]$ *are normalizable. More precisely, if a proof P in* NPB$[l]$ *is given, then there is a normal proof Q such that* $\mathrm{oa}(Q) = \mathrm{oa}(P)$ *and* $\mathrm{end}(Q) = \mathrm{end}(P)$.

Proof. Suppose $\mathrm{oa}(P) = \Gamma$ and $\mathrm{end}(P) = \{\beta\}$. By Theorem 7.4.5 (1), the sequent $\Gamma \Rightarrow \beta$ is provable in PB$[l]$. By Theorem 7.3.5, the sequent $\Gamma \Rightarrow \beta$ is also provable in PB$[l]$ − (cut). Then, by Theorem 7.4.5 (2), there is a normal proof Q in NPB$[l]$ such that $\mathrm{oa}(Q) = \mathrm{oa}(P)$ and $\mathrm{end}(Q) = \mathrm{end}(P)$. ∎

Next, we discuss NIB$[l]$. The proof of the equivalence and normalization theorems are omitted.

Definition 7.4.7 (NIB[l]) NIB[l] *is obtained from* NPB[l] *by deleting the inference rules concerning* $\sim$ *and adding the inference rule of the form:*

$$\frac{\mathrm{X}^i\bot}{\alpha}\ (\bot E).$$

The notions of proof, reduction, etc. are defined similarly as for NPB[l].

Theorem 7.4.8 (Equivalence between NIB[l] **and** IB[l]**)** *We have the following.*

1. *If P is a proof in* NIB[l] *such that* $\mathrm{oa}(P) = \Gamma$ *and* $\mathrm{end}(P) = \{\beta\}$, *then the sequent* $\Gamma \Rightarrow \beta$ *is provable in* IB[l].

2. *If a sequent* $\Gamma \Rightarrow \beta$ *is provable in* IB[l] $-$ (cut), *then there is a proof* Q *in* NIB[l] *which satisfies the following conditions:* (a) $\mathrm{oa}(Q) = \Gamma$, (b) $\mathrm{end}(Q) = \{\beta\}$, *and* (c) Q *is normal.*

Theorem 7.4.9 (Normalization for NIB[l]**)** *All proofs in* NIB[l] *are normalizable. More precisely, if a proof P in* NIB[l] *is given, then there is a normal proof Q such that* $\mathrm{oa}(Q) = \mathrm{oa}(P)$ *and* $\mathrm{end}(Q) = \mathrm{end}(P)$.

7.5. Display calculus

7.5.1. Non-paraconsistent case

In this subsection, we present a display sequent calculus δIB[l] for IB[l]. In comparison to the sequent calculus from Section 7.2, δIB[l] has some advantages from a philosophical point of view, see also [23, 67, 199]. In particular, if the introduction rules of a sequent calculus are viewed as meaning assignments, then the sequent calculus from Section 7.2 is *holistic* in the sense that it assigns a meaning to the operators X^i only in combination with *each* of the other object language connectives. By suitably generalizing the notion of a sequent and exploiting the fact that (i) $\wedge$ and $\rightarrow$, (ii) G and P ("sometimes in the past"), (iii) H ("always in the past") and F, and (iv) X^i and E^i ("i steps earlier") form residuated pairs, it is possible to state introduction rules for the connectives in such a way that every operation is introduced as the main connective of a single-antecedent (single-succedent) conclusion sequent. Moreover, the right and left introduction rules exhibit only one occurrence of the operation and no occurrence of another connective from the object language. Furthermore, the interpretation of some *structural* connectives in the display calculus as backward-looking temporal operators in either antecedent or succedent position allows one to add introduction rules with the just mentioned property also for the backward-looking modalities. Certain properties of the assumed temporal order such as the boundedness of the time domain can then be expressed by purely structural sequent rules not exhibiting any operations of the logical object language.

In ordinary sequent calculi, the comma, ",", may be seen as a context-sensitive structural connective. It is to be understood as conjunction in antecedent position and as disjunction in succedent position of a sequent. In δIB$[l]$ we shall use one binary operation and certain unary operations as structural connectives. A sequent is an expression of the shape $\Delta \Rightarrow \Gamma$, where Δ and Γ are structures (or "Gentzen terms"). We assume the empty structure **I**, and the set of structures is inductively defined from a set *Atom* of atomic formulas as follows:

$$\begin{array}{rl} \text{formulas:} & \alpha \in \mathit{Form(Atom)} \\ \text{structures} & \Delta \in \mathit{Struc(Form)} \\ \Delta ::= & A \mid \mathbf{I} \mid (\Delta;\Delta) \mid \lhd\Delta \mid \rhd\Delta \mid \ltimes\Delta \mid \rtimes\Delta. \end{array}$$

In antecedent position, ; is to be interpreted as conjunction and in succedent position as implication. In antecedent position, $\lhd$ is to be read as P and in succedent position as G, whereas $\rhd$ is to be understood as F in antecedent position and as H in succedent position. The structure $\ltimes^i\Delta$ for any $i \in \omega$ is inductively defined by $(\ltimes^0\Delta := \Delta)$ and $(\ltimes^{n+1}\Delta := \ltimes^n \ltimes \Delta)$. Similarly, $(\rtimes^0\Delta := \Delta)$ and $(\rtimes^{n+1}\Delta := \rtimes^n \rtimes \Delta)$. In succedent position $\ltimes^i$ means E^i and in antecedent position it means X^i. In succedent position $\rtimes^i$ means X^i and in antecedent position it means E^i.

The suggested interpretation of the structural connectives justifies a number of "display postulates" (*dp*) (we omit outer brackets):[4]

$$\frac{\Gamma \Rightarrow \Delta;\Sigma}{\dfrac{\Delta;\Gamma \Rightarrow \Sigma}{\Delta \Rightarrow \Gamma;\Sigma}} \qquad \frac{\Delta \Rightarrow \Gamma;\Sigma}{\dfrac{\Delta;\Gamma \Rightarrow \Sigma}{\Gamma \Rightarrow \Delta;\Sigma}}$$

$$\frac{\lhd\Delta \Rightarrow \Gamma}{\Delta \Rightarrow \lhd\Gamma} \qquad \frac{\Delta \Rightarrow \lhd\Gamma}{\lhd\Delta \Rightarrow \Gamma} \qquad \frac{\rhd\Delta \Rightarrow \Gamma}{\Delta \Rightarrow \rhd\Gamma} \qquad \frac{\Delta \Rightarrow \rhd\Gamma}{\rhd\Delta \Rightarrow \Gamma}$$

$$\frac{\ltimes^i\Delta \Rightarrow \Gamma}{\Delta \Rightarrow \ltimes^i\Gamma} \qquad \frac{\Delta \Rightarrow \ltimes^i\Gamma}{\ltimes^i\Delta \Rightarrow \Gamma} \qquad \frac{\rtimes^i\Delta \Rightarrow \Gamma}{\Delta \Rightarrow \rtimes^i\Gamma} \qquad \frac{\Delta \Rightarrow \rtimes^i\Gamma}{\rtimes^i\Delta \Rightarrow \Gamma}$$

$$\frac{\Delta \Rightarrow \Gamma}{\Delta \Rightarrow \ltimes^i \rtimes^i \Gamma} \qquad \frac{\Delta \Rightarrow \ltimes^i \rtimes^i \Gamma}{\Delta \Rightarrow \Gamma} \qquad \frac{\Delta \Rightarrow \Gamma}{\Delta \Rightarrow \rtimes^i \ltimes^i \Gamma} \qquad \frac{\Delta \Rightarrow \rtimes^i \ltimes^i \Gamma}{\Delta \Rightarrow \Gamma}$$

Moreover, we assume initial sequents $p \Rightarrow p$, a cut-rule:

[4] Note that the following display postulates are derivable:

$$\frac{\Delta \Rightarrow \alpha \quad \alpha \Rightarrow \Gamma}{\Delta \Rightarrow \Gamma}\ (cut)$$

rules which govern the empty structure:

$$\begin{array}{c} \Delta; \mathbf{I} \Rightarrow \Gamma \\ \hline \Delta \Rightarrow \Gamma \\ \hline \mathbf{I}; \Delta \Rightarrow \Gamma \end{array} \qquad \begin{array}{c} \mathbf{I}; \Delta \Rightarrow \Gamma \\ \hline \Delta \Rightarrow \Gamma \\ \hline \Delta; \mathbf{I} \Rightarrow \Gamma \end{array}$$

and versions of the standard left structural rules from ordinary Gentzen calculi, weakening, exchange, and contraction, together with associativity:

$$\frac{\Delta \Rightarrow \Gamma}{\Delta; \Sigma \Rightarrow \Gamma}\ (lm) \qquad \frac{\Delta; \Sigma \Rightarrow \Gamma}{\Sigma; \Delta \Rightarrow \Gamma}\ (le)$$

$$\frac{\Delta; \Delta \Rightarrow \Gamma}{\Delta \Rightarrow \Gamma}\ (lc) \qquad \frac{(\Delta; \Gamma); \Sigma \Rightarrow \Theta}{\Delta; (\Gamma; \Sigma) \Rightarrow \Theta}\ (la)$$

We also assume further structural rules, for any $k \in \omega_l$, to express the boundedness of the temporal order (rules (b), (b')), to capture the interaction between the temporal operators (rules (lg) – (rf)), and to capture part of the interaction between X^i and $\to$ (rule $\star$):

$$\frac{\Delta \Rightarrow \ltimes^i \Gamma; \rtimes^i \Sigma}{\Delta \Rightarrow \rtimes^i (\Gamma; \Sigma)}\ \star$$

$$\frac{\Delta \Rightarrow \Gamma}{\Delta \Rightarrow \ltimes^{l+m} \rtimes^l \Gamma}\ (b) \qquad \frac{\Delta \Rightarrow \Gamma}{\rtimes^{l+m} \ltimes^l \Delta \Rightarrow \Gamma}\ (b')$$

$$\frac{\Delta \Rightarrow \triangleleft \Gamma}{\Delta \Rightarrow \ltimes^i \rtimes^{i+k} \Gamma}\ (lg) \qquad \frac{\{\Delta \Rightarrow \rtimes^{i+j} \Gamma\}\ j \in \omega_l}{\triangleleft \rtimes^i \Delta \Rightarrow \Gamma}\ (rg)$$

$$\frac{\{\ltimes^{i+j} \Delta \Rightarrow \Gamma\}\ j \in \omega_l}{\Delta \Rightarrow \triangleright \ltimes^i \Gamma}\ (lf) \qquad \frac{\triangleright \Delta \Rightarrow \Gamma}{\rtimes^i \ltimes^{i+k} \Delta \Rightarrow \Gamma}\ (rf)$$

Definition 7.5.1 *The display sequent calculus* δIB$[l]$ *consists of the above sequent rules together with the following right and left introduction rules:*[5]

$$\frac{}{\bot \Rightarrow \Delta}\ (\bot\text{left})$$

$$\frac{\Delta \Rightarrow \alpha \quad \Gamma \Rightarrow \beta}{\Delta; \Gamma \Rightarrow (\alpha \wedge \beta)}\ (\wedge\text{right}) \qquad \frac{\alpha; \beta \Rightarrow \Delta}{(\alpha \wedge \beta) \Rightarrow \Delta}\ (\wedge\text{left})$$

$$\frac{\Delta \Rightarrow \Gamma}{\ltimes^i \rtimes^i \Delta \Rightarrow \Gamma} \qquad \frac{\ltimes^i \rtimes^i \Delta \Rightarrow \Gamma}{\Delta \Rightarrow \Gamma} \qquad \frac{\Delta \Rightarrow \Gamma}{\rtimes^i \ltimes^i \Delta \Rightarrow \Gamma} \qquad \frac{\rtimes^i \ltimes^i \Delta \Rightarrow \Gamma}{\Delta \Rightarrow \Gamma}$$

[5] We use the same names for the logical inference rules as in the standard-style IB$[l]$.

$$\frac{\Delta \Rightarrow \alpha}{\Delta \Rightarrow (\alpha \vee \beta)} \quad \frac{\Delta \Rightarrow \beta}{\Delta \Rightarrow (\alpha \vee \beta)} \ (\vee\text{right}) \qquad \frac{\alpha \Rightarrow \Delta \quad \beta \Rightarrow \Delta}{(\alpha \vee \beta) \Rightarrow \Delta} \ (\vee\text{left})$$

$$\frac{\Delta \Rightarrow \alpha; \beta}{\Delta \Rightarrow (\alpha \to \beta)} \ (\to \text{right}) \qquad \frac{\Delta \Rightarrow \alpha \quad \beta \Rightarrow \Gamma}{(\alpha \to \beta) \Rightarrow \Delta; \Gamma} \ (\to \text{left})$$

$$\frac{\lhd\Delta \Rightarrow \alpha}{\Delta \Rightarrow \mathrm{G}\alpha} \ (\text{Gright}) \qquad \frac{\alpha \Rightarrow \Delta}{\mathrm{G}\alpha \Rightarrow \lhd\Delta} \ (\text{Gleft})$$

$$\frac{\Delta \Rightarrow \alpha}{\rhd\Delta \Rightarrow \mathrm{F}\alpha} \ (\text{Fright}) \qquad \frac{\alpha \Rightarrow \rhd\Delta}{\mathrm{F}\alpha \Rightarrow \Delta} \ (\text{Fleft})$$

$$\frac{\rtimes^i\Delta \Rightarrow \alpha}{\Delta \Rightarrow \mathrm{X}^i\alpha} \ (\mathrm{X}^i\text{right}) \qquad \frac{\alpha \Rightarrow \ltimes^i\Delta}{\mathrm{X}^i\alpha \Rightarrow \Delta} \ (\mathrm{X}^i\text{left})$$

Proposition 7.5.2 *In* $\delta\mathrm{IB}[l]$, $\alpha \Rightarrow \alpha$ *is provable for every formula* α.

Proof. By induction on α. ∎

We take up the earlier example of a natural deduction proof and present a proof in $\delta\mathrm{IB}[l]$ of the temporal induction axiom for the case $l = 1$. We first present a proof Π_1 of $\mathrm{G}(\alpha \to \mathrm{X}\alpha) \Rightarrow \mathrm{X}(\alpha \to \mathrm{X}\alpha)$.

$$\begin{array}{c}
\alpha \to \mathrm{X}\alpha \Rightarrow \alpha \to \mathrm{X}\alpha \\ \hline
\mathrm{G}(\alpha \to \mathrm{X}\alpha) \Rightarrow \lhd(\alpha \to \mathrm{X}\alpha) \\ \hline
\mathrm{G}(\alpha \to \mathrm{X}\alpha) \Rightarrow \rtimes(\alpha \to \mathrm{X}\alpha) \\ \hline
\rtimes\mathrm{G}(\alpha \to \mathrm{X}\alpha) \Rightarrow \alpha \to \mathrm{X}\alpha \\ \hline
\mathrm{G}(\alpha \to \mathrm{X}\alpha) \Rightarrow \mathrm{X}(\alpha \to \mathrm{X}\alpha)
\end{array} \ (lg)$$

Next, we give a proof Π_2 of $\alpha; \mathrm{X}(\alpha \to \mathrm{X}\alpha) \Rightarrow \mathrm{X}\alpha$.

$$\begin{array}{c}
\alpha \Rightarrow \alpha \quad \mathrm{X}\alpha \Rightarrow \mathrm{X}\alpha \\ \hline
\alpha \to \mathrm{X}\alpha \Rightarrow \alpha; \mathrm{X}\alpha \\ \hline
\alpha \to \mathrm{X}\alpha \Rightarrow \ltimes \ltimes \rtimes(\alpha; \mathrm{X}\alpha) \\ \hline
\mathrm{X}(\alpha \to \mathrm{X}\alpha) \Rightarrow \ltimes \rtimes (\alpha; \mathrm{X}\alpha) \\ \hline
\mathrm{X}(\alpha \to \mathrm{X}\alpha) \Rightarrow \alpha; \mathrm{X}\alpha \\ \hline
\alpha; \mathrm{X}(\alpha \to \mathrm{X}\alpha) \Rightarrow \mathrm{X}\alpha
\end{array} \ (b)$$

We combine Π_1 and Π_2 to obtain a proof Π_3 of $\alpha \wedge \mathrm{G}(\alpha \to \mathrm{X}\alpha) \Rightarrow \mathrm{X}\alpha$.

$$\dfrac{\dfrac{\Pi_1 \qquad \dfrac{\dfrac{\dfrac{\alpha \Rightarrow \alpha \quad \mathrm{X}(\alpha \to \mathrm{X}\alpha) \Rightarrow \mathrm{X}(\alpha \to \mathrm{X}\alpha)}{\alpha; \mathrm{X}(\alpha \to \mathrm{X}\alpha) \Rightarrow \alpha \wedge \mathrm{X}(\alpha \to \mathrm{X}\alpha)} \quad \dfrac{\Pi_2}{\alpha \wedge \mathrm{X}(\alpha \to \mathrm{X}\alpha) \Rightarrow \mathrm{X}\alpha}}{\alpha; \mathrm{X}(\alpha \to \mathrm{X}\alpha)) \Rightarrow \mathrm{X}\alpha}}{\mathrm{X}(\alpha \to \mathrm{X}\alpha) \Rightarrow \alpha; \mathrm{X}\alpha}}{\dfrac{\mathrm{G}(\alpha \to \mathrm{X}\alpha) \Rightarrow \alpha; \mathrm{X}\alpha}{\alpha; \mathrm{G}(\alpha \to \mathrm{X}\alpha) \Rightarrow \mathrm{X}\alpha}}}{\alpha \wedge \mathrm{G}(\alpha \to \mathrm{X}\alpha) \Rightarrow \mathrm{X}\alpha}$$

We can now use Π_3 in a proof of the induction axiom (for the case $l = 1$.).

$$\dfrac{\dfrac{\dfrac{\dfrac{\alpha \Rightarrow \alpha}{\alpha; \mathrm{G}(\alpha{\to}\mathrm{X}\alpha) \Rightarrow \alpha}}{\alpha \wedge \mathrm{G}(\alpha{\to}\mathrm{X}\alpha) \Rightarrow \alpha} \qquad \dfrac{\Pi_3 \quad \dfrac{\dfrac{\alpha \Rightarrow \alpha}{\alpha \Rightarrow \ltimes \rtimes \alpha}}{\mathrm{X}\alpha \Rightarrow \rtimes\alpha}}{\alpha \wedge \mathrm{G}(\alpha{\to}\mathrm{X}\alpha) \Rightarrow \rtimes\alpha}}{\lhd(\alpha \wedge \mathrm{G}(\alpha{\to}\mathrm{X}\alpha)) \Rightarrow \alpha}\,(rg)}{\alpha \wedge \mathrm{G}(\alpha{\to}\mathrm{X}\alpha) \Rightarrow \mathrm{G}\alpha}$$

If two sequents are interderivable by means of the display postulates, the sequents are said to be *display equivalent*. In display logic, any substructure of a given sequent s may be displayed as either the entire antecedent or the entire succedent of a display equivalent sequent s'. In order to state this claim precisely, we need the notions of antecedent parts and succedent parts of a sequent. A succedent part of a sequent $\Delta \Rightarrow \Gamma$ is a certain occurrence of a substructure of Γ. Suppose Σ occurs as a substructure of Γ. Then this occurrence of Σ is said to be a succedent part of the sequent $\Delta \Rightarrow \Gamma$ iff

1. $\Gamma \equiv \Sigma$, or
2. $\Gamma \equiv \Gamma_1; \Gamma_2$ and Σ is a succedent part of $\Delta; \Gamma_1 \Rightarrow \Gamma_2$, or
3. $\Gamma \equiv \sharp\Gamma_1$, and Σ is a succedent part of $\sharp\Delta \Rightarrow \Gamma_1$, $\sharp \in \{\rhd, \lhd, \rtimes^i, \ltimes^i\}$.

An antecedent part of $s \equiv \Delta \Rightarrow \Gamma$ is either an occurrence of a substructure of Δ or an occurrence of a substructure of Γ that is not a succedent part of s.

Theorem 7.5.3 (Display property for δIB$[l]$, Belnap [23]) *For every sequent s and every antecedent part (succedent part) Δ of s there exists a sequent s' display equivalent to s such that Δ is the entire antecedent (succedent) of s'.*[6]

Theorem 7.5.4 (Cut-elimination for δIB[l], Belnap [23]) *Every proof of a sequent $\Delta \Rightarrow \Gamma$ in δIB[l], can be converted into a cut-free proof of $\Delta \Rightarrow \Gamma$ in δIB[l].*

Proof. This follows from Belnap's cut-elimination theorem for properly displayable logics. The calculus δIB[l] satisfies Belnap's conditions (C1) – (C8) and hence is properly displayable. ∎

The context-sensitive reading of the structural connectives is made explicit by the following translation from sequents into formulas, where $\top$ is defined as $\alpha \to \alpha$ for some fixed atomic formula α.

Definition 7.5.5 *If $\Delta \Rightarrow \Gamma$ is a sequent then its translation $\tau(\Delta \Rightarrow \Gamma)$ is the formula $\tau_1(\Delta) \to \tau_2(\Gamma)$, where the translations τ_1 and τ_2 from structures into formulas are inductively defines as follows:*

(1) *If Σ is a formula α, then $\tau_1(\Sigma) \equiv \tau_2(\Sigma) := \alpha$.*

(2) *If $\Sigma \equiv \mathbf{I}$, then $\tau_1(\Sigma) := \top$; $\tau_2(\Sigma) := \bot$.*

(3) $\tau_1(\Sigma;\Theta) := \tau_1(\Sigma) \wedge \tau_1(\Theta)$; $\tau_2(\Sigma;\Theta) := \tau_1(\Sigma) \to \tau_2(\Theta)$.

(4) $\tau_1(\lhd\Sigma) := \mathrm{P}\tau_1(\Sigma)$; $\tau_2(\lhd\Sigma) := \mathrm{G}\tau_2(\Sigma)$.

(5) $\tau_1(\rhd\Sigma) := \mathrm{H}\tau_1(\Sigma)$; $\tau_2(\rhd\Sigma) := \mathrm{F}\tau_2(\Sigma)$.

(6) $\tau_1(\ltimes^i\Sigma) := \mathrm{X}^i\tau_1(\Sigma)$; $\tau_2(\ltimes^i\Sigma) := \mathrm{E}^i\tau_2(\Sigma)$.

(7) $\tau_1(\rtimes^i\Sigma) := \mathrm{E}^i\tau_1(\Sigma)$; $\tau_2(\rtimes^i\Sigma) := \mathrm{X}^i\tau_2(\Sigma)$.

Theorem 7.5.6 (Soundness of δIB[l]) *If δIB[l] $\vdash \Delta \Rightarrow \Gamma$, then $\tau(\Delta \Rightarrow \Gamma)$ is valid in the class of all Kripke frames.*

Proof. The evaluation conditions for formulas $\mathrm{E}\alpha$ are: $(x,i) \models \mathrm{E}\alpha$ iff $(x,i-1) \models \alpha$ if $i > 0$ and otherwise $(x,0) \models \alpha$. For formulas $\mathrm{P}\alpha$ we have: $(x,i) \models \mathrm{P}\alpha$ iff $\exists j \in N_l$ [$j \leq i$ and $(x,j) \models \alpha$] if $i > 0$ and otherwise $(x,0) \models \alpha$. By induction on proofs in δIB[l], it can be shown that if δIB[l] $\vdash \Delta \Rightarrow \Gamma$, then for every Kripke frame and valuation $\models$, $\tau(\Delta \Rightarrow \Gamma)$ is true at every state x and every moment i. Thus, $\forall x$ $(x,0) \models \tau(\Delta \Rightarrow \Gamma)$. ∎

If $\Delta \equiv \langle \alpha_1, \alpha_2, ..., \alpha_n \rangle$ $(1 \leq n)$, let $\Delta^\wedge$ and $\Delta^\vee$ stand for $\alpha_1 \wedge \alpha_2 \wedge \cdots \wedge \alpha_n$ and $\alpha_1 \vee \alpha_2 \vee \cdots \vee \alpha_n$, respectively. If Δ is the empty sequence, let $\Delta^\wedge$ and $\Delta^\vee$ stand for $\top$ and $\bot$, respectively.

6 An elegant method for proving the display property can be found in [168].

Theorem 7.5.7 (Completeness of δIB[l]) *If the sequent $\Delta \Rightarrow \Gamma$ is provable in the sequent calculus for* IB[l] *from Section 7.2, then $\Delta^\wedge \Rightarrow \Gamma^\vee$ is provable in δIB[l].*

Proof. By induction on proofs in the standard sequent system for IB[l]. Initial sequents: By Proposition 7.5.2 and

$$\dfrac{\bot \Rightarrow \ltimes^i \bot}{X^i \bot \Rightarrow \bot}$$

(cut); the cases of the other structural rules are obvious:

$$\dfrac{\dfrac{\Gamma\wedge \Rightarrow \alpha \qquad \dfrac{\dfrac{\dfrac{\dfrac{\alpha \Rightarrow \alpha \quad \Sigma^\wedge \Rightarrow \Sigma^\wedge}{\alpha; \Sigma^\wedge \Rightarrow \alpha \wedge \Sigma^\wedge} \quad \alpha \wedge \Sigma^\wedge \Rightarrow \Delta^\vee}{\alpha; \Sigma^\wedge \Rightarrow \Delta^\vee}}{\alpha \Rightarrow \Sigma^\wedge; \Delta^\vee}}{\alpha \Rightarrow \Sigma^\wedge \rightarrow \Delta^\vee}}{\Gamma^\wedge \Rightarrow \Sigma^\wedge \rightarrow \Delta^\vee} \qquad \dfrac{\Sigma^\wedge \Rightarrow \Sigma\wedge \quad \Delta^\vee \Rightarrow \Delta^\vee}{\Sigma\wedge \rightarrow \Delta^\vee \Rightarrow \Sigma\wedge; \Delta^\vee}}{\dfrac{\dfrac{\Gamma^\wedge \Rightarrow \Sigma^\wedge; \Delta^\vee}{\Gamma^\wedge; \Sigma^\wedge \Rightarrow \Delta^\vee}}{\Gamma^\wedge \wedge \Sigma^\wedge \Rightarrow \Delta^\vee}}$$

($\rightarrow$ left): By the induction hypothesis, the sequents $\Gamma^\wedge \Rightarrow \mathrm{X}^i\alpha$ and $\mathrm{X}^i\beta \wedge \Sigma^\wedge \Rightarrow \Delta^\vee$ are provable in δIB[l]. If $\mathrm{X}^i(\alpha \rightarrow \beta) \Rightarrow \mathrm{X}^i\alpha \rightarrow \mathrm{X}^i\beta$ is provable in δIB[l], we can proceed as follows:

$$\dfrac{\dfrac{\mathrm{X}^i(\alpha \rightarrow \beta) \Rightarrow \mathrm{X}^i\alpha \rightarrow \mathrm{X}^i\beta \qquad \dfrac{\Gamma^\wedge \Rightarrow \mathrm{X}^i\alpha \qquad \dfrac{\dfrac{\dfrac{\mathrm{X}^i\beta \Rightarrow \mathrm{X}^i\beta \quad \Sigma^\wedge \Rightarrow \Sigma^\wedge}{\mathrm{X}^i\beta; \Sigma^\wedge \Rightarrow \mathrm{X}^i\beta \wedge \Sigma^\wedge} \quad \mathrm{X}^i\beta \wedge \Sigma^\wedge \Rightarrow \Delta^\vee}{\mathrm{X}^i\beta; \Sigma^\wedge \Rightarrow \Delta^\vee}}{\mathrm{X}^i\beta \Rightarrow \Sigma^\wedge; \Delta^\vee}}{\mathrm{X}^i\alpha \rightarrow \mathrm{X}^i\beta \Rightarrow \Gamma^\wedge; (\Sigma^\wedge; \Delta^\vee)}}{\mathrm{X}^i(\alpha \rightarrow \beta) \Rightarrow \Gamma^\wedge; (\Sigma^\wedge; \Delta^\vee)}}{\dfrac{\vdots}{\mathrm{X}^i(\alpha \rightarrow \beta) \wedge \Gamma^\wedge \wedge \Sigma^\wedge \Rightarrow \Delta^\vee}}$$

Thus, it remains to show that $\mathrm{X}^i(\alpha \rightarrow \beta) \Rightarrow \mathrm{X}^i\alpha \rightarrow \mathrm{X}^i\beta$ is probable in δIB[l]. To this end we will combine the following four proofs, $\Pi_1 - \Pi_4$.

Π_1:

$$\dfrac{\dfrac{\dfrac{\dfrac{\alpha \rightarrow \beta \Rightarrow \alpha \rightarrow \beta}{\alpha \rightarrow \beta \Rightarrow \ltimes^i \rtimes^i (\alpha \rightarrow \beta)}}{\mathrm{X}^i(\alpha \rightarrow \beta) \Rightarrow \rtimes^i(\alpha \rightarrow \beta)}}{\mathrm{X}^i(\alpha \rightarrow \beta); \mathrm{X}^i\alpha \Rightarrow \rtimes^i(\alpha \rightarrow \beta)}}{\rtimes^i(\mathrm{X}^i(\alpha \rightarrow \beta); \mathrm{X}^i\alpha) \Rightarrow (\alpha \rightarrow \beta)}$$

Π_2:

$$\begin{array}{c}
\alpha \Rightarrow \alpha \\ \hline
\alpha \Rightarrow \ltimes^i \rtimes^i \alpha \\ \hline
\mathrm{X}^i\alpha \Rightarrow \rtimes^i(\alpha \to \beta) \\ \hline
\mathrm{X}^i(\alpha \to \beta); \mathrm{X}^i\alpha \Rightarrow \rtimes^i\alpha \\ \hline
\rtimes^i(\mathrm{X}^i(\alpha \to \beta); \mathrm{X}^i\alpha) \Rightarrow \alpha
\end{array}$$

Π_3:

$$\begin{array}{c}
\Pi_1 \qquad\qquad \Pi_2 \\ \hline
\rtimes^i(\mathrm{X}^i(\alpha \to \beta); \mathrm{X}^i\alpha); \rtimes^i(\mathrm{X}^i(\alpha \to \beta); \mathrm{X}^i\alpha) \Rightarrow (\alpha \to \beta) \wedge \alpha \\ \hline
\rtimes^i(\mathrm{X}^i(\alpha \to \beta); \mathrm{X}^i\alpha) \Rightarrow (\alpha \to \beta) \wedge \alpha \\ \hline
\mathrm{X}^i(\alpha \to \beta); \mathrm{X}^i\alpha \Rightarrow \rtimes^i((\alpha \to \beta) \wedge \alpha)
\end{array}$$

Π_4:

$$\begin{array}{c}
\alpha \Rightarrow \alpha \qquad \beta \Rightarrow \beta \\ \hline
\alpha \to \beta \Rightarrow \alpha; \beta \\ \hline
(\alpha \to \beta); \alpha \Rightarrow \beta \\ \hline
(\alpha \to \beta) \wedge \alpha \Rightarrow \beta \\ \hline
(\alpha \to \beta) \wedge \alpha \Rightarrow \ltimes^i \rtimes^i \beta \\ \hline
\ltimes^i((\alpha \to \beta) \wedge \alpha) \Rightarrow \rtimes^i\beta
\end{array}$$

We obtain the following proof:

$$\begin{array}{c}
\Pi_3 \qquad \Pi_4 \\ \hline
(\mathrm{X}^i(\alpha \to \beta); \mathrm{X}^i\alpha) \Rightarrow \rtimes^i\beta \\ \hline
\rtimes^i(\mathrm{X}^i(\alpha \to \beta); \mathrm{X}^i\alpha) \Rightarrow \beta \\ \hline
(\mathrm{X}^i(\alpha \to \beta); \mathrm{X}^i\alpha) \Rightarrow \mathrm{X}^i\beta \\ \hline
\mathrm{X}^i(\alpha \to \beta) \Rightarrow \mathrm{X}^i\alpha; \mathrm{X}^i\beta \\ \hline
\mathrm{X}^i(\alpha \to \beta) \Rightarrow \mathrm{X}^i\alpha \to \mathrm{X}^i\beta
\end{array}$$

($\to$ right): Here the essential step is to show that $\mathrm{X}^i\alpha \to \mathrm{X}^i\beta \Rightarrow \mathrm{X}^i(\alpha \to \beta)$ is probable in $\delta\mathrm{IB}[l]$. We use the structural sequent rule $\star$:

$$\begin{array}{c}
\begin{array}{c}
\alpha \Rightarrow \alpha \\ \hline
\rtimes^i \ltimes^i \alpha \to \alpha \\ \hline
\ltimes^i\alpha \to \mathrm{X}^i\alpha
\end{array}
\qquad
\begin{array}{c}
\beta \Rightarrow \beta \\ \hline
\beta \Rightarrow \ltimes^i \rtimes^i \beta \\ \hline
\mathrm{X}^i\beta \Rightarrow \rtimes^i\alpha
\end{array} \\ \hline
\mathrm{X}^i\alpha \to \mathrm{X}^i\beta \Rightarrow \ltimes^i\alpha; \rtimes^i\beta \\ \hline
\mathrm{X}^i\alpha \to \mathrm{X}^i\beta \Rightarrow \rtimes^i(\alpha; \beta) \;\star \\ \hline
\rtimes^i(\mathrm{X}^i\alpha \to \mathrm{X}^i\beta) \Rightarrow \alpha; \beta \\ \hline
\rtimes^i(\mathrm{X}^i\alpha \to \mathrm{X}^i\beta) \Rightarrow \alpha \to \beta \\ \hline
\mathrm{X}^i\alpha \to \mathrm{X}^i\beta \Rightarrow \mathrm{X}^i(\alpha \to \beta)
\end{array}$$

($\wedge$left1, 2): The proof uses (cut), (lm), and the provability of $\ltimes^i\gamma \Rightarrow \mathrm{X}^i\gamma$.
($\wedge$right): The proof uses (cut), (lc), and the provability of $\mathrm{X}^i\gamma \Rightarrow \rtimes^i\gamma$.
($\vee$left): The proof uses (cut) and the provability of $\ltimes^i\gamma \Rightarrow \mathrm{X}^i\gamma$.
($\vee$right1, 2): The proof makes use of (cut) and the provability of $\mathrm{X}^i\gamma \Rightarrow \rtimes^i\gamma$.
(Xleft): The proof makes use of (cut) and the provability of $\mathrm{X}^{l+m}\gamma \Rightarrow \mathrm{X}^l\gamma$ by means of the boundedness rule (b).
(Xright): We may use (cut) and the provability of $\mathrm{X}^l\gamma \Rightarrow \mathrm{X}^{l+m}\gamma$ by means of the boundedness rule (b').
(Gleft): The proof makes use of (cut) and the following subproof:

$$\begin{array}{c}
\alpha \Rightarrow \alpha \\ \hline
\mathrm{G}\alpha \Rightarrow \lhd\alpha \\ \hline
\mathrm{G}\alpha \Rightarrow \ltimes^i \rtimes^{i+k} \alpha \\ \hline
\vdots \\ \hline
\mathrm{G}\alpha \Rightarrow \ltimes^i \mathrm{X}^{i+k}\alpha \\ \hline
\mathrm{X}^i\mathrm{G}\alpha \Rightarrow \mathrm{X}^{i+k}\alpha
\end{array}
\quad \begin{array}{l} (lg) \\ (dp) \end{array}$$

(Gright):

$$\begin{array}{c}
\{\Gamma^\wedge \Rightarrow \rtimes^{i+j}\alpha\}\ j \in \omega_l \\ \hline
\lhd \rtimes^i \Gamma^\wedge \Rightarrow \alpha \\ \hline
\rtimes^i\Gamma\wedge \Rightarrow \mathrm{G}\alpha \\ \hline
\Gamma^\wedge \Rightarrow \mathrm{X}^i\mathrm{G}\alpha
\end{array}
\quad (rg)$$

(Fleft):

$$\begin{array}{c}
\{\ltimes^{i+j}\alpha \Rightarrow \Gamma^\wedge; \Delta^\vee\}\ j \in \omega_l \\ \hline
\alpha \Rightarrow \rhd \rtimes^i (\Gamma^\wedge; \Delta^\vee) \\ \hline
\mathrm{F}\alpha \Rightarrow \rtimes^i(\Gamma^\wedge; \Delta^\vee) \\ \hline
\mathrm{X}^i\mathrm{F}\alpha \Rightarrow \Gamma^\wedge; \Delta^\vee \\ \hline
\mathrm{X}^i\mathrm{F}\alpha; \Gamma^\wedge \Rightarrow \Delta^\vee \\ \hline
\mathrm{X}^i\mathrm{F}\alpha \wedge \Gamma^\wedge \Rightarrow \Delta^\vee
\end{array}
\quad (lf)$$

(Fright):

$$\dfrac{\Gamma^\wedge \Rightarrow \mathrm{X}^{i+k}\alpha \qquad
\begin{array}{c}
\alpha \Rightarrow \alpha \\ \hline
\rhd\alpha \Rightarrow \mathrm{F}\alpha \\ \hline
\rtimes^i \ltimes^{i+k} \alpha \Rightarrow \mathrm{F}\alpha \\ \hline
\ltimes^{i+k}\alpha \Rightarrow \mathrm{X}^i\mathrm{F}\alpha \\ \hline
\alpha \Rightarrow \ltimes^{i+k}\mathrm{X}^i\mathrm{F}\alpha \\ \hline
\mathrm{X}^{i+k}\alpha \Rightarrow \mathrm{X}^i\mathrm{F}\alpha
\end{array}\ (rf)}{\Gamma^\wedge \Rightarrow \mathrm{X}^i\mathrm{F}\alpha}$$

∎

Due to the presence of both the forward-looking *and* the backward-looking structural connectives, introduction rules for the backward-looking counterparts of IB[l]'s temporal operators are easily available:

$$\frac{\triangleright\Delta \Rightarrow \alpha}{\Delta \Rightarrow \mathrm{H}\alpha}\ (\mathrm{Hright}) \qquad \frac{\alpha \Rightarrow \Delta}{\mathrm{H}\alpha \Rightarrow \triangleright\Delta}\ (\mathrm{Hleft})$$

$$\frac{\Delta \Rightarrow \alpha}{\triangleleft\Delta \Rightarrow \mathrm{P}\alpha}\ (\mathrm{Pright}) \qquad \frac{\alpha \Rightarrow \triangleleft\Delta}{\mathrm{P}\alpha \Rightarrow \Delta}\ (\mathrm{Pleft})$$

$$\frac{\ltimes^i\Delta \Rightarrow \alpha}{\Delta \Rightarrow \mathrm{E}^i\alpha}\ (\mathrm{E}^i\mathrm{right}) \qquad \frac{\alpha \Rightarrow \rtimes^i\Delta}{\mathrm{E}^i\alpha \Rightarrow \Delta}\ (\mathrm{E}^i\mathrm{left})$$

7.5.2. Paraconsistent case

A sound and complete display calculus δPB[l] for PB[l] can be obtained in a natural and straightforward way, see also the display calculi presented in [209] for extensions of Heyting-Brouwer logic by strong negation. Again, the inferential understanding of X^i as laid down by the introduction rules is (basically) non-holistic. It is only the meaning of $\sim$ that is specified in combination with each of the other object language connectives.

Definition 7.5.8 *The display calculus δPB[l] is obtained from δIB[l] by removing* ($\bot$left) *and adding initial sequents $\sim p \Rightarrow \sim p$ together with the following sequent rules:*

$$\frac{\Delta \Rightarrow \sim\alpha}{\Delta \Rightarrow \sim(\alpha \wedge \beta)}\ \frac{\Delta \Rightarrow \sim\beta}{\Delta \Rightarrow \sim(\alpha \wedge \beta)}\ (\sim\wedge\,\mathrm{right}) \qquad \frac{\sim\alpha \Rightarrow \Delta \quad \sim\beta \Rightarrow \Delta}{\sim(\alpha \wedge \beta) \Rightarrow \Delta}\ (\sim\wedge\,\mathrm{left})$$

$$\frac{\Delta \Rightarrow \sim\alpha \quad \Gamma \Rightarrow \sim\beta}{\Delta;\Gamma \Rightarrow \sim(\alpha \vee \beta)}\ (\sim\vee\,\mathrm{right}) \qquad \frac{\sim\alpha;\sim\beta \Rightarrow \Delta}{\sim(\alpha \vee \beta) \Rightarrow \Delta}\ (\sim\vee\,\mathrm{left})$$

$$\frac{\Delta \Rightarrow \alpha \quad \Gamma \Rightarrow \sim\beta}{\Delta;\Gamma \Rightarrow \sim(\alpha \to \beta)}\ (\sim\to\mathrm{right}) \qquad \frac{\alpha;\sim\beta \Rightarrow \Delta}{\sim(\alpha \to \beta) \Rightarrow \Delta}\ (\sim\to\mathrm{left})$$

$$\frac{\Delta \Rightarrow \sim\alpha}{\triangleright\Delta \Rightarrow \sim\mathrm{G}\alpha}\ (\sim\mathrm{Gright}) \qquad \frac{\sim\alpha \Rightarrow \triangleright\Delta}{\sim\mathrm{G}\alpha \Rightarrow \Delta}\ (\sim\mathrm{Gleft})$$

$$\frac{\triangleleft\Delta \Rightarrow \sim\alpha}{\Delta \Rightarrow \sim\mathrm{F}\alpha}\ (\sim\mathrm{Fright}) \qquad \frac{\sim\alpha \Rightarrow \Delta}{\sim\mathrm{F}\alpha \Rightarrow \triangleleft\Delta}\ (\sim\mathrm{Fleft})$$

$$\frac{\Delta \Rightarrow \rtimes^i\sim\alpha}{\Delta \Rightarrow \sim\mathrm{X}^i\alpha}\ (\sim\mathrm{X}^i\mathrm{right}) \qquad \frac{\ltimes^i\sim\alpha \Rightarrow \Delta}{\sim\mathrm{X}^i\alpha \Rightarrow \Delta}\ (\sim\mathrm{X}^i\mathrm{left})$$

Note that for every formula α, the sequent $\alpha \Rightarrow \alpha$ is provable in δPB$[l]$. Moreover, the modifications leading from δIB$[l]$ to δPB$[l]$ clearly do not spoil the display property.

Theorem 7.5.9 (Display property for δPB$[l]$) *For every sequent s and every antecedent part (succedent part) Δ of s, there exists a sequent s' display equivalent to s such that Δ is the entire antecedent (succedent) of s'.*

Theorem 7.5.10 (Soundness of δPB$[l]$) *If $\delta\mathrm{PB}[l] \vdash \Delta \Rightarrow \Gamma$, then $\tau(\Delta \Rightarrow \Gamma)$ is p-valid in the class of all Kripke models for δPB$[l]$.*

Proof. The verification conditions for formulas Pα and Eα are analogous to the intuitionistic case. The refutation conditions are: $(x,i) \models^- \mathrm{E}\alpha$ iff $(x,i-1) \models^- \alpha$ if $i > 0$ and otherwise $(x,0) \models^- \alpha$; $(x,i) \models^- \mathrm{P}\alpha$ iff $\exists j \in N_l\ [j \leq i$ and $(x,j) \models^- \alpha]$ if $i > 0$ and otherwise $(x,0) \models^- \alpha$. By induction on proofs in δPB$[l]$, it can be shown that if $\delta\mathrm{PB}[l] \vdash \Delta \Rightarrow \Gamma$, then for every Kripke frame and valuations $\models^+$ and $\models^-$, $(x,i) \models^+ \tau(\Delta \Rightarrow \Gamma)$ for every state x and moment i. ∎

Theorem 7.5.11 (Completeness of δPB$[l]$) *If the sequent $\Delta \Rightarrow \gamma$ is provable in the sequent calculus for* PB$[l]$ *from Section 7.3, then $\Delta^\wedge \Rightarrow \gamma$ is provable in δPB$[l]$.*

Proof. By induction on proofs in the standard sequent system for PB$[l]$. We consider here just the rules for strongly negated temporal operators. (${\sim}$Gleft): By the induction hypothesis, for every $j \in \omega_l$, $\mathrm{X}^{i+j}{\sim}\alpha \wedge \Gamma^\wedge \Rightarrow \gamma$ is provable. It can easily be shown that $\ltimes^{i+j}{\sim}\alpha \Rightarrow \wedge\Gamma^\wedge; \gamma$ is provable, for every $j \in \omega_l$. Then we have the following proof:

$$\begin{array}{cl}
\{\ltimes^{i+j}{\sim}\alpha \Rightarrow \wedge\Gamma^\wedge;\gamma\}\ j \in \omega_l & \\
\hline
{\sim}\alpha \Rightarrow \triangleright \ltimes^{i}(\Gamma^\wedge;\gamma) & (lf) \\
\hline
{\sim}\mathrm{G}\alpha \Rightarrow \ltimes^{i}(\Gamma^\wedge;\gamma) & \\
\hline
\mathrm{X}^i{\sim}\mathrm{G}\alpha \Rightarrow \Gamma^\wedge;\gamma & \\
\hline
\vdots & \\
\hline
\mathrm{X}^i{\sim}\mathrm{G}\alpha \wedge \Gamma^\wedge \Rightarrow \gamma &
\end{array}$$

(${\sim}$Gright): We may use (cut) and the following proof:

$$\begin{array}{cl}
{\sim}\alpha \Rightarrow {\sim}\alpha & \\
\hline
\triangleleft{\sim}\alpha \Rightarrow {\sim}\mathrm{G}\alpha & \\
\hline
{\sim}\alpha \Rightarrow \triangleleft{\sim}\mathrm{G}\alpha & \\
\hline
{\sim}\alpha \Rightarrow \ltimes^{i+k} \rtimes^{i+k} {\sim}\mathrm{G}\alpha & (lg) \\
\hline
\vdots & (dp) \\
\hline
\mathrm{X}^{i+k}{\sim}\alpha \Rightarrow \mathrm{X}^i{\sim}\mathrm{G}\alpha &
\end{array}$$

(∼Fleft): Use (*lg*). (∼Fleft): Use (*rg*). (∼Xleft); (∼Xright): We may use the rules (∼X^iright) and (∼X^ileft). ∎

Note that one may describe a so-called "full circle" through the different proof systems for IB[l] and PB[l] to show their mutual equivalence.

The proof of Belnap's [23] general cut-elimination theorem cannot be applied to δPB[l], because δPB[l] fails to satisfy his condition (C1). This condition guarantees the subformula-property as a corollary of cut-elimination and says that each formula which is a constituent of some premise of a sequent rule is a subformula of the conclusion sequent. The calculus δPB[l] satisfies, however, a negation normal form theorem. The provability of the following sequents:

$$\begin{gathered}
{\sim}{\sim}\alpha \Leftrightarrow \alpha \\
{\sim}(\alpha \wedge \beta) \Leftrightarrow {\sim}\alpha \vee {\sim}\beta \\
{\sim}(\alpha \vee \beta) \Leftrightarrow {\sim}\alpha \wedge {\sim}\beta \\
{\sim}(\alpha \to \beta) \Leftrightarrow \alpha \wedge {\sim}\beta \\
{\sim}\mathrm{G}\alpha \Leftrightarrow \mathrm{F}{\sim}\alpha \\
{\sim}\mathrm{F}\alpha \Leftrightarrow \mathrm{G}{\sim}\alpha \\
{\sim}\mathrm{X}^i\alpha \Leftrightarrow \mathrm{X}^i{\sim}\alpha
\end{gathered}$$

induces the definition of a function nnf on the set of $\mathcal{L}_{\mathrm{PB}[l]}$-formulas such that for every $\mathcal{L}_{\mathrm{PB}[l]}$-formula α, $nnf(\alpha)$ is a formula containing ∼ at most in front of atomic formulas.[7]

Proposition 7.5.12 *For every* $\alpha \in \mathcal{L}_{\mathrm{PB}[l]}$, $\delta\mathrm{PB}[l] \vdash \alpha \Rightarrow nnf(\alpha)$ *and* $\delta\mathrm{PB}[l] \vdash nnf(\alpha) \Rightarrow \alpha$.

Proof. By induction on α. ∎

If s is a sequent, let $(s)'$ be the result of replacing every $\mathcal{L}_{\mathrm{PB}[l]}$-formula α in s by $nnf(\alpha)$. If δPB[l] is restricted to formulas in negation normal form, Belnap's proof can be applied.

Theorem 7.5.13 *If* $\delta\mathrm{PB}[l] \vdash \Delta \Rightarrow \Gamma$, *then there is a cut-free proof of* $(\Delta \Rightarrow \Gamma)'$ *in* δPB[l].

Introduction rules for the strong negation of formulas Pα, Hα, and $\mathrm{E}^i\alpha$ are readily available:

[7] Note, however, that the Replacement Theorem does not hold for PB[l]. Although ${\sim}(\alpha \to \beta)$ and $(\alpha \wedge {\sim}\beta)$ are provably equivalent, $(\alpha \to \beta)$ and ${\sim}(\alpha \wedge {\sim}\beta)$ are not.

$$\frac{\Delta \Rightarrow {\sim}\alpha}{\lhd\Delta \Rightarrow {\sim}\mathrm{H}\alpha}\ ({\sim}\mathrm{Hright}) \qquad \frac{{\sim}\alpha \Rightarrow \rhd\Delta}{{\sim}\mathrm{H}\alpha \Rightarrow \Delta}\ ({\sim}\mathrm{Hleft})$$

$$\frac{\rhd\Delta \Rightarrow {\sim}\alpha}{\Delta \Rightarrow {\sim}\mathrm{P}\alpha}\ ({\sim}\mathrm{Pright}) \qquad \frac{{\sim}\alpha \Rightarrow \Delta}{{\sim}\mathrm{P}\alpha \Rightarrow \rhd\Delta}\ ({\sim}\mathrm{Pleft})$$

$$\frac{\ltimes^i\Delta \Rightarrow {\sim}\alpha}{\Delta \Rightarrow {\sim}\mathrm{E}^i\alpha}\ ({\sim}\mathrm{E}^i\mathrm{right}) \qquad \frac{{\sim}\alpha \Rightarrow \rtimes^i\Delta}{{\sim}\mathrm{E}^i\alpha \Rightarrow \Delta}\ {\sim}(\mathrm{E}^i\mathrm{left})$$

7.6. Remarks

In this section, we suggest to construct some infinite time domain versions of the proposed systems. Infinite (unbounded) versions of IB$[l]$ and PB$[l]$ can naturally be considered. Let IB$[\omega]$ and PB$[\omega]$ be obtained from IB$[l]$ and PB$[l]$, respectively, by deleting (Xleft) and (Xright), and replacing ω_l by ω. Embedding theorems of IB$[\omega]$ and PB$[\omega]$ into *infinitary* intuitionistic logic (see [80]) and *infinitary* Nelson's paraconsistent logic, respectively, can be obtained, and then the cut-elimination theorems for IB$[\omega]$ and PB$[\omega]$ may also be obtained using these embedding theorems. Also, normalizable natural deduction systems NIB$[\omega]$ and NPB$[\omega]$ can be obtained. Unbounded display calculi δIB$[\omega]$ and δPB$[\omega]$ can be defined from δIB$[l]$ and δPB$[l]$, respectively, by deleting the boundedness rules (b) and (b'), and replacing ω_l by ω. However, the Kripke completeness theorems for these systems may not be shown for the natural Kripke semantics with the following valuation conditions:

1. $(x,i) \models \mathrm{X}\alpha$ iff $(x,i+1) \models \alpha$,
2. $(x,i) \models \mathrm{G}\alpha$ iff $\forall j \in N\ [i \leq j$ implies $(x,j) \models \alpha]$,
3. $(x,i) \models \mathrm{F}\alpha$ iff $\exists j \in N\ [i \leq j$ and $(x,j) \models \alpha]$
4. (and analogously for $\models^+$),
5. $(x,i) \models^- \mathrm{X}\alpha$ iff $(x,i+1) \models^- \alpha$,
6. $(x,i) \models^- \mathrm{G}\alpha$ iff $\exists j \in N_l\ [i \leq j$ and $(x,j) \models^- \alpha]$,
7. $(x,i) \models^- \mathrm{F}\alpha$ iff $\forall j \in N_l\ [i \leq j$ implies $(x,j) \models^- \alpha]$.

The reason of the failure of the completeness proof is that in the proof of Lemmas 7.2.17 and 7.3.14, the cases for $\alpha \equiv \mathrm{G}\beta$ and $\alpha \equiv \mathrm{F}\beta$, respectively, require the following facts:

$$\vdash \mathrm{X}^i\beta, \mathrm{X}^{i+1}\beta, \mathrm{X}^{i+2}\beta, ..., \infty \Rightarrow \mathrm{X}^i\mathrm{G}\beta$$

where the antecedent is an infinite sequence of formulas, and

$$\vdash \mathrm{X}^i\mathrm{F}\beta \Rightarrow \mathrm{X}^i\beta \vee \mathrm{X}^{i+1}\beta \vee \mathrm{X}^{i+2}\beta \vee \cdots \infty$$

where the succedent is an infinite disjunction of formulas. Thus, in order to obtain completeness, the notion of a sequent should be extended to encompass infinite sequents that permit us to have infinite antecedents. Moreover, infinite disjunctions should be allowed, and the treatment of unbounded display calculi would require infinite conjunctions as well. By imposing these modifications, the completeness theorems can be obtained, but the corresponding logics may differ from IB[ω], PB[ω], δIB[ω], and δPB[ω]. Moreover, the cut-elimination theorems for the corresponding sequent calculi with these modifications cannot be shown. The reason of the failure of the cut-elimination proof is that the infinite antecedents require an infinite version of the cut rule of the form:

$$\frac{\Gamma \Rightarrow \alpha \quad \Delta \Rightarrow \Sigma}{\Gamma, \Delta^* \Rightarrow \Sigma}\ (\omega\text{−cut})$$

where Δ contains α (it can appear infinitely many times), and Δ^* is obtained from Δ by deleting all occurrences of α.

8. Paraconsistent classical temporal logics

Inconsistency-tolerant reasoning and paraconsistent logic are of growing importance not only in Knowledge Representation, AI and other areas of Computer Science, but also in Philosophical Logic. In this chapter, a new logic, paraconsistent linear-time temporal logic (PLTL), is obtained semantically from the linear-time temporal logic LTL by adding a paraconsistent negation. Some theorems for embedding PLTL into LTL are proved, and PLTL is shown to be decidable. A Gentzen-type sequent calculus PLT_ω for PLTL is introduced, and the completeness and cut-elimination theorems for this calculus are proved. In addition, a display calculus $\delta\mathrm{PLT}_\omega$ for PLTL is defined.[1]

8.1. Introduction

Linear-time temporal logic, LTL, is known to be one of the most useful temporal logics for verifying time-dependent and concurrent systems [77, 152]. On the other hand, as we have emphasized in Chapter 2 *Nelson's paraconsistent four-valued logic with strong negation*, N4, is known to be one of the most important base logics for paraconsistent or inconsistency-tolerant reasoning [6, 134, 138, 196]. Whereas LTL is an extension of *classical* logic, N4 is an extension of positive *intuitionistic* logic. In Chapter 7 bounded versions $\mathrm{IB}[l]$ and $\mathrm{PB}[l]$ of LTL based on intuitionistic logic and on N4, respectively, have been presented and investigated. The time domain of $\mathrm{IB}[l]$ and $\mathrm{PB}[l]$ is bounded by a fixed positive integer l; nevertheless, all the typical temporal axioms of LTL can be proved.

Also in Chapter 7, completeness (with respect to Kripke semantics), cut-elimination, normalization (with respect to natural deduction) and decidability theorems for $\mathrm{IB}[l]$ and $\mathrm{PB}[l]$ are proved. Moreover, unbounded linear-time versions $\mathrm{IB}[\omega]$ and $\mathrm{PB}[\omega]$ of $\mathrm{IB}[l]$ and $\mathrm{PB}[l]$, respectively, are discussed. It is explained that embedding theorems of $\mathrm{IB}[\omega]$ and $\mathrm{PB}[\omega]$ into *infinitary* intuitionistic logic (see [80]) and *infinitary* N4,

[1] This chapter is based on [102].

respectively, can be obtained and that the cut-elimination theorems for IB[ω] and PB[ω] may be proved using these embedding theorems. The completeness theorems for these systems, however, are not known to hold for what seems to be the natural Kripke semantics for these logics, cf. Chapter 7.

The aim of the present chapter is to define and study an *unbounded* linear-time temporal logic by adding to LTL a paraconsistent negation similar to the strong negation in N4,[2] such that (i) time-dependent *and* inconsistency-tolerant reasoning can be represented within a single logic, and (ii) decidability, completeness and cut-elimination are satisfied. For this aim, a new logic, *paraconsistent linear-time temporal logic* (PLTL), is obtained semantically from the semantical definition of LTL. A *semantical embedding theorem* of PLTL into LTL is proved by constructing a standard single-consequence model of LTL from a paraconsistent dual-consequence model of PLTL, and vice versa. By using this embedding theorem, PLTL is shown to be decidable. Next, a Gentzen-type sequent calculus PLT_ω for PLTL is introduced, and a *syntactical embedding theorem* of this calculus into its paraconsistent negationless subsystem is proved. By using this embedding theorem, the cut-elimination theorem for PLT_ω is shown. The completeness theorem for it is proved by combining both the semantical and syntactical embedding theorems.

From the philosophical point of view of a proof-theoretic semantics, the sequent system PLT_ω has the disadvantage of introducing both the next-time operator X and the strong negation $\sim$ only in combination with the other connectives of PLTL. As meaning assignments the rules of PLT_ω are therefore *holistic*, specifying the meaning of strong negation and X always along with the meaning of some other logical operation. We will present a display sequent calculus $\delta\mathrm{PLT}_\omega$ for PLTL that is less holistic insofar as only the rules for strong negation refer to more than one connective of PLTL.

A remarkable feature of PLTL is that PLTL has the paraconsistent negation connective $\sim$ similar to the strong negation connective in N4. The paraconsistent (strong) negation connective $\sim$, which was first introduced by Nelson in [134], has been studied by many researchers and has been applied in several non-classical logics (see, e.g., [140, 142, 196] and the references therein). One reason why $\sim$ is considered is that it may be added in such a way that the extended logics satisfy the property of *paraconsistency*. A semantic consequence relation $\models$ is called paraconsistent with respect to a negation connective $\sim$ if there are

[2] This paraconsistent negation is also similar to the negation in Dunn and Belnap's *useful four-valued logic* [7]. Whereas N4 contains intuitionistic implication, the logic PLTL defined in the present chapter contains classical implication.

formulas α, β such that not $\{\alpha, \sim\alpha\} \models \beta$. In the case of a linear-time temporal logic this means that there is a model M and a position i of a sequence $\sigma = t_0, t_1, t_2, ...$ of time-points in M with not $[(M, i) \models (\alpha \wedge \sim\alpha) \rightarrow \beta]$. It is known that logical systems with paraconsistency can deal with inconsistency-tolerant and uncertainty reasoning more appropriately than systems which are non-paraconsistent. For example, we do not desire that $(s(x) \wedge \sim s(x)) \rightarrow d(x)$ is satisfied for any symptom s and disease d where $\sim s(x)$ means "person x does not have symptom s" and $d(x)$ means "person x suffers from disease d", because there may be situations that support the truth of both $s(a)$ and $\sim s(a)$ for some individual a but do not support the truth of $d(a)$. For more information on paraconsistency see [154, 157] and the references therein.

The advantages of PLTL can then be summarized as follows. PLTL is decidable. Cut-elimination and completeness theorems hold for the proof system PLT_ω. The decidability of the validity, satisfiability and model-checking problems of PLTL is especially useful as a basis for computer science applications. For instance, by the semantical embedding theorem of PLTL into LTL, the existing LTL model-checking algorithms are applicable to PLTL. Existing application areas of both LTL and N4 can thus be merged or integrated based on PLTL, without loss of any theoretical foundations. In addition, the proposed embedding-based method can give a simple and uniform proof of the decidability, cut-elimination and completeness theorems for PLTL and PLT_ω. The proposed method is considerably simpler than other proposals such as directly extending proof methods for LTL, and may be widely applicable to other related temporal logics. Moreover, a display calculus δPLT_ω for PLTL can be defined. In this way, we also obtain a display calculus for LTL. This calculus for LTL is new.

8.2. Paraconsistent linear-time temporal logic

In this section, firstly, we present a semantical definition of LTL, and secondly, we introduce PLTL by extending LTL with a paraconsistent negation.

Formulas of LTL are constructed from countably many propositional variables, $\rightarrow$ (implication), $\wedge$ (conjunction), $\vee$ (disjunction), $\neg$ (negation), X (next), G (globally) and F (eventually). Lower-case letters $p, q, ...$ are used to denote propositional variables, and Greek lower-case letters $\alpha, \beta, ...$ are used to denote formulas. We write $A \equiv B$ to indicate the syntactical identity between A and B. The symbol ω is used to represent the set of natural numbers. Lower-case letters i, j and k are used to denote any natural numbers. The symbol $\geq$ or $\leq$ is used to represent the standard linear order on ω.

Definition 8.2.1 (LTL) *Let S be a non-empty set of states. A structure $M := (\sigma, I)$ is a* model *if*

1. *σ is an infinite sequence $s_0, s_1, s_2, ...$ of states in S,*

2. *I is a mapping from the set Φ of propositional variables to the power set of S.*

An evaluation relation $(M, i) \models \alpha$ for any formula α, where M is a model (σ, I) and i ($\in \omega$) represents some position within σ, is defined inductively by

1. *for any $p \in \Phi$, $(M, i) \models p$ iff $s_i \in I(p)$,*

2. *$(M, i) \models \alpha \wedge \beta$ iff $(M, i) \models \alpha$ and $(M, i) \models \beta$,*

3. *$(M, i) \models \alpha \vee \beta$ iff $(M, i) \models \alpha$ or $(M, i) \models \beta$,*

4. *$(M, i) \models \alpha \rightarrow \beta$ iff $(M, i) \models \alpha$ implies $(M, i) \models \beta$,*

5. *$(M, i) \models \neg\alpha$ iff not-$[(M, i) \models \alpha]$,*

6. *$(M, i) \models \mathrm{X}\alpha$ iff $(M, i+1) \models \alpha$,*

7. *$(M, i) \models \mathrm{G}\alpha$ iff $\forall j \geq i[(M, j) \models \alpha]$,*

8. *$(M, i) \models \mathrm{F}\alpha$ iff $\exists j \geq i[(M, j) \models \alpha]$.*

A formula α is satisfiable *in* LTL *iff $(M, 0) \models \alpha$ for some model $M := (\sigma, I)$. A formula α is* valid *in* LTL *iff $(M, 0) \models \alpha$ for any model $M := (\sigma, I)$.*

Formulas of PLTL are constructed from countably many propositional variables, $\rightarrow$, $\wedge$, $\vee$, $\neg$, X, G, F and $\sim$ (paraconsistent negation). The formulation of PLTL uses two evaluation relations $\models^+$ and $\models^-$. The intuitive interpretations of $\models^+$ and $\models^-$ are "verification" (or "support of truth") and "refutation" (or "falsification","support of falsity"), respectively.

Definition 8.2.2 (PLTL) *Let S be a non-empty set of states. A structure $M := (\sigma, I^+, I^-)$ is a* paraconsistent model *if*

1. *σ is an infinite sequence $s_0, s_1, s_2, ...$ of states in S,*

2. *I^+ and I^- are mappings from the set Φ of propositional variables to the power set of S.*

Evaluation relations $(M,i) \models^+ \alpha$ and $(M,i) \models^- \alpha$ for any formula α, where M is a paraconsistent model (σ, I^+, I^-) and i $(\in \omega)$ represents some position within σ, are defined inductively by

1. *for any $p \in \Phi$, $(M,i) \models^+ p$ iff $s_i \in I^+(p)$,*
2. *for any $p \in \Phi$, $(M,i) \models^- p$ iff $s_i \in I^-(p)$,*
3. $(M,i) \models^+ \alpha \wedge \beta$ *iff* $(M,i) \models^+ \alpha$ *and* $(M,i) \models^+ \beta$,
4. $(M,i) \models^+ \alpha \vee \beta$ *iff* $(M,i) \models^+ \alpha$ *or* $(M,i) \models^+ \beta$,
5. $(M,i) \models^+ \alpha \rightarrow \beta$ *iff* $(M,i) \models^+ \alpha$ *implies* $(M,i) \models^+ \beta$,
6. $(M,i) \models^+ \neg\alpha$ *iff not-*$[(M,i) \models^+ \alpha]$,
7. $(M,i) \models^+ \mathrm{X}\alpha$ *iff* $(M,i+1) \models^+ \alpha$,
8. $(M,i) \models^+ \mathrm{G}\alpha$ *iff* $\forall j \geq i[(M,j) \models^+ \alpha]$,
9. $(M,i) \models^+ \mathrm{F}\alpha$ *iff* $\exists j \geq i[(M,j) \models^+ \alpha]$,
10. $(M,i) \models^+ {\sim}\alpha$ *iff* $(M,i) \models^- \alpha$,
11. $(M,i) \models^- {\sim}\alpha$ *iff* $(M,i) \models^+ \alpha$,
12. $(M,i) \models^- \alpha \wedge \beta$ *iff* $(M,i) \models^- \alpha$ *or* $(M,i) \models^- \beta$,
13. $(M,i) \models^- \alpha \vee \beta$ *iff* $(M,i) \models^- \alpha$ *and* $(M,i) \models^- \beta$,
14. $(M,i) \models^- \alpha \rightarrow \beta$ *iff* $(M,i) \models^+ \alpha$ *and* $(M,i) \models^- \beta$,
15. $(M,i) \models^- \neg\alpha$ *iff not-*$[(M,i) \models^- \alpha]$,
16. $(M,i) \models^- \mathrm{X}\alpha$ *iff* $(M,i+1) \models^- \alpha$,
17. $(M,i) \models^- \mathrm{G}\alpha$ *iff* $\exists j \geq i[(M,j) \models^- \alpha]$,
18. $(M,i) \models^- \mathrm{F}\alpha$ *iff* $\forall j \geq i[(M,j) \models^- \alpha]$.

The falsification conditions for classical negation may be felt to be in need of some justification. Suppose that a is a person who is neither rich nor poor and that, as a matter of fact, no one is both rich and poor. Let p stand for the claim that a is poor and r for the claim that a is rich. Intuitively, a state definitely verifies p iff it falsifies r, and vice versa. Suppose now that $\neg p$ is indeed falsified at a state i in model M:

$(M,i) \models^- \neg p$. This should mean that it is verified at i that p is poor or neither poor or rich. But this is the case iff r is not verified at i, which means that p is not falsified at i.

Definition 8.2.3 *A formula α is* satisfiable *in* PLTL *iff* $(M,0) \models^+ \alpha$ *for some paraconsistent model* $M := (\sigma, I^+, I^-)$. *A formula α is* valid *in* PLTL *iff* $(M,0) \models^+ \alpha$ *for any paraconsistent model* $M := (\sigma, I^+, I^-)$.

Note that the relation $\models^+$ of PLTL includes the relation $\models$ of LTL, and hence PLTL is an extension of LTL.

An expression $\alpha \leftrightarrow \beta$ means $(\alpha \rightarrow \beta) \wedge (\beta \rightarrow \alpha)$.

Proposition 8.2.4 *The following formulas are valid in* PLTL, *for any formulas α and β:*

1. $\sim\sim\alpha \leftrightarrow \alpha$,
2. $\sim(\alpha \wedge \beta) \leftrightarrow \sim\alpha \vee \sim\beta$,
3. $\sim(\alpha \vee \beta) \leftrightarrow \sim\alpha \wedge \sim\beta$,
4. $\sim(\alpha \rightarrow \beta) \leftrightarrow \alpha \wedge \sim\beta$,
5. $\sim\neg\alpha \leftrightarrow \neg\sim\alpha$,
6. $\sim \mathrm{X}\alpha \leftrightarrow \mathrm{X}\sim\alpha$,
7. $\sim \mathrm{F}\alpha \leftrightarrow \mathrm{G}\sim\alpha$,
8. $\sim \mathrm{G}\alpha \leftrightarrow \mathrm{F}\sim\alpha$.

The operators F and G are thus duals of each other not only with respect to $\neg$ but also with respect to $\sim$, and X is a self dual not only with respect to $\neg$ but also with respect to $\sim$. The negations $\neg$ and $\sim$ are self-duals with respect to $\sim$ and $\neg$, respectively.

Definition 8.2.5 *A formula α is said to be in classical (strong) negation normal form iff α contains $\neg$ ($\sim$) only in front of propositional variables.*

In view of Definition 8.2.2 and Proposition 8.2.4 it is clear that PLTL has the negation normal form property with respect to both of its negations.

Observation 8.2.6 *For every formula α, there is a formula α^* (α^{**}) such that $\alpha \leftrightarrow \alpha^*$ ($\alpha \leftrightarrow \alpha^{**}$) is valid in* PLTL *and α^* (α^{**}) is in classical (strong) negation normal form.*

For each $i \in \omega$ and each formula α, we can take one of the following four cases: (1) α is verified at i, i.e., $(M, i) \models^+ \alpha$, (2) α is falsified at i, i.e., $(M, i) \models^- \alpha$, (3) α is both verified and falsified at i, and (4) α is neither verified nor falsified at i. Thus, PLTL may be regarded as a four-valued logic.

Assume a paraconsistent model $M = (\sigma, I^+, I^-)$ such that $s_i \in I^+(p)$, $s_i \in I^-(p)$ and $s_i \notin I^+(q)$ for a pair of distinct propositional variables p and q. Then, $(M, i) \models^+ (p \wedge {\sim}p) \rightarrow q$ does not hold, and hence $\models^+$ in PLTL is paraconsistent with respect to $\sim$.

8.3. Embedding and decidability

In the following, we introduce a translation of PLTL into LTL, and by using this translation, we show a semantical embedding theorem of PLTL into LTL. A similar translation has been used by Vorob'ev [193], Gurevich [73], and Rautenberg [164] to embed Nelson's three-valued constructive logic [6, 134] into intuitionistic logic. Although the proposed translation is a straightforward extension of the traditional translation concerning Nelson's logic, the way (proof) of showing the (semantical) embedding theorem of PLTL into LTL is a new technical contribution developed in this chapter. The semantical embedding theorem is used to show the decidability and completeness theorems for PLTL.

Definition 8.3.1 *Let Φ be a non-empty set of propositional variables and Φ' be the set $\{p' \mid p \in \Phi\}$ of propositional variables. The language $\mathcal{L}^{\sim}$ (the set of formulas) of* PLTL *is defined using Φ, $\sim$, $\neg, \rightarrow, \wedge, \vee$,* F*,* G *and* X*. The language $\mathcal{L}$ of* LTL *is obtained from $\mathcal{L}^{\sim}$ by adding Φ' and deleting $\sim$.*

A mapping f from $\mathcal{L}^{\sim}$ to $\mathcal{L}$ is defined inductively by

1. *for any $p \in \Phi$, $f(p) := p$ and $f({\sim}p) := p' \in \Phi'$,*
2. *$f(\alpha \sharp \beta) := f(\alpha) \sharp f(\beta)$ where $\sharp \in \{\wedge, \vee, \rightarrow\}$,*
3. *$f(\sharp\alpha) := \sharp f(\alpha)$ where $\sharp \in \{\neg, \mathrm{X}, \mathrm{F}, \mathrm{G}\}$,*
4. *$f({\sim}{\sim}\alpha) := f(\alpha)$,*
5. *$f({\sim}\sharp\alpha) := \sharp f({\sim}\alpha)$ where $\sharp \in \{\neg, \mathrm{X}\}$,*
6. *$f({\sim}(\alpha \wedge \beta)) := f({\sim}\alpha) \vee f({\sim}\beta)$,*
7. *$f({\sim}(\alpha \vee \beta)) := f({\sim}\alpha) \wedge f({\sim}\beta)$,*
8. *$f({\sim}(\alpha \rightarrow \beta)) := f(\alpha) \wedge f({\sim}\beta)$,*

9. $f(\sim\mathrm{F}\alpha) := \mathrm{G}f(\sim\alpha)$,
10. $f(\sim\mathrm{G}\alpha) := \mathrm{F}f(\sim\alpha)$.

Lemma 8.3.2 *Let f be the mapping defined in Definition 8.3.1, and S be a non-empty set of states. For any paraconsistent model $M := (\sigma, I^+, I^-)$ of PLTL, any evaluation relations $\models^+$ and $\models^-$ on M, and any state s_i in σ, there exist a model $N := (\sigma, I)$ of LTL and an evaluation relation $\models$ on N such that for any formula α in $\mathcal{L}^{\sim}$,*

1. $(M, i) \models^+ \alpha$ *iff* $(N, i) \models f(\alpha)$,
2. $(M, i) \models^- \alpha$ *iff* $(N, i) \models f(\sim\alpha)$.

Proof. Let Φ be a non-empty set of propositional variables and Φ' be the set $\{p' \mid p \in \Phi\}$. Suppose that M is a paraconsistent model (σ, I^+, I^-) where

I^+ and I^- are mappings from Φ to the power set of S.

Suppose that N is a model (σ, I) where

I is a mapping from $\Phi \cup \Phi'$ to the power set of S.

Suppose moreover that M and N satisfy the following conditions: for any s_i in σ and any $p \in \Phi$,

1. $s_i \in I^+(p)$ iff $s_i \in I(p)$,
2. $s_i \in I^-(p)$ iff $s_i \in I(p')$.

The lemma is then proved by (simultaneous) induction on the complexity of α.

• Base step:

Case $\alpha \equiv p \in \Phi$: For (1), we obtain: $(M, i) \models^+ p$ iff $s_i \in I^+(p)$ iff $s_i \in I(p)$ iff $(N, i) \models p$ iff $(N, i) \models f(p)$ (by the definition of f). For (2), we obtain: $(M, i) \models^- p$ iff $s_i \in I^-(p)$ iff $s_i \in I(p')$ iff $(N, i) \models p'$ iff $(N, i) \models f(\sim p)$ (by the definition of f).

• Induction step: We show some cases.

Case $\alpha \equiv \beta \wedge \gamma$: For (1), we obtain: $(M, i) \models^+ \beta \wedge \gamma$ iff $(M, i) \models^+ \beta$ and $(M, i) \models^+ \gamma$ iff $(N, i) \models f(\beta)$ and $(N, i) \models f(\gamma)$ (by induction hypothesis for 1) iff $(N, i) \models f(\beta) \wedge f(\gamma)$ iff $(N, i) \models f(\beta \wedge \gamma)$ (by the definition of f). For (2), we obtain: $(M, i) \models^- \beta \wedge \gamma$ iff $(M, i) \models^- \beta$ or $(M, i) \models^- \gamma$ iff $(N, i) \models f(\sim\beta)$ or $(N, i) \models f(\sim\gamma)$ (by induction hypothesis for 2) iff $(N, i) \models f(\sim\beta) \vee f(\sim\gamma)$ iff $(N, i) \models f(\sim(\beta \wedge \gamma))$ (by the definition of f).

Case $\alpha \equiv \beta{\rightarrow}\gamma$: For (1), we obtain: $(M,i) \models^{+} \beta{\rightarrow}\gamma$ iff $(M,i) \models^{+} \beta$ implies $(M,i) \models^{+} \gamma$ iff $(N,i) \models f(\beta)$ implies $(N,i) \models f(\gamma)$ (by induction hypothesis for 1) iff $(N,i) \models f(\beta){\rightarrow}f(\gamma)$ iff $(N,i) \models f(\beta{\rightarrow}\gamma)$ (by the definition of f). For (2), we obtain: $(M,i) \models^{-} \beta{\rightarrow}\gamma$ iff $(M,i) \models^{+} \beta$ and $(M,i) \models^{-} \gamma$ iff $(N,i) \models f(\beta)$ and $(N,i) \models f({\sim}\gamma)$ (by induction hypothesis for 1 and 2) iff $(N,i) \models f(\beta) \wedge f({\sim}\gamma)$ iff $(N,i) \models f({\sim}(\beta{\rightarrow}\gamma))$ (by the definition of f).

Case $\alpha \equiv {\sim}\beta$: For (1), we obtain: $(M,i) \models^{+} {\sim}\beta$ iff $(M,i) \models^{-} \beta$ iff $(N,i) \models f({\sim}\beta)$ (by induction hypothesis for 2). For (2), we obtain: $(M,i) \models^{-} {\sim}\beta$ iff $(M,i) \models^{+} \beta$ iff $(N,i) \models f(\beta)$ (by induction hypothesis for 1) iff $(N,i) \models f({\sim\sim}\beta)$ (by the definition of f).

Case $\alpha \equiv \mathrm{X}\beta$: For (1), we obtain: $(M,i) \models^{+} \mathrm{X}\beta$ iff $(M,i{+}1) \models^{+} \beta$ iff $(N,i+1) \models f(\beta)$ (by induction hypothesis for 1) iff $(N,i) \models \mathrm{X}f(\beta)$ iff $(N,i) \models f(\mathrm{X}\beta)$ (by the definition of f). For (2), we obtain: $(M,i) \models^{-} \mathrm{X}\beta$ iff $(M,i+1) \models^{-} \beta$ iff $(N,i+1) \models f({\sim}\beta)$ (by induction hypothesis for 2) iff $(N,i) \models \mathrm{X}f({\sim}\beta)$ iff $(N,i) \models f({\sim}\mathrm{X}\beta)$ (by the definition of f).

Case $\alpha \equiv \mathrm{G}\beta$: For (1), we obtain: $(M,i) \models^{+} \mathrm{G}\beta$ iff $\forall j \geq i[(M,j) \models^{+} \beta]$ iff $\forall j \geq i[(N,j) \models f(\beta)]$ (by induction hypothesis for 1) iff $(N,i) \models \mathrm{G}f(\beta)$ iff $(N,i) \models f(\mathrm{G}\beta)$ (by the definition of f). For (2), we obtain: $(M,i) \models^{-} \mathrm{G}\beta$ iff $\exists j \geq i[(M,j) \models^{-} \beta]$ iff $\exists j \geq i[(N,j) \models f({\sim}\beta)]$ (by induction hypothesis for 2) iff $(N,i) \models \mathrm{F}f({\sim}\beta)$ iff $(N,i) \models f({\sim}\mathrm{G}\beta))$ (by the definition of f). ∎

Lemma 8.3.3 *Let f be the mapping defined in Definition 8.3.1, and S be a non-empty set of states. For any model $N := (\sigma, I)$ of* LTL, *any evaluation relation $\models$ on N, and any state s_i in σ, there exist a paraconsistent model $M := (\sigma, I^{+}, I^{-})$ of* PLTL *and evaluation relations $\models^{+}$ and $\models^{-}$ on M such that for any formula α in $\mathcal{L}^{\sim}$,*

1. $(N,i) \models f(\alpha)$ *iff* $(M,i) \models^{+} \alpha$,
2. $(N,i) \models f({\sim}\alpha)$ *iff* $(M,i) \models^{-} \alpha$.

Proof. Similar to the proof of Lemma 8.3.2. ∎

Theorem 8.3.4 (Semantical embedding) *Let f be the mapping defined in Definition 8.3.1. For any formula α, α is valid in* PLTL *iff $f(\alpha)$ is valid in* LTL.

Proof. By Lemmas 8.3.2 and 8.3.3. ∎

Theorem 8.3.5 (Decidability) PLTL *is decidable.*

Proof. By decidability of LTL, for each α, it is possible to decide if $f(\alpha)$ is valid in LTL. Then, by Theorem 8.3.4, PLTL is decidable. ∎

Theorem 8.3.5 shows that the validity problem of PLTL is decidable. Similarly, we can also show that both the satisfiability and model checking problems of PLTL are decidable.

8.4. Cut-elimination and completeness

Greek capital letters $\Gamma, \Delta, ...$ are used to represent finite (possibly empty) sets of formulas, and $\mathrm{X}\Gamma$ is used to denote the set $\{\mathrm{X}\gamma \mid \gamma \in \Gamma\}$. The expression $\mathrm{X}^i\alpha$ for any $i \in \omega$ is defined inductively by $\mathrm{X}^0\alpha \equiv \alpha$ and $\mathrm{X}^{n+1}\alpha \equiv \mathrm{X}^n\mathrm{X}\alpha$. An expression of the form $\Gamma \Rightarrow \Delta$ is called a *sequent*. An expression $L \vdash S$ is used to denote the fact that a sequent S is provable in a sequent calculus L.

Kawai's sequent calculus LT_ω [108] for LTL is presented below, cf. also [91].

Definition 8.4.1 ($\mathbf{LT}_\omega$) *The initial sequents of* LT_ω *are of the following form, for any propositional variable* p:

$$\mathrm{X}^i p \Rightarrow \mathrm{X}^i p.$$

The structural rules of LT_ω *are of the form:*

$$\frac{\Gamma \Rightarrow \Delta, \alpha \quad \alpha, \Sigma \Rightarrow \Pi}{\Gamma, \Sigma \Rightarrow \Delta, \Pi}\ (\text{cut}) \quad \frac{\Gamma \Rightarrow \Delta}{\alpha, \Gamma \Rightarrow \Delta}\ (\text{we-left}) \quad \frac{\Gamma \Rightarrow \Delta}{\Gamma \Rightarrow \Delta, \alpha}\ (\text{we-right}).$$

The logical inference rules of LT_ω *are of the form:*

$$\frac{\Gamma \Rightarrow \Sigma, \mathrm{X}^i\alpha \quad \mathrm{X}^i\beta, \Delta \Rightarrow \Pi}{\mathrm{X}^i(\alpha{\to}\beta), \Gamma, \Delta \Rightarrow \Sigma, \Pi}\ ({\to}\text{left}) \quad \frac{\mathrm{X}^i\alpha, \Gamma \Rightarrow \Delta, \mathrm{X}^i\beta}{\Gamma \Rightarrow \Delta, \mathrm{X}^i(\alpha{\to}\beta)}\ ({\to}\text{right})$$

$$\frac{\mathrm{X}^i\alpha, \Gamma \Rightarrow \Delta}{\mathrm{X}^i(\alpha \wedge \beta), \Gamma \Rightarrow \Delta}\ (\wedge\text{left1}) \quad \frac{\mathrm{X}^i\beta, \Gamma \Rightarrow \Delta}{\mathrm{X}^i(\alpha \wedge \beta), \Gamma \Rightarrow \Delta}\ (\wedge\text{left2})$$

$$\frac{\Gamma \Rightarrow \Delta, \mathrm{X}^i\alpha \quad \Gamma \Rightarrow \Delta, \mathrm{X}^i\beta}{\Gamma \Rightarrow \Delta, \mathrm{X}^i(\alpha \wedge \beta)}\ (\wedge\text{right}) \quad \frac{\mathrm{X}^i\alpha, \Gamma \Rightarrow \Delta \quad \mathrm{X}^i\beta, \Gamma \Rightarrow \Delta}{\mathrm{X}^i(\alpha \vee \beta), \Gamma \Rightarrow \Delta}\ (\vee\text{left})$$

$$\frac{\Gamma \Rightarrow \Delta, \mathrm{X}^i\alpha}{\Gamma \Rightarrow \Delta, \mathrm{X}^i(\alpha \vee \beta)}\ (\vee\text{right1}) \quad \frac{\Gamma \Rightarrow \Delta, \mathrm{X}^i\beta}{\Gamma \Rightarrow \Delta, \mathrm{X}^i(\alpha \vee \beta)}\ (\vee\text{right2})$$

$$\frac{\Gamma \Rightarrow \Delta, \mathrm{X}^i\alpha}{\mathrm{X}^i\neg\alpha, \Gamma \Rightarrow \Delta}\ (\neg\text{left}) \quad \frac{\mathrm{X}^i\alpha, \Gamma \Rightarrow \Delta}{\Gamma \Rightarrow \Delta, \mathrm{X}^i\neg\alpha}\ (\neg\text{right})$$

$$\frac{X^{i+k}\alpha, \Gamma \Rightarrow \Delta}{X^i G\alpha, \Gamma \Rightarrow \Delta} \text{ (Gleft)} \quad \frac{\{\ \Gamma \Rightarrow \Delta, X^{i+j}\alpha\ \}_{j\in\omega}}{\Gamma \Rightarrow \Delta, X^i G\alpha} \text{ (Gright)}$$

$$\frac{\{\ X^{i+j}\alpha, \Gamma \Rightarrow \Delta\ \}_{j\in\omega}}{X^i F\alpha, \Gamma \Rightarrow \Delta} \text{ (Fleft)} \quad \frac{\Gamma \Rightarrow \Delta, X^{i+k}\alpha}{\Gamma \Rightarrow \Delta, X^i F\alpha} \text{ (Fright).}$$

Note that (Gright) and (Fleft) have infinitely many premises. The sequents of the form: $X^i\alpha \Rightarrow X^i\alpha$ for any formula α are provable in cut-free LT_ω. This fact can be proved by induction on the complexity of α. The cut-elimination and completeness theorems for LT_ω were proved by Kawai in [108].

We now introduce a sequent calculus PLT_ω for PLTL.

Definition 8.4.2 ($\mathbf{PLT}_\omega$) PLT_ω *is obtained from* LT_ω *by adding the initial sequents of the following form, for any propositional variable* p*:*

$$X^i{\sim}p \Rightarrow X^i{\sim}p,$$

and adding the logical inference rules of the form:

$$\frac{X^i\alpha, \Gamma \Rightarrow \Delta}{X^i{\sim}{\sim}\alpha, \Gamma \Rightarrow \Delta} \text{ (}{\sim}{\sim}\text{left)} \quad \frac{\Gamma \Rightarrow \Delta, X^i\alpha}{\Gamma \Rightarrow \Delta, X^i{\sim}{\sim}\alpha} \text{ (}{\sim}{\sim}\text{right)}$$

$$\frac{X^i\alpha, \Gamma \Rightarrow \Delta}{X^i{\sim}(\alpha{\rightarrow}\beta), \Gamma \Rightarrow \Delta} \text{ (}{\sim}{\rightarrow}\text{left1)} \quad \frac{X^i{\sim}\beta, \Gamma \Rightarrow \Delta}{X^i{\sim}(\alpha{\rightarrow}\beta), \Gamma \Rightarrow \Delta} \text{ (}{\sim}{\rightarrow}\text{left2)}$$

$$\frac{\Gamma \Rightarrow \Delta, X^i\alpha \quad \Gamma \Rightarrow \Delta, X^i{\sim}\beta}{\Gamma \Rightarrow \Delta, X^i{\sim}(\alpha{\rightarrow}\beta)} \text{ (}{\sim}{\rightarrow}\text{right)}$$

$$\frac{X^i{\sim}\alpha, \Gamma \Rightarrow \Delta \quad X^i{\sim}\beta, \Gamma \Rightarrow \Delta}{X^i{\sim}(\alpha \wedge \beta), \Gamma \Rightarrow \Delta} \text{ (}{\sim} \wedge \text{ left)}$$

$$\frac{\Gamma \Rightarrow \Delta, X^i{\sim}\alpha}{\Gamma \Rightarrow \Delta, X^i{\sim}(\alpha \wedge \beta)} \text{ (}{\sim} \wedge \text{ right1)} \quad \frac{\Gamma \Rightarrow \Delta, X^i{\sim}\beta}{\Gamma \Rightarrow \Delta, X^i{\sim}(\alpha \wedge \beta)} \text{ (}{\sim} \wedge \text{ right2)}$$

$$\frac{X^i{\sim}\alpha, \Gamma \Rightarrow \Delta}{X^i{\sim}(\alpha \vee \beta), \Gamma \Rightarrow \Delta} \text{ (}{\sim} \vee \text{ left1)} \quad \frac{X^i{\sim}\beta, \Gamma \Rightarrow \Delta}{X^i{\sim}(\alpha \vee \beta), \Gamma \Rightarrow \Delta} \text{ (}{\sim} \vee \text{ left2)}$$

$$\frac{\Gamma \Rightarrow \Delta, X^i{\sim}\alpha \quad \Gamma \Rightarrow \Delta, X^i{\sim}\beta}{\Gamma \Rightarrow \Delta, X^i{\sim}(\alpha \vee \beta)} \text{ (}{\sim} \vee \text{ right)}$$

$$\frac{\Gamma \Rightarrow \Delta, X^i{\sim}\alpha}{X^i{\sim}\neg\alpha, \Gamma \Rightarrow \Delta} \text{ (}{\sim}\neg\text{left)} \quad \frac{X^i{\sim}\alpha, \Gamma \Rightarrow \Delta}{\Gamma \Rightarrow \Delta, X^i{\sim}\neg\alpha} \text{ (}{\sim}\neg\text{right)}$$

$$\frac{\{\ X^{i+j}{\sim}\alpha, \Gamma \Rightarrow \Delta\ \}_{j\in\omega}}{X^i{\sim}G\alpha, \Gamma \Rightarrow \Delta} \text{ (}{\sim}\text{Gleft)} \quad \frac{\Gamma \Rightarrow \Delta, X^{i+k}{\sim}\alpha}{\Gamma \Rightarrow \Delta, X^i{\sim}G\alpha} \text{ (}{\sim}\text{Gright)}$$

$$\frac{\mathrm{X}^{i+k}{\sim}\alpha, \Gamma \Rightarrow \Delta}{\mathrm{X}^{i}{\sim}\mathrm{F}\alpha, \Gamma \Rightarrow \Delta}\ ({\sim}\text{Fleft}) \quad \frac{\{\ \Gamma \Rightarrow \Delta, \mathrm{X}^{i+j}{\sim}\alpha\ \}_{j\in\omega}}{\Gamma \Rightarrow \Delta, \mathrm{X}^{i}{\sim}\mathrm{F}\alpha}\ ({\sim}\text{Fright})$$

$$\frac{\mathrm{X}^{i}{\sim}\alpha, \Gamma \Rightarrow \Delta}{{\sim}\mathrm{X}^{i}\alpha, \Gamma \Rightarrow \Delta}\ ({\sim}\text{Xleft}) \quad \frac{\Gamma \Rightarrow \Delta, \mathrm{X}^{i}{\sim}\alpha}{\Gamma \Rightarrow \Delta, {\sim}\mathrm{X}^{i}\alpha}\ ({\sim}\text{Xright}).$$

The sequents of the form: $\mathrm{X}^i\alpha \Rightarrow \mathrm{X}^i\alpha$ for any formula α are provable in cut-free PLT_ω. The rule of the form:

$$\frac{\Gamma \Rightarrow \Delta}{\mathrm{X}\Gamma \Rightarrow \mathrm{X}\Delta}\ (\text{Xregu})$$

is admissible in cut-free PLT_ω.

As an example, we show that the sequent $\alpha, \mathrm{G}(\alpha{\rightarrow}\mathrm{X}\alpha) \Rightarrow \mathrm{G}\alpha$ is provable in cut-free PLT_ω:

$$\frac{\begin{array}{c}\vdots \\ \{\alpha, \mathrm{G}(\alpha{\rightarrow}\mathrm{X}\alpha) \Rightarrow \mathrm{X}^k\alpha\}_{k\in\omega}\end{array}}{\alpha, \mathrm{G}(\alpha{\rightarrow}\mathrm{X}\alpha) \Rightarrow \mathrm{G}\alpha}\ (\text{Gright})$$

where $\vdash \alpha, \mathrm{G}(\alpha{\rightarrow}\mathrm{X}\alpha) \Rightarrow \mathrm{X}^k\alpha$ for any $k \in \omega$ is shown by mathematical induction on k as follows: the base step is obvious, and the induction step can be shown by

$$\dfrac{\dfrac{\begin{array}{c}\vdots\ \text{ind.hyp.} \\ \alpha, \mathrm{G}(\alpha{\rightarrow}\mathrm{X}\alpha) \Rightarrow \mathrm{X}^k\alpha\end{array} \quad \mathrm{X}^{k+1}\alpha \Rightarrow \mathrm{X}^{k+1}\alpha}{\alpha, \mathrm{G}(\alpha{\rightarrow}\mathrm{X}\alpha), \mathrm{X}^k(\alpha{\rightarrow}\mathrm{X}\alpha) \Rightarrow \mathrm{X}^{k+1}\alpha}\ ({\rightarrow}\text{left})}{\alpha, \mathrm{G}(\alpha{\rightarrow}\mathrm{X}\alpha), \mathrm{G}(\alpha{\rightarrow}\mathrm{X}\alpha) \Rightarrow \mathrm{X}^{k+1}\alpha}\ (\text{Gleft})$$

An expression $f(\Gamma)$ denotes the result of replacing every occurrence of a formula α in Γ by an occurrence of $f(\alpha)$.

Theorem 8.4.3 (Syntactical embedding) *Let Γ and Δ be sets of formulas in $\mathcal{L}^{\sim}$, and f be the mapping defined in Definition 8.3.1. Then:*

1. $\mathrm{PLT}_\omega \vdash \Gamma \Rightarrow \Delta$ *iff* $\mathrm{LT}_\omega \vdash f(\Gamma) \Rightarrow f(\Delta)$.

2. $\mathrm{PLT}_\omega - (\text{cut}) \vdash \Gamma \Rightarrow \Delta$ *iff* $\mathrm{LT}_\omega - (\text{cut}) \vdash f(\Gamma) \Rightarrow f(\Delta)$.

Proof. Since claim (2) can be obtained as a subproof of (1), we show only (1) in the following.

• $(\Longrightarrow)$: By induction on the proofs P of $\Gamma \Rightarrow \Delta$ in PLT_ω. We distinguish the cases according to the last inference of P, and show some cases.

Case ($X^i{\sim}p \Rightarrow X^i{\sim}p$): The last inference of P is of the form: $X^i{\sim}p \Rightarrow X^i{\sim}p$. In this case, we obtain the required fact $\mathrm{LT}_\omega \vdash f(X^i{\sim}p) \Rightarrow f(X^i{\sim}p)$, since $f(X^i{\sim}p)$ coincides with $X^i p'$ by the definition of f.

Case (${\sim}{\rightarrow}$left): The last inference of P is of the form:

$$\frac{\Gamma \Rightarrow \Delta, X^i\alpha \quad \Gamma \Rightarrow \Delta, X^i{\sim}\beta}{\Gamma \Rightarrow \Delta, X^i{\sim}(\alpha{\rightarrow}\beta)}\ ({\sim}{\rightarrow}\text{left}).$$

By induction hypothesis, we have: $\mathrm{LT}_\omega \vdash f(\Gamma) \Rightarrow f(\Delta), f(X^i\alpha)$ and $\mathrm{LT}_\omega \vdash f(\Gamma) \Rightarrow f(\Delta), f(X^i{\sim}\beta)$ where $f(X^i\alpha)$ and $f(X^i{\sim}\beta)$ respectively coincide with $X^i f(\alpha)$ and $X^i f({\sim}\beta)$ by the definition of f. Then, we obtain:

$$\frac{\begin{array}{c}\vdots \\ f(\Gamma) \Rightarrow f(\Delta), X^i f(\alpha)\end{array} \quad \begin{array}{c}\vdots \\ f(\Gamma) \Rightarrow f(\Delta), X^i f({\sim}\beta)\end{array}}{f(\Gamma) \Rightarrow f(\Delta), X^i(f(\alpha) \wedge f({\sim}\beta))}\ (\wedge\text{left})$$

where $X^i(f(\alpha) \wedge f({\sim}\beta))$ coincides with $f(X^i{\sim}(\alpha{\rightarrow}\beta))$ by the definition of f.

Case (${\sim}$Gleft): The last inference of P is of the form:

$$\frac{\{\ X^{i+j}{\sim}\alpha, \Gamma \Rightarrow \Delta\ \}_{j\in\omega}}{X^i{\sim}G\alpha, \Gamma \Rightarrow \Delta}\ ({\sim}\text{Gleft}).$$

By induction hypothesis, we have: $\mathrm{LT}_\omega \vdash f(X^{i+j}{\sim}\alpha), f(\Gamma) \Rightarrow f(\Delta)$ for any $j \in \omega$, where $f(X^{i+j}{\sim}\alpha)$ coincides with $X^{i+j} f({\sim}\alpha)$ by the definition of f. Then, we obtain:

$$\frac{\begin{array}{c}\vdots \\ \{\ X^{i+j} f({\sim}\alpha), f(\Gamma) \Rightarrow f(\Delta)\ \}_{j\in\omega}\end{array}}{X^i F f({\sim}\alpha), f(\Gamma) \Rightarrow f(\Delta)}\ (\text{Fleft})$$

where $X^i F f({\sim}\alpha)$ coincides with $f(X^i{\sim}G\alpha)$ by the definition of f.

Case (${\sim}$Xleft): The last inference of P is of the form:

$$\frac{X^i{\sim}\alpha, \Gamma \Rightarrow \Delta}{{\sim}X^i\alpha, \Gamma \Rightarrow \Delta}\ ({\sim}\text{Xleft}).$$

By induction hypothesis, we have: $\mathrm{LT}_\omega \vdash f(X^i{\sim}\alpha), f(\Gamma) \Rightarrow f(\Delta)$ where $f(X^i{\sim}\alpha)$ coincides with $f({\sim}X^i\alpha)$ by the definition of f.

• ($\Longleftarrow$) : By induction on the proofs Q of $f(\Gamma) \Rightarrow f(\Delta)$ in LT_ω. We distinguish the cases according to the last inference of Q, and show only the following cases.

Case (Gleft): The last inference of Q is (Gleft).
Subcase (1): The last inference of Q is of the form:

$$\frac{X^{i+k}f(\alpha), f(\Gamma) \Rightarrow f(\Delta)}{X^{i}Gf(\alpha), f(\Gamma) \Rightarrow f(\Delta)} \text{ (Gleft)}$$

where $X^{i+k}f(\alpha)$ and $X^{i}Gf(\alpha)$ respectively coincide with $f(X^{i+k}\alpha)$ and $f(X^{i}G\alpha)$ by the definition of f. By induction hypothesis, we have: $\text{PLT}_\omega \vdash X^{i+k}\alpha, \Gamma \Rightarrow \Delta$, and hence obtain the required fact:

$$\frac{\begin{array}{c}\vdots\\ X^{i+k}\alpha, \Gamma \Rightarrow \Delta\end{array}}{X^{i}G\alpha, \Gamma \Rightarrow \Delta} \text{ (Gleft).}$$

Subcase (2): The last inference of Q is of the form:

$$\frac{X^{i+k}f({\sim}\alpha), f(\Gamma) \Rightarrow f(\Delta)}{X^{i}Gf({\sim}\alpha), f(\Gamma) \Rightarrow f(\Delta)} \text{ (Gleft)}$$

where $X^{i+k}f({\sim}\alpha)$ and $X^{i}Gf({\sim}\alpha)$ respectively coincide with $f(X^{i+k}{\sim}\alpha)$ and $f(X^{i}{\sim}F\alpha)$ by the definition of f. By induction hypothesis, we have: $\text{PLT}_\omega \vdash X^{i+k}{\sim}\alpha, \Gamma \Rightarrow \Delta$, and hence obtain the required fact:

$$\frac{\begin{array}{c}\vdots\\ X^{i+k}{\sim}\alpha, \Gamma \Rightarrow \Delta\end{array}}{X^{i}{\sim}F\alpha, \Gamma \Rightarrow \Delta} \text{ (}{\sim}\text{Fleft).}$$

Case (cut): The last inference of Q is of the form:

$$\frac{f(\Gamma_1) \Rightarrow f(\Delta_1), \beta \quad \beta, f(\Gamma_2) \Rightarrow f(\Delta_2)}{f(\Gamma_1), f(\Gamma_2) \Rightarrow f(\Delta_1), f(\Delta_2)} \text{ (cut).}$$

Since β is in $\mathcal{L}$, we have the fact $\beta = f(\beta)$. This fact can be shown by induction on β. Then, by induction hypothesis, we have: $\text{PLT}_\omega \vdash \Gamma_1 \Rightarrow \Delta_1, \beta$ and $\text{PLT}_\omega \vdash \beta, \Gamma_2 \Rightarrow \Delta_2$. We then obtain the required fact: $\text{PLT}_\omega \vdash \Gamma_1, \Gamma_2 \Rightarrow \Delta_1, \Delta_2$ by using (cut) in PLT_ω. ∎

Theorem 8.4.4 (Cut-elimination) *The rule* (cut) *is admissible in cut-free* PLT_ω.

Proof. Suppose $\text{PLT}_\omega \vdash \Gamma \Rightarrow \Delta$. Then, we have $\text{LT}_\omega \vdash f(\Gamma) \Rightarrow f(\Delta)$ by Theorem 8.4.3 (1), and hence $\text{LT}_\omega - (\text{cut}) \vdash f(\Gamma) \Rightarrow f(\Delta)$ by the cut-elimination theorem for LT_ω. By Theorem 8.4.3 (2), we obtain $\text{PLT}_\omega - (\text{cut}) \vdash \Gamma \Rightarrow \Delta$. ∎

Theorem 8.4.5 (Completeness) *For any formula* α, $\mathrm{PLT}_\omega \vdash \Rightarrow \alpha$ *iff* α *is valid in* PLTL.

Proof. $\mathrm{PLT}_\omega \vdash \Rightarrow \alpha$ iff $\mathrm{LT}_\omega \vdash \Rightarrow f(\alpha)$ (by Theorem 8.4.3) iff $f(\alpha)$ is valid in LTL (by the completeness theorem for LTL) iff α is valid in PLTL (by Theorem 8.3.4). ∎

8.5. Display calculus

We will define a display sequent calculus $\delta\mathrm{PLT}_\omega$ for PLTL. This calculus differs from the display calculi for the constructive bounded linear-time temporal logics defined in [100] in the assumptions concerning the structural language of sequents. Whereas in the display calculi in [100] a single binary structure connective is used in the introduction rules for conjunction and *intuitionistic implication*, in $\delta\mathrm{PLT}_\omega$ a single binary structure connective is used in the introduction rules for conjunction and *disjunction*. Moreover, in $\delta\mathrm{PLT}_\omega$ a unary structure connective $*$ is utilized in the introduction rules for classical negation, and no structural assumptions are required to capture the boundedness of the temporal order. The presence of $*$ makes it possible to reduce the number of structure connectives used in the introduction rules for the temporal operators in comparison with the display calculi for constructive bounded linear-time temporal logics.

One remarkable difference between the sequent systems PLT_ω and $\delta\mathrm{PLT}_\omega$ is that the introduction rules for the next-time operator X in $\delta\mathrm{PLT}_\omega$ do not exhibit any other logical operations except of strong negation, whereas the rules for X in PLT_ω exhibit all the remaining connectives. This property of $\delta\mathrm{PLT}_\omega$ may be seen as an advantage from the perspective of a proof-theoretic interpretation of the logical operations, cf. [23, 199]. If the introduction rules of a sequent calculus are regarded as meaning assignments, then the sequent rules of $\delta\mathrm{PLT}_\omega$ isolate the meaning of X to a greater extent than the sequent rules of PLT_ω. The idea behind the treatment of X in $\delta\mathrm{PLT}_\omega$ is to observe that the operations X and E ("one step earlier") form a residuated pair. The rules for the other temporal operators exploit the fact that G and P ("sometimes in the past") form a residuated pair. The introduction rules for X come with the unary structure connective $\odot$, and the introduction rules for G and F come with the unary structure connective $\bullet$. The assumptions concerning the structure connectives make it possible to state the introduction rules for the connectives in such a way that every operation in the language without strong negation is introduced as the main connective of a single-antecedent (single-succedent) conclusion sequent.

Al already explained in Chapter 7, in standard sequent systems, the comma, ",", may be seen as a context-sensitive *polyvalent* structure connective. It is to be understood as conjunction in antecedent position and as disjunction in succedent position of a sequent. In $\delta\mathrm{PLT}_\omega$ the comma is replaced by the *binary* structure connective $\circ$. Moreover, we will use the unary structure operations $*$, $\bullet$ and $\odot$ to state introduction rules for the one-place connectives of PLTL distinct from $\sim$. The structure connectives are again used to build up more complex structures (or Gentzen terms) from the formulas and the empty structure **I**. We use Greek capital letters with and without subscripts to denote structures. A sequent is an expression of the shape $\Delta \Rightarrow \Gamma$, where Δ and Γ are structures. The set of structures is inductively defined from a countable set *Atom* of propositional variables as follows:

$$\begin{array}{rl} \text{formulas:} & \alpha \in \mathit{Form(Atom)} \\ \text{structures} & \Delta \in \mathit{Struc(Form)} \\ \Delta ::= & \alpha \mid \mathbf{I} \mid (\Delta \circ \Delta) \mid *\Delta \mid \bullet\Delta \mid \odot\Delta. \end{array}$$

In antecedent position, $\circ$ is thus to be interpreted as conjunction and in succedent position as disjunction. In antecedent position, $\bullet$ is to be read as P and in succedent position as G. The structure $\odot^i\Delta$ for any $i \in \omega$ is inductively defined by $(\odot^0\Delta := \Delta)$ and $(\odot^{n+1}\Delta := \odot^n \odot \Delta)$. In succedent position $\odot^i$ is to be understood as X^i and in antecedent position as E^i ("n steps in the past"). The unary $*$ is a shift-operation used to introduce classical negation. Let $\Delta_1 \Rightarrow \Gamma_2 \dashv\vdash \Delta_2 \Rightarrow \Gamma_2$ be a notation for the pair of rules:

$$\frac{\Delta_2 \Rightarrow \Gamma_2}{\Delta_1 \Rightarrow \Gamma_1} \quad \text{and} \quad \frac{\Delta_1 \Rightarrow \Gamma_1}{\Delta_2 \Rightarrow \Gamma_2}.$$

The inferential behaviour of the structure connectives is codified by the following basic "display postulates" (*dp*) of $\delta\mathrm{PLT}_\omega$ (we omit outer brackets):

1. $\Delta \circ \Gamma \Rightarrow \Theta \dashv\vdash \Delta \Rightarrow \Theta \circ *\Gamma \dashv\vdash \Gamma \Rightarrow *\Delta \circ \Theta$

2. $\Delta \Rightarrow \Gamma \circ \Theta \dashv\vdash \Delta \circ *\Theta \Rightarrow \Gamma \dashv\vdash *\Gamma \circ \Delta \Rightarrow \Theta$

3. $\Delta \Rightarrow \Gamma \dashv\vdash *\Gamma \Rightarrow *\Delta \dashv\vdash \Delta \Rightarrow **\Gamma$

4. $\Delta \Rightarrow \bullet\Gamma \dashv\vdash \bullet\Delta \Rightarrow \Gamma$

5. $\Delta \Rightarrow \odot\Gamma \dashv\vdash \odot\Delta \Rightarrow \Gamma$

6. $\Delta \Rightarrow * \odot^i * \odot^i \Gamma \dashv\vdash \Delta \Rightarrow \Gamma$

7. $* \odot^i * \odot^i \Delta \Rightarrow \Gamma \dashv\vdash \Delta \Rightarrow \Gamma$.

If two sequents are interderivable by means of display postulates, then these sequents are said to be *structurally* or *display equivalent.* The following pairs of sequents are display equivalent:

1. $\Delta \circ \Gamma \Rightarrow \Theta$, $*\Theta \Rightarrow *\Gamma \circ *\Delta$
2. $\Delta \Rightarrow \Gamma \circ \Theta$, $*\Theta \circ *\Gamma \Rightarrow *\Delta$
3. $\Delta \Rightarrow \Gamma$, $** \Delta \Rightarrow \Gamma$
4. $*\Delta \Rightarrow \Gamma$, $*\Gamma \Rightarrow \Delta$
5. $\Delta \Rightarrow *\Gamma$, $\Gamma \Rightarrow *\Delta$
6. $\Delta \Rightarrow * \odot^i * \odot^i \Gamma$, $\odot^i * \odot^i * \Delta \Rightarrow \Gamma$
7. $* \odot^i * \odot^i \Delta \Rightarrow \Gamma$, $\Delta \Rightarrow \odot^i * \odot^i * \Gamma$
8. $\Delta \Rightarrow \Gamma$, $\odot^i * \odot^i * \Delta \Rightarrow \Gamma$
9. $\Delta \Rightarrow \Gamma$, $\Delta \Rightarrow \odot^i * \odot^i * \Gamma$.

Moreover, we assume initial sequents $p \Rightarrow p$ and $\sim p \Rightarrow \sim p$, and a cut-rule:

$$\frac{\Delta \Rightarrow \alpha \quad \alpha \Rightarrow \Gamma}{\Delta \Rightarrow \Gamma} \ (cut).$$

Display logic has been developed by Nuel Belnap [23] in order to obtain a general and flexible proof-theoretical framework, in particular a framework for presenting all kinds of *substructural* logics and modal logics based on substructural logics by means of a systematic combination of structural sequent rules together with a fixed set of introduction rules for the logical operations. Since PLTL is an extension of *classical* propositional logic, a less fine-grained distinction of structural sequent rules suffices in comparison with what is needed in the more general context of substructural subsystems of classical propositional logic. We assume the following structural rules for the empty structure **I**:

$$\frac{\Delta \Rightarrow \Gamma}{\Delta \circ \mathbf{I} \Rightarrow \Gamma} \quad \frac{\Delta \circ \mathbf{I} \Rightarrow \Gamma}{\Delta \Rightarrow \Gamma} \quad \frac{\Delta \Rightarrow \Gamma}{\Delta \Rightarrow \mathbf{I} \circ \Gamma} \quad \frac{\Delta \Rightarrow \mathbf{I} \circ \Gamma}{\Delta \Rightarrow \Gamma}$$

and versions of the standard structural rules from ordinary Gentzen calculi, weakening (monotonicity), exchange, and contraction, together with associativity:

$$\frac{\Delta \Rightarrow \Gamma}{\Delta \circ \Sigma \Rightarrow \Gamma}\ (lm) \qquad \frac{\Delta \Rightarrow \Gamma}{\Delta \Rightarrow \Gamma \circ \Sigma}\ (rm)$$

$$\frac{\Delta \circ \Sigma \Rightarrow \Gamma}{\Sigma \circ \Delta \Rightarrow \Gamma}\ (le) \qquad \frac{\Delta \Rightarrow \Sigma \circ \Gamma}{\Delta \Rightarrow \Gamma \circ \Sigma}\ (re)$$

$$\frac{\Delta \circ \Delta \Rightarrow \Gamma}{\Delta \Rightarrow \Gamma}\ (lc) \qquad \frac{\Delta \Rightarrow \Gamma \circ \Gamma}{\Delta \Rightarrow \Gamma}\ (rc)$$

$$\frac{(\Delta \circ \Gamma) \circ \Sigma \Rightarrow \Theta}{\Delta \circ (\Gamma \circ \Sigma) \Rightarrow \Theta}\ (la) \qquad \frac{\Theta \Rightarrow (\Delta \circ \Gamma) \circ \Sigma}{\Theta \Rightarrow \Delta \circ (\Gamma \circ \Sigma)}\ (ra).$$

We also assume some structural rules to capture the interaction between the temporal operators, namely, for any $i \in \omega$:

$$\frac{\Delta \Rightarrow \bullet\Gamma}{\Delta \Rightarrow \odot^{i}\Gamma}\ (Lg) \qquad \frac{\{\Delta \Rightarrow \odot^{i+j}\Gamma\}_{j\in\omega}}{\bullet \odot^{i} \Delta \Rightarrow \Gamma}\ (Rg)$$

$$\frac{\{\odot^{i+j}\Delta \Rightarrow \Gamma\}_{j\in\omega}}{\Delta \Rightarrow \odot^{i} \bullet \Gamma}\ (Lf) \qquad \frac{* \bullet * \odot^{i} \Delta \Rightarrow \Gamma}{\Delta \Rightarrow \Gamma}\ (Rf).$$

Definition 8.5.1 (δPLT$_\omega$) *The sequent calculus $\delta\mathrm{PLT}_\omega$ consists of the display postulates, the above initial sequents, cut-rule and additional structural rules plus the following right and left introduction rules:*

$$\frac{\Delta \Rightarrow *\alpha}{\Delta \Rightarrow \neg\alpha}\ (\neg\text{Right}) \qquad \frac{*\alpha \Rightarrow \Delta}{\neg\alpha \Rightarrow \Delta}\ (\neg\text{Left})$$

$$\frac{\Delta \circ \alpha \Rightarrow \beta}{\Delta \Rightarrow (\alpha \rightarrow \beta)}\ (\rightarrow\text{Right}) \qquad \frac{\Delta \Rightarrow \alpha \quad \beta \Rightarrow \Gamma}{(\alpha \rightarrow \beta) \Rightarrow *\Delta \circ \Gamma}\ (\rightarrow\text{Left})$$

$$\frac{\Delta \Rightarrow \alpha \quad \Gamma \Rightarrow \beta}{\Delta \circ \Gamma \Rightarrow (\alpha \wedge \beta)}\ (\wedge\text{Right}) \qquad \frac{\alpha \circ \beta \Rightarrow \Delta}{(\alpha \wedge \beta) \Rightarrow \Delta}\ (\wedge\text{Left})$$

$$\frac{\Delta \Rightarrow \alpha \circ \beta}{\Delta \Rightarrow (\alpha \vee \beta)}\ (\vee\text{Right}) \qquad \frac{\alpha \Rightarrow \Delta \quad \beta \Rightarrow \Gamma}{(\alpha \vee \beta) \Rightarrow \Delta \circ \Gamma}\ (\vee\text{Left})$$

$$\frac{\bullet\Delta \Rightarrow \alpha}{\Delta \Rightarrow \mathrm{G}\alpha}\ (\text{GRight}) \qquad \frac{\alpha \Rightarrow \Delta}{\mathrm{G}\alpha \Rightarrow \bullet\Delta}\ (\text{GLeft})$$

$$\frac{\Delta \Rightarrow \alpha}{* \bullet *\Delta \Rightarrow \mathrm{F}\alpha}\ (\text{FRight}) \qquad \frac{* \bullet *\alpha \Rightarrow \Delta}{\mathrm{F}\alpha \Rightarrow \Delta}\ (\text{FLeft})$$

$$\frac{\odot^i \Delta \Rightarrow \alpha}{\Delta \Rightarrow \mathrm{X}^i \alpha}\ (\mathrm{X}^i\mathrm{Right}) \qquad \frac{\alpha \Rightarrow \Delta}{\mathrm{X}^i \alpha \Rightarrow \odot^i \Delta}\ (\mathrm{X}^i\mathrm{Left})$$

$$\frac{\Delta \Rightarrow *{\sim}\alpha}{\Delta \Rightarrow {\sim}\neg\alpha}\ ({\sim}\neg\mathrm{Right}) \qquad \frac{*{\sim}\alpha \Rightarrow \Delta}{{\sim}\neg\alpha \Rightarrow \Delta}\ ({\sim}\neg\mathrm{Left})$$

$$\frac{\Delta \Rightarrow \alpha \quad \Gamma \Rightarrow {\sim}\beta}{\Delta \circ \Gamma \Rightarrow {\sim}(\alpha \to \beta)}\ ({\sim} \to \mathrm{Right}) \qquad \frac{\alpha \circ {\sim}\beta \Rightarrow \Delta}{{\sim}(\alpha \to \beta) \Rightarrow \Delta}\ ({\sim} \to \mathrm{Left})$$

$$\frac{\Delta \Rightarrow {\sim}\alpha \circ {\sim}\beta}{\Delta \Rightarrow {\sim}(\alpha \wedge \beta)}\ ({\sim} \wedge \mathrm{Right}) \qquad \frac{{\sim}\alpha \Rightarrow \Delta \quad {\sim}\beta \Rightarrow \Gamma}{{\sim}(\alpha \wedge \beta) \Rightarrow \Delta \circ \Gamma}\ ({\sim} \wedge \mathrm{Left})$$

$$\frac{\Delta \Rightarrow {\sim}\alpha \quad \Gamma \Rightarrow {\sim}\beta}{\Delta \circ \Gamma \Rightarrow {\sim}(\alpha \vee \beta)}\ ({\sim} \vee \mathrm{Right}) \qquad \frac{{\sim}\alpha \circ {\sim}\beta \Rightarrow \Delta}{{\sim}(\alpha \vee \beta) \Rightarrow \Delta}\ ({\sim} \vee \mathrm{Left})$$

$$\frac{\Delta \Rightarrow {\sim}\alpha}{* \bullet * \Delta \Rightarrow {\sim}\mathrm{G}\alpha}\ ({\sim}\mathrm{GRight}) \qquad \frac{* \bullet *{\sim}\alpha \Rightarrow \Delta}{{\sim}\mathrm{G}\alpha \Rightarrow \Delta}\ ({\sim}\mathrm{GLeft})$$

$$\frac{\bullet\Delta \Rightarrow {\sim}\alpha}{\Delta \Rightarrow {\sim}\mathrm{F}\alpha}\ ({\sim}\mathrm{FRight}) \qquad \frac{{\sim}\alpha \Rightarrow \Delta}{{\sim}\mathrm{F}\alpha \Rightarrow \bullet\Delta}\ ({\sim}\mathrm{FLeft})$$

$$\frac{\odot^i \Delta \Rightarrow {\sim}\alpha}{\Delta \Rightarrow {\sim}\mathrm{X}^i \alpha}\ ({\sim}\mathrm{X}^i\mathrm{Right}) \qquad \frac{{\sim}\alpha \Rightarrow \Delta}{{\sim}\mathrm{X}^i \alpha \Rightarrow \odot^i \Delta}\ ({\sim}\mathrm{X}^i\mathrm{Left}).$$

A formula α is provable in $\delta\mathrm{PLT}_\omega$ iff the sequent $\mathbf{I} \Rightarrow \alpha$ is provable in $\delta\mathrm{PLT}_\omega$.

By induction on α it can be shown that $\alpha \Rightarrow \alpha$ is provable in cut-free $\delta\mathrm{PLT}_\omega$ for every formula α.

The following proof is a simple example of a derivation in $\delta\mathrm{PLT}_\omega$:

$$\cfrac{\cfrac{\cfrac{\cfrac{\cfrac{\cfrac{\cfrac{\alpha \Rightarrow \alpha}{\odot^i * \odot^i * \alpha \Rightarrow \alpha}\ (dp)}{* \odot^i * \alpha \Rightarrow \mathrm{X}^i \alpha}}{*\mathrm{X}^i \alpha \Rightarrow \odot^i * \alpha}}{\neg\mathrm{X}^i \alpha \Rightarrow \odot^i * \alpha}}{\odot^i \neg\mathrm{X}^i \alpha \Rightarrow *\alpha}}{\odot^i \neg\mathrm{X}^i \alpha \Rightarrow \neg\alpha}}{\neg\mathrm{X}^i \alpha \Rightarrow \mathrm{X}^i \neg\alpha}$$

As further examples of derivations in $\delta\mathrm{PLT}_\omega$, we consider a proof of the reflexivity axiom $\mathrm{G}\alpha \to \alpha$ and a proof of the time induction axiom $(\alpha \wedge \mathrm{G}(\alpha{\to}\mathrm{X}\alpha)) \to \mathrm{G}\alpha$. The first example illustrates the use of rule (Lg) for the case $(i = 0)$:

$$\dfrac{\dfrac{\dfrac{\dfrac{\alpha \Rightarrow \alpha}{\mathrm{G}\alpha \Rightarrow \bullet\alpha}\,(Lg)}{\mathrm{G}\alpha \Rightarrow \alpha}}{\mathbf{I} \circ \mathrm{G}\alpha \Rightarrow \alpha}}{\mathbf{I} \Rightarrow \mathrm{G}\alpha \to \alpha}$$

The second example illustrates the use of rule (Rg):

$$\dfrac{\dfrac{\dfrac{\dfrac{\dfrac{\{(\alpha \wedge \mathrm{G}(\alpha \to \mathrm{X}\alpha)) \Rightarrow \odot^i\alpha\}_{j\in\omega}}{\bullet(\alpha \wedge \mathrm{G}(\alpha \to \mathrm{X}\alpha)) \Rightarrow \alpha}\,(Rg)}{(\alpha \wedge \mathrm{G}(\alpha \to \mathrm{X}\alpha)) \Rightarrow \mathrm{G}\alpha}}{(\alpha \wedge \mathrm{G}(\alpha \to \mathrm{X}\alpha)) \circ \mathbf{I} \Rightarrow \mathrm{G}\alpha}}{\mathbf{I} \circ (\alpha \wedge \mathrm{G}(\alpha \to \mathrm{X}\alpha)) \Rightarrow \mathrm{G}\alpha}}{\mathbf{I} \Rightarrow (\alpha \wedge \mathrm{G}(\alpha \to \mathrm{X}\alpha)) \to \mathrm{G}\alpha}$$

The sequents $(\alpha \wedge \mathrm{G}(\alpha \to \mathrm{X}\alpha)) \Rightarrow \odot^i\alpha$ $(i \in \omega)$ are provable in $\delta\mathrm{PLT}_\omega$. The case $i = 0$ is straightforward. For the remaining cases we may assume as induction hypothesis that $\delta\mathrm{PLT}_\omega \vdash (\alpha \wedge \mathrm{G}(\alpha \to \mathrm{X}\alpha)) \Rightarrow \odot^k\alpha$ and must show that $(\alpha \wedge \mathrm{G}(\alpha \to \mathrm{X}\alpha)) \Rightarrow \odot^{k+1}\alpha$ is provable in $\delta\mathrm{PTL}_\omega$. Note that $\delta\mathrm{PTL}_\omega \vdash \alpha \wedge G(\alpha \to \mathrm{X}\alpha) \Rightarrow \mathrm{X}\alpha$ and that $\delta\mathrm{PTL}_\omega \vdash \alpha \wedge G(\alpha \to \mathrm{X}\alpha) \Rightarrow \mathrm{XG}(\alpha \to \mathrm{X}\alpha)$. We have:

$$\dfrac{\dfrac{\odot(\alpha \wedge \mathrm{G}(\alpha \to \mathrm{X}\alpha)) \Rightarrow \alpha \wedge \mathrm{G}(\alpha \to \mathrm{X}\alpha) \quad \alpha \wedge \mathrm{G}(\alpha \to \mathrm{X}\alpha) \Rightarrow \odot^k\alpha}{\odot(\alpha \wedge \mathrm{G}(\alpha \to \mathrm{X}\alpha)) \Rightarrow \odot^k\alpha}}{\alpha \wedge \mathrm{G}(\alpha \to \mathrm{X}\alpha) \Rightarrow \odot\odot^k\,\alpha}$$

The left premise sequent is provable in $\delta\mathrm{PTL}_\omega$ as follows:

$$\dfrac{\dfrac{\alpha \wedge \mathrm{G}(\alpha \to \mathrm{X}\alpha) \Rightarrow \mathrm{X}(\alpha \wedge \mathrm{G}(\alpha \to \mathrm{X}\alpha)) \quad S}{\alpha \wedge \mathrm{G}(\alpha \to \mathrm{X}\alpha) \Rightarrow \odot(\alpha \wedge \mathrm{G}(\alpha \to \mathrm{X}\alpha))}}{\odot(\alpha \wedge \mathrm{G}(\alpha \to \mathrm{X}\alpha)) \Rightarrow \alpha \wedge \mathrm{G}(\alpha \to \mathrm{X}\alpha)}$$

where the sequent $S = \mathrm{X}(\alpha \wedge \mathrm{G}(\alpha \to \mathrm{X}\alpha)) \Rightarrow \odot(\alpha \wedge \mathrm{G}(\alpha \to \mathrm{X}\alpha))$ is provable in $\delta\mathrm{PTL}_\omega$, and where the left premise sequent of the above derivation is derivable from provable sequents as follows:

$$\dfrac{\dfrac{\dfrac{\alpha \wedge \mathrm{G}(\alpha \to \mathrm{X}\alpha) \Rightarrow \mathrm{X}\alpha \quad \alpha \wedge \mathrm{G}(\alpha \to \mathrm{X}\alpha) \Rightarrow \mathrm{XG}(\alpha \to \mathrm{X}\alpha)}{(\alpha \wedge \mathrm{G}(\alpha \to \mathrm{X}\alpha)) \circ (\alpha \wedge \mathrm{G}(\alpha \to \mathrm{X}\alpha)) \Rightarrow \mathrm{X}\alpha \wedge \mathrm{XG}(\alpha \to \mathrm{X}\alpha)}}{\alpha \wedge \mathrm{G}(\alpha \to \mathrm{X}\alpha) \Rightarrow \mathrm{X}\alpha \wedge \mathrm{XG}(\alpha \to \mathrm{X}\alpha)} \quad S'}{\alpha \wedge \mathrm{G}(\alpha \to \mathrm{X}\alpha) \Rightarrow \mathrm{X}(\alpha \wedge \mathrm{G}(\alpha \to \mathrm{X}\alpha))}$$

with $S' = X\alpha \wedge XG(\alpha \to X\alpha) \Rightarrow X(\alpha \wedge G(\alpha \to X\alpha))$.

In display logic, any substructure of a given sequent s may be displayed as either the entire antecedent or the entire succedent of a display equivalent sequent s'. In order to state this claim precisely, we need the notions of antecedent parts and succedent parts of a sequent. In order to state this fact precisely, we need some definitions. An occurrence of a substructure in a given structure is said to be positive (negative) if it is in the scope of an even (uneven) number of $*$'s. An antecedent (succedent) part of a sequent $\Delta \Rightarrow \Gamma$ is a positive occurrence of a substructure of Δ or a negative occurrence of a substructure of Γ (a positive occurrence of a substructure of Γ or a negative occurrence of a substructure of Δ).

Theorem 8.5.2 (Display property for δPLT$_\omega$, Belnap [23]) *For every sequent s and every antecedent part (succedent part) Δ of s there exists a sequent s' display equivalent to s such that Δ is the entire antecedent (succedent) of s'.*[3]

The context-sensitive reading of the structural connectives is made explicit by the following translation from sequents into formulas of the language of PLTL enhanced by E, where $\top$ is defined as $p \to p$ and $\bot$ is defined as $\neg(p \to p)$, for some fixed atomic formula p.

Definition 8.5.3 *If $\Delta \Rightarrow \Gamma$ is a display sequent, then its translation $\tau(\Delta \Rightarrow \Gamma)$ is the formula $\tau_1(\Delta) \to \tau_2(\Gamma)$, where the translations τ_1 and τ_2 from structures into formulas are inductively defines as follows:*

1. *If Σ is a formula α, then $\tau_1(\Sigma) \equiv \tau_2(\Sigma) := \alpha$.*
2. *If $\Sigma \equiv \mathbf{I}$, then $\tau_1(\Sigma) := \top$; $\tau_2(\Sigma) := \bot$.*
3. *$\tau_1(\Sigma \circ \Theta) := \tau_1(\Sigma) \wedge \tau_1(\Theta)$; $\tau_2(\Sigma \circ \Theta) := \tau_2(\Sigma) \vee \tau_2(\Theta)$.*
4. *$\tau_1(\bullet\Sigma) := P\tau_1(\Sigma)$; $\tau_2(\bullet\Sigma) := G\tau_2(\Sigma)$.*
5. *$\tau_1(*\Sigma) := \neg\tau_2(\Sigma)$; $\tau_2(*\Sigma) := \neg\tau_1(\Sigma)$.*
6. *$\tau_1(\odot^i\Sigma) := E^i\tau_1(\Sigma)$; $\tau_2(\odot^i\Sigma) := X^i\tau_2(\Sigma)$.*

Definition 8.5.4 *Let $M := (\sigma, I^+, I^-)$ be a paraconsistent model.*

1. *$(M, i) \models^+ E\alpha$ iff $[(M, i-1) \models^+ \alpha$ if $i > 0$ and otherwise $(M, 0) \models^+ \alpha]$*

[3] An elegant method for proving the display property can be found in [168], see also [199].

2. *$(M,i) \models^- \mathrm{E}\alpha$ iff $[(M,i-1) \models^- \alpha$ and otherwise $(M,0) \models^- \alpha]$*

3. *$(M,i) \models^+ \mathrm{P}\alpha$ iff $[\exists j \in \omega\ [j \leq i$ and $(x,j) \models^+ \alpha]$ if $i > 0$ and otherwise $(M,0) \models \alpha]$*

4. *$(M,i) \models^- \mathrm{P}\alpha$ iff $[\exists j \in \omega\ [j \leq i$ and $(M,j) \models^- \alpha]$ if $i > 0$ and otherwise $(M,0) \models^- \alpha]$.*

A formula α in the language of PLTL *extended by* E *and* P *is said to be strongly valid iff for every paraconsistent model $M := (\sigma, I^+, I^-)$ and every position i in σ, $(M,i) \models^+ \alpha$.*

Note that the addition of E and P to PLTL is conservative.

If $\Delta = \{\alpha_1, \alpha_2, ..., \alpha_n\}$ $(1 \leq n)$, let $\Delta^\wedge$ and $\Delta^\vee$ stand for $\alpha_1 \wedge \alpha_2 \wedge \cdots \wedge \alpha_n$ and $\alpha_1 \vee \alpha_2 \vee \cdots \vee \alpha_n$, respectively. If Δ is the empty sequence, let $\Delta^\wedge$ and $\Delta^\vee$ stand for $\top$ and $\bot$, respectively.

Theorem 8.5.5 (Equivalence) *Let Δ and Γ be finite sets of* PLTL-*formulas. Then* $\mathrm{PLT}_\omega \vdash \Delta \Rightarrow \Gamma$ *iff* $\delta\mathrm{PLT}_\omega \vdash \Delta^\wedge \Rightarrow \Gamma^\vee$.

Proof. ($\Longleftarrow$): We show by induction on proofs in $\delta\mathrm{PLT}_\omega$ that for any structures Σ, Θ the following holds: If $\delta\mathrm{PLT}_\omega \vdash \Sigma \Rightarrow \Theta$ then $\tau(\Sigma \Rightarrow \Theta)$ is strongly valid. Hence $\delta\mathrm{PLT}_\omega \vdash \Delta^\wedge \Rightarrow \Gamma^\vee$, implies that $\Delta^\wedge \rightarrow \Gamma^\vee$ is valid in PLTL. By Theorem 8.4.5 $\mathrm{PLT}_\omega \vdash\ \Rightarrow \Delta^\wedge \rightarrow \Gamma^\vee$, and by using (cut), we obtain $\mathrm{PLT}_\omega \vdash \Delta \Rightarrow \Gamma$. We here present one case of the inductive proof. Suppose that the last rule applied in the proof of $\Delta \Rightarrow \Gamma$ is (Rg):

$$\frac{\{\Delta' \Rightarrow \odot^{i+j}\Gamma\}_{j\in\omega}}{\bullet \odot^i \Delta' \Rightarrow \Gamma}$$

By the induction hypothesis: (*) for every $j \in \omega$, $\Delta'^\wedge \rightarrow \mathrm{X}^{i+j}\Gamma^\vee$ is strongly valid. Suppose $\tau_1(\bullet \odot^i \Delta'^\wedge)$ is verified at position n in model M: $(M,n) \models^+ \mathrm{PE}^i\tau_1(\Delta'^\wedge)$. Then $(\exists k \leq n)$ $(M,k-i) \models^+ \tau_1(\Delta'^\wedge)$. By (*), $(M,k-i) \models^+ \mathrm{X}^{i+j}\Gamma^\vee$ for every $j \in \omega$. Hence $(M,n) \models^+ \tau_2(\Gamma^\vee)$.
($\Longrightarrow$): By induction on proofs in PLT_ω. For initial sequents, the claim is obvious.
(cut): By the induction hypothesis, the sequents $\Gamma^\wedge \Rightarrow \Delta^\vee \vee \alpha$ and

$\alpha \wedge \Sigma^\wedge \Rightarrow \Pi^\vee$ are provable in $\delta\mathrm{PLT}_\omega$.

$$\dfrac{\dfrac{\dfrac{\dfrac{\dfrac{\Gamma^\wedge \Rightarrow \Delta^\vee \vee \alpha \quad \dfrac{\Delta^\vee \Rightarrow \Delta^\vee \quad \alpha \Rightarrow \alpha}{\Delta^\vee \vee \alpha \Rightarrow \Delta^\vee \circ \alpha}}{\Gamma^\wedge \Rightarrow \Delta^\vee \circ \alpha}}{\Gamma^\wedge \circ *\Delta^\vee \Rightarrow \alpha} \quad \dfrac{\dfrac{\dfrac{\alpha \Rightarrow \alpha \quad \Sigma^\wedge \Rightarrow \Sigma^\wedge}{\alpha \circ \Sigma^\wedge \Rightarrow \alpha \wedge \Sigma^\wedge} \quad \alpha \wedge \Sigma^\wedge \Rightarrow \Pi^\vee}{\alpha \circ \Sigma^\wedge \Rightarrow \Pi^\vee}}{\alpha \Rightarrow *\Sigma^\wedge \circ \Pi^\vee}}{\Gamma^\wedge \circ *\Delta^\vee \Rightarrow *\Sigma^\wedge \circ \Pi^\vee}}{\Gamma^\wedge \circ \Sigma^\wedge \Rightarrow \Delta^\vee \circ \Pi^\vee}\ (dp)^*}{\Gamma^\wedge \circ \Sigma^\wedge \Rightarrow \Delta^\vee \vee \Pi^\vee}}{\Gamma^\wedge \wedge \Sigma^\wedge \Rightarrow \Delta^\vee \vee \Pi^\vee}$$

where $(dp)^*$ indicates the repeated use of display postulates.
The cases of the other structural rules of PLT_ω are obvious.
($\to$ left): By the induction hypothesis, the sequents $\Gamma^\wedge \Rightarrow \Sigma^\vee \vee \mathrm{X}^i\alpha$ and $\mathrm{X}^i\beta \wedge \Delta^\wedge \Rightarrow \Pi^\vee$ are provable in $\delta\mathrm{PLT}_\omega$. Therefore, by applying (*cut*) and display postulates, the sequents $*\Sigma^\vee \circ \Gamma^\wedge \Rightarrow \mathrm{X}^i\alpha$ and $\mathrm{X}^i\beta \Rightarrow *\Delta^\wedge \circ \Pi^\vee$ can be proved. We then have the following derivation of $(\mathrm{X}^i(\alpha \to \beta) \wedge \Gamma^\wedge) \wedge \Delta^\wedge \Rightarrow \Sigma^\vee \vee \Pi^\vee$:

$$\dfrac{\dfrac{\dfrac{\dfrac{\dfrac{\dfrac{\mathrm{X}^i(\alpha \to \beta) \Rightarrow \mathrm{X}^i\alpha \to \mathrm{X}^i\beta \quad \dfrac{*\Sigma^\vee \circ \Gamma^\wedge \Rightarrow \mathrm{X}^i\alpha \quad \mathrm{X}^i\beta \Rightarrow *\Delta^\wedge \circ \Pi^\vee}{\mathrm{X}^i\alpha \to \mathrm{X}^i\beta \Rightarrow *(*\Sigma^\vee \circ \Gamma^\wedge) \circ (*\Delta^\wedge \circ \Pi^\vee)}}{\mathrm{X}^i(\alpha \to \beta) \Rightarrow *(*\Sigma^\vee \circ \Gamma^\wedge) \circ (*\Delta^\wedge \circ \Pi^\vee)}}{\mathrm{X}^i(\alpha \to \beta) \circ \Gamma^\wedge \Rightarrow *\Delta^\wedge \circ (\Sigma^\vee \circ \Pi^\vee)}\ (dp)^*}{\mathrm{X}^i(\alpha \to \beta) \wedge \Gamma^\wedge \Rightarrow *\Delta^\wedge \circ (\Sigma^\vee \circ \Pi^\vee)}}{(\mathrm{X}^i(\alpha \to \beta) \wedge \Gamma^\wedge) \circ \Delta^\wedge \Rightarrow \Sigma^\vee \circ \Pi^\vee}}{(\mathrm{X}^i(\alpha \to \beta) \wedge \Gamma^\wedge) \wedge \Delta^\wedge \Rightarrow \Sigma^\vee \circ \Pi^\vee}}{(\mathrm{X}^i(\alpha \to \beta) \wedge \Gamma^\wedge) \wedge \Delta^\wedge \Rightarrow \Sigma^\vee \vee \Pi^\vee}$$

It remains to prove the sequent $\mathrm{X}^i(\alpha \to \beta) \Rightarrow \mathrm{X}^i\alpha \to \mathrm{X}^i\beta$ in $\delta\mathrm{PLT}_\omega$:

$$\dfrac{\dfrac{\dfrac{\dfrac{\dfrac{\dfrac{\dfrac{\dfrac{\dfrac{\dfrac{\dfrac{\alpha \Rightarrow \alpha}{\mathrm{X}^i\alpha \Rightarrow \odot^i\alpha}}{(\mathrm{X}^i(\alpha \to \beta) \circ \mathrm{X}^i\alpha) \Rightarrow \odot^i\alpha}}{\odot^i(\mathrm{X}^i(\alpha \to \beta) \circ \mathrm{X}^i\alpha) \Rightarrow \alpha} \quad \beta \Rightarrow \beta}{(\alpha \to \beta) \Rightarrow * \odot^i (\mathrm{X}^i(\alpha \to \beta) \circ \mathrm{X}^i\alpha) \circ \beta}}{\mathrm{X}^i(\alpha \to \beta) \Rightarrow \odot^i(* \odot^i (\mathrm{X}^i(\alpha \to \beta) \circ \mathrm{X}^i\alpha) \circ \beta)}}{\mathrm{X}^i(\alpha \to \beta) \circ \mathrm{X}^i\alpha \Rightarrow \odot^i(* \odot^i (\mathrm{X}^i(\alpha \to \beta) \circ \mathrm{X}^i\alpha) \circ \beta)}\ (lm)}{\odot^i(\mathrm{X}^i(\alpha \to \beta) \circ \mathrm{X}^i\alpha) \Rightarrow * \odot^i (\mathrm{X}^i(\alpha \to \beta) \circ \mathrm{X}^i\alpha) \circ \beta}}{\odot^i(\mathrm{X}^i(\alpha \to \beta) \circ \mathrm{X}^i\alpha) \circ \odot^i(\mathrm{X}^i(\alpha \to \beta) \circ \mathrm{X}^i\alpha) \Rightarrow \beta}}{\odot^i(\mathrm{X}^i(\alpha \to \beta) \circ \mathrm{X}^i\alpha) \Rightarrow \beta}}{\mathrm{X}^i(\alpha \to \beta) \circ \mathrm{X}^i\alpha \Rightarrow \mathrm{X}^i\beta}}{\mathrm{X}^i(\alpha \to \beta) \Rightarrow \mathrm{X}^i\alpha \to \mathrm{X}^i\beta}$$

($\rightarrow$ right): By the induction hypothesis, $X^i\alpha \wedge \Gamma^\wedge \Rightarrow \Delta^\vee \vee X^i\beta$ is provable in $\delta\mathrm{PLT}_\omega$. Using (*cut*) it is easy to show that $X^i\alpha \circ \Gamma^\wedge \Rightarrow \Delta^\vee \circ X^i\beta$ is provable in $\delta\mathrm{PLT}_\omega$. We then have the following derivation of $\Gamma^\wedge \Rightarrow \Delta^\vee \vee X^i(\alpha \rightarrow \beta)$:

$$\dfrac{\dfrac{\dfrac{\dfrac{\dfrac{X^i\alpha \circ \Gamma^\wedge \Rightarrow \Delta^\vee \circ X^i\beta}{\Gamma^\wedge \circ *\Delta^\vee \circ *X^i\beta \Rightarrow *X^i\alpha}\,(dp)^*}{\Gamma^\wedge \circ *\Delta^\vee \circ *X^i\beta \Rightarrow \neg X^i\alpha}}{\Gamma^\wedge \circ *\Delta^\vee \Rightarrow \neg X^i\alpha \circ X^i\beta}}{\Gamma^\wedge \circ *\Delta^\vee \Rightarrow \neg X^i\alpha \vee X^i\beta} \qquad \neg X^i\alpha \vee X^i\beta \Rightarrow X^i(\alpha \rightarrow \beta)}{\dfrac{\dfrac{\Gamma^\wedge \circ *\Delta^\vee \Rightarrow X^i(\alpha \rightarrow \beta)}{\Gamma^\wedge \Rightarrow \Delta^\vee \circ X^i(\alpha \rightarrow \beta)}}{\Gamma^\wedge \Rightarrow \Delta^\vee \vee X^i(\alpha \rightarrow \beta)}}$$

We are done if we can show that $\neg X^i\alpha \vee X^i\beta \Rightarrow X^i(\alpha \rightarrow \beta)$ is provable in $\delta\mathrm{PLT}_\omega$:

$$\dfrac{\dfrac{\dfrac{\dfrac{\dfrac{\dfrac{\dfrac{\dfrac{\dfrac{\alpha \Rightarrow \alpha}{\odot^i * \odot^i * \alpha \Rightarrow \alpha}}{* \odot^i *\alpha \Rightarrow X^i\alpha}}{*X^i\alpha \Rightarrow \odot^i * \alpha}}{\odot^i * X^i\alpha \Rightarrow *\alpha}}{\odot^i * X^i\alpha \Rightarrow *\alpha \circ \beta}}{*X^i\alpha \Rightarrow \odot^i(*\alpha \circ \beta)}}{\neg X^i\alpha \Rightarrow \odot^i(*\alpha \circ \beta)} \quad \dfrac{\dfrac{\beta \Rightarrow \beta}{\beta \Rightarrow *\alpha \circ \beta}}{X^i\beta \Rightarrow \odot^i(*\alpha \circ \beta)}}{\dfrac{\dfrac{\dfrac{\dfrac{\dfrac{\neg X^i\alpha \vee X^i\beta \Rightarrow \odot^i(*\alpha \circ \beta) \circ \odot^i(*\alpha \circ \beta)}{\neg X^i\alpha \vee X^i\beta \Rightarrow \odot^i(*\alpha \circ \beta)}}{\odot^i(\neg X^i\alpha \vee X^i\beta) \Rightarrow *\alpha \circ \beta}}{\odot^i(\neg X^i\alpha \vee X^i\beta) \circ \alpha \Rightarrow \beta}}{\odot^i(\neg X^i\alpha \vee X^i\beta) \Rightarrow (\alpha \rightarrow \beta)}}{\neg X^i\alpha \vee X^i\beta \Rightarrow X^i(\alpha \rightarrow \beta)}}$$

($\wedge$left1), ($\wedge$left2): The crucial step is to show that $X^i(\alpha \wedge \beta) \Rightarrow X^i\alpha$ and $X^i(\alpha \wedge \beta) \Rightarrow X^i\beta$ are provable in $\delta\mathrm{PLT}_\omega$, which is simple.
($\wedge$right): The crucial step is to show that $X^i\alpha \wedge X^i\beta \Rightarrow X^i(\alpha \wedge \beta)$ is

provable in $\delta\mathrm{PLT}_\omega$:

$$\dfrac{\dfrac{\dfrac{\dfrac{\dfrac{\dfrac{\dfrac{\alpha\Rightarrow\alpha}{\mathrm{X}^i\alpha\Rightarrow\odot^i\alpha}}{\mathrm{X}^i\alpha\circ\mathrm{X}^i\beta\Rightarrow\odot^i\alpha}}{\odot^i(\mathrm{X}^i\alpha\circ\mathrm{X}^i\beta)\Rightarrow\alpha}\quad\dfrac{\dfrac{\dfrac{\beta\Rightarrow\beta}{\mathrm{X}^i\beta\Rightarrow\odot^i\beta}}{\mathrm{X}^i\alpha\circ\mathrm{X}^i\beta\Rightarrow\odot^i\beta}}{\odot^i(\mathrm{X}^i\alpha\circ\mathrm{X}^i\beta)\Rightarrow\beta}}{\odot^i(\mathrm{X}^i\alpha\circ\mathrm{X}^i\beta)\circ\odot^i(\mathrm{X}^i\alpha\circ\mathrm{X}^i\beta)\Rightarrow\alpha\wedge\beta}}{\odot^i(\mathrm{X}^i\alpha\circ\mathrm{X}^i\beta)\Rightarrow\alpha\wedge\beta}}{\mathrm{X}^i\alpha\circ\mathrm{X}^i\beta\Rightarrow\mathrm{X}^i(\alpha\wedge\beta)}}{\mathrm{X}^i\alpha\wedge\mathrm{X}^i\beta\Rightarrow\mathrm{X}^i(\alpha\wedge\beta)}$$

($\vee$left), ($\vee$right1), ($\vee$right2): analogous to the cases for $\wedge$.
($\neg$left): We may assume that $\Gamma^\wedge\Rightarrow\Delta^\vee\vee\mathrm{X}^i\alpha$ is provable in $\delta\mathrm{PLT}_\omega$. Then also $\Gamma^\wedge\Rightarrow\Delta^\vee\circ\mathrm{X}^i\alpha$ is derivable, and we have:

$$\dfrac{\dfrac{\dfrac{\dfrac{\dfrac{\dfrac{\dfrac{\dfrac{\dfrac{\dfrac{\dfrac{\dfrac{\Gamma^\wedge\Rightarrow\Delta^\vee\circ\mathrm{X}^i\alpha}{*\mathrm{X}^i\alpha\circ\Gamma^\wedge\Rightarrow\Delta^\vee}}{*\mathrm{X}^i\alpha\Rightarrow*\Gamma^\wedge\circ\Delta^\vee}}{*(*\Gamma^\wedge\circ\Delta^\vee)\Rightarrow**\mathrm{X}^i\alpha}}{*(*\Gamma^\wedge\circ\Delta^\vee)\Rightarrow\mathrm{X}^i\alpha}\quad\dfrac{\alpha\Rightarrow\alpha}{\mathrm{X}^i\alpha\Rightarrow\odot^i\alpha}}{*(*\Gamma^\wedge\circ\Delta^\vee)\Rightarrow\odot\alpha}}{\odot*(*\Gamma^\wedge\circ\Delta^\vee)\Rightarrow\alpha}}{*\alpha\Rightarrow*\odot*(*\Gamma^\wedge\circ\Delta^\vee)}}{\neg\alpha\Rightarrow*\odot*(*\Gamma^\wedge\circ\Delta^\vee)}}{\mathrm{X}^i\neg\alpha\Rightarrow\odot*\odot*(*\Gamma^\wedge\circ\Delta^\vee)}}{\mathrm{X}^i\neg\alpha\Rightarrow*\Gamma^\wedge\circ\Delta^\vee}\,(dp)^*}{\mathrm{X}^i\neg\alpha\circ\Gamma^\wedge\Rightarrow\Delta^\vee}}{\mathrm{X}^i\neg\alpha\wedge\Gamma^\wedge\Rightarrow\Delta^\vee}$$

($\neg$right): By the induction hypothesis, $\mathrm{X}^i\alpha\wedge\Gamma^\wedge\Rightarrow\Delta^\vee$ is provable in $\delta\mathrm{PLT}_\omega$. Then also $\mathrm{X}^i\circ\wedge\Gamma^\wedge\Rightarrow\Delta^\vee$ is provable in $\delta\mathrm{PLT}_\omega$. Note first that $*\mathrm{X}^i\neg\alpha\Rightarrow\mathrm{X}^i\alpha$ is provable in $\delta\mathrm{PLT}_\omega$ (left derivation); then we can infer $\Gamma^\wedge\Rightarrow\Delta^\vee\vee\mathrm{X}^i\neg\alpha$ from $\mathrm{X}^i\alpha\circ\Gamma^\wedge\Rightarrow\Delta^\vee$ in $\delta\mathrm{PLT}_\omega$ (right derivation):

$$\dfrac{\dfrac{\dfrac{\dfrac{\dfrac{\dfrac{\dfrac{\alpha\Rightarrow\alpha}{\alpha\Rightarrow*\odot^i*\odot^i\alpha}}{\odot^i*\odot^i\alpha\Rightarrow*\alpha}}{\odot^i*\odot^i\alpha\Rightarrow\neg\alpha}}{*\odot^i\alpha\Rightarrow\mathrm{X}^i\neg\alpha}}{*\mathrm{X}^i\neg\alpha\Rightarrow\odot^i\alpha}}{\odot^i*\mathrm{X}^i\neg\alpha\Rightarrow\alpha}}{*\mathrm{X}^i\neg\alpha\Rightarrow\mathrm{X}^i\alpha}\qquad\dfrac{\dfrac{\dfrac{*\mathrm{X}^i\neg\alpha\Rightarrow\mathrm{X}^i\alpha\quad\dfrac{\mathrm{X}^i\alpha\circ\Gamma^\wedge\Rightarrow\Delta^\vee}{\mathrm{X}^i\alpha\Rightarrow*\Gamma^\wedge\circ\Delta^\vee}}{*\mathrm{X}^i\neg\alpha\Rightarrow*\Gamma^\wedge\circ\Delta^\vee}\,(dp)^*}{\Gamma^\wedge\Rightarrow\Delta^\vee\circ\mathrm{X}^i\neg\alpha}}{\Gamma^\wedge\Rightarrow\Delta^\vee\vee\mathrm{X}^i\neg\alpha}$$

(Gleft): Here the essential step is to show that $X^i G\alpha \Rightarrow X^{i+k}\alpha$ is provable in $\delta\mathrm{PLT}_\omega$:

$$\dfrac{\dfrac{\dfrac{\dfrac{\dfrac{\dfrac{\alpha \Rightarrow \alpha}{G\alpha \Rightarrow \bullet\alpha}}{G\alpha \Rightarrow \odot^k\alpha}\,(Lg)}{X^iG\alpha \Rightarrow \odot^{i+k}\alpha}}{\odot^{i+k}X^iG\alpha \Rightarrow \alpha}}{X^iG\alpha \Rightarrow X^{i+k}\alpha}}{}$$

(Gright): By the induction hypothesis, $\Gamma^\wedge \Rightarrow \Delta^\vee \vee X^{i+j}\alpha$ is provable in $\delta\mathrm{PLT}_\omega$ for every $j \in \omega$. By applying display postulates, for every $j \in \omega$ it can be shown that $\Gamma^\wedge \circ *\Delta^\vee \Rightarrow X^{i+j}\alpha$ is provable in $\delta\mathrm{PLT}_\omega$. By infinitely many applications of (*cut*), $\Gamma^\wedge \circ *\Delta^\vee \Rightarrow \odot^{i+j}\alpha$ is provable in $\delta\mathrm{PLT}_\omega$. Then we have:

$$\dfrac{\dfrac{\dfrac{\dfrac{\dfrac{\{\Gamma^\wedge \circ *\Delta^\vee \Rightarrow \odot^{i+j}\alpha\}_{j\in\omega}}{\bullet \odot^i (\Gamma^\wedge \circ *\Delta^\vee) \Rightarrow \alpha}\,(Rg)}{\odot^i(\Gamma^\wedge \circ *\Delta^\vee) \Rightarrow G\alpha}}{\Gamma^\wedge \circ *\Delta^\vee \Rightarrow X^iG\alpha}}{\Gamma^\wedge \Rightarrow \Delta^\vee \circ X^iG\alpha}}{\Gamma^\wedge \Rightarrow \Delta^\vee \vee X^iG\alpha}$$

(Fleft): By the induction hypothesis, $X^{i+j}\alpha \wedge \Gamma^\wedge \Rightarrow \Delta^\vee$ is provable in $\delta\mathrm{PLT}_\omega$ for every $j \in \omega$. By infinitely many applications of (*cut*), $X^{i+j}\alpha \circ \Gamma^\wedge \Rightarrow \Delta^\vee$ is provable in $\delta\mathrm{PLT}_\omega$ for each $j \in \omega$. Then we have the derivation in Figure 8.1, where the dots indicate infinitely many applications of sequent rules.

(Fright): The crucial step is to show that $X^{i+k}\alpha \Rightarrow X^iF\alpha$ is provable in $\delta\mathrm{PLT}_\omega$:

$$\dfrac{\dfrac{\dfrac{\dfrac{\dfrac{\dfrac{\dfrac{\alpha \Rightarrow \alpha}{X^k\alpha \Rightarrow \odot^k\alpha}}{\odot^kX^k\alpha \Rightarrow \alpha}}{* \bullet * \odot^k X^k\alpha \Rightarrow F\alpha}\,(Rf)}{X^k\alpha \Rightarrow F\alpha}}{X^{i+k}\alpha \Rightarrow \odot^iF\alpha}}{\odot^iX^{i+k}\alpha \Rightarrow F\alpha}}{X^{i+k}\alpha \Rightarrow X^iF\alpha}$$

The rules involving $\sim$ can be dealt with in a similar way. ∎

$$
\begin{array}{cl}
\{\mathrm{X}^{i+j}\alpha \circ \Gamma^{\wedge} \Rightarrow \Delta^{\vee}\}_{j\in\omega} & \\ \hline
\vdots & \\ \hline
\{\mathrm{X}^{i+j}\alpha \Rightarrow *\Gamma^{\wedge} \circ \Delta^{\vee}\}_{j\in\omega} & \\ \hline
\vdots & \\ \hline
\{*(*\Gamma^{\wedge} \circ \Delta^{\vee}) \Rightarrow *\mathrm{X}^{i+j}\alpha\}_{j\in\omega} & \\ \hline
\vdots & \\ \hline
\{*(*\Gamma^{\wedge} \circ \Delta^{\vee}) \Rightarrow \neg\mathrm{X}^{i+j}\alpha\}_{j\in\omega} & \\ \hline
\vdots & \\ \hline
\{*(*\Gamma^{\wedge} \circ \Delta^{\vee}) \Rightarrow \mathrm{X}^{i+j}\neg\alpha\}_{j\in\omega} & \\ \hline
\vdots & \\ \hline
\{*(*\Gamma^{\wedge} \circ \Delta^{\vee}) \Rightarrow \odot^{i+j}\neg\alpha\}_{j\in\omega} & \\ \hline
\vdots & \\ \hline
\{\odot^{i+j} * (*\Gamma^{\wedge} \circ \Delta^{\vee}) \Rightarrow \neg\alpha\}_{j\in\omega} & \\ \hline
\vdots & \\ \hline
\{\odot^{i+j} * (*\Gamma^{\wedge} \circ \Delta^{\vee}) \Rightarrow *\alpha\}_{j\in\omega} & \\ \hline
(\Gamma^{\wedge} \circ \Delta^{\vee}) \Rightarrow \odot^{i} \bullet *\alpha & (Lf) \\ \hline
\odot^{i} * (*\Gamma^{\wedge} \circ \Delta^{\vee}) \Rightarrow \bullet * \alpha & \\ \hline
* \bullet *\alpha \Rightarrow * \odot^{i} *(*\Gamma^{\wedge} \circ \Delta^{\vee}) & \\ \hline
\mathrm{F}\alpha \Rightarrow * \odot^{i} *(*\Gamma^{\wedge} \circ \Delta^{\vee}) & \\ \hline
\mathrm{X}^{i}\mathrm{F}\alpha \Rightarrow \odot^{i} * \odot^{i} * (*\Gamma^{\wedge} \circ \Delta^{\vee}) & \\ \hline
\mathrm{X}^{i}\mathrm{F}\alpha \Rightarrow *\Gamma^{\wedge} \circ \Delta^{\vee} & (dp) \\ \hline
\mathrm{X}^{i}\mathrm{F}\alpha \circ \Gamma^{\wedge} \Rightarrow \Delta^{\vee} & \\ \hline
\mathrm{X}^{i}\mathrm{F}\alpha \wedge \Gamma^{\wedge} \Rightarrow \Delta^{\vee} &
\end{array}
$$

Figure 8.1.: (Fleft) case.

Part IV.

Paraconsistent substructural logics

9. Paraconsistent classical substructural logics

A general Gentzen-style framework for handling both bilattice (or strong) negation and usual negation is introduced based on the characterization of negation by a modal-like operator. This framework is regarded as an extension, generalization or refinement of not only bilattice logics and logics with strong negation, but also traditional logics including classical logic LK, classical modal logic S4 and classical linear logic CL. Cut-elimination theorems are proved for a variety of proposed sequent calculi including CLS (a conservative extension of CL) and $\mathrm{CLS_{cw}}$ (a conservative extension of some bilattice logics, LK and S4). Completeness theorems are given for these calculi with respect to phase semantics, for SLK (a conservative extension and fragment of LK and $\mathrm{CLS_{cw}}$, respectively) with respect to a classical-like semantics, and for SS4 (a conservative extension and fragment of S4 and $\mathrm{CLS_{cw}}$, respectively) with respect to a Kripke-type semantics. The proposed framework allows for an embedding of the proposed calculi into LK, S4 and CL.[1]

9.1. Introduction

Recall that the notion of a *bilattice* as a generalization of Belnap and Dunn's four-valued logic [21, 47] was first introduced by Ginsberg [63] as a tool for knowledge representation in AI, and has since been further applied by Fitting as a framework for logic programming semantics [57]. Theories and applications of various kinds of bilattices have been proposed by many researchers, as discussed in detail in a comprehensive review by Gargov [62].

Moreover, we already pointed out that logics of logical bilattices were introduced by Arieli and Avron [12, 13], and were formulated using Gentzen-type cut-free sequent calculi that correspond to certain bilattices. A number of bilattice logics or four-valued logics, which are natural extensions of Belnap's four-valued logic, have also been introduced and studied comprehensively in [159]. This gives a number of cut-free

[1] This chapter is based on [88].

sequent calculi with not only bilattice negation but also usual classical negation. An alternative approach to bilattice logic was also proposed by Schöter [172], called "evidential bilattice logic", which has a number of applications in knowledge representation and natural language semantics, but does not have a sequent calculus formulation. The logics which were investigated by Arieli and Avron [12, 13] and also in [159] have a novel kind of bilattice negation that is known to have the desirable feature of paraconsistency.

As we noted repeatedly, the notion of *strong negation* was first introduced by Nelson [134], and we pointed out that a number of studies in connection with logics with strong negation have since been developed by many logicians and computer scientists (see e.g., [6, 194, 196]). In particular, intuitionistic substructural logics with strong negation were presented by Wansing [196], and further extended by Kamide [82]. These intuitionistic substructural logics have a characteristic property called *constructible falsity*, and can be applied to logic programming languages and concurrent process calculi. The main difference between bilattice negation and strong negation is the base logic: Although the bilattice logics by Arieli and Avron [12, 13] and in [159] are based on classical (substructural) positive logics, various versions of strong negation logics are based on intuitionistic (substructural) positive logics. Thus, the kind of negation analogous to bilattice negation is called strong negation in this chapter.

The motivation of the study presented in this chapter is to give a general and widely applicable Gentzen-type framework for handling both strong negation and usual classical negation. Various kinds of cut-free sequent calculi (or extended classical substructural logics) are introduced based on the concept of strong negation as a modal-like operator. In these sequent calculi, a kind of strong negation is introduced as a new modal-like operator, representing conservative extensions of known logics including some of Arieli and Avron's bilattice logics, some of the bilattice logics in [159], classical linear logic CL [64], classical logic LK and the modal logic S4. The idea of controlling or characterizing negation using a modal-like operator is regarded as an analogue of the idea in linear logics to control structural rules by exponential modal operators. A basic sequent calculus CLS (classical linear logic with strong negation), an extended classical linear logic with two kinds of negation, is introduced as an example of using such negation. This extended logic is considered to be a resource-conscious refinement of Arieli and Avron's bilattice logics, the bilattice logics in [159], and the strong-negation logics mentioned above because CLS has not only paraconsistency and constructible falsity but also resource-sensitivity.

This chapter is then organized as follows. In Section 9.2, some classical substructural logics with two kinds of negation (including CLS and an extended classical logic $\mathrm{CLS_{cw}}$) are introduced as two-sided sequent calculi, and the relationships between these proposed logics, Arieli and Avron's logics, the logics from [159], and other logics are clarified. A faithful embedding of the proposed logics into logics without strong negation (i.e., standard logics such as S4, LK and CL) is also presented.

In Section 9.3, a novel sequent calculus SLK is introduced as a conservative extension of LK and a fragment of $\mathrm{CLS_{cw}}$. The completeness theorem (with respect to a simple classical-like semantics) for SLK is proved using Maehara's decomposition method that allows the cut-elimination theorem to be derived at the same time.

In Section 9.4, a sequent calculus SS4, also a fragment of $\mathrm{CLS_{cw}}$ and a conservative extension of S4 and LK, is introduced, and the completeness theorem (with respect to a Kripke semantics) for SS4 is presented.

In Section 9.5, using a modified version of the phase semantics proof method introduced by Okada [144], the completeness theorems and cut-elimination theorems are proved uniformly for one-sided versions of the proposed sequent calculi including CLS and $\mathrm{CLS_{cw}}$.

In Section 9.6, some remarks on two versions of cut-free one-sided calculi for the proposed calculi, called *subformula calculi* and *dual calculi*, are given. An intuitive meaning of strong negation is clarified using these calculi.

9.2. Two-sided calculus

We again first introduce the language used in this chapter. *Formulas* are constructed from propositional variables, constants $\mathbf{1}, \bot$ (multiplicatives), $\mathbf{0}, \top$ (additives), $\rightarrow$ (implication), $\wedge$ (additive conjunction), $*$ (fusion or multiplicative conjunction), $\vee$ (additive disjunction), $+$ (fission, par or multiplicative disjunction), $\cdot^{\perp}$ (negation), $\sim$ (strong negation) and modal operators ! (of course), ? (why not). Lower-case letters $p, q, ...$ are used to denote propositional variables, Greek lower-case letters $\alpha, \beta, ...$ are used to denote formulas, and Greek capital letters $\Gamma, \Delta, ...$ are used to represent finite (possibly empty) multisets of formulas. $!\Gamma$ ($?\Gamma$ or $\sim\Gamma$) denotes the multiset $\{!\gamma \mid \gamma \in \Gamma\}$ ($\{?\gamma \mid \gamma \in \Gamma\}$ or $\{\sim\gamma \mid \gamma \in \Gamma\}$ respectively). A *sequent* of two-sided calculi is an expression of the form $\Gamma \Rightarrow \Delta$. The symbol $\equiv$ is used to denote equality as sequences (or multisets) of symbols. The expression of the form $L \vdash \Gamma \Rightarrow \Delta$ means that the sequent $\Gamma \Rightarrow \Delta$ is provable in a two-sided sequent calculus L. We will sometimes omit L in this expression. Since all logics discussed in this chapter are formulated as sequent calculi, we will occasionally identify a sequent calculus with the logic determined by it.

We define a basic two-sided sequent calculus CLS for the classical linear logic with strong negation.

Definition 9.2.1 (CLS) *The initial sequents of* CLS *are of the forms:*

$$\alpha \Rightarrow \alpha \qquad \Rightarrow \mathbf{1} \qquad \bot \Rightarrow \qquad \Gamma \Rightarrow \Delta, \top \qquad \mathbf{0}, \Gamma \Rightarrow \Delta$$

$$\sim\mathbf{1} \Rightarrow \qquad \Rightarrow \sim\bot \qquad \sim\top, \Gamma \Rightarrow \Delta \qquad \Gamma \Rightarrow \Delta, \sim\mathbf{0}$$

The rules of inferences of CLS *are of the forms:*

$$\frac{\Delta \Rightarrow \Pi, \alpha \quad \alpha, \Sigma \Rightarrow \Gamma}{\Delta, \Sigma \Rightarrow \Pi, \Gamma} \text{ (cut)} \qquad \frac{\Gamma \Rightarrow \Delta}{\mathbf{1}, \Gamma \Rightarrow \Delta} \text{ (}\mathbf{1}\text{we)} \qquad \frac{\Gamma \Rightarrow \Delta}{\Gamma \Rightarrow \Delta, \bot} \text{ (}\bot\text{we)}$$

$$\frac{\Gamma \Rightarrow \Delta, \alpha \quad \beta, \Sigma \Rightarrow \Pi}{\alpha \to \beta, \Gamma, \Sigma \Rightarrow \Delta, \Pi} \text{ (}\to\text{left)} \qquad \frac{\alpha, \Gamma \Rightarrow \Delta, \beta}{\Gamma \Rightarrow \Delta, \alpha \to \beta} \text{ (}\to\text{right)}$$

$$\frac{\alpha, \beta, \Gamma \Rightarrow \Delta}{\alpha * \beta, \Gamma \Rightarrow \Delta} \text{ (}*\text{left)} \qquad \frac{\Gamma \Rightarrow \Sigma, \alpha \quad \Delta \Rightarrow \Pi, \beta}{\Gamma, \Delta \Rightarrow \Sigma, \Pi, \alpha * \beta} \text{ (}*\text{right)}$$

$$\frac{\alpha, \Gamma \Rightarrow \Delta}{\alpha \wedge \beta, \Gamma \Rightarrow \Delta} \text{ (}\wedge\text{left1)} \qquad \frac{\beta, \Gamma \Rightarrow \Delta}{\alpha \wedge \beta, \Gamma \Rightarrow \Delta} \text{ (}\wedge\text{left2)}$$

$$\frac{\Gamma \Rightarrow \Delta, \alpha \quad \Gamma \Rightarrow \Delta, \beta}{\Gamma \Rightarrow \Delta, \alpha \wedge \beta} \text{ (}\wedge\text{right)} \qquad \frac{\alpha, \Gamma \Rightarrow \Delta \quad \beta, \Gamma \Rightarrow \Delta}{\alpha \vee \beta, \Gamma \Rightarrow \Delta} \text{ (}\vee\text{left)}$$

$$\frac{\Gamma \Rightarrow \Delta, \alpha}{\Gamma \Rightarrow \Delta, \alpha \vee \beta} \text{ (}\vee\text{right1)} \qquad \frac{\Gamma \Rightarrow \Delta, \beta}{\Gamma \Rightarrow \Delta, \alpha \vee \beta} \text{ (}\vee\text{right2)}$$

$$\frac{\alpha, \Gamma \Rightarrow \Sigma \quad \beta, \Delta \Rightarrow \Pi}{\alpha + \beta, \Gamma, \Delta \Rightarrow \Sigma, \Pi} \text{ (+left)} \qquad \frac{\Gamma \Rightarrow \Pi, \alpha, \beta}{\Gamma \Rightarrow \Pi, \alpha + \beta} \text{ (+right)}$$

$$\frac{\Gamma \Rightarrow \Delta, \alpha}{\alpha^{\bot}, \Gamma \Rightarrow \Delta} \text{ (}\bot\text{left)} \qquad \frac{\alpha, \Gamma \Rightarrow \Delta}{\Gamma \Rightarrow \Delta, \alpha^{\bot}} \text{ (}\bot\text{right)}$$

$$\frac{\alpha, \Gamma \Rightarrow \Delta}{\sim\sim\alpha, \Gamma \Rightarrow \Delta} \text{ (}\sim\text{left)} \qquad \frac{\Gamma \Rightarrow \Delta, \alpha}{\Gamma \Rightarrow \Delta, \sim\sim\alpha} \text{ (}\sim\text{right)}$$

$$\frac{\Gamma \Rightarrow \Delta}{\sim\bot, \Gamma \Rightarrow \Delta} \text{ (}\sim\bot\text{we)} \qquad \frac{\Gamma \Rightarrow \Delta}{\Gamma \Rightarrow \Delta, \sim\mathbf{1}} \text{ (}\sim\mathbf{1}\text{we)}$$

$$\frac{\Gamma \Rightarrow \Delta, \sim\alpha \quad \sim\beta, \Sigma \Rightarrow \Pi}{\sim(\alpha \to \beta), \Gamma, \Sigma \Rightarrow \Delta, \Pi} \text{ (}\sim\to\text{left)} \qquad \frac{\sim\alpha, \Gamma \Rightarrow \Delta, \sim\beta}{\Gamma \Rightarrow \Delta, \sim(\alpha \to \beta)} \text{ (}\sim\to\text{right)}$$

$$\frac{\sim\alpha, \sim\beta, \Gamma \Rightarrow \Delta}{\sim(\alpha * \beta), \Gamma \Rightarrow \Delta} \text{ (}\sim *\text{ left)} \qquad \frac{\Gamma \Rightarrow \Sigma, \sim\alpha \quad \Delta \Rightarrow \Pi, \sim\beta}{\Gamma, \Delta \Rightarrow \Sigma, \Pi, \sim(\alpha * \beta)} \text{ (}\sim *\text{ right)}$$

$$\frac{\sim\alpha, \Gamma \Rightarrow \Delta \quad \sim\beta, \Gamma \Rightarrow \Delta}{\sim(\alpha \wedge \beta), \Gamma \Rightarrow \Delta} \text{ (}\sim \wedge\text{ left)}$$

$$\frac{\Gamma \Rightarrow \Delta, {\sim}\alpha}{\Gamma \Rightarrow \Delta, {\sim}(\alpha \wedge \beta)} \ ({\sim} \wedge \text{right1}) \qquad \frac{\Gamma \Rightarrow \Delta, {\sim}\beta}{\Gamma \Rightarrow \Delta, {\sim}(\alpha \wedge \beta)} \ ({\sim} \wedge \text{right2})$$

$$\frac{{\sim}\alpha, \Gamma \Rightarrow \Delta}{{\sim}(\alpha \vee \beta), \Gamma \Rightarrow \Delta} \ ({\sim} \vee \text{left1}) \qquad \frac{{\sim}\beta, \Gamma \Rightarrow \Delta}{{\sim}(\alpha \vee \beta), \Gamma \Rightarrow \Delta} \ ({\sim} \vee \text{left2})$$

$$\frac{\Gamma \Rightarrow \Delta, {\sim}\alpha \quad \Gamma \Rightarrow \Delta, {\sim}\beta}{\Gamma \Rightarrow \Delta, {\sim}(\alpha \vee \beta)} \ ({\sim} \vee \text{right})$$

$$\frac{{\sim}\alpha, \Gamma \Rightarrow \Sigma \quad {\sim}\beta, \Delta \Rightarrow \Pi}{{\sim}(\alpha + \beta), \Gamma, \Delta \Rightarrow \Sigma, \Pi} \ ({\sim} + \text{left}) \qquad \frac{\Gamma \Rightarrow \Pi, {\sim}\alpha, {\sim}\beta}{\Gamma \Rightarrow \Pi, {\sim}(\alpha + \beta)} \ ({\sim} + \text{right})$$

$$\frac{\Gamma \Rightarrow \Delta, {\sim}\alpha}{{\sim}(\alpha^{\perp}), \Gamma \Rightarrow \Delta} \ ({\sim}\perp\text{left}) \qquad \frac{{\sim}\alpha, \Gamma \Rightarrow \Delta}{\Gamma \Rightarrow \Delta, {\sim}(\alpha^{\perp})} \ ({\sim}\perp\text{right})$$

$$\frac{\alpha, \Gamma \Rightarrow \Delta}{!\alpha, \Gamma \Rightarrow \Delta} \ (!\text{left}) \qquad \frac{!\Gamma_1, {\sim}!\Gamma_2 \Rightarrow ?\Delta_1, {\sim}?\Delta_2, \alpha}{!\Gamma_1, {\sim}!\Gamma_2 \Rightarrow ?\Delta_1, {\sim}?\Delta_2, !\alpha} \ (!\text{right})$$

$$\frac{\alpha, !\Gamma_1, {\sim}!\Gamma_2 \Rightarrow ?\Delta_1, {\sim}?\Delta_2}{?\alpha, !\Gamma_1, {\sim}!\Gamma_2 \Rightarrow ?\Delta_1, {\sim}?\Delta_2} \ (?\text{left}) \qquad \frac{\Gamma \Rightarrow \Delta, \alpha}{\Gamma \Rightarrow \Delta, ?\alpha} \ (?\text{right})$$

$$\frac{!\alpha, !\alpha, \Gamma \Rightarrow \Delta}{!\alpha, \Gamma \Rightarrow \Delta} \ (!\text{co}) \qquad \frac{\Gamma \Rightarrow \Delta, ?\alpha, ?\alpha}{\Gamma \Rightarrow \Delta, ?\alpha} \ (?\text{co})$$

$$\frac{\Gamma \Rightarrow \Delta}{!\alpha, \Gamma \Rightarrow \Delta} \ (!\text{we}) \qquad \frac{\Gamma \Rightarrow \Delta}{\Gamma \Rightarrow \Delta, ?\alpha} \ (?\text{we})$$

$$\frac{{\sim}\alpha, \Gamma \Rightarrow \Delta}{{\sim}!\alpha, \Gamma \Rightarrow \Delta} \ ({\sim}!\text{left}) \qquad \frac{!\Gamma_1, {\sim}!\Gamma_2 \Rightarrow ?\Delta_1, {\sim}?\Delta_2, {\sim}\alpha}{!\Gamma_1, {\sim}!\Gamma_2 \Rightarrow ?\Delta_1, {\sim}?\Delta_2, {\sim}!\alpha} \ ({\sim}!\text{right})$$

$$\frac{{\sim}\alpha, !\Gamma_1, {\sim}!\Gamma_2 \Rightarrow ?\Delta_1, {\sim}?\Delta_2}{{\sim}?\alpha, !\Gamma_1, {\sim}!\Gamma_2 \Rightarrow ?\Delta_1, {\sim}?\Delta_2} \ ({\sim}?\text{left}) \qquad \frac{\Gamma \Rightarrow \Delta, {\sim}\alpha}{\Gamma \Rightarrow \Delta, {\sim}?\alpha} \ ({\sim}?\text{right})$$

$$\frac{{\sim}!\alpha, {\sim}!\alpha, \Gamma \Rightarrow \Delta}{{\sim}!\alpha, \Gamma \Rightarrow \Delta} \ ({\sim}!\text{co}) \qquad \frac{\Gamma \Rightarrow \Delta, {\sim}?\alpha, {\sim}?\alpha}{\Gamma \Rightarrow \Delta, {\sim}?\alpha} \ ({\sim}?\text{co})$$

$$\frac{\Gamma \Rightarrow \Delta}{{\sim}!\alpha, \Gamma \Rightarrow \Delta} \ ({\sim}!\text{we}) \qquad \frac{\Gamma \Rightarrow \Delta}{\Gamma \Rightarrow \Delta, {\sim}?\alpha} \ ({\sim}?\text{we})$$

We remark that the exchange rules are omitted since we have agreed that the antecedents and succedents of the sequents in this system are multisets.

We consider the following structural rules:

$$\frac{\alpha, \alpha, \Pi \Rightarrow \Delta}{\alpha, \Pi \Rightarrow \Delta} \ (\text{co-left}) \qquad \frac{\Gamma \Rightarrow \Delta, \alpha, \alpha}{\Gamma \Rightarrow \Delta, \alpha} \ (\text{co-right})$$

$$\frac{\Pi \Rightarrow \Delta}{\alpha, \Pi \Rightarrow \Delta} \ (\text{we-left}) \qquad \frac{\Gamma \Rightarrow \Delta}{\Gamma \Rightarrow \Delta, \alpha} \ (\text{we-right})$$

We define the calculi:

1. $\text{CLS}_{\text{w}} = \text{CLS}+(\text{we-left})+(\text{we-right})$,

2. $\mathrm{CLS_c} = \mathrm{CLS}$+(co-left)+(co-right),

3. $\mathrm{CLS_{cw}} = \mathrm{CLS_w}$+(co-left)+(co-right).

The $\{!, ?, \cdot^{\perp}, \to\}$-free part of $\mathrm{CLS_{cw}}$ is equivalent to GBL(4), which is a logic of logical bilattices by Arieli and Avron, and the constant-free fragment of GBL(4) is called GBL (see [12, 13]). The $\{\wedge, \vee, \sim\}$-part of CLS is Avron's basic system BS [16]. The rules ($\sim\perp$left) and ($\sim\perp$right) are from [159], and the $\{!, ?, \to\}$-free part of $\mathrm{CLS_{cw}}$ is a logic from [159], referred to here as PL. The $\{\wedge, \vee, \sim\}$-part of $\mathrm{CLS_{cw}}$ is Belnap and Dunn's four-valued logic, called BL here . The $\{\wedge, \vee, \to, \cdot^{\perp}\}$-part of $\mathrm{CLS_{cw}}$ is equivalent to a sequent calculus for classical logic without constants, and is called LK in this chapter. The $\{\sim, *, +, \mathit{constants}\}$-free part of $\mathrm{CLS_{cw}}$ is equivalent to a sequent calculus for the bi-modal S4 logic without constants, and is called Bi-S4 here. The $\sim$-free part of CLS is the classical linear logic CL.[2]

For any formulas α and β, the expression $\alpha \leftrightarrow \beta$ means that both the sequents $\alpha \Rightarrow \beta$ and $\beta \Rightarrow \alpha$ are provable in CLS. Then, we have the following laws with respect to $\sim$:[3]

1. $\sim(\alpha \to \beta) \leftrightarrow (\sim\alpha \to \sim\beta)$,

2. $\sim(\alpha^{\perp}) \leftrightarrow (\sim\alpha)^{\perp}$,

3. $\sim(\alpha \wedge \beta) \leftrightarrow \sim\alpha \vee \sim\beta$,

4. $\sim(\alpha \vee \beta) \leftrightarrow \sim\alpha \wedge \sim\beta$,

5. $\sim(\alpha * \beta) \leftrightarrow \sim\alpha * \sim\beta$,

6. $\sim(\alpha + \beta) \leftrightarrow \sim\alpha + \sim\beta$,

7. $\sim(!\alpha) \leftrightarrow\ !(\sim\alpha)$,

8. $\sim(?\alpha) \leftrightarrow\ ?(\sim\alpha)$.

The self-duality laws $\sim(!\alpha) \leftrightarrow\ !(\sim\alpha)$ and $\sim(?\alpha) \leftrightarrow\ ?(\sim\alpha)$, which correspond to postulating the inference rules concerning $\{\sim, !, ?\}$, seem to be unusual, because the standard duality laws are $\sim(!\alpha) \leftrightarrow\ ?(\sim\alpha)$ and $\sim(?\alpha) \leftrightarrow\ !(\sim\alpha)$, which are modal versions of the laws $\sim(\alpha \wedge \beta) \leftrightarrow$

[2] Strictly speaking, in the case of Bi-S4 and CL, the rules (!right) and (?left) need to modified by deleting $\sim!\Gamma_2$ and $\sim?\Delta_2$.

[3] Of course, we also have the standard laws (in the classical linear logic) with respect to $\cdot^{\perp}$: $(\alpha \wedge \beta)^{\perp} \leftrightarrow \alpha^{\perp} \vee \beta^{\perp}$, $(\alpha \vee \beta)^{\perp} \leftrightarrow \alpha^{\perp} \wedge \beta^{\perp}$, $(\alpha * \beta)^{\perp} \leftrightarrow \alpha^{\perp} + \beta^{\perp}$, $(\alpha + \beta)^{\perp} \leftrightarrow \alpha^{\perp} * \beta^{\perp}$, $(\alpha \to \beta)^{\perp} \leftrightarrow \alpha^{\perp} + \beta$, $(!\alpha)^{\perp} \leftrightarrow ?(\alpha^{\perp})$ and $(?\alpha)^{\perp} \leftrightarrow !(\alpha^{\perp})$.

$\sim\alpha \vee \sim\beta$ and $\sim(\alpha \vee \beta) \leftrightarrow \sim\alpha \wedge \sim\beta$, respectively.[4] However, since the operators ! and ? are known to be infinite versions of the multiplicative connectives $*$ and $+$, respectively, the self-duality laws with respect to ! and ? are regarded as natural modal versions of the laws $\sim(\alpha * \beta) \leftrightarrow \sim\alpha * \sim\beta$ and $\sim(\alpha + \beta) \leftrightarrow \sim\alpha + \sim\beta$, respectively. The standard duality laws can also be adopted for the underlying logic as a natural choice. But, in this chapter, we adopt the self-duality laws with motivations of linear logic, e.g., concurrency theory. As mentioned in [82], a formula $\sim\alpha$ can be given a concurrency-theoretic reading as "action (or process) α is stopped or suspended". For example, the concurrency-theoretic meaning of $\sim(!\alpha) \leftrightarrow !(\sim\alpha)$ is that the fact "infinitely duplicated actions $!\alpha$ are stopped" is equivalent to the fact "stop action $\sim\alpha$ is executed for each (infinitely) duplicated action α".

The law $\sim(\alpha \rightarrow \beta) \leftrightarrow (\sim\alpha \rightarrow \sim\beta)$, which corresponds to the inference rules ($\sim\rightarrow$left) and ($\sim\rightarrow$right), means that the strong negation connective $\sim$ is introduced as a modal-like operator, since the modal logic K has the law $K(\alpha \rightarrow \beta) \rightarrow (K\alpha \rightarrow K\beta)$ with the modal operator K. On the other hand, the law $(K\alpha \rightarrow K\beta) \rightarrow K(\alpha \rightarrow \beta)$ is not a law of K, and hence $\sim$ is different from the K-type operator. The operator $\sim$ is, indeed, a kind of *spatial modal operators* introduced in [86] rather than the standard modal operator. We explain the analogy between $\sim$ and these spatial modal operators. Let S be a non-empty set of locations (or spaces). Then, the spatial modal operators $[l_i]$ ($l_i \in S$), which are introduced in [86] based on a linear logic, are, roughly speaking, axiomatized as follows: for any $l, k \in S$,

1. $[l](\alpha \rightarrow \beta) \leftrightarrow [l]\alpha \rightarrow [l]\beta$,
2. $[l](\alpha * \beta) \leftrightarrow [l]\alpha * [l]\beta$,
3. $[l]!\alpha \leftrightarrow ![l]\alpha$,
4. $[l][k]\alpha \leftrightarrow [k]\alpha$,

$$\frac{\alpha}{[l]\alpha} \qquad \frac{\forall s \in S\ ([s]\alpha)}{\alpha}.$$

In this axiomatization, the law $[l](\alpha \rightarrow \beta) \leftrightarrow [l]\alpha \rightarrow [l]\beta$ is the same setting as $\sim(\alpha \rightarrow \beta) \leftrightarrow \sim\alpha \rightarrow \sim\beta$. The axiomatization derives a possible reading of a formula $[l]\alpha$ as "α is true at location l". Analogously, we may read $\sim\alpha$ as "α is negated at a location". A possible reading of $\sim(\alpha \rightarrow \beta) \leftrightarrow \sim\alpha \rightarrow \sim\beta$ is thus "if $\alpha \rightarrow \beta$ is negated at a location,

[4] As a motivation of intuitionistic modal logic with strong negation, to analyze the duality laws is known to be an important issue [141].

then α is negated at the location implies β is negated at the location, and vice versa". Furthermore, a special case of the spatial operators is regarded as the involution operator which appears in the algebraic structures of *involutive quantales* introduced by Mulvey and Pelletier. Roughly speaking, the involution operator $\cdot^{\bullet}$ in a logic corresponding to the involutive quantales is considered to be a single domain version (i.e. S is singleton) of the spatial operators (see e.g. [85]). The setting of $\sim(\alpha\rightarrow\beta) \leftrightarrow \sim\alpha\rightarrow\sim\beta$ for $\sim$ is thus also the same as that for $\cdot^{\bullet}$. In addition, another possible reading of $\sim(\alpha\rightarrow\beta) \leftrightarrow \sim\alpha\rightarrow\sim\beta$ is motivated in concurrency theory: "to stop the action $\alpha\rightarrow\beta$ (first α invokes, then β invokes)" is equivalent to "first α halts, then β halts".

The law $\sim(\alpha\rightarrow\beta) \leftrightarrow (\sim\alpha\rightarrow\sim\beta)$ corresponds to postulating ($\sim\rightarrow$left) and ($\sim\rightarrow$right), although, other settings are considered. For example, we can consider the following inference rules corresponding to $\sim(\alpha\rightarrow\beta) \leftrightarrow \alpha * \sim\beta$.

$$\frac{\alpha, \sim\beta, \Gamma \Rightarrow \Delta}{\sim(\alpha\rightarrow\beta), \Gamma \Rightarrow \Delta} \qquad \frac{\Gamma \Rightarrow \Sigma, \alpha \quad \Delta \Rightarrow \Pi, \sim\beta}{\Gamma, \Delta \Rightarrow \Sigma, \Pi, \sim(\alpha\rightarrow\beta)}$$

A number of $\{\text{constants}, !, ?, \cdot^{\perp}\}$-free sequent calculi that have these inference rules instead of ($\sim\rightarrow$left) and ($\sim\rightarrow$right) are discussed in [12, 81, 159].[5] The following inference rules were presented by Crolard [38].

$$\frac{\alpha, \Gamma \Rightarrow \Delta, \beta}{\alpha - \beta, \Gamma \Rightarrow \Delta} \qquad \frac{\Gamma \Rightarrow \Sigma, \alpha \quad \beta, \Delta \Rightarrow \Pi}{\Gamma, \Delta \Rightarrow \Sigma, \Pi, \alpha - \beta}$$

where the connective "$-$" denotes the "subtraction operator". Since these rules by Crolard are similar to the rules presented above, we can approximately regard $\sim(\alpha\rightarrow\beta)$ as "subtraction", see also Chapters 2 and refchap10. Alternative inference rules concerning $\sim$ and $\rightarrow$ have also been proposed in [12]. These inference rules can provide many different meanings for knowledge representation in AI.

Note that we can also consider the following inference rules corresponding to the axiom scheme $\sim(\alpha\rightarrow\beta) \leftrightarrow (\alpha\rightarrow\sim\beta)$, which was introduced by Wansing [204] in order to axiomatize a basic system of *connexive modal logic*.

$$\frac{\alpha, \Gamma \Rightarrow \Delta, \sim\beta}{\Gamma \Rightarrow \Delta, \sim(\alpha\rightarrow\beta)} \qquad \frac{\Gamma \Rightarrow \Sigma, \alpha \quad \sim\beta, \Delta \Rightarrow \Pi}{\sim(\alpha\rightarrow\beta), \Gamma, \Delta \Rightarrow \Sigma, \Pi}$$

[5] These calculi have the cut-elimination property, and the inference rules corresponding to $\sim(\alpha\rightarrow\beta) \leftrightarrow \alpha + \sim\beta$ can be considered. The proposed framework can deal with such variations in a similar way.

The setting of the rules ($\sim\to$left) and ($\sim\to$right) provides very simple and uniform versions of one-sided calculi, phase semantics and two-valued-like semantics, as well as Kripke semantics. This is an advantage of introducing the setting.

Various other negation rules can be considered. For example, the following inference rules by Arieli and Avron [12], called the inference rules of *conflation*,[6] can be introduced.

$$\frac{\Gamma \Rightarrow \Delta, \alpha}{\sim(\alpha^{\perp}), \Gamma \Rightarrow \Delta} \qquad \frac{\alpha, \Gamma \Rightarrow \Delta}{\Gamma \Rightarrow \Delta, \sim(\alpha^{\perp})}$$

These rules correspond to the law $\sim(\alpha^{\perp}) \leftrightarrow \alpha^{\perp}$. The following inference rules can also be considered.

$$\frac{\alpha, \Gamma \Rightarrow \Delta}{\sim(\alpha^{\perp}), \Gamma \Rightarrow \Delta} \qquad \frac{\Gamma \Rightarrow \Delta, \alpha}{\Gamma \Rightarrow \Delta, \sim(\alpha^{\perp})}$$

These rules correspond to the law $\sim(\alpha^{\perp}) \leftrightarrow \alpha$. We can deal with these kinds of negation using the proposed framework in a similar way with some appropriate modifications, however, we will discuss only the first setting in detail here.

The following theorem is derived from the phase semantics completeness proof in a later section.

Theorem 9.2.2 (Cut-elimination) *Let L be* CLS, CLS_{w}, CLS_{c} *or* CLS_{cw}. *The rule* (cut) *is admissible in cut-free L.*

As a corollary of Theorem 9.2.2, we can obtain various basic results.

Corollary 9.2.3 (Conservativity) *Let* Bi-S4, LK, CL, GB(4), PK, BL *and* BS *be the sequent calculi of bi-modal logic, classical logic, classical linear logic, Arieli and Avron's bilattice logic, the bilattice logic from [159], Belnap and Dunn's four-valued logic, and Avron's basic system, respectively. Then:*

1. CLS_{cw} *is a conservative extension of* Bi-S4, LK, GB(4), PK *and* BL.

2. CLS *is a conservative extension of* CL *and* BS.

Since this corollary shows the importance of discussing the logics CLS_{cw} and CLS, we will focus on here the logics CLS and CLS_{cw}, along with two remarkable fragments of CLS_{cw}: SLK and SS4.

[6] Strictly speaking, in [12], $\sim$ is $-$ (conflation), and $\cdot^{\perp}$ is $\sim$ (bilattice negation).

Paraconsistency is usually defined with respect to consequence relations [156]. Here, however, it is defined with respect to sequents. Let $\neg$ be a negation connective. A sequent calculus L is called *explosive* with respect to $\neg$ if for any formulas α and β, the sequent $\alpha, \neg\alpha \Rightarrow \beta$ is provable in L. It is called *paraconsistent* with respect to $\neg$ if it is not explosive with respect to $\neg$.

Corollary 9.2.4 (Paraconsistency) *We have the following.*

1. $\mathrm{CLS_{cw}}$ *is paraconsistent with respect to* $\sim$.

2. $\mathrm{CLS_{cw}}$ *is explosive with respect to* $\cdot^{\perp}$.

The subsystems of $\mathrm{CLS_{cw}}$ discussed above have the same property.

The following property, called "constructible falsity", is an important property for logics with strong negation (see, e.g., [134, 196]).

Corollary 9.2.5 (Constructible falsity) *For any formulas* α *and* β, *if* $\Rightarrow \sim(\alpha \wedge \beta)$ *is provable in* CLS, *then so is* $\Rightarrow \sim\alpha$ *or* $\Rightarrow \sim\beta$.

We explain some of the differences between $\sim$ and $\cdot^{\perp}$ as follows. Let p and q be distinct propositional variables. The sequents $p, \sim p \Rightarrow q$ and $\Rightarrow p \vee \sim p$ are not provable in $\mathrm{CLS_{cw}}$, but $p, p^{\perp} \Rightarrow q$ and $\Rightarrow p \vee p^{\perp}$ are provable in $\mathrm{CLS_{cw}}$. Similarly, $p, \sim p \Rightarrow ?q$ and $\Rightarrow ?(p \vee \sim p)$ are not provable in CLS, but $p, p^{\perp} \Rightarrow ?q$ and $\Rightarrow ?(p \vee p^{\perp})$ are provable in CLS.

Next we give an embedding of CLS into CL.

Definition 9.2.6 *We fix a set* $PROP$ *of propositional variables, used as a component of the language of* CLS, *and define the set* $PROP' := \{p' \mid p \in PROP\}$ *of propositional variables. The language* L_{CLS} *of* CLS *is defined by using* $PROP$, $\mathbf{1}, \top, \perp, \mathbf{0}, \rightarrow, \wedge, \vee, *, +, !, ?, \cdot^{\perp}$ *and* $\sim$. *The language* L_{CL} *of* CL *is obtained from* L_{CLS} *by adding* $PROP'$ *and by deleting* $\sim$.

A mapping f *from* L_{CLS} *to* L_{CL} *is defined inductively as follows:*

1. $f(p) := p$ *and* $f(\sim p) := p' \in PROP'$ *for any* $p \in PROP$,

2. $f(\mathbf{1}) := \mathbf{1}$,

3. $f(\mathbf{0}) := \mathbf{0}$,

4. $f(\top) := \top$,

5. $f(\perp) := \perp$,

6. $f(\alpha \circ \beta) := f(\alpha) \circ f(\beta)$ *where* $\circ \in \{\rightarrow, *, +, \wedge, \vee\}$,

7. $f(\alpha^{\perp}) := (f(\alpha))^{\perp}$, $f(\circ\alpha) := \circ f(\alpha)$ *where* $\circ \in \{!, ?\}$,

8. $f(\sim\sim\alpha) := f(\alpha)$,

9. $f(\sim\mathbf{1}) := \perp$,

10. $f(\sim\mathbf{0}) := \top$,

11. $f(\sim\top) := \mathbf{0}$,

12. $f(\sim\perp) := \mathbf{1}$,

13. $f(\sim(\alpha \circ \beta)) := f(\sim\alpha) \circ f(\sim\beta)$ *where* $\circ \in \{\rightarrow, *, +\}$,

14. $f(\sim(\alpha \wedge \beta)) := f(\sim\alpha) \vee f(\sim\beta)$,

15. $f(\sim(\alpha \vee \beta)) := f(\sim\alpha) \wedge f(\sim\beta)$,

16. $f(\sim(\alpha^{\perp})) := (f(\sim\alpha))^{\perp}$,

17. $f(\sim(\circ\alpha)) := \circ f(\sim\alpha)$ *where* $\circ \in \{!, ?\}$.

Let Γ be a multiset of formulas in L_{CLS}. Then, $f(\Gamma)$ denotes the result of replacing every occurrence of a formula α in Γ by an occurrence of $f(\alpha)$.

Proposition 9.2.7 *Let* Γ *and* Δ *be a multisets of formulas in* L_{CLS}. *Then:*

$$\mathrm{CLS} \vdash \Gamma \Rightarrow \Delta \ \textit{iff} \ \mathrm{CL} \vdash f(\Gamma) \Rightarrow f(\Delta).$$

With some appropriate modifications, this type of embedding is applicable to other calculi discussed in this chapter.

9.3. Non-modal version and completeness

In the following, we consider formulas constructed from propositional variables, $\perp, \top, \rightarrow, \wedge, \vee, \cdot^{\perp}$ and $\sim$. The following discussion is mainly based on the textbook [146].

Definition 9.3.1 (SLK) *A sequent calculus* SLK *is obtained from the* $\{\rightarrow, \wedge, \vee, \cdot^{\perp}, \sim\}$*-fragment of* $\mathrm{CLS_{cw}}$ *by adding the initial sequents of the* *forms* $(\bot \Rightarrow)$, $(\Rightarrow \top)$, $(\sim\top \Rightarrow)$ *and* $(\Rightarrow \sim\bot)$.[7]

Let Γ be a multiset $\{\alpha_1, ..., \alpha_m\}$ $(m \geq 0)$. Then Γ^* is defined as $\alpha_1 \vee \cdots \vee \alpha_m$ if $m \geq 1$, and $\bot$ if $m = 0$. Also Γ_* is defined as $\alpha_1 \wedge \cdots \wedge \alpha_m$ if $m \geq 1$, and $\top$ if $m = 0$. In the following discussion, the commutativity of $\wedge$ or $\vee$ is assumed. We have the following fact: for any formulas $\alpha_1, ..., \alpha_m, \beta_1, ..., \beta_n$, the sequent $\alpha_1, ..., \alpha_m \Rightarrow \beta_1, ..., \beta_n$ is provable in SLK if and only if so is $\alpha_1 \wedge \cdots \wedge \alpha_m \Rightarrow \beta_1 \vee \cdots \vee \beta_n$. Also, in the sequent expression of the form $\Gamma \Rightarrow \Delta$, the expressions Γ and Δ are considered as sets of formulas.

Definition 9.3.2 *Valuations v^+ and v^- are mappings from the set of all propositional variables to the set $\{t, f\}$. These valuations v^+ and v^- are extended to mappings from the set of all formulas to $\{t, f\}$ by*

1. $v^+(\top) = t$ *and* $v^+(\bot) = f$,
2. $v^+(\alpha \wedge \beta) = t$ *iff* $v^+(\alpha) = v^+(\beta) = t$,
3. $v^+(\alpha \vee \beta) = t$ *iff* $v^+(\alpha) = t$ *or* $v^+(\beta) = t$,
4. $v^+(\alpha \rightarrow \beta) = t$ *iff* $v^+(\alpha) = f$ *or* $v^+(\beta) = t$,
5. $v^+(\alpha^\bot) = t$ *iff* $v^+(\alpha) = f$,
6. $v^+(\sim\alpha) = t$ *iff* $v^-(\alpha) = t$,
7. $v^-(\top) = f$ *and* $v^-(\bot) = t$,
8. $v^-(\alpha \wedge \beta) = t$ *iff* $v^-(\alpha) = t$ *or* $v^-(\beta) = t$,
9. $v^-(\alpha \vee \beta) = t$ *iff* $v^-(\alpha) = v^-(\beta) = t$,
10. $v^-(\alpha \rightarrow \beta) = t$ *iff* $v^-(\alpha) = f$ *or* $v^-(\beta) = t$,
11. $v^-(\alpha^\bot) = t$ *iff* $v^-(\alpha) = f$,
12. $v^-(\sim\alpha) = t$ *iff* $v^+(\alpha) = t$.

[7] We remark that in $\mathrm{CLS_{cw}}$, the formulas $\mathbf{1}, \top, \sim\bot$ and $\sim\mathbf{0}$ are logically equivalent, and also so are $\mathbf{0}, \bot, \sim\top$ and $\sim\mathbf{1}$. Thus, in $\mathrm{CLS_{cw}}$, we can adopt the initial sequents (w.r.t. the constants) of the forms $(\Rightarrow \top)$, $(\bot \Rightarrow)$, $(\sim\top \Rightarrow)$, $(\Rightarrow \sim\bot)$, and can delete the rules (**1**we), ($\bot$we), ($\sim\bot$we), ($\sim$**1**we), and the initial sequents of the forms $(\Gamma \Rightarrow \Delta, \top)$, $(\mathbf{0}, \Gamma \Rightarrow \Delta)$, $(\sim\mathbf{1} \Rightarrow)$, $(\sim\top, \Gamma \Rightarrow \Delta)$, $(\Gamma \Rightarrow \Delta, \sim\mathbf{0})$. Hence, SLK is in fact the $\{\top, \bot, \rightarrow, \wedge, \vee, \cdot^\bot, \sim\}$-fragment of $\mathrm{CLS_{cw}}$.

A formula α *is called a* tautology *if* $v^+(\alpha) = t$ *holds for any valuations* v^+ *and* v^-. *A sequent of the form* $\Gamma \Rightarrow \Delta$ *is called a tautology if so is the formula* $\Gamma_* \rightarrow \Delta^*$.

Theorem 9.3.3 (Soundness of SLK) *For any sequent* S, *if* SLK $\vdash S$, *then* S *is a tautology.*

In the following, we will prove the completeness theorem (Theorem 9.3.8) by using Maehara's method, , which is presented, for example, in [146].

Definition 9.3.4 *A* decomposition *of a sequent* S *is defined as of the form* S' *or* $S'; S''$ *by*

(1a) $\Gamma \Rightarrow \Delta, \alpha \,;\, \Gamma \Rightarrow \Delta, \beta$ *is a decomposition of* $\Gamma \Rightarrow \Delta, \alpha \wedge \beta$,

(1b) $\alpha, \beta, \Gamma \Rightarrow \Delta$ *is a decomposition of* $\alpha \wedge \beta, \Gamma \Rightarrow \Delta$,

(2a) $\Gamma \Rightarrow \Delta, \alpha, \beta$ *is a decomposition of* $\Gamma \Rightarrow \Delta, \alpha \vee \beta$,

(2b) $\alpha, \Gamma \Rightarrow \Delta \,;\, \beta, \Gamma \Rightarrow \Delta$ *is a decomposition of* $\alpha \vee \beta, \Gamma \Rightarrow \Delta$,

(3a) $\alpha, \Gamma \Rightarrow \Delta, \beta$ *is a decomposition of* $\Gamma \Rightarrow \Delta, \alpha \rightarrow \beta$,

(3b) $\Gamma \Rightarrow \Delta, \alpha \,;\, \beta, \Gamma \Rightarrow \Delta$ *is a decomposition of* $\alpha \rightarrow \beta, \Gamma \Rightarrow \Delta$,

(4a) $\alpha, \Gamma \Rightarrow \Delta$ *is a decomposition of* $\Gamma \Rightarrow \Delta, \alpha^{\perp}$,

(4b) $\Gamma \Rightarrow \Delta, \alpha$ *is a decomposition of* $\alpha^{\perp}, \Gamma \Rightarrow \Delta$,

(5a) $\alpha, \Gamma \Rightarrow \Delta$ *is a decomposition of* ${\sim\sim}\alpha, \Gamma \Rightarrow \Delta$,

(5b) $\Gamma \Rightarrow \Delta, \alpha$ *is a decomposition of* $\Gamma \Rightarrow \Delta, {\sim\sim}\alpha$,

(6a) $\Gamma \Rightarrow \Delta, {\sim}\alpha, {\sim}\beta$ *is a decomposition of* $\Gamma \Rightarrow \Delta, {\sim}(\alpha \wedge \beta)$,

(6b) ${\sim}\alpha, \Gamma \Rightarrow \Delta \,;\, {\sim}\beta, \Gamma \Rightarrow \Delta$ *is a decomposition of* ${\sim}(\alpha \wedge \beta), \Gamma \Rightarrow \Delta$,

(7a) $\Gamma \Rightarrow \Delta, {\sim}\alpha \,;\, \Gamma \Rightarrow \Delta, {\sim}\beta$ *is a decomposition of* $\Gamma \Rightarrow \Delta, {\sim}(\alpha \vee \beta)$,

(7b) ${\sim}\alpha, {\sim}\beta, \Gamma \Rightarrow \Delta$ *is a decomposition of* ${\sim}(\alpha \vee \beta), \Gamma \Rightarrow \Delta$,

(8a) ${\sim}\alpha, \Gamma \Rightarrow \Delta, {\sim}\beta$ *is a decomposition of* $\Gamma \Rightarrow \Delta, {\sim}(\alpha \rightarrow \beta)$,

(8b) $\Gamma \Rightarrow \Delta, {\sim}\alpha \,;\, {\sim}\beta, \Gamma \Rightarrow \Delta$ *is a decomposition of* ${\sim}(\alpha \rightarrow \beta), \Gamma \Rightarrow \Delta$,

(9a) ${\sim}\alpha, \Gamma \Rightarrow \Delta$ *is a decomposition of* $\Gamma \Rightarrow \Delta, {\sim}(\alpha^{\perp})$,

(9b) $\Gamma \Rightarrow \Delta, {\sim}\alpha$ *is a decomposition of* ${\sim}(\alpha^{\perp}), \Gamma \Rightarrow \Delta$.

A *decomposition tree of* S is a tree which expresses a process of some repeated decomposition of S. In other words, a decomposition tree corresponds to a bottom up proof search tree. We remark that any repeated decomposition process terminates by the definition of decomposition. A *complete decomposition tree of* S is a decomposition tree of S in which all the formulas occurring in all the leaves of the tree are one of the following forms: $p, {\sim}p, \top, \bot, {\sim}\top$ and ${\sim}\bot$.

Lemma 9.3.5 *Let* S_1 *or* $S_1 \ ; \ S_2$ *be a decomposition of* S. *If* S *is a tautology, then so are* S_1 *and* S_2.

Proof. We show only (8b).

(8b): Suppose that ${\sim}(\alpha{\to}\beta) \wedge \Gamma_*{\to}\Delta^*$ is a tautology. First, we show that $\Gamma_*{\to}\Delta^* \vee {\sim}\alpha$ is a tautology. Suppose that (1) $v^+(\Gamma_*) = t$. We show $v^+(\Delta^* \vee {\sim}\alpha) = t$. If $v^+({\sim}\alpha) = t$, then $v^+(\Delta^* \vee {\sim}\alpha) = t$. Thus, suppose that $v^+({\sim}\alpha) = v^-(\alpha) = f$. Then, $v^-(\alpha{\to}\beta) = t$, and hence (2) $v^+({\sim}(\alpha{\to}\beta)) = v^-(\alpha{\to}\beta) = t$. On the other hand, we have (3) $v^+({\sim}(\alpha{\to}\beta) \wedge \Gamma_*{\to}\Delta^*) = t$ by the hypothesis. Thus, we obtain $v^+(\Delta^*) = t$ by (1), (2) and (3). Therefore $v^+(\Delta^* \vee {\sim}\alpha) = t$. Second, we show that ${\sim}\beta \wedge \Gamma_*{\to}\Delta^*$ is a tautology. Suppose that (4) $v^+({\sim}\beta \wedge \Gamma_*) = t$. Then, (5) $v^+({\sim}\beta) = v^-(\beta) = t$ and (6) $v^+(\Gamma_*) = t$. By (5), we have $v^-(\alpha{\to}\beta) = t$, and hence (7) $v^+({\sim}(\alpha{\to}\beta)) = v^-(\alpha{\to}\beta) = t$. By (3), (6) and (7), we obtain the required fact $v^+(\Delta^*) = t$. ∎

Lemma 9.3.6 *We have the following.*

1. *Suppose that each* α_i *or* β_j *in* $\{\alpha_1, ..., \alpha_m, \beta_1, ..., \beta_n\}$ *is a propositional constant, a propositional variable or the formula of the form* ${\sim}\gamma$ *where* γ *is a propositional variable or constant. Then, the sequent* $\alpha_1, ..., \alpha_m \Rightarrow \beta_1, ..., \beta_n$ *is a tautology if and only if* (a) *there is* α_i $(i \leq m)$ *such that* $\alpha_i \equiv \bot$ *or* $\alpha_i \equiv {\sim}\top$, (b) *there is* β_i $(i \leq n)$ *such that* $\beta_i \equiv \top$ *or* $\beta_i \equiv {\sim}\bot$, *or* (c) *there are* α_i $(i \leq m)$ *and* β_j $(j \leq n)$ *such that* $\alpha_i \equiv \beta_j$.
2. *The sequents of the forms* $(\alpha, \Gamma \Rightarrow \Delta, \alpha)$, $(\bot, \Gamma \Rightarrow \Delta)$, $(\Gamma \Rightarrow \Delta, \top)$, $({\sim}\top, \Gamma \Rightarrow \Delta)$ *and* $(\Gamma \Rightarrow \Delta, {\sim}\bot)$ *are provable in cut-free* SLK.

Lemma 9.3.7 *Let* S_1 *(or* $S_1 \ ; \ S_2$*) be a decomposition of* S. *If* S_1 *(or* S_2*) is provable in cut-free* SLK, *then so is* S.

Theorem 9.3.8 (Completeness) *For any sequent* S, *if* S *is a tautology, then* SLK$-$(cut) $\vdash S$.

Proof. Suppose that a sequent S is a tautology. Then all the leaves of a complete decomposition tree of S are tautologies by using Lemma 9.3.5 repeatedly. Then, these leaves are provable in cut-free SLK by Lemma 9.3.6 (1) (2). By using Lemma 9.3.7 repeatedly for the complete decomposition tree of S, all the sequents in the tree are provable in cut-free SLK. Therefore, in particular, S is provable in cut-free SLK. ∎

Combining this theorem with the soundness theorem, we can obtain the cut-elimination theorem for SLK.[8]

Theorem 9.3.9 (Cut-elimination) *The rule* (cut) *is admissible in cut-free* SLK.

9.4. Modal version and completeness

In the following, we consider formulas constructed from propositional variables, $\rightarrow, \wedge, \vee, \cdot^{\perp}, \sim$ and $\heartsuit$. Also, $\heartsuit\Gamma$ denotes the multiset $\{\heartsuit\gamma \mid \gamma \in \Gamma\}$. Suppose that Γ is a multiset $\{\alpha_1, ..., \alpha_m\}$ $(m \geq 0)$, and that p is a fixed propositional variable. Then Γ^* is defined as $\alpha_1 \vee \cdots \vee \alpha_m$ if $m \geq 1$, and $(p\rightarrow p)^{\perp}$ if $m = 0$. Also Γ_* is defined as $\alpha_1 \wedge \cdots \wedge \alpha_m$ if $m \geq 1$, and $p\rightarrow p$ if $m = 0$. In the following discussion, we consult [146].

Definition 9.4.1 (SS4) *A sequent calculus* SS4 *is obtained from the* $\{\top, \bot\}$*-free* SLK[9] *by adding the inference rules of the forms:*

$$\frac{\alpha, \Gamma \Rightarrow \Delta}{\heartsuit\alpha, \Gamma \Rightarrow \Delta}\ (\heartsuit\text{left}), \qquad \frac{\heartsuit\Gamma, \sim\heartsuit\Delta \Rightarrow \alpha}{\heartsuit\Gamma, \sim\heartsuit\Delta \Rightarrow \heartsuit\alpha}\ (\heartsuit\text{right}),$$

$$\frac{\sim\alpha, \Gamma \Rightarrow \Delta}{\sim\heartsuit\alpha, \Gamma \Rightarrow \Delta}\ (\sim\heartsuit\text{left}), \qquad \frac{\heartsuit\Gamma, \sim\heartsuit\Delta \Rightarrow \sim\alpha}{\heartsuit\Gamma, \sim\heartsuit\Delta \Rightarrow \sim\heartsuit\alpha}\ (\sim\heartsuit\text{right}).$$

We have the following.

Theorem 9.4.2 (Cut-elimination for SS4) *The rule* (cut) *is admissible in cut-free* SS4.

Definition 9.4.3 *A structure $\langle M, R\rangle$ is called a Kripke frame if*

1. *M is a non-empty set and*

2. *R is a transitive and reflexive binary relation on M.*

[8] The cut-elimination theorem for SLK is also an immediate consequence of that for CLS_{cw}.

[9] We can consider the logic SS4 with the addition of $\top$ and $\bot$, however, for the sake of simplicity of the discussion, we adopt the $\{\top, \bot\}$-less logic. We can also introduce alternative modal logics based on the standard logics such as K, KT and S5, and can prove the completeness theorems for such logics, with some appropriate modifications of the present framework.

Valuations v^+ *and* v^- *are mappings from the set of all propositional variables to the power set of* M. *These valuations* v^+ *and* v^- *are extended to mappings from the set of all formulas to the power set of* M *by*

1. $(x \models^+ p$ *iff* $x \in v^+(p))$ *and* $(x \models^- p$ *iff* $x \in v^-(p))$ *for any propositional variable* p,
2. $x \models^+ \alpha \wedge \beta$ *iff* $x \models^+ \alpha$ *and* $x \models^+ \beta$,
3. $x \models^+ \alpha \vee \beta$ *iff* $x \models^+ \alpha$ *or* $x \models^+ \beta$,
4. $x \models^+ \alpha \rightarrow \beta$ *iff not* $(x \models^+ \alpha)$ *or* $x \models^+ \beta$,
5. $x \models^+ \alpha^\perp$ *iff not* $(x \models^+ \alpha)$,
6. $x \models^+ \heartsuit\alpha$ *iff* $\forall y \in M\ (xRy \Longrightarrow y \models^+ \alpha)$,
7. $x \models^+ {\sim}\alpha$ *iff* $x \models^- \alpha$,
8. $x \models^- \alpha \wedge \beta$ *iff* $x \models^- \alpha$ *or* $x \models^- \beta$,
9. $x \models^- \alpha \vee \beta$ *iff* $x \models^- \alpha$ *and* $x \models^- \beta$,
10. $x \models^- \alpha \rightarrow \beta$ *iff not* $(x \models^- \alpha)$ *or* $x \models^- \beta$,
11. $x \models^- \alpha^\perp$ *iff not* $(x \models^- \alpha)$,
12. $x \models^- \heartsuit\alpha$ *iff* $\forall y \in M\ (xRy \Longrightarrow y \models^- \alpha)$,
13. $x \models^- {\sim}\alpha$ *iff* $x \models^+ \alpha$.

A formula α *is* valid in a Kripke frame $\langle M, R\rangle$ *if* $x \models^+ \alpha$ *holds for any valuations* $\models^+$, $\models^-$ *and any* $x \in M$. *A sequent* $\Gamma \Rightarrow \Delta$ *is valid in a Kripke frame if so is the formula* $\Gamma_* \rightarrow \Delta^*$.

Theorem 9.4.4 (Soundness of SS4**)** *For any sequent* S, *if* SS4 $\vdash S$, *then* S *is valid in any Kripke frame.*

We now prove the completeness theorem (Theorem 9.4.11).

Definition 9.4.5 *Let* Φ *be the set of all formulas of* SS4, *and* $U, V \subseteq \Phi$. *A pair* (U, V) *is called* consistent *if for any* $\alpha_1, ..., \alpha_m \in U$ *and any* $\beta_1, ..., \beta_n \in V$ *(*m *and* n *are arbitrary finite fixed integers and* $m, n \geq 0$*), the sequent* $\alpha_1, ..., \alpha_m \Rightarrow \beta_1, ..., \beta_n$ *is not provable in* SS4. *A pair* (U, V) *is called* maximal consistent *if* (1) (U, V) *is consistent and* (2) $U \cup V = \Phi$.

Lemma 9.4.6 *If a pair (U_0, V_0) is consistent, then there are $U, V \in \Phi$ such that $U_0 \subseteq U$, $V_0 \subseteq V$ and (U, V) is maximal consistent.*

Definition 9.4.7 *We define a canonical model $\langle M_L, R_L, \models_L^+, \models_L^- \rangle$ in the following. We define $M_L := \{U(\subseteq \Phi) \mid (U, \Phi - U)$ is maximal consistent$\}$ and $U_{\heartsuit} := \{\alpha \mid \heartsuit\alpha \in U\} \cup \{{\sim}\alpha \mid {\sim}\heartsuit\alpha \in U\}$ for any $U \in M_L$. For any $U_1, U_2 \in M_L$, $U_1 R_L U_2$ is defined as $(U_1)_{\heartsuit} \subseteq (U_2)_{\heartsuit}$. For any $U \in M_L$ and any propositional variable p, $U \models_L^+ p$ is defined as $p \in U$ and $U \models_L^- p$ is defined as ${\sim}p \in U$.*

Lemma 9.4.8 *Let $U \in M_L$.*

1. *If $\alpha_1, ..., \alpha_m \in U$ and $\mathrm{SS4} \vdash \alpha_1, ..., \alpha_m \Rightarrow \beta$, then $\beta \in U$.*
2. *For any formula α, either $\alpha \in U$ or $\alpha^{\perp} \in U$.*

By using this lemma, we can show the following.

Corollary 9.4.9 *Let $U \in M_L$.*

1. *$\alpha \wedge \beta \in U$ iff $\alpha \in U$ and $\beta \in U$.*
2. *$\alpha \vee \beta \in U$ iff $\alpha \in U$ or $\beta \in U$.*
3. *$\alpha \rightarrow \beta \in U$ iff $\alpha \notin U$ or $\beta \in U$.*
4. *$\alpha^{\perp} \in U$ iff $\alpha \notin U$.*
5. *$\heartsuit\alpha \in U$ iff $\forall W \in M_L\ [U R_L W \Longrightarrow \alpha \in W]$.*
6. *$\alpha \in U$ iff ${\sim}{\sim}\alpha \in U$.*
7. *${\sim}(\alpha \wedge \beta) \in U$ iff ${\sim}\alpha \in U$ or ${\sim}\beta \in U$.*
8. *${\sim}(\alpha \vee \beta) \in U$ iff ${\sim}\alpha \in U$ and ${\sim}\beta \in U$.*
9. *${\sim}(\alpha \rightarrow \beta) \in U$ iff ${\sim}\alpha \notin U$ or ${\sim}\beta \in U$.*
10. *${\sim}(\alpha^{\perp}) \in U$ iff ${\sim}\alpha \notin U$.*
11. *${\sim}\heartsuit\alpha \in U$ iff $\forall W \in M_L\ [U R_L W \Longrightarrow {\sim}\alpha \in W]$.*

Proof. We show only (11).

(11): ($\Longrightarrow$) Suppose ${\sim}\heartsuit\alpha \in U$, $U R_L W$ and $W \in M_L$. Then we have ${\sim}\alpha \in U_{\heartsuit} \subseteq W_{\heartsuit}$, and hence $\heartsuit{\sim}\alpha \in W$ or ${\sim}\heartsuit\alpha \in W$. In the former case, by using Lemma 9.4.8 (1) and the fact $\vdash \heartsuit{\sim}\alpha \Rightarrow {\sim}\alpha$, we obtain ${\sim}\alpha \in W$. In the latter case, by using Lemma 9.4.8 (1) and the fact $\vdash$

$\sim\heartsuit\alpha \Rightarrow \sim\alpha$, we obtain $\sim\alpha \in W$. ($\Longleftarrow$) We show the contraposition. Suppose $\sim\heartsuit\alpha \notin U$. Then (*): $(U_\heartsuit, \{\sim\alpha\})$ is consistent (this is proved later). By using Lemma 9.4.6, we have that there is a maximal consistent pair (W, V) such that $U_\heartsuit \subseteq W$ and $\sim\alpha \in V$. Then we have $W \in M_L$ and $\sim\alpha \notin W$. Moreover we have (**): $U_\heartsuit \subseteq W$ implies $U_\heartsuit \subseteq W_\heartsuit$ (this is proved later). Therefore we have the required fact that there is $W \in M_L$ such that UR_LW and $\sim\alpha \notin W$.

We show the remained fact (*). Suppose that $(U_\heartsuit, \{\sim\alpha\})$ is not consistent. Then there are $\beta_1, ..., \beta_n, \sim\delta_1, ..., \sim\delta_o \in U_\heartsuit$ (and $\heartsuit\beta_1, ..., \heartsuit\beta_n, \sim\heartsuit\delta_1, ..., \sim\heartsuit\delta_o \in U$) such that $\vdash \beta_1, ..., \beta_n, \sim\delta_1, ..., \sim\delta_o \Rightarrow \sim\alpha$. Applying ($\heartsuit$left), ($\sim\heartsuit$left), ($\sim\heartsuit$right), we obtain $\vdash \heartsuit\beta_1, ..., \heartsuit\beta_n, \sim\heartsuit\delta_1, ..., \sim\heartsuit\delta_o \Rightarrow \sim\heartsuit\alpha$. By Lemma 9.4.8 (1), and $\heartsuit\beta_1, ..., \heartsuit\beta_n, \sim\delta_1, ..., \sim\heartsuit\delta_o \in U$, we obtain $\sim\heartsuit\alpha \in U$. This contradicts for the assumption $\sim\heartsuit\alpha \notin U$. Therefore $(U_\heartsuit, \{\sim\alpha\})$ is consistent.

We show the remained fact that (**): $U_\heartsuit \subseteq W$ implies $U_\heartsuit \subseteq W_\heartsuit$. Suppose $\gamma \in U_\heartsuit$. Then $\heartsuit\gamma \in U$ or ($\gamma \equiv \sim\beta$ and $\sim\heartsuit\beta \in U$). By Lemma 9.4.8 and the facts $\vdash \heartsuit\gamma \Rightarrow \heartsuit\heartsuit\gamma$ and $\vdash \sim\heartsuit\beta \Rightarrow \heartsuit\sim\heartsuit\beta$, we obtain $\heartsuit\heartsuit\gamma \in U$ or $\heartsuit\sim\heartsuit\beta \in U$. In the latter case, by using Lemma 9.4.8 (1) and the fact $\vdash \heartsuit\sim\heartsuit\beta \Rightarrow \heartsuit\heartsuit\sim\beta$, we also obtain $\heartsuit\heartsuit\gamma \in U$. Thus, we have $\heartsuit\gamma \in U_\heartsuit \subseteq W$ by the assumption. Therefore $\gamma \in W_\heartsuit$. ∎

By using Corollary 9.4.9, we can prove the following.

Lemma 9.4.10 *For any $U \in M_L$ and any formula α,*

1. $U \models^+_L \alpha$ *iff* $\alpha \in U$,
2. $U \models^-_L \alpha$ *iff* $\sim\alpha \in U$.

Proof. By (simultaneous) induction on the complexity of the formula α. We show only the case $\alpha \equiv \heartsuit\beta$ for (2) as follows: $U \models^-_L \heartsuit\beta$ iff $\forall W \in M_L$ $[(U)_\heartsuit \subseteq (W)_\heartsuit \Longrightarrow W \models^-_L \beta]$ iff $\forall W \in M_L$ $[(U)_\heartsuit \subseteq (W)_\heartsuit \Longrightarrow \sim\beta \in W]$ (by the induction hypothesis for (2)) iff $\sim\heartsuit\beta \in U$ (by Corollary 9.4.9 (11)). ∎

Now, we show the completeness theorem by using the canonical model.

Theorem 9.4.11 (Completeness) *For any sequent S, if S is valid in any Kripke frame, then* SS4 $\vdash S$.

Proof. Suppose that a sequent $\Gamma \Rightarrow \Delta$ is valid in every Kripke frame. Moreover, suppose that $\Gamma \Rightarrow \Delta$ is not provable in SS4. Let $\Gamma \equiv \{\alpha_1, ..., \alpha_m\}$, $\Delta \equiv \{\beta_1, ..., \beta_n\}$ and $m, n \geq 0$. Then, the pair $(\{\alpha_1, ..., \alpha_m\},$

$\{\beta_1, ..., \beta_n\})$ is consistent. By using Lemma 9.4.6, we have that there is a maximal consistent pair (U, V) such that $\{\alpha_1, ..., \alpha_m\} \subseteq U$ and $\{\beta_1, ..., \beta_n\} \subseteq V$. Then, we have $U \in M_L$. Taking the statement (1) in Lemma 9.4.10, we obtain that $U \models_L^+ \alpha_i$ $(i = 1, ..., m)$ and not $[U \models_L^+ \beta_j]$ $(j = 1, ..., n)$. Hence, we have that not $[U \models_L^+ (\alpha_1 \wedge \cdots \wedge \alpha_m) \rightarrow (\beta_1 \vee \cdots \vee \beta_n)]$. This contradicts for the hypothesis. Therefore SS4 $\vdash \Gamma \Rightarrow \Delta$. ∎

9.5. One-sided calculi, cut-elimination and completeness

For the sake of simplicity of the completeness and cut-elimination proofs, we introduce one-sided calculi for CLS, CLS_w, CLS_c and CLS_cw. These calculi and the two-sided calculi are essentially the same thing by assuming the De Morgan and other laws w.r.t. $\cdot^\perp$ (see [64]). We thus use the same names CLS, CLS_w, CLS_c and CLS_cw for the one-sided calculi. An expression $\vdash \Gamma$ is a sequent of one-sided calculi where Γ denotes a multiset of formulas. We remark that $\Gamma \Rightarrow \Delta$ of the two-sided calculi corresponds to $\vdash \Gamma^\perp, \Delta$ in the one-sided calculi where $\Gamma^\perp \equiv \{\gamma_1^\perp, \cdots, \gamma_n^\perp\}$ if $\Gamma \equiv \{\gamma_1, \cdots, \gamma_n\}$. Moreover, note that if the cut-elimination theorem holds for a one-sided calculus, then the theorem also holds for the corresponding two-sided calculus.

Definition 9.5.1 (One-sided CLS) *We define a one-sided calculus* CLS. *The initial sequents of* CLS *are of the forms:*

$$\vdash \alpha, \alpha^\perp \qquad \vdash \mathbf{1} \qquad \vdash \Gamma, \top \qquad \vdash \Gamma, {\sim}\mathbf{0} \qquad \vdash {\sim}\bot$$

The rules of inferences of CLS *are of the forms:*

$$\frac{\vdash \Gamma, \alpha \quad \vdash \Delta, \alpha^\perp}{\vdash \Gamma, \Delta}\ (\text{cut}) \qquad \frac{\vdash \Gamma}{\vdash \Gamma, \bot}\ (\bot) \qquad \frac{\vdash \Gamma}{\vdash \Gamma, {\sim}\mathbf{1}}\ ({\sim}\mathbf{1})$$

$$\frac{\vdash \Gamma, \alpha \quad \vdash \Delta, \beta}{\vdash \Gamma, \Delta, \alpha * \beta}\ (*) \qquad \frac{\vdash \Gamma, \alpha, \beta}{\vdash \Gamma, \alpha + \beta}\ (+)$$

$$\frac{\vdash \Gamma, \alpha \quad \vdash \Gamma, \beta}{\vdash \Gamma, \alpha \wedge \beta}\ (\wedge) \qquad \frac{\vdash \Gamma, \alpha}{\vdash \Gamma, \alpha \vee \beta}\ (\vee 1) \qquad \frac{\vdash \Gamma, \beta}{\vdash \Gamma, \alpha \vee \beta}\ (\vee 2)$$

$$\frac{\vdash \Gamma, \alpha}{\vdash \Gamma, {\sim}{\sim}\alpha}\ ({\sim}) \qquad \frac{\vdash \Gamma, {\sim}\alpha \quad \vdash \Delta, {\sim}\beta}{\vdash \Gamma, \Delta, {\sim}(\alpha * \beta)}\ ({\sim}*)$$

$$\frac{\vdash \Gamma, {\sim}\alpha, {\sim}\beta}{\vdash \Gamma, {\sim}(\alpha + \beta)}\ ({\sim}+) \qquad \frac{\vdash \Gamma, {\sim}\alpha \quad \vdash \Gamma, {\sim}\beta}{\vdash \Gamma, {\sim}(\alpha \vee \beta)}\ ({\sim}\vee)$$

$$\frac{\vdash \Gamma, \sim\alpha}{\vdash \Gamma, \sim(\alpha \wedge \beta)} (\sim\wedge 1) \quad \frac{\vdash \Gamma, \sim\beta}{\vdash \Gamma, \sim(\alpha \wedge \beta)} (\sim\wedge 2)$$

$$\frac{\vdash ?\Gamma_1, \sim ?\Gamma_2, \alpha}{\vdash ?\Gamma_1, \sim ?\Gamma_2, !\alpha} (!) \quad \frac{\vdash \Gamma, \alpha}{\vdash \Gamma, ?\alpha} (?) \quad \frac{\vdash \Gamma, ?\alpha, ?\alpha}{\vdash \Gamma, ?\alpha} (?\text{co}) \quad \frac{\vdash \Gamma}{\vdash \Gamma, ?\alpha} (?\text{we})$$

$$\frac{\vdash ?\Gamma_1, \sim ?\Gamma_2, \sim\alpha}{\vdash ?\Gamma_1, \sim ?\Gamma_2, \sim !\alpha} (\sim !) \quad \frac{\vdash \Gamma, \sim\alpha}{\vdash \Gamma, \sim ?\alpha} (\sim ?)$$

$$\frac{\vdash \Gamma, \sim ?\alpha, \sim ?\alpha}{\vdash \Gamma, \sim ?\alpha} (\sim ?\text{co}) \quad \frac{\vdash \Gamma}{\vdash \Gamma, \sim ?\alpha} (\sim ?\text{we})$$

We define the structural rules:

$$\frac{\vdash \Gamma, \alpha, \alpha}{\vdash \Gamma, \alpha} (\text{co}) \qquad \frac{\vdash \Gamma}{\vdash \Gamma, \alpha} (\text{we})$$

and the calculi:

1. CLS_w = CLS+(we),
2. CLS_c = CLS+(co),
3. CLS_cw = CLS+(co)+(we).

We remark that the one-sided CL of the original classical linear logic is obtained from the one-sided CLS by deleting the initial sequents and rules with respect to $\sim$, and by deleting $\sim ?\Gamma_2$ in the rule (!).

Next, we define a phase semantics for these logics. The difference between such semantics and the original semantics is only the definition of the valuations: whereas the original semantics has a valuation ϕ, our semantics has two kinds of valuations ϕ^+ and ϕ^-, where ϕ^+ is the same as ϕ.

Definition 9.5.2 *Let $\langle M, \cdot, 1\rangle$ be a commutative monoid with the unit 1. If $X, Y \subseteq M$, we define $X \circ Y := \{x \cdot y \mid x \in X \text{ and } y \in Y\}$.*[10] *A* phase space *is a structure $\langle M, \hat{\bot}, \hat{I}\rangle$ where $\hat{\bot}$ is a fixed subset of M. For $X \subseteq M$, we define $X^{\hat{\bot}} := \{y \mid \forall x \in X\ (x \cdot y \in \hat{\bot})\}$. We define $\hat{I} := \{x \in M \mid x \cdot x = x\} \cap \hat{\bot}^{\hat{\bot}}$. X ($\subseteq M$) is called a* fact *if $X^{\hat{\bot}\hat{\bot}} = X$. The set of facts is denoted by D_M.*

[10] We remark that the operation $\circ$ is commutative and associative, and has the monotonicity property: $X_1 \subseteq Y_1$ and $X_2 \subseteq Y_2$ imply $X_1 \circ X_2 \subseteq Y_1 \circ Y_2$ for any X_1, X_2, Y_1, Y_2 ($\subseteq M$).

Definition 9.5.3 *A phase space* $\langle M, \hat{\bot}, \hat{I}\rangle$ *is called a* weakening phase space *if the following weakening-condition holds:* $\hat{\bot}^{\hat{\bot}} = M$. *A phase space* $\langle M, \hat{\bot}, \hat{I}\rangle$ *is called a* contraction phase space *if the following contraction-condition holds:* $z \cdot x \cdot x \in \hat{\bot}$ *implies* $z \cdot x \in \hat{\bot}$ *for all* $x, z \in M$. *A phase space* $\langle M, \hat{\bot}, \hat{I}\rangle$ *is called a* contraction-weakening phase space *if both the contraction-condition and the weakening-condition hold.*

Proposition 9.5.4 *Let* $X, Y \subseteq M$.

1. $X \subseteq Y^{\hat{\bot}}$ *iff* $X \circ Y \subseteq \hat{\bot}$.
2. *If* $X \subseteq Y$ *then* $X \circ Y^{\hat{\bot}} \subseteq \hat{\bot}$.
3. $X \circ X^{\hat{\bot}} \subseteq \hat{\bot}$.
4. *If* $X \subseteq Y$ *then* $Y^{\hat{\bot}} \subseteq X^{\hat{\bot}}$.
5. *If* $X \subseteq Y$ *then* $X^{\hat{\bot}\hat{\bot}} \subseteq Y^{\hat{\bot}\hat{\bot}}$.
6. $X \subseteq X^{\hat{\bot}\hat{\bot}}$.
7. $(X^{\hat{\bot}\hat{\bot}})^{\hat{\bot}\hat{\bot}} \subseteq X^{\hat{\bot}\hat{\bot}}$.
8. $X^{\hat{\bot}\hat{\bot}} \circ Y^{\hat{\bot}\hat{\bot}} \subseteq (X \circ Y)^{\hat{\bot}\hat{\bot}}$.
9. $x \in X^{\hat{\bot}}$ *iff* $\{x\} \circ X \subseteq \hat{\bot}$.
10. *If* $X \circ Y \subseteq \hat{\bot}$ *then* $X \circ Y^{\hat{\bot}\hat{\bot}} \subseteq \hat{\bot}$.

Proposition 9.5.5 *Let* $X, Y \subseteq M$.

1. $X^{\hat{\bot}}$ *is a fact.*
2. $X^{\hat{\bot}\hat{\bot}}$ *is the smallest fact that includes* X.
3. *if* X *and* Y *are facts, then so is* $X \cap Y$.

Definition 9.5.6 *Let* $A, B \subseteq M$. *We define the following operators and constants:*

1. $\hat{\bot} := \{1\}^{\hat{\bot}}$,
2. $\hat{\mathbf{1}} := \hat{\bot}^{\hat{\bot}} = \{1\}^{\hat{\bot}\hat{\bot}}$,
3. $\hat{\top} := M = \emptyset^{\hat{\bot}}$,
4. $\hat{\mathbf{0}} := \hat{\top}^{\hat{\bot}} = M^{\hat{\bot}} = \emptyset^{\hat{\bot}\hat{\bot}}$,

5. $A \mathbin{\hat{\wedge}} B := A \cap B$,

6. $A \mathbin{\hat{\vee}} B := (A \cup B)^{\hat{\bot}\hat{\bot}}$,

7. $A \mathbin{\hat{*}} B := (A \circ B)^{\hat{\bot}\hat{\bot}}$,

8. $A \mathbin{\hat{+}} B := (A^{\hat{\bot}} \circ B^{\hat{\bot}})^{\hat{\bot}}$,

9. $\hat{!}A := (A \cap \hat{I})^{\hat{\bot}\hat{\bot}}$,

10. $\hat{?}A := (A^{\hat{\bot}} \cap \hat{I})^{\hat{\bot}}$.

We can show that, by Proposition 9.5.5, the constants defined above are facts and the operators defined above are closed under D_M.

Definition 9.5.7 *A valuation ϕ^+ (ϕ^-) on a phase space $\langle M, \hat{\bot}, \hat{I}\rangle$ is a mapping which assigns a fact to each propositional variables. Each valuation ϕ^+ (ϕ^-) can be extended to a mapping $\cdot^+$ ($\cdot^-$) from the set Φ of all formulas to D_M by*

1. $p^+ := \phi^+(p)$ *for any propositional variable* p,

2. $\bot^+ := \hat{\bot}$,

3. $\mathbf{1}^+ := \hat{\mathbf{1}}$,

4. $\top^+ := \hat{\top}$,

5. $\mathbf{0}^+ := \hat{\mathbf{0}}$,

6. $(\alpha^{\bot})^+ := (\alpha^+)^{\hat{\bot}}$,

7. $(\alpha \wedge \beta)^+ := \alpha^+ \mathbin{\hat{\wedge}} \beta^+$,

8. $(\alpha \vee \beta)^+ := \alpha^+ \mathbin{\hat{\vee}} \beta^+$,

9. $(\alpha * \beta)^+ := \alpha^+ \mathbin{\hat{*}} \beta^+$,

10. $(\alpha + \beta)^+ := \alpha^+ \mathbin{\hat{+}} \beta^+$,

11. $(!\alpha)^+ := \hat{!}(\alpha^+)$,

12. $(?\alpha)^+ := \hat{?}(\alpha^+)$,

13. $({\sim}\alpha)^+ := \alpha^-$,

14. $p^- := \phi^-(p)$ *for any propositional variable* p,

15. $\bot^- := \hat{\mathbf{1}}$,

16. $\mathbf{1}^- := \hat{\bot}$,

17. $\top^- := \hat{\mathbf{0}}$,

18. $\mathbf{0}^- := \hat{\top}$,

19. $(\alpha^\bot)^- := (\alpha^-)^{\hat{\bot}}$,

20. $(\alpha \wedge \beta)^- := \alpha^- \hat{\vee} \beta^-$,

21. $(\alpha \vee \beta)^- := \alpha^- \hat{\wedge} \beta^-$,

22. $(\alpha * \beta)^- := \alpha^- \hat{*} \beta^-$,

23. $(\alpha + \beta)^- := \alpha^- \hat{+} \beta^-$,

24. $(!\alpha)^- := \hat{!}(\alpha^-)$,

25. $(?\alpha)^- := \hat{?}(\alpha^-)$,

26. $(\sim\alpha)^- := \alpha^+$.

We call the values α^+ *and* α^- *the* inner-values *of* α $(\in \Phi)$.

Definition 9.5.8 $\langle M, \hat{\bot}, \hat{I}, \phi^+, \phi^- \rangle$ *is a* phase (weakening phase, contraction phase or contraction-weakening phase) model *if* $\langle M, \hat{\bot}, \hat{I} \rangle$ *is a phase (weakening phase, contraction phase or contraction-weakening phase respectively) space and* ϕ^+ *and* ϕ^- *are valuations on* $\langle M, \hat{\bot}, \hat{I} \rangle$. *A sequent* $\vdash \alpha$ *is* true *in a phase (weakening phase, contraction phase or contraction-weakening phase) model* $\langle M, \hat{\bot}, \hat{I}, \phi^+, \phi^- \rangle$ *if* $\alpha^{+\hat{\bot}} \subseteq \hat{\bot}$ *(or equivalently* $1 \in \alpha^+$*), and* valid *in a phase (weakening phase, contraction phase or contraction-weakening phase respectively) space* $\langle M, \hat{\bot}, \hat{I} \rangle$ *if it is true for any valuations* ϕ^+ *and* ϕ^- *on the phase space. A sequent* $\vdash \alpha_1, \cdots, \alpha_n$ *is* true *in a phase (weakening phase, contraction phase or contraction-weakening phase) model* $\langle M, \hat{\bot}, \hat{I}, \phi^+, \phi^- \rangle$ *if* $\vdash \alpha_1 + \cdots + \alpha_n$ *is true in the model, and* valid *in a phase (weakening phase, contraction phase or contraction-weakening phase respectively) space* $\langle M, \hat{\bot}, \hat{I} \rangle$ *if it is true for any valuation* ϕ^+ *and* ϕ^- *on the phase space.*

Theorem 9.5.9 (Soundness) *Let* L_1, L_2, L_3 *and* L_4 *be* CLS, $\mathrm{CLS_w}$, $\mathrm{CLS_c}$ *and* $\mathrm{CLS_{cw}}$ *respectively. If a sequent* S *is provable in* L_1 *(*L_2*,* L_3 *or* L_4*) then the sequent* S *is valid for any phase (weakening phase, contraction phase or contraction-weakening phase respectively) space.*

Theorem 9.5.10 (Completeness) *Let L_1, L_2, L_3 and L_4 be* CLS, CLS_w, CLS_c *and* CLS_cw *respectively. If a sequent S is valid for any phase (weakening phase, contraction phase or contraction-weakening phase respectively) space, then the sequent S is cut-free provable in L_1 (L_2, L_3 or L_4 respectively).*

In the following, we only prove this theorem for CLS. To prove this theorem, we construct a canonical phase model $\langle M, \hat{\bot}, \hat{I}, \phi^+, \phi^- \rangle$. Here M is the set of all multisets of formulas where multiple occurrence of a formula of the form $?\alpha$ (or $\sim ?\alpha$) in the multisets counts only once. $\langle M, \cdot, 1 \rangle$ is a commutative monoid where $\Delta \cdot \Gamma := \Delta \cup \Gamma$ (the multiset union) for all $\Delta, \Gamma \in M$, and 1 ($\in M$) is $\varnothing$ (the empty multiset). For any formula α, we define $[\alpha]_+ := \{\Delta \mid \vdash_{cf} \Delta, \alpha\}$ and $[\alpha]_- := \{\Delta \mid \vdash_{cf} \Delta, \sim\alpha\}$ where $\vdash_{cf} \Delta, \alpha$ means that $\vdash \Delta, \alpha$ is cut-free provable. We call $[\alpha]_*$ ($* \in \{+, -\}$) the *outer-values* of α ($\in \Phi$). We define $\hat{\bot} := [\bot]_+ = \{\Delta \mid \vdash_{cf} \Delta\}$. Moreover $\hat{I}$ is defined as $\{?\Gamma \cup \sim ?\Delta \in 2^\Phi \mid \Gamma \cup \Delta \in M\}$.

We show that the set $\hat{I}$ is equivalent to the set $\dot{I} = \{\Delta \in M \mid \Delta \cup \Delta = \Delta\} \cap \hat{\bot}^{\hat{\bot}}$. First, we show $\hat{I} \subseteq \dot{I}$. Suppose $\Sigma \in \hat{I}$. Then, this means that Σ is a set of formulas of the forms $?\gamma$ and $\sim ?\beta$, because the multiple occurrence of a formula of the form $?\gamma$ or $\sim ?\gamma$ in the multisets in M counts only once. We can verify $\Sigma \in M$ and $\Sigma \cup \Sigma = \Sigma$, because Σ is a set and $\cup$ is regarded as the set union. We can also show $\Sigma \in \hat{\bot}^{\hat{\bot}} = \{\Delta \mid \forall \Gamma \in \hat{\bot}\ (\Delta \cup \Gamma \in \hat{\bot})\}$, i.e., $\forall \Gamma\ [\vdash_{cf} \Gamma \Longrightarrow\ \vdash_{cf} \Gamma, \Sigma]$, by using (?we) and/or ($\sim$?we). Thus, we have $\Sigma \in \dot{I}$. Next, we show $\dot{I} \subseteq \hat{I}$. Suppose $\Sigma \in \dot{I}$. Then, we have $\Sigma \in M$, $\Sigma \cup \Sigma = \Sigma$ and $\forall \Gamma\ [\vdash_{cf} \Gamma \Longrightarrow\ \vdash_{cf} \Gamma, \Sigma]$. These mean that Σ is a set of formulas of the forms $?\gamma$ and $\sim ?\beta$, and hence $\Sigma \in \hat{I}$. Therefore $\dot{I} = \hat{I}$.

Using $M, \hat{\bot}$ and $\hat{I}$ defined above, we have the fact that $\langle M, \hat{\bot}, \hat{I} \rangle$ is a phase space. The valuations ϕ^+ and ϕ^- of the canonical model are defined as $\phi^+(p) := [p]_+$ and $\phi^-(p) := [p]_-$ for any propositional variable p.

To prove the completeness theorem, we must prove some lemmas.

Lemma 9.5.11 *Let α be any formula.*

1. *if $\alpha^+ \subseteq [\alpha]_+$ then $\{\alpha\} \in \alpha^{+\hat{\bot}}$.*

2. *if $\alpha^- \subseteq [\alpha]_-$ then $\{\sim\alpha\} \in \alpha^{-\hat{\bot}}$.*

Lemma 9.5.12 *Let α be any formula.*

1. $[\alpha]_+^{\hat{\bot}\hat{\bot}} = [\alpha]_+$.

2. $[\alpha]_-^{\hat{\perp}\hat{\perp}} = [\alpha]_-$.

3. $[\sim\alpha]_+ = [\alpha]_-$.

4. $[\sim\alpha]_- = [\alpha]_+$.

Proof. We show only (4).

(4): We first show $[\alpha]_+ \subseteq [\sim\alpha]_-$. Suppose $\Delta \in [\alpha]_+$, i.e., $\vdash_{cf} \alpha, \Delta$. Then, by using the rule $(\sim)$, we obtain $\vdash_{cf} \sim\sim\alpha, \Delta$, and hence $\Delta \in [\sim\alpha]_-$. Next, we show the converse, i.e., $\Delta \in [\sim\alpha]_-$ implies $\Delta \in [\alpha]_+$. To prove the fact, we must use the fact that the rule

$$\frac{\vdash \Gamma, \sim\sim\alpha}{\vdash \Gamma, \alpha}\ (\sim^{-1})$$

is admissible in cut-free CLS. This fact can be proved easily by induction on the cut-free proof in CLS. Now suppose $\Delta \in [\sim\alpha]_-$, i.e., $\vdash_{cf} \sim\sim\alpha, \Delta$. By applying $(\sim^{-1})$ to $\vdash_{cf} \sim\sim\alpha, \Delta$, we obtain $\vdash_{cf} \alpha, \Delta$, and hence $\Delta \in [\alpha]_+$. ∎

By Lemmas 9.5.11 and 9.5.12, we can prove the following main lemma. The lemma directly implies the completeness theorem. If the sequent $\vdash \alpha$ is true, then $\emptyset \in \alpha^+$. On the other hand $\alpha^+ \subseteq [\alpha]_+$, and hence $\emptyset \in [\alpha]_+$. This means that $\vdash \alpha$ is cut-free provable.

Lemma 9.5.13 *Let α be any formula.*

1. $\alpha^+ \subseteq [\alpha]_+$,

2. $\alpha^- \subseteq [\alpha]_-$.

Proof. We prove this lemma by (simultaneous) induction on the complexity of the formula α.

• Base step: Obvious by the definitions.

• Induction step for (1): We show only the following case. The other cases are the same as that in [144].

(Case $\alpha \equiv \sim\beta$ for (1)): We show $(\sim\beta)^+ \subseteq [\sim\beta]_+$. Suppose $\Gamma \in (\sim\beta)^+$, i.e., $\Gamma \in \beta^-$. Then we obtain $\Gamma \in \beta^- \subseteq [\beta]_- = [\sim\beta]_+$ by the induction hypothesis for (2) and Lemma 9.5.12 (3).

• Induction step for (2): We show some cases.

(Case $\alpha \equiv p^\perp$ for (2) where p is a propositional variable):[11]

[11] We do not have to consider the other cases such as $\alpha \equiv \gamma^\perp$ where γ is any formula except propositional variable. Because we have the De Morgan and other laws for $\cdot^\perp$. For example, in the case $\alpha \equiv (\sim\beta)^\perp$, we consider $(\sim\beta)^{\perp -} \subseteq [(\sim\beta)^\perp]_-$. Then it is enough to consider $(\sim(\beta^\perp))^- \subseteq [\sim(\beta^\perp)]_-$ by the law $(\sim\beta)^\perp \leftrightarrow \sim(\beta^\perp)$. Here taking $\gamma \equiv \beta^\perp$, we see $(\sim\gamma)^- \subseteq [\sim\gamma]_-$. Therefore this case is included the case $\alpha \equiv \sim\gamma$ for (2).

We show $p^{\perp -} \subseteq [p^{\perp}]_{-}$. Suppose $\Gamma \in p^{-\hat{\perp}}$. Then we have $\Gamma \in [p]_{-}^{\hat{\perp}}$ by the definition $p^{-} := [p]_{-}$, and hence we have (*): $\forall \Pi$ [$\Pi \in [p]_{-}$ implies $\vdash_{cf} \Gamma, \Pi$]. Now we have $\{(\sim p)^{\perp}\} \in [p]_{-}$, i.e., $\vdash_{cf} (\sim p)^{\perp}, \sim p$, as an initial sequent. Thus, we obtain $\vdash_{cf} \Gamma, \sim(p^{\perp})$ by (*) and the law $(\sim p)^{\perp} \leftrightarrow \sim(p^{\perp})$. Therefore $\Gamma \in [p^{\perp}]_{-}$.

(Case $\alpha \equiv \beta \wedge \gamma$ for (2)): We show $(\beta \wedge \gamma)^{-} \subseteq [\beta \wedge \gamma]_{-}$. Suppose $\Gamma \in (\beta \wedge \gamma)^{-} = \beta^{-} \hat{\vee} \gamma^{-} = (\beta^{-} \cup \gamma^{-})^{\hat{\perp}\hat{\perp}}$. Now we have $\beta^{-} \subseteq [\beta]_{-}$ and $\gamma^{-} \subseteq [\gamma]_{-}$ by the induction hypothesis. Suppose (*): $[\beta]_{-} \cup [\gamma]_{-} \subseteq [\beta \wedge \gamma]_{-}$ (this will be proved). Then we have $\beta^{-} \cup \gamma^{-} \subseteq [\beta]_{-} \cup [\gamma]_{-} \subseteq [\beta \wedge \gamma]_{-}$, and hence (**): $\beta^{-} \cup \gamma^{-} \subseteq [\beta \wedge \gamma]_{-}$. By the hypothesis, (**), Proposition 9.5.4 (5) and Lemma 9.5.12 (2), we obtain $\Gamma \in (\beta^{-} \cup \gamma^{-})^{\hat{\perp}\hat{\perp}} \subseteq [\beta \wedge \gamma]_{-}^{\hat{\perp}\hat{\perp}} = [\beta \wedge \gamma]_{-}$. This means $\Gamma \in [\beta \wedge \gamma]_{-}$. We show the remained fact (*). Suppose $\Delta \in [\beta]_{-} \cup [\gamma]_{-}$. Then we have $\vdash_{cf} \Delta, \sim\beta$ or $\vdash_{cf} \Delta, \sim\gamma$. By applying ($\sim\wedge$1) or ($\sim\wedge$2), we obtain $\vdash_{cf} \Delta, \sim(\beta \wedge \gamma)$, and hence $\Delta \in [\beta \wedge \gamma]_{-}$.

(Case $\alpha \equiv !\beta$ for (2)): We show $(!\beta)^{-} \subseteq [!\beta]^{-}$. Suppose (*): $\Gamma \in (!\beta)^{-} = \hat{!}(\beta^{-}) = (\beta^{-} \cap \hat{I})^{\hat{\perp}\hat{\perp}}$. Also, suppose (**): $\beta^{-} \cap \hat{I} \subseteq [!\beta]_{-}$ (this will be proved). By (*) and (**), Proposition 9.5.4 (5) and Lemma 9.5.12 (2), we obtain $\Gamma \in (\beta^{-} \cap \hat{I})^{\hat{\perp}\hat{\perp}} \subseteq [!\beta]_{-}^{\hat{\perp}\hat{\perp}} = [!\beta]_{-}$. Therefore $\Gamma \in [!\beta]_{-}$. Next, we show the remained fact (**). Suppose $\Delta \in \beta^{-} \cap \hat{I}$, i.e., $\Delta \in \beta^{-}$ and $\Delta \in \hat{I}$. By the induction hypothesis $\beta^{-} \subseteq [\beta]_{-}$, we have $\Delta \in [\beta]_{-}$, and hence (1): $\vdash_{cf} \Delta, \sim\beta$. On the other hand, $\Delta \in \hat{I} = \{?\Sigma \cup \sim ?\Pi \in 2^{\Phi} \mid \Sigma \cup \Pi \in M\}$ and hence (2): Δ is of the form $?\Delta_1 \cup \sim ?\Delta_2$. By applying ($\sim$!) to (1), we obtain $\vdash_{cf} ?\Delta_1, \sim ?\Delta_2, \sim !\beta$, i.e., $\vdash_{cf} \Delta, \sim !\beta$. This means $\Delta \in [!\beta]_{-}$. ∎

By combining this main lemma with the soundness theorem, we can derive the cut-elimination theorem for (one-sided and two-sided) CLS (i.e., Theorem 9.2.2). Of course, we can also show the cut-elimination and completeness theorems for $\mathrm{CLS_w}$, $\mathrm{CLS_c}$ and $\mathrm{CLS_{cw}}$ with some appropriate modifications of the proof for CLS.

9.6. Remarks

We can introduce two alternative cut-free sequent calculi for the one-sided CLS, which are regarded as classical versions of the calculi proposed in [82]. These calculi can be modified for the other logics mentioned in this chapter, and can also be obtained for the two-sided versions in a similar way. Thus, we only focus on the case of the one-sided CLS.

First, we introduce a subformula calculus SC, which has the subformula property . The sequents of SC are of the forms $\vdash \Gamma : \Delta$ where Γ, Δ

are multisets of formulas. We call Γ and Δ in the expression *negative (or refutability) context* and *positive (or provability) context* respectively. The sequent $\vdash \Gamma : \Delta$ in SC intuitively means $\vdash {\sim}\Gamma, \Delta$ in CLS.

Definition 9.6.1 (SC) *We define a (one-sided) subformula calculus* SC*. The initial sequents of* SC *are of the forms:*

$$\vdash \varnothing : \alpha, \alpha^{\perp} \quad \vdash \alpha, \alpha^{\perp} : \varnothing \quad \vdash \varnothing : \mathbf{1}$$

$$\vdash \Gamma_1 : \Gamma_2, \top \quad \vdash \mathbf{0}, \Gamma_1 : \Gamma_2 \quad \vdash \bot : \varnothing$$

The inference rules of SC *are of the forms:*

$$\frac{\vdash \alpha, \Gamma_1 : \Gamma_2 \quad \vdash \alpha^{\perp}, \Delta_1 : \Delta_2}{\vdash \Gamma_1, \Delta_1 : \Gamma_2, \Delta_2} \quad \frac{\vdash \Gamma_1 : \Gamma_2, \alpha \quad \vdash \Delta_1 : \Delta_2, \alpha^{\perp}}{\vdash \Gamma_1, \Delta_1 : \Gamma_2, \Delta_2}$$

$$\frac{\vdash \Gamma : \Delta, \alpha}{\vdash {\sim}\alpha, \Gamma : \Delta} \quad \frac{\vdash \alpha, \Gamma : \Delta}{\vdash \Gamma : \Delta, {\sim}\alpha}$$

$$\frac{\vdash \Gamma : \Delta}{\vdash \Gamma : \Delta, \bot} \quad \frac{\vdash \Gamma_1 : \Gamma_2, \alpha \quad \vdash \Delta_1 : \Delta_2, \beta}{\vdash \Gamma_1, \Delta_1 : \Gamma_2, \Delta_2, \alpha * \beta}$$

$$\frac{\vdash \Gamma : \Delta, \alpha, \beta}{\vdash \Gamma : \Delta, \alpha + \beta} \quad \frac{\vdash \Gamma : \Delta, \alpha \quad \vdash \Gamma : \Delta, \beta}{\vdash \Gamma : \Delta, \alpha \wedge \beta}$$

$$\frac{\vdash \Gamma : \Delta, \alpha}{\vdash \Gamma : \Delta, \alpha \vee \beta} \quad \frac{\vdash \Gamma : \Delta, \beta}{\vdash \Gamma : \Delta, \alpha \vee \beta}$$

$$\frac{\vdash ?\Gamma_1 : ?\Gamma_2, \alpha}{\vdash ?\Gamma_1 : ?\Gamma_2, !\alpha} \quad \frac{\vdash \Gamma : \Delta, \alpha}{\vdash \Gamma : \Delta, ?\alpha} \quad \frac{\vdash \Gamma : \Delta, ?\alpha, ?\alpha}{\vdash \Gamma : \Delta, ?\alpha}$$

$$\frac{\vdash \Gamma : \Delta}{\vdash \Gamma : \Delta, ?\alpha} \quad \frac{\vdash \Gamma : \Delta}{\vdash \mathbf{1}, \Gamma : \Delta}$$

$$\frac{\vdash \alpha, \Gamma_1 : \Gamma_2 \quad \vdash \beta, \Delta_1 : \Delta_2}{\vdash \alpha * \beta, \Gamma_1, \Delta_1 : \Gamma_2, \Delta_2} \quad \frac{\vdash \alpha, \beta, \Gamma : \Delta}{\vdash \alpha + \beta, \Gamma : \Delta}$$

$$\frac{\vdash \alpha, \Gamma : \Delta \quad \vdash \beta, \Gamma : \Delta}{\vdash \alpha \vee \beta, \Gamma : \Delta} \quad \frac{\vdash \alpha, \Gamma : \Delta}{\vdash \alpha \wedge \beta, \Gamma : \Delta}$$

$$\frac{\vdash \beta, \Gamma : \Delta}{\vdash \alpha \wedge \beta, \Gamma : \Delta} \quad \frac{\vdash \alpha, ?\Gamma_1 : ?\Gamma_2}{\vdash !\alpha, ?\Gamma_1 : ?\Gamma_2} \quad \frac{\vdash \alpha, \Gamma : \Delta}{\vdash ?\alpha, \Gamma : \Delta}$$

$$\frac{\vdash ?\alpha, ?\alpha, \Gamma : \Delta}{\vdash ?\alpha, \Gamma : \Delta} \quad \frac{\vdash \Gamma : \Delta}{\vdash ?\alpha, \Gamma : \Delta}$$

We can obtain the equivalence between SC and CLS by (1) if $\vdash \Gamma : \Delta$ is provable in SC, then $\vdash {\sim}\Gamma, \Delta$ is provable in CLS, and (2) if $\vdash {\sim}\Gamma, \Delta$ is cut-free provable in CLS, then $\vdash \Gamma : \Delta$ is cut-free provable in SC. We can also obtain the cut-elimination and subformula properties for SC.

Next, we introduce a dual calculus DC. A sequent of the form $\vdash^{+} \Gamma$ is called a *positive (or provability) sequent*, and a sequent of the form $\vdash^{-} \Gamma$ is called a *negative (or refutability) sequent*.

Definition 9.6.2 (DC) *We define a (one-sided) dual calculus* DC. *The initial sequents of* DC *are of the forms:*

$$\vdash^{+} \alpha, \alpha^{\perp} \qquad \vdash^{+} \mathbf{1} \qquad \vdash^{+} \Gamma, \top \qquad \vdash^{-} \alpha, \alpha^{\perp} \qquad \vdash^{-} \Gamma, \mathbf{0} \qquad \vdash^{-} \perp$$

The inference rules of DC *are of the forms:*

$$\frac{\vdash^{-} \Gamma, \alpha \quad \vdash^{-} \Delta, \alpha^{\perp}}{\vdash^{-} \Gamma, \Delta} \qquad \frac{\vdash^{+} \Gamma, \alpha \quad \vdash^{+} \Delta, \alpha^{\perp}}{\vdash^{+} \Gamma, \Delta} \qquad \frac{\vdash^{-} {\sim}\Gamma, \Delta}{\vdash^{+} \Gamma, {\sim}\Delta} \qquad \frac{\vdash^{+} {\sim}\Gamma, \Delta}{\vdash^{-} \Gamma, {\sim}\Delta}$$

$$\frac{\vdash^{+} \Gamma}{\vdash^{+} \Gamma, \perp} \qquad \frac{\vdash^{+} \Gamma, \alpha \quad \vdash^{+} \Delta, \beta}{\vdash^{+} \Gamma, \Delta, \alpha * \beta} \qquad \frac{\vdash^{+} \Gamma, \alpha, \beta}{\vdash^{+} \Gamma, \alpha + \beta} \qquad \frac{\vdash^{+} \Gamma, \alpha \quad \vdash^{+} \Gamma, \beta}{\vdash^{+} \Gamma, \alpha \wedge \beta}$$

$$\frac{\vdash^{+} \Gamma, \alpha}{\vdash^{+} \Gamma, \alpha \vee \beta} \qquad \frac{\vdash^{+} \Gamma, \beta}{\vdash^{+} \Gamma, \alpha \vee \beta} \qquad \frac{\vdash^{+} ?\Gamma_1, {\sim}?\Gamma_2, \alpha}{\vdash^{+} ?\Gamma_1, {\sim}?\Gamma_2, !\alpha}$$

$$\frac{\vdash^{+} \Gamma, \alpha}{\vdash^{+} \Gamma, ?\alpha} \qquad \frac{\vdash^{+} \Gamma, ?\alpha, ?\alpha}{\vdash^{+} \Gamma, ?\alpha}$$

$$\frac{\vdash^{+} \Gamma}{\vdash^{+} \Gamma, ?\alpha} \qquad \frac{\vdash^{-} \Gamma}{\vdash^{-} \Gamma, \mathbf{1}} \qquad \frac{\vdash^{-} \Gamma, \alpha \quad \vdash^{-} \Delta, \beta}{\vdash^{-} \Gamma, \Delta, \alpha * \beta}$$

$$\frac{\vdash^{-} \Gamma, \alpha, \beta}{\vdash^{-} \Gamma, \alpha + \beta} \qquad \frac{\vdash^{-} \Gamma, \alpha \quad \vdash^{-} \Gamma, \beta}{\vdash^{-} \Gamma, \alpha \vee \beta}$$

$$\frac{\vdash^{-} \Gamma, \alpha}{\vdash^{-} \Gamma, \alpha \wedge \beta} \qquad \frac{\vdash^{-} \Gamma, \beta}{\vdash^{-} \Gamma, \alpha \wedge \beta} \qquad \frac{\vdash^{-} ?\Gamma_1, {\sim}?\Gamma_2, \alpha}{\vdash^{-} ?\Gamma_1, {\sim}?\Gamma_2, !\alpha}$$

$$\frac{\vdash^{-} \Gamma, \alpha}{\vdash^{-} \Gamma, ?\alpha} \qquad \frac{\vdash^{-} \Gamma, ?\alpha, ?\alpha}{\vdash^{-} \Gamma, ?\alpha} \qquad \frac{\vdash^{-} \Gamma}{\vdash^{-} \Gamma, ?\alpha}$$

We can obtain the equivalence between DC and CLS by (1) if $\vdash^{*} \Gamma$ ($* \in \{+, -\}$) is provable in DC, then the sequent $\vdash \Gamma$ is provable in CLS if $* = +$, or the sequent $\vdash {\sim}\Gamma$ is provable in CLS if $* = -$, and (2) if $\vdash \Gamma$ is cut-free provable in CLS, then the sequent $\vdash^{+} \Gamma$ is cut-free provable in DC. We can also obtain the cut-elimination property for DC.

10. Paraconsistent intuitionistic linear logics

The extended intuitionistic linear logic with strong negation, investigated in [196, 197] and referred to as WILL in [82] and elsewhere, may be regarded as a resource-conscious refinement of Nelson's constructive logics with strong negation. In this chapter, (1) the completeness theorem with respect to phase semantics is proved for WILL using a method that simultaneously allows one to derive the cut-elimination theorem, (2) a simple correspondence between the class of Petri nets with inhibitor arcs and a fragment of WILL is obtained using a Kripke semantics, (3) a cut-free sequent calculus for WILL, called twist calculus, is presented, and (4) new applications of WILL in medical diagnosis and electric circuit theory are proposed. Strong negation in WILL is found to be expressible as a resource-conscious operation of refutability, and is shown to correspond to inhibitor arcs in Petri net theory.[1]

10.1. Introduction

This chapter discusses the role of strong negation in both pure theoretical foundations and some realistic applications. To consider this issue, a phase semantics and a sequent calculus for an underlying logic with strong negation are introduced in order to allow a fundamental analysis of strong negation. The theory of Petri nets with inhibitor arcs, the theory of inconsistency-tolerant medical diagnosis, and the theory of electric circuits are then discussed as realistic applications of strong negation. Before the detailed discussion, linear logics, logics with strong negation, and Petri net theory are briefly explained below.

Linear logics [64] have been widely researched as tools for resource-sensitive reasoning, as frameworks for logic programming languages [75], and as logical foundations of concurrency theory including π-calculus specifications [20], process-algebraic treatments [43], process-calculus interpretations [2, 143] and Petri net interpretations. In particular, *intuitionistic linear logic* is known as a good candidate for obtaining the

[1] This chapter rests upon [92].

logical foundations of Petri net theory.

Logics with strong negation were first introduced by Nelson [134]. The extended intuitionistic (non-modal) linear logic with strong negation, called WILL here, was introduced by Wansing as a generalization (or resource-conscious refinement) of Nelson's logics, and various kinds of substructural logics including WILL and their informational interpretations using Došen's groupoid models have been studied [197]. An extension of WILL with the exponential operator ! has also been proposed and discussed [82]. These kinds of logics with strong negation have a number of useful sequent calculi, as well as a number of complete semantics such as Kripke-type semantics, with many applications including the process calculus, logic programming languages, non-monotonic reasoning and the analysis of the knowability paradox (see e.g., [4, 83, 194, 202]).

Strong negation in WILL has the desirable feature of paraconsistency, which is known to be useful to discuss inconsistency-tolerant reasoning and non-monotonic reasoning in AI [41]. Strong negation in WILL also has an important characteristic property called *constructible falsity* [134]. This property can deal with *inexact predicates*, which can express imperfect or incomplete information in an empirical domain [194]. Thus, WILL can express not only paraconsistency and constructible falsity, but also resource-sensitivity.

Petri nets were already conceived by C.A. Petri in 1939 and later presented in his seminal Ph.D. thesis, . Both the theory and the applications of his model have flourished in concurrency theory (see e.g., [165, 166]). The relationships between Petri nets and intuitionistic linear logics have been studied by many computer scientists and logicians [55, 79, 106, 107, 116, 117, 118, 126, 143]. For example, a category theoretical investigation of such a relationship was given by Martí-Oliet and Meseguer [126], a purely syntactical approach using Horn linear logic was established by Kanovich [106, 107], a naïve phase model for a certain class of Petri nets was given by Okada [143], and various Petri net interpretations of linear logic using quantale models have been obtained by Ishihara and Hiraishi [79], Engberg and Winskel [55], Larchey-Wendling and Galmiche [116, 117] and Lilius [118].

Petri nets with inhibitor arcs were first introduced by Kosaraju [110] to show the limitation of the usual Petri nets. The class of Petri nets with inhibitor arcs is known as a proper extension of the class of usual Petri nets. In such Petri nets, XOR (exclusive-or) transitions can be expressed, Turing machines can be encoded, and the problem of common channel with priority can be solved [151]. Extended or modified versions of Petri nets with inhibitor arcs have been studied by many researchers (e.g., [109, 175, 177]).

The content of this chapter is then summarized as follows. In Section 10.2, WILL and its characteristic properties such as constructible falsity and paraconsistency are reviewed. Admissibility of the rule of *weak contraposition* is also discussed. In addition, a faithful embedding of WILL into ILL (the original non-modal intuitionistic linear logic) is obtained.

In Section 10.3, a new cut-free sequent calculus for WILL, called *twist calculus*, is presented in order to consider the derivability of a new rule "*quasi weak contraposition*". This calculus provides an intuitive meaning for strong negation, i.e., strong negation can represent refutability. This calculus is also a modification of the dual calculus discussed in [82].

In Section 10.4, a phase semantic completeness and cut-elimination proof for WILL is given through a modification of Okada's method [144], representing the main contribution to the theoretical treatment in this chapter. The difference between the present semantics and that for the original linear logic is the use of a negative valuation v^- to represent the refutability of formulas.

In Section 10.5, a simple Petri net interpretation for WL (a fragment of WILL) is given based on a simple Kripke semantics with a partially ordered commutative monoid as the base frame. Using this interpretation, the correspondence between the class of Petri nets with inhibitor arcs and WL is shown, i.e., the reachability in a Petri net with inhibitor arcs corresponds to the provability in WL. Some examples such as a Petri net with an infinite reachability tree and a Petri net expressing an XOR transition are also presented using the proposed interpretation, and it is shown that strong negation is closely related to inhibitor arcs. This result using Kripke semantics is a new approach to obtain the logical foundation of Petri net theory, and is also applicable to a fragment of ILL.

In Section 10.6, an application of WILL in medical diagnosis is presented. In this example, the notion of paraconsistency discussed in Section 10.2 is reconsidered from a realistic point of view. The comparison of strong negation and usual linear negation is also discussed.

In Section 10.7, an application of WILL in the theory of electric circuits is discussed. In this example, strong negation is closely related to the direction of the flow of electric current.

Finally, some remarks are given in Section 10.8.

10.2. Sequent calculus

Prior to a more detailed discussion, the language and notion used in this chapter are introduced. *Formulas* are constructed from propositional variables, propositional constants $\mathbf{1}, \top$, and $\bot$, $\rightarrow$ (implication), $\wedge$ (conjunction), $*$ (fusion), $\vee$ (disjunction) and $\sim$ (strong negation). This chapter adopts the notation for constants $\mathbf{1}, \top$ and $\bot$ as presented in [187], differing from that in [64] ($\wedge, \vee$, and $*$ correspond to $\&, \oplus$ and $\otimes$ in [64]). Lower case letters $p, q, ...$ are used for propositional variables, lower case Greek letters $\alpha, \beta, ...$ are used to denote formulas, and Greek capital letters $\Gamma, \Delta, ...$ are used to represent finite (possibly empty) sequences of formulas, and $[\Gamma]$ denotes the multiset consists of all elements of a sequence Γ of formulas. $\sim\Gamma$ denotes the sequence $\langle \sim\gamma \mid \gamma \in \Gamma \rangle$. A *sequent* is an expression of the form $\Gamma \Rightarrow \alpha$, where α is non-empty (i.e. a single formula). The symbol $\equiv$ is used to denote equality of sequences (or multisets) of symbols. Δ^* denotes $\delta_1 * \cdots * \delta_n$ if $\Delta \equiv \langle \delta_1, ..., \delta_n \rangle$ $(1 \leq n)$, and denotes $\mathbf{1}$ if Δ is empty. If a sequent S is provable in a sequent system L, then we write $L \vdash S$, or also $\vdash S$ by omitting L. Parentheses for $*$ are omitted because $*$ is associative (i.e. $\vdash \alpha * (\beta * \gamma) \Rightarrow (\alpha * \beta) * \gamma$ and $\vdash (\alpha * \beta) * \gamma \Rightarrow \alpha * (\beta * \gamma)$ for any formulas α, β, γ). Since all logics discussed in this chapter are formulated as sequent calculi, a sequent calculus will occasionally be identified with the logic determined by it.

First, we introduce ILL (intuitionistic linear logic without exponential !).

Definition 10.2.1 (ILL) *The initial sequents of* ILL *are of the forms:*

$$\alpha \Rightarrow \alpha \qquad \Rightarrow \mathbf{1} \qquad \Gamma \Rightarrow \top \qquad \bot, \Delta \Rightarrow \gamma.$$

The structural rules of ILL *are of the forms:*

$$\frac{\Gamma \Rightarrow \alpha \quad \alpha, \Sigma \Rightarrow \gamma}{\Gamma, \Sigma \Rightarrow \gamma} \text{ (cut)} \qquad \frac{\Gamma, \beta, \alpha, \Delta \Rightarrow \gamma}{\Gamma, \alpha, \beta, \Delta \Rightarrow \gamma} \text{ (ex)}.$$

The inference rules of ILL *are of the forms:*

$$\frac{\Gamma \Rightarrow \gamma}{\mathbf{1}, \Gamma \Rightarrow \gamma} \text{ (}\mathbf{1}\text{we)}$$

$$\frac{\Gamma \Rightarrow \alpha \quad \beta, \Sigma \Rightarrow \gamma}{\alpha \rightarrow \beta, \Gamma, \Sigma \Rightarrow \gamma} \text{ (}\rightarrow\text{left)} \qquad \frac{\Gamma, \alpha \Rightarrow \beta}{\Gamma \Rightarrow \alpha \rightarrow \beta} \text{ (}\rightarrow\text{right)}$$

$$\frac{\alpha, \beta, \Delta \Rightarrow \gamma}{\alpha * \beta, \Delta \Rightarrow \gamma} \text{ (}*\text{left)} \qquad \frac{\Gamma \Rightarrow \alpha \quad \Delta \Rightarrow \beta}{\Gamma, \Delta \Rightarrow \alpha * \beta} \text{ (}*\text{right)}$$

$$\frac{\alpha, \Delta \Rightarrow \gamma}{\alpha \wedge \beta, \Delta \Rightarrow \gamma} \ (\wedge\text{left1}) \qquad \frac{\beta, \Delta \Rightarrow \gamma}{\alpha \wedge \beta, \Delta \Rightarrow \gamma} \ (\wedge\text{left2})$$

$$\frac{\Gamma \Rightarrow \alpha \quad \Gamma \Rightarrow \beta}{\Gamma \Rightarrow \alpha \wedge \beta} \ (\wedge\text{right}) \qquad \frac{\alpha, \Delta \Rightarrow \gamma \quad \beta, \Delta \Rightarrow \gamma}{\alpha \vee \beta, \Delta \Rightarrow \gamma} \ (\vee\text{left})$$

$$\frac{\Gamma \Rightarrow \alpha}{\Gamma \Rightarrow \alpha \vee \beta} \ (\vee\text{right1}) \qquad \frac{\Gamma \Rightarrow \beta}{\Gamma \Rightarrow \alpha \vee \beta} \ (\vee\text{right2}).$$

Definition 10.2.2 (WILL) *The logic* WILL *(an extension of ILL with strong negation) is obtained from* ILL *by adding the initial sequents and inference rules of the forms:*

$$\sim\mathbf{1}, \Gamma \Rightarrow \gamma \qquad \sim\top, \Gamma \Rightarrow \gamma \qquad \Gamma \Rightarrow \sim\bot$$

$$\frac{\Gamma \Rightarrow \alpha}{\Gamma \Rightarrow \sim\sim\alpha} \ (\sim\text{right}) \qquad \frac{\alpha, \Delta \Rightarrow \gamma}{\sim\sim\alpha, \Delta \Rightarrow \gamma} \ (\sim\text{left})$$

$$\frac{\sim\beta, \alpha, \Delta \Rightarrow \gamma}{\sim(\alpha\to\beta), \Delta \Rightarrow \gamma} \ (\sim\to\text{left}) \qquad \frac{\Gamma \Rightarrow \sim\beta \quad \Delta \Rightarrow \alpha}{\Gamma, \Delta \Rightarrow \sim(\alpha\to\beta)} \ (\sim\to\text{right})$$

$$\frac{\sim\alpha, \sim\beta, \Delta \Rightarrow \gamma}{\sim(\alpha * \beta), \Delta \Rightarrow \gamma} \ (\sim * \text{left}) \qquad \frac{\Gamma \Rightarrow \sim\alpha \quad \Delta \Rightarrow \sim\beta}{\Gamma, \Delta \Rightarrow \sim(\alpha * \beta)} \ (\sim * \text{right})$$

$$\frac{\sim\alpha, \Delta \Rightarrow \gamma \quad \sim\beta, \Delta \Rightarrow \gamma}{\sim(\alpha \wedge \beta), \Delta \Rightarrow \gamma} \ (\sim \wedge \text{left})$$

$$\frac{\Gamma \Rightarrow \sim\alpha}{\Gamma \Rightarrow \sim(\alpha \wedge \beta)} \ (\sim \wedge \text{right1}) \qquad \frac{\Gamma \Rightarrow \sim\beta}{\Gamma \Rightarrow \sim(\alpha \wedge \beta)} \ (\sim \wedge \text{right2})$$

$$\frac{\sim\alpha, \Delta \Rightarrow \gamma}{\sim(\alpha \vee \beta), \Delta \Rightarrow \gamma} \ (\sim \vee \text{left1}) \qquad \frac{\sim\beta, \Delta \Rightarrow \gamma}{\sim(\alpha \vee \beta), \Delta \Rightarrow \gamma} \ (\sim \vee \text{left2})$$

$$\frac{\Gamma \Rightarrow \sim\alpha \quad \Gamma \Rightarrow \sim\beta}{\Gamma \Rightarrow \sim(\alpha \vee \beta)} \ (\sim \vee \text{right}).$$

Note that the rules ($\sim\to$left) and ($\sim\to$right) correspond to the axiom scheme $\sim(\alpha\to\beta) \leftrightarrow \alpha * \sim\beta$ by assuming the presence of (ex). Since the absence of the contraction and weakening rules results in the distinction between $*$ and $\wedge$, it can naturally be discussed on the rules of the forms:

$$\frac{\alpha, \Gamma \Rightarrow \gamma}{\sim(\alpha\to\beta), \Gamma \Rightarrow \gamma} \qquad \frac{\sim\beta, \Gamma \Rightarrow \gamma}{\sim(\alpha\to\beta), \Gamma \Rightarrow \gamma} \qquad \frac{\Gamma \Rightarrow \alpha \quad \Gamma \Rightarrow \sim\beta}{\Gamma \Rightarrow \sim(\alpha\to\beta)}$$

which correspond to the axiom scheme $\sim(\alpha\to\beta) \leftrightarrow \alpha \wedge \sim\beta$. We can consider the system obtained from WILL by adding these rules, and can obtain some good properties, such as cut-elimination and embedding, for the extended system. But, the following discussion focuses only on WILL.

The following theorem is presented in [197].

Theorem 10.2.3 (Cut-elimination) *The rule* (cut) *is admissible in cut-free* WILL.

In Section 10.6, we will obtain an alternative semantic proof of this theorem. Using Theorem 10.2.3, we can derive the following.

Corollary 10.2.4 WILL *is decidable and a conservative extension of* ILL.

A logic L is called *explosive* if for any formulas α and β, the sequent $\alpha, {\sim}\alpha \Rightarrow \beta$ is provable in L. L is called *paraconsistent* if L is not explosive.

Corollary 10.2.5 (Paraconsistency) WILL *is paraconsistent.*

The following is a characteristic property of logics with strong negation.

Corollary 10.2.6 (Constructible falsity) *If* WILL $\vdash \; \Rightarrow {\sim}(\alpha \wedge \beta)$, *then either* WILL $\vdash \; \Rightarrow {\sim}\alpha$ *or* WILL $\vdash \; \Rightarrow {\sim}\beta$.

In addition, we consider the rule, called *weak contraposition*, of the form:

$$\frac{\alpha \Rightarrow \beta}{{\sim}\beta \Rightarrow {\sim}\alpha}\ (\text{w-con}).$$

This rule will be discussed from a theoretical point of view (Section 10.3) and from an application point of view (Section 10.6).

We have the following by restricting the language of WILL.

Proposition 10.2.7 (Admissibility of weak contraposition) *Let* α, β *and* Γ *be formulas and a sequence of formulas respectively which are constructed only from* $\wedge, \vee, {\sim}$ *and propositional variables. Then the rule* (w-con) *is admissible in cut-free* WILL.

Proof. To prove this proposition, we remark that the rules

$$\frac{\Gamma \Rightarrow {\sim}{\sim}\alpha}{\Gamma \Rightarrow \alpha}\ ({\sim}\text{right}^{-1}) \qquad \frac{{\sim}{\sim}\alpha, \Gamma \Rightarrow \gamma}{\alpha, \Gamma \Rightarrow \gamma}\ ({\sim}\text{left}^{-1})$$

are admissible in cut-free WILL.

Then, we prove this proposition by induction on the cut-free proof P of $\alpha \Rightarrow \beta$ in WILL. We distinguish cases according to the last inferences of P. By the assumption, it is enough to consider the rule (ex) and the rules concerning the connectives $\wedge, \vee$ and $\sim$. Here, we only show the following case.

Case ($\sim\wedge$left): The last inference of P is of the form:

$$\frac{\sim\alpha_1 \Rightarrow \beta \quad \sim\alpha_2 \Rightarrow \beta}{\sim(\alpha_1 \wedge \alpha_2) \Rightarrow \beta} \ (\sim\wedge\text{left})$$

where $\alpha \equiv \sim(\alpha_1 \wedge \alpha_2)$. By the hypothesis of induction, we have that $\vdash \sim\beta \Rightarrow \sim\sim\alpha_1$ and $\vdash \sim\beta \Rightarrow \sim\sim\alpha_2$. Then we obtain the required fact:

$$\cfrac{\cfrac{\cfrac{\sim\beta \Rightarrow \sim\sim\alpha_1}{\sim\beta \Rightarrow \alpha_1} \ (\sim\text{right}^{-1}) \quad \cfrac{\sim\beta \Rightarrow \sim\sim\alpha_2}{\sim\beta \Rightarrow \alpha_2} \ (\sim\text{right}^{-1})}{\sim\beta \Rightarrow \alpha_1 \wedge \alpha_2}}{\sim\beta \Rightarrow \sim\sim(\alpha_1 \wedge \alpha_2)} \ .$$

■

This property cannot be proved for WILL in a similar way if the language includes the full connectives $\rightarrow, \mathbf{1}, \top, \bot$ and $*$. However, we have not yet found a counterexample to the non-admissibility of (w-con). On the other hand, in the next section, we can derive the fact that some analogous rules for (w-con) are derivable in a twist calculus for WILL.

Next, we give a faithful embedding of WILL into ILL.

Definition 10.2.8 *We fix a set PROP of propositional variables used as components of the language of* WILL, *and define the set* $PROP' := \{p' \mid p \in PROP\}$ *of propositional variables. The language* L_{WILL} *of* WILL *is defined using PROP,* $\mathbf{1}, \top, \bot, \wedge, \vee, *, \rightarrow$ *and* $\sim$. *The language* L_{ILL} *of* ILL *is obtained from* L_{WILL} *by adding PROP′ and deleting* $\sim$.

A mapping f from L_{WILL} to L_{ILL} is defined inductively as follows.

1. $f(p) := p$ *and* $f(\sim p) := p' \in PROP'$ *for any* $p \in PROP$,
2. $f(\mathbf{1}) := \mathbf{1}$,
3. $f(\top) := \top$,
4. $f(\bot) := \bot$,
5. $f(\alpha \circ \beta) := f(\alpha) \circ f(\beta)$ *where* $\circ \in \{*, \wedge, \vee, \rightarrow\}$,
6. $f(\sim\mathbf{1}) := \bot$,
7. $f(\sim\top) := \bot$,

8. $f(\sim\bot) := \top$,

9. $f(\sim\sim\alpha) := f(\alpha)$,

10. $f(\sim(\alpha * \beta)) := f(\sim\alpha) * f(\sim\beta)$,

11. $f(\sim(\alpha\rightarrow\beta)) := f(\alpha) * f(\sim\beta)$,

12. $f(\sim(\alpha \wedge \beta)) := f(\sim\alpha) \vee f(\sim\beta)$,

13. $f(\sim(\alpha \vee \beta)) := f(\sim\alpha) \wedge f(\sim\beta)$.

Let Γ be a sequence of formulas in L_{WILL}. Then, $f(\Gamma)$ denotes the result of replacing every occurrence of a formula α in Γ by an occurrence of $f(\alpha)$.

Proposition 10.2.9 *Let Γ be a sequence of formulas in L_{WILL}. Then:*

$$\text{WILL} \vdash \Gamma \Rightarrow \gamma \textit{ iff } \text{ILL} \vdash f(\Gamma) \Rightarrow f(\gamma).$$

Proof. The case ($\Longleftarrow$) is obvious. Thus, we show the case ($\Longrightarrow$) by induction on the proof P of WILL. We distinguish cases according to the last inference of P. We only show the following case.

Case ($\sim\rightarrow$right): The last inference rule of P is of the form:

$$\frac{\sim\beta, \alpha, \Delta \Rightarrow \gamma}{\sim(\alpha\rightarrow\beta), \Delta \Rightarrow \gamma} \ (\sim\rightarrow\text{right}).$$

By the hypothesis of induction, we have that ILL $\vdash f(\sim\beta), f(\alpha), f(\Delta) \Rightarrow f(\gamma)$, and hence we obtain that ILL $\vdash f(\alpha) * f(\sim\beta), f(\Delta) \Rightarrow f(\gamma)$ by using the rules (ex) and ($*$left). Therefore we have the required fact that ILL $\vdash f(\sim(\alpha\rightarrow\beta)), f(\Delta) \Rightarrow f(\gamma)$ by the definition $f(\sim(\alpha\rightarrow\beta)) = f(\alpha) * f(\sim\beta)$. ∎

10.3. Twist calculus

We introduce a new sequent calculus TWILL, called *twist calculus*. In the following, a sequent of the form $\Gamma \Rightarrow^{+} \gamma$ is called a *positive sequent*, which corresponds to "*provability*", and a sequent of the form $\gamma \Rightarrow^{-} \Gamma$ is called a *negative sequent*, which corresponds to "*refutability*". In the following definitions, γ in expression $\Gamma \Rightarrow^{+} \gamma$ or $\gamma \Rightarrow^{-} \Gamma$ for any Γ is a single formula.

Definition 10.3.1 (TWILL) *The initial sequents of* TWILL *are of the forms:*

$$\alpha \Rightarrow^+ \alpha \qquad \Rightarrow^+ \mathbf{1} \qquad \Gamma \Rightarrow^+ \top \qquad \bot, \Delta \Rightarrow^+ \gamma$$

$$\alpha \Rightarrow^- \alpha \qquad \gamma \Rightarrow^- \Gamma, \mathbf{1} \qquad \gamma \Rightarrow^- \Gamma, \top \qquad \bot \Rightarrow^- \Gamma.$$

The specific inference rules of TWILL *are of the forms:*

$$\frac{\gamma \Rightarrow^- {\sim}\Gamma, \Delta}{\Gamma, {\sim}\Delta \Rightarrow^+ {\sim}\gamma}\ (-/+1) \qquad \frac{{\sim}\gamma \Rightarrow^- {\sim}\Gamma, \Delta}{\Gamma, {\sim}\Delta \Rightarrow^+ \gamma}\ (-/+2)$$

$$\frac{{\sim}\Gamma, \Delta \Rightarrow^+ \gamma}{{\sim}\gamma \Rightarrow^- \Gamma, {\sim}\Delta}\ (+/-1) \qquad \frac{{\sim}\Gamma, \Delta \Rightarrow^+ {\sim}\gamma}{\gamma \Rightarrow^- \Gamma, {\sim}\Delta}\ (+/-2).$$

The cut rules of TWILL *are of the forms:*

$$\frac{\Gamma \Rightarrow^+ \alpha \quad \alpha, \Sigma \Rightarrow^+ \gamma}{\Gamma, \Sigma \Rightarrow^+ \gamma}\ (+\text{cut}) \qquad \frac{\gamma \Rightarrow^- \alpha, \Gamma \quad \alpha \Rightarrow^- \Sigma}{\gamma \Rightarrow^- \Gamma, \Sigma}\ (-\text{cut}).$$

The positive inference rules of TWILL *are obtained from the inference rules of the* $\sim$*-free part of cut-free* WILL *by replacing* $\Rightarrow$ *by* $\Rightarrow^+$.

The negative rules of TWILL *are of the forms:*

$$\frac{\gamma \Rightarrow^- \Delta, \beta, \alpha, \Gamma}{\gamma \Rightarrow^- \Delta, \alpha, \beta, \Gamma}\ (-\text{ex})$$

$$\frac{\gamma \Rightarrow^- \beta, {\sim}\alpha, \Delta}{\gamma \Rightarrow^- \alpha{\rightarrow}\beta, \Delta}\ (-{\rightarrow}\text{right}) \qquad \frac{\beta \Rightarrow^- \Gamma \quad {\sim}\alpha \Rightarrow^- \Delta}{\alpha{\rightarrow}\beta \Rightarrow^- \Gamma, \Delta}\ (-{\rightarrow}\text{left})$$

$$\frac{\gamma \Rightarrow^- \alpha, \beta, \Delta}{\gamma \Rightarrow^- \alpha * \beta, \Delta}\ (- * \text{right}) \qquad \frac{\alpha \Rightarrow^- \Gamma \quad \beta \Rightarrow^- \Delta}{\alpha * \beta \Rightarrow^- \Gamma, \Delta}\ (- * \text{left})$$

$$\frac{\gamma \Rightarrow^- \alpha, \Delta \quad \gamma \Rightarrow^- \beta, \Delta}{\gamma \Rightarrow^- \alpha \wedge \beta, \Delta}\ (- \wedge \text{right})$$

$$\frac{\alpha \Rightarrow^- \Gamma}{\alpha \wedge \beta \Rightarrow^- \Gamma}\ (- \wedge \text{left1}) \qquad \frac{\beta \Rightarrow^- \Gamma}{\alpha \wedge \beta \Rightarrow^- \Gamma}\ (- \wedge \text{left2})$$

$$\frac{\gamma \Rightarrow^- \alpha, \Delta}{\gamma \Rightarrow^- \alpha \vee \beta, \Delta}\ (- \vee \text{right1}) \qquad \frac{\gamma \Rightarrow^- \beta, \Delta}{\gamma \Rightarrow^- \alpha \vee \beta, \Delta}\ (- \vee \text{right2})$$

$$\frac{\alpha \Rightarrow^- \Gamma \quad \beta \Rightarrow^- \Gamma}{\alpha \vee \beta \Rightarrow^- \Gamma}\ (- \vee \text{left}).$$

Theorem 10.3.2 (Equivalence between TWILL **and** WILL**)** *Let* Γ *be a sequence of formulas,* γ *be a formula. Then:*

1. *If* TWILL $\vdash \Gamma \Rightarrow^+ \gamma$ *then* WILL $\vdash \Gamma \Rightarrow \gamma$*, and if* TWILL $\vdash \gamma \Rightarrow^- \Gamma$ *then* WILL $\vdash {\sim}\Gamma \Rightarrow {\sim}\gamma$.

2. *If* WILL$-$(cut) $\vdash \Gamma \Rightarrow \gamma$ *then* TWILL$-$(+cut)$-$($-$cut) $\vdash \Gamma \Rightarrow^{+} \gamma$.

Proof. First we show (1) by induction on the proof P of $\Gamma \Rightarrow^{+} \gamma$ or $\gamma \Rightarrow^{-} \Gamma$ in TWILL. We distinguish cases according to the last inference in P. We only show the following cases.

Case (+/$-$1): The last inference of P is

$$\frac{\sim\Gamma, \Delta \Rightarrow^{+} \gamma}{\sim\gamma \Rightarrow^{-} \Gamma, \sim\Delta}\ (+/-1).$$

By the induction hypothesis, $\sim\Gamma, \Delta \Rightarrow \gamma$ is provable in WILL. Then we obtain the required fact that $\sim\Gamma, \sim\sim\Delta \Rightarrow \sim\sim\gamma$ is provable in WILL using ($\sim$left) and ($\sim$right).

Case ($-$$\to$right): The last inference of P is

$$\frac{\gamma \Rightarrow^{-} \beta, \sim\alpha, \Delta}{\gamma \Rightarrow^{-} \alpha\to\beta, \Delta}\ (-{\to}\text{right}).$$

By the induction hypothesis, $\sim\beta, \sim\sim\alpha, \sim\Delta \Rightarrow \sim\gamma$ is provable in WILL. Then we obtain the required fact:

$$\cfrac{\cfrac{\cfrac{\alpha \Rightarrow \alpha}{\alpha \Rightarrow \sim\sim\alpha}\ (\sim\text{right}) \qquad \sim\beta, \sim\sim\alpha, \sim\Delta \Rightarrow \sim\gamma}{\sim\beta, \alpha, \sim\Delta \Rightarrow \sim\gamma}\ (\text{cut})}{\sim(\alpha\to\beta), \sim\Delta \Rightarrow \sim\gamma}\ (\sim{\to}\text{left}).$$

Next we show (2) by induction on the cut-free proof P of $\Gamma \Rightarrow \gamma$ in WILL. We distinguish cases according to the last inference in P. We only show the following case.

Case ($\sim$$\to$left): The last inference of P is

$$\frac{\sim\beta, \alpha, \Delta \Rightarrow \gamma}{\sim(\alpha\to\beta), \Delta \Rightarrow \gamma}\ (\sim{\to}\text{left}).$$

By the induction hypothesis, $\sim\beta, \alpha, \Delta \Rightarrow^{+} \gamma$ is cut-free provable in TWILL. Then we obtain

$$\cfrac{\cfrac{\cfrac{\sim\beta, \alpha, \Delta \Rightarrow^{+} \gamma}{\sim\gamma \Rightarrow^{-} \beta, \sim\alpha, \sim\Delta}\ (+/-1)}{\sim\gamma \Rightarrow^{-} \alpha\to\beta, \sim\Delta}\ (-{\to}\text{right})}{\sim(\alpha\to\beta), \Delta \Rightarrow^{+} \gamma}\ (-/+2).$$

∎

Theorem 10.3.3 (Cut-elimination) *The rules* (+cut) *and* ($-$cut) *are admissible in cut-free* TWILL.

Proof. Suppose that a sequent $\Gamma \Rightarrow^{+} \gamma$ or $\gamma \Rightarrow^{-} \Gamma$ is provable in TWILL. Then, by Theorem 10.3.2 (1), the sequent $\Gamma \Rightarrow \gamma$ is provable in WILL in the former case, or $\sim\Gamma \Rightarrow \sim\gamma$ is provable in WILL in the latter case. Hence the sequent $\Gamma \Rightarrow \gamma$ or $\sim\Gamma \Rightarrow \sim\gamma$ is cut-free provable in WILL by Theorem 10.2.3. If $\Gamma \Rightarrow \gamma$ is cut-free provable in WILL then $\Gamma \Rightarrow^{+} \gamma$ is cut-free provable in TWILL by Theorem 10.3.2 (2). If $\sim\Gamma \Rightarrow \sim\gamma$ is cut-free provable in WILL then $\sim\Gamma \Rightarrow^{+} \sim\gamma$ is cut-free provable in TWILL by Theorem 10.3.2 (2), and hence $\gamma \Rightarrow^{-} \Gamma$ is cut-free provable in TWILL by using (+/−2). ∎

Obviously, we have the following proposition.

Proposition 10.3.4 (Derivability of quasi weak contraposition) *The following rules, called* quasi weak contraposition rules, *are derivable in* TWILL.

$$\frac{\alpha \Rightarrow^{+} \beta}{\sim\beta \Rightarrow^{-} \sim\alpha} \qquad \frac{\alpha \Rightarrow^{-} \beta}{\sim\beta \Rightarrow^{+} \sim\alpha}.$$

10.4. Phase semantics and completeness

Definition 10.4.1 *An* intuitionistic phase space[2] *is a structure* $\langle \mathbf{M}, cl\rangle$ *satisfying the following conditions:*

1. $\mathbf{M} := \langle M, \cdot, 1\rangle$ *is a commutative monoid with the identity* 1,
2. *cl is a closure operation on the power set* $P(M)$ *of* M *such that, for any* $X, Y \in P(M)$,

   C1: $X \subseteq cl(X)$,

   C2: $cl cl(X) \subseteq cl(X)$,

   C3: $X \subseteq Y$ *implies* $cl(X) \subseteq cl(Y)$,

   C4: $cl(X) \circ cl(Y) \subseteq cl(X \circ Y)$

   where the operation $\circ$ *is defined as* $X \circ Y := \{x \cdot y \mid x \in X \text{ and } y \in Y\}$ *for any* $X, Y \in P(M)$.

Definition 10.4.2 *We define constants and operations on* $P(M)$ *as follows: for any* $X, Y \in P(M)$,

[2] The term "intuitionistic phase space" is borrowed from [144]. Since the phase semantics can be adapted to other substructural logics and to intuitionistic logic, this term may give rise to the confusion that the intuitionistic phase space is for intuitionistic logic. But, in this chapter, we only discuss the linear logic base, and hence the term "intuitionistic" means "intuitionistic linear".

1. $\dot{\mathbf{1}} := cl\{1\}$,
2. $\dot{\top} := M$,
3. $\dot{\bot} := cl(\emptyset)$,
4. $X \dot{\rightarrow} Y := \{y \mid \forall x \in X\ (x \cdot y \in Y)\}$,
5. $X \dot{\wedge} Y := X \cap Y$,
6. $X \dot{\vee} Y := cl(X \cup Y)$,
7. $X \dot{*} Y := cl(X \circ Y)$.

We define $D := \{X \in P(M) \mid X = cl(X)\}$. *Then*

$$\mathbf{D} := \langle D, \dot{\rightarrow}, \dot{*}, \dot{\wedge}, \dot{\vee}, \dot{\mathbf{1}}, \dot{\top}, \dot{\bot} \rangle$$

is called an intuitionistic phase structure.

We remark that the following conditions hold: for any $X, X', Y, Y', Z \in P(M)$,

1. $X \subseteq Y \dot{\rightarrow} Z$ iff $X \circ Y \subseteq Z$, and
2. $X \subseteq X'$ and $Y \subseteq Y'$ imply $X \circ Y \subseteq X' \circ Y'$ and $X' \dot{\rightarrow} Y \subseteq X \dot{\rightarrow} Y'$.

We remark that D is closed under the operations $\dot{\rightarrow}, \dot{*}, \dot{\wedge}$ and $\dot{\vee}$, and $\dot{\mathbf{1}}, \dot{\top}, \dot{\bot} \in D$.

Definition 10.4.3 Valuations v^+ *and* v^- *on an intuitionistic phase structure* $\mathbf{D} := \langle D, \dot{\rightarrow}, \dot{*}, \dot{\wedge}, \dot{\vee}, \dot{\mathbf{1}}, \dot{\top}, \dot{\bot} \rangle$ *are mappings from the set of all propositional variables to* D. *Then,* v^+ *and* v^- *are extended to mappings from the set* Φ *of all formulas to* D *by*

1. $v^+(\mathbf{1}) := \dot{\mathbf{1}}$,
2. $v^+(\top) := \dot{\top}$,
3. $v^+(\bot) := \dot{\bot}$,
4. $v^+(\alpha \wedge \beta) := v^+(\alpha)\ \dot{\wedge}\ v^+(\beta)$,
5. $v^+(\alpha \vee \beta) := v^+(\alpha)\ \dot{\vee}\ v^+(\beta)$,
6. $v^+(\alpha * \beta) := v^+(\alpha)\ \dot{*}\ v^+(\beta)$,
7. $v^+(\alpha \rightarrow \beta) := v^+(\alpha)\ \dot{\rightarrow}\ v^+(\beta)$,

8. $v^+(\sim\alpha) := v^-(\alpha)$,
9. $v^-(\mathbf{1}) := \dot{\bot}$,
10. $v^-(\top) := \dot{\bot}$,
11. $v^-(\bot) := \dot{\top}$,
12. $v^-(\alpha \wedge \beta) := v^-(\alpha) \dot{\vee} v^-(\beta)$,
13. $v^-(\alpha \vee \beta) := v^-(\alpha) \dot{\wedge} v^-(\beta)$,
14. $v^-(\alpha * \beta) := v^-(\alpha) \dot{*} v^-(\beta)$,
15. $v^-(\alpha \rightarrow \beta) := v^+(\alpha) \dot{*} v^-(\beta)$,
16. $v^-(\sim\alpha) := v^+(\alpha)$.

An intuitive meaning of these valuations is that v^+ and v^- respectively correspond to *provability* and *refutability*.

Definition 10.4.4 *An* intuitionistic phase model *for* WILL *is a structure* $\langle \mathbf{D}, v^+, v^- \rangle$ *such that* $\mathbf{D}$ *is an intuitionistic phase structure, and* v^+ *and* v^- *are valuations. A formula* α *is* true *in an intuitionistic phase model* $\langle \mathbf{D}, v^+, v^- \rangle$ *for* WILL *if* $\dot{\mathbf{1}} \subseteq v^+(\alpha)$ *(or equivalently* $1 \in v^+(\alpha)$*) holds, and* valid *in an intuitionistic phase structure* $\mathbf{D}$ *if it is true for any valuations* v^+ *and* v^- *on the intuitionistic phase structure. A sequent* $\alpha_1, \cdots, \alpha_n \Rightarrow \beta$ *(or* $\Rightarrow \beta$*) is* true *in an intuitionistic phase model* $\langle \mathbf{D}, v^+, v^- \rangle$ *for* WILL *if the formula* $\alpha_1 * \cdots * \alpha_n \rightarrow \beta$ *(or* β*) is true in it, and valid in an intuitionistic phase structure if so is* $\alpha_1 * \cdots * \alpha_n \rightarrow \beta$ *(or* β*).*

Then, we can prove the following.

Theorem 10.4.5 (Soundness) *Let* C *be a class of intuitionistic phase structures,* $L(C) := \{S \mid$ *a sequent* S *is valid in all intuitionistic phase structures of* $C\}$*, and* $L := \{S \mid \text{WILL} \vdash S\}$*. Then,* $L \subseteq L(C)$*.*

Next we show the completeness theorem for WILL.

Definition 10.4.6 *An expression* $[\Gamma]$ *represents the multiset consists of all elements of a sequence* Γ *of formulas. We define a commutative monoid* $\langle M, \cdot, 1 \rangle$ *as follows:*

1. $M := \{[\Gamma] \mid [\Gamma]$ *is a finite multiset of formulas*$\}$,

2. $[\Gamma] \cdot [\Delta] := [\Gamma, \Delta]$ *(the multiset union),*

3. $1 := [\,]$ *(the empty multiset).*

We define the following: for any formula α,

1. $\|\alpha\|^+ := \{[\Gamma] \mid \vdash_{cf} \Gamma \Rightarrow \alpha\}$,

2. $\|\alpha\|^- := \{[\Gamma] \mid \vdash_{cf} \Gamma \Rightarrow {\sim}\alpha\}$

where $\vdash_{cf}$ *means "cut-free provable in* WILL*". We remark that the following fact holds:*

$$\|\alpha\|^+ = \|{\sim}\alpha\|^-$$

for any formula α. *This fact is verified using the rules* (~right) *and* (~right^{-1}).[3] *We then define*

$$D := \{X \mid X = \bigcap_{i \in I} \|\alpha_i\|^+\} = \{X \mid X = \bigcap_{i \in I} \|\beta_i\|^-\}$$

for arbitrary indexing set I, *and arbitrary formula* α_i *and* $\beta_i \equiv {\sim}\alpha_i$. *Then we define*

$$cl(X) := \bigcap\{Y \in D \mid X \subseteq Y\}.$$

We define the following constants and operations on $P(M)$: *for any* $X, Y \in P(M)$,

1. $\dot{1} := cl\{1\}$,

2. $\dot{\top} := M$,

3. $\dot{\bot} := cl(\varnothing)$,

4. $X \dot{\rightarrow} Y := \{[\Delta] \mid \forall\, [\Gamma] \in X\ ([\Gamma, \Delta] \in Y)\}$,

5. $X \dot{\wedge} Y := X \cap Y$,

6. $X \dot{\vee} Y := cl(X \cup Y)$,

7. $X \dot{*} Y := cl(X \circ Y)$ *where* $X \circ Y := \{[\Gamma, \Delta] \mid [\Gamma] \in X$ *and* $[\Delta] \in Y\}$.

Valuations v^+ *and* v^- *are mappings from the set of all propositional variables to* D *such that*

1. $v^+(p) := \|p\|^+$,

[3] For the admissibility of (~right^{-1}), see the proof of Proposition 10.2.7.

2. $v^-(p) := \|p\|^-$

for any propositional variable p.

We have the following: for any $X, Y, Z \in P(M)$,

$$X \circ Y \subseteq Z \text{ iff } X \subseteq Y \dot{\to} Z.$$

We remark that D is closed under arbitrary $\bigcap$.

Lemma 10.4.7 *Let D be defined above and* $D_c := \{X \in P(M) \mid X = cl(X)\}$. *Then* $D = D_c$.

Proof. First, we show $D_c \subseteq D$. Suppose $X \in D_c$. Then $X = cl(X) = \bigcap\{Y \in D \mid X \subseteq Y\} \in D$. Next, we show $D \subseteq D_c$. Suppose $X \in D$. We show $X \in D_c$, that is, $X = \bigcap\{Y \in D \mid X \subseteq Y\}$. To show this, it is sufficient to prove that

1. $X \subseteq \{[\Gamma] \mid \forall\, W\ [W \in D \text{ and } X \subseteq W \text{ imply } [\Gamma] \in W]\}$,
2. $\{[\Gamma] \mid \forall\, W\ [W \in D \text{ and } X \subseteq W \text{ imply } [\Gamma] \in W]\} \subseteq X$.

First, we show (1). Suppose $[\Delta] \in X$ and assume $W \in D$ and $X \subseteq W$ for any W. Then we have $[\Delta] \in X \subseteq W$. Next we show (2). Suppose $[\Delta] \in \{[\Gamma] \mid \forall\, W\ [W \in D \text{ and } X \subseteq W \text{ imply } [\Gamma] \in W]\}$. By the assumption $X \in D$ and the fact that $X \subseteq X$, we have $[\Delta] \in X$. ∎

Lemma 10.4.8 *For any* $X \subseteq M$ *and any* $Y \in D$, *we have* $X \dot{\to} Y \in D$.

Proof. Before the proof, we remark that the rules

$$\frac{\Gamma \Rightarrow \alpha \to \beta}{\Gamma, \alpha \Rightarrow \beta}\ (\to\text{right}^{-1}) \qquad \frac{\alpha * \beta, \Gamma \Rightarrow \gamma}{\alpha, \beta, \Gamma \Rightarrow \gamma}\ (*\text{left}^{-1})$$

are admissible in cut-free WILL. We set $\Gamma^* \equiv \varnothing$ if Γ is empty, and otherwise $\Gamma^* \equiv \Gamma^{\star}$.

Suppose $X \subseteq M$ and $Y \in D$. We have:

$X \dot{\to} Y$

$= X \dot{\to} \bigcap_{i \in I} \|\alpha_i\|^+$

$= \{[\Delta] \mid \forall [\Gamma] \in X\ ([\Delta, \Gamma] \in \{[\Pi] \mid \forall i \in I([\Pi] \in \|\alpha_i\|^+)\})\}$

$= \{[\Delta] \mid \forall [\Gamma] \in X(\forall i \in I(\vdash_{cf} \Delta, \Gamma \Rightarrow \alpha_i))\}$

$= \{[\Delta] \mid \forall [\Gamma] \in X(\forall i \in I(\vdash_{cf} \Delta \Rightarrow \Gamma^* \to \alpha_i))\}$ (by using $(*\text{left})$, $(*\text{left}^{-1})$, $(\to\text{right})$ and $(\to\text{right}^{-1})$)

$= \{[\Delta] \mid \forall [\Gamma] \in X(\forall i \in I([\Delta] \in \|\Gamma^* \to \alpha_i\|^+))\}$

$= \bigcap\{\|\Gamma^* \to \alpha_i\|^+ \mid i \in I \text{ and } [\Gamma] \in X\} \in D.$

∎

Then, we can show the following.

Proposition 10.4.9 *The structure* $\mathbf{D} := \langle D, \dot{\rightarrow}, \dot{*}, \dot{\wedge}, \dot{\vee}, \dot{\mathbf{1}}, \dot{\top}, \dot{\bot} \rangle$ *defined above forms an intuitionistic phase structure for* WILL.

Proof. We can verify that D is closed under $\dot{\rightarrow}, \dot{*}, \dot{\wedge}$ and $\dot{\vee}$. In particular, for $\dot{\rightarrow}$, we use Lemma 10.4.8. The fact $\dot{\mathbf{1}}, \dot{\top}, \dot{\bot} \in D$ is obvious. We can verify that the conditions C1—C4 for closure operation hold for this structure. The conditions C1—C3 are obvious. We only show C4: $cl(X) \circ cl(Y) \subseteq cl(X \circ Y)$ for any $X, Y \in P(M)$. We have $X \circ Y \subseteq cl(X \circ Y)$ by the condition C1, and hence $X \subseteq Y \dot{\rightarrow} cl(X \circ Y)$. Moreover, by the condition C3, we have $cl(X) \subseteq cl(Y \dot{\rightarrow} cl(X \circ Y))$. Here, $cl(X \circ Y) \in D$ and by Lemma 10.4.8, we have $Y \dot{\rightarrow} cl(X \circ Y) \in D$. Thus, we obtain

$$cl(X) \subseteq cl(Y \dot{\rightarrow} cl(X \circ Y)) = Y \dot{\rightarrow} cl(X \circ Y)$$

by Lemma 10.4.7. Therefore we obtain (*): $cl(X) \circ Y \subseteq cl(X \circ Y)$ for any $X, Y \in P(M)$. By applying the fact (*) twice and Lemma 10.4.7, we have[4]

$$cl(X) \circ cl(Y) \subseteq cl(cl(X) \circ Y) \subseteq cl(cl(X \circ Y)) = cl(X \circ Y).$$

∎

We then obtain the following key lemma.

Lemma 10.4.10 *For any formula* α*, we have*

1. $[\alpha] \in v^{+}(\alpha) \subseteq \|\alpha\|^{+}$,
2. $[{\sim}\alpha] \in v^{-}(\alpha) \subseteq \|\alpha\|^{-}$.

Proof. We can prove this lemma by (simultaneous) induction on the complexity of α. We show only some cases for the induction step for (2).

Case ($\alpha \equiv {\sim}\beta$ for (2)): First we show $[{\sim}{\sim}\beta] \in v^{-}({\sim}\beta)$. By the induction hypothesis for (1), we have

$$[\beta] \in v^{+}(\beta) = \bigcap_{i \in I} \|\delta_i\|^{+} = \{[\Delta] \mid \forall i \in I([\Delta] \in \|\delta_i\|^{+})\}.$$

Thus we obtain:

[4] We assume the commutativity of $\circ$.

$\forall i \in I\ ([\beta] \in \|\delta_i\|^+)$
iff $\forall i \in I\ (\vdash_{cf} \beta \Rightarrow \delta_i)$
implies $\forall i \in I\ (\vdash_{cf} {\sim}{\sim}\beta \Rightarrow \delta_i)$ (by (${\sim}$left))
iff $[{\sim}{\sim}\beta] \in v^+(\beta) = v^-({\sim}\beta)$.

Next we show $v^-({\sim}\beta) \subseteq \|{\sim}\beta\|^-$. Suppose $[\Gamma] \in v^-({\sim}\beta)$. We have $[\Gamma] \in v^+(\beta) \subseteq \|\beta\|^+$ by the induction hypothesis for (1). This means $\vdash_{cf} \Gamma \Rightarrow \beta$, and hence we obtain $\vdash_{cf} \Gamma \Rightarrow {\sim}{\sim}\beta$ by (${\sim}$right). Therefore $[\Gamma] \in \|{\sim}\beta\|^-$.

Case ($\alpha \equiv \beta{\rightarrow}\gamma$ for (2)): We show $[{\sim}(\beta{\rightarrow}\gamma)] \in v^-(\beta{\rightarrow}\gamma) \subseteq \|\beta{\rightarrow}\gamma\|^-$. First, we show $[{\sim}(\beta{\rightarrow}\gamma)] \in v^-(\beta{\rightarrow}\gamma)$, that is,

$[{\sim}(\beta{\rightarrow}\gamma)] \in v^-(\beta{\rightarrow}\gamma)$
iff $[{\sim}(\beta{\rightarrow}\gamma)] \in v^+(\beta) \mathbin{\dot{*}} v^-(\gamma)$
iff $[{\sim}(\beta{\rightarrow}\gamma)] \in cl(v^+(\beta) \circ v^-(\gamma))$
iff $[{\sim}(\beta{\rightarrow}\gamma)] \in \bigcap\{Y \in D \mid v^+(\beta) \circ v^-(\gamma) \subseteq Y\}$
iff $\forall W\ [W \in D$ and $v^+ \circ v^-(\gamma) \subseteq W$ imply $[{\sim}(\beta{\rightarrow}\gamma)] \in W]$.

We show this. Suppose $W \in D$ and $v^+(\beta) \circ v^-(\gamma) \subseteq W$. By the induction hypothesis, we have $[\beta] \in v^+(\beta)$ and $[{\sim}\gamma] \in v^-(\gamma)$. Hence, we have

$$[\beta, {\sim}\gamma] \in v^+(\beta) \circ v^-(\gamma) \subseteq W = \bigcap_{i \in I} \|\delta_i\|^+ \in D.$$

Thus, we obtain $[\beta, {\sim}\gamma] \in \bigcap_{i \in I} \|\delta_i\|^+ = \{[\Delta] \mid \forall i \in I([\Delta] \in \|\delta_i\|^+)\}$, that is, $\forall i \in I(\vdash_{cf} \beta, {\sim}\gamma \Rightarrow \delta_i)$. Then we have $\forall i \in I(\vdash_{cf} {\sim}(\beta{\rightarrow}\gamma) \Rightarrow \delta_i)$ by (${\sim}{\rightarrow}$left). Therefore $[{\sim}(\beta{\rightarrow}\gamma)] \in \bigcap_{i \in I} \|\delta_i\|^+ = W$.

Second, we show $v^-(\beta{\rightarrow}\gamma) \subseteq \|\beta{\rightarrow}\gamma\|^-$. Suppose $[\Gamma] \in v^-(\beta{\rightarrow}\gamma)$. We show $[\Gamma] \in \|\beta{\rightarrow}\gamma\|^-$. For the assumption, we have

$[\Gamma] \in v^-(\beta{\rightarrow}\gamma)$
iff $[\Gamma] \in v^+(\beta) \mathbin{\dot{*}} v^-(\gamma)$
iff $[\Gamma] \in cl(v^+(\beta) \circ v^-(\gamma))$
iff $[\Gamma] \in \bigcap\{Y \in D \mid v^+(\beta) \circ v^-(\gamma) \subseteq Y\}$
iff $\forall W\ [W \in D$ and $v^+(\beta) \circ v^-(\gamma) \subseteq W$ imply $[\Gamma] \in W]$.

For this, if $W = \|\beta{\rightarrow}\gamma\|^-$, then $[\Gamma] \in \|\beta{\rightarrow}\gamma\|^-$. Thus, it is sufficient to prove that $v^+(\beta) \circ v^-(\gamma) \subseteq \|\beta{\rightarrow}\gamma\|^-$. Now, we prove this. Suppose $[\Delta] \in v^+(\beta) \circ v^-(\gamma)$. Then $[\Delta] \equiv [\Delta_1, \Delta_2]$, $[\Delta_1] \in v^+(\beta)$ and $[\Delta_2] \in v^-(\gamma)$. By the induction hypothesis, we have $[\Delta_1] \in v^+(\beta) \subseteq \|\beta\|^+$ and $[\Delta_2] \in v^-(\gamma) \subseteq \|\gamma\|^-$, and hence $\vdash_{cf} \Delta_1 \Rightarrow \beta$ and $\vdash_{cf} \Delta_2 \Rightarrow {\sim}\gamma$. By applying (${\sim}{\rightarrow}$right) to these, we have $\vdash_{cf} \Delta \Rightarrow {\sim}(\beta{\rightarrow}\gamma)$. This means $[\Delta] \in \|\beta{\rightarrow}\gamma\|^-$.

Case ($\alpha \equiv \beta \wedge \gamma$ for (2)): First, we show $[{\sim}(\beta\wedge\gamma)] \in v^-(\beta\wedge\gamma)$, that is, $[{\sim}(\beta \wedge \gamma)] \in v^-(\beta \wedge \gamma) = v^-(\beta)\ \dot{\vee}\ v^-(\gamma) = cl(v^-(\beta) \cup v^-(\gamma)) = \cap\{Y \in D \mid v^-(\beta) \cup v^-(\gamma) \subseteq Y\}$. Thus, we show

$$\forall W[W \in D \text{ and } v^-(\beta) \cup v^-(\gamma) \subseteq W \text{ imply } [{\sim}(\beta \wedge \gamma)] \in W].$$

Suppose $W \in D$ and $v^-(\beta) \cup v^-(\gamma) \subseteq W$, and the induction hypothesis $[{\sim}\beta] \in v^-(\beta)$ and $[{\sim}\gamma] \in v^-(\gamma)$. Then, we have

$$[{\sim}\gamma], [{\sim}\beta] \in v^-(\beta) \cup v^-(\gamma) \subseteq W = \bigcap_{i\in I} \|\delta_i\|^+ = \{[\Delta] \mid \forall i \in I([\Delta] \in \|\delta_i\|^+)\}$$

and hence $\forall i \in I(\vdash_{cf} {\sim}\beta \Rightarrow \delta_i$ and $\vdash_{cf} {\sim}\gamma \Rightarrow \delta_i)$. Thus we obtain $\forall i \in I(\vdash_{cf} {\sim}(\beta \wedge \gamma) \Rightarrow \delta_i)$ by (${\sim}\wedge$left). This means $[{\sim}(\beta \wedge \gamma)] \in \bigcap_{i\in I} \|\delta_i\|^+ = W$.

Second, we show $v^-(\beta \wedge \gamma) \subseteq \|\beta \wedge \gamma\|^-$. Suppose $[\Gamma] \in v^-(\beta \wedge \gamma)$. Then we have $[\Gamma] \in cl(v^-(\beta) \cup v^-(\gamma))$, that is,

$$\forall W[W \in D \text{ and } v^-(\beta) \cup v^-(\gamma) \subseteq W \text{ imply } [\Gamma] \in W].$$

We take $\|\beta \wedge \gamma\|^-$ for W. If we can show $v^-(\beta) \cup v^-(\gamma) \subseteq \|\beta \wedge \gamma\|^-$, then $[\Gamma] \in \|\beta \wedge \gamma\|^-$. Thus we prove this. Suppose $[\Delta] \in v^-(\beta) \cup v^-(\gamma)$. Then, $[\Delta] \in v^-(\beta) \cup v^-(\gamma) \subseteq \|\beta\|^- \cup \|\gamma\|^-$ by the induction hypothesis, and hence we obtain $[\Delta] \in \|\beta\|^-$ or $[\Delta] \in \|\gamma\|^-$, that is, $\vdash_{cf} \Delta \Rightarrow {\sim}\beta$ or $\vdash_{cf} \Delta \Rightarrow {\sim}\gamma$. For both cases, we can obtain $\vdash_{cf} \Delta \Rightarrow {\sim}(\beta \wedge \gamma)$ by (${\sim}\wedge$right1) or (${\sim}\wedge$right2). This means $[\Delta] \in \|\beta \wedge \gamma\|^-$. ∎

By using this key lemma, we can obtain the completeness theorem for WILL as follows. If formula α is true, then $[\ ] \in v^+(\alpha)$. On the other hand $v^+(\alpha) \subseteq \|\alpha\|^+$, and hence $[\ ] \in \|\alpha\|^+$, which means "α is cut-free provable". By combining this with the soundness theorem, we also obtain the cut-elimination theorem for WILL, i.e. Theorem 10.2.3.

Theorem 10.4.11 (Completeness) *Let C be a class of intuitionistic phase structures, $L(C) := \{S \mid$ a sequent S is valid in all intuitionistic phase structures of $C\}$, and $L := \{S \mid$ WILL $\vdash S\}$. Then, $L(C) \subseteq L$.*

10.5. Petri net interpretation

First, we define the notion of a Kripke model for the logic WL, which is the $\{\to, *, \mathbf{1}, \sim\}$-fragment of WILL, and demonstrate the completeness theorem. The Kripke models are based on partially ordered commutative monoids, which are discussed by Ono and Komori [147]. We have already obtained this completeness result for the extended WL with ! [82], and hence we omit the proof.

Definition 10.5.1 *A* Kripke frame *is a structure* $\langle M, \cdot, \varepsilon, \omega, \geq\rangle$ *satisfying the following conditions:*

1. $\langle M, \cdot, \varepsilon\rangle$ *is a commutative monoid with the identity* ε*, and* $\omega \in M$ *with the condition:*

   C0*:* $\omega \cdot x = \omega$ *for all* $x \in M$*,*

2. $\langle M, \geq\rangle$ *is a partially ordered set with the greatest element* ω *with respect to* $\geq$*,*

3. $\cdot$ *is monotonic with respect to* $\geq$*, that is,*

   C1*:* $y \geq z$ *implies* $x \cdot y \geq x \cdot z$ *for all* $x, y, z \in M$*.*

We can derive the following monotonicity condition:

C1′: $y_1 \geq z_1$ and $y_2 \geq z_2$ imply $y_1 \cdot y_2 \geq z_1 \cdot z_2$ for all $y_1, y_2, z_1, z_2 \in M$.

Definition 10.5.2 Valuations $\models^+$ *and* $\models^-$ *on a Kripke frame* $\langle M, \cdot, \varepsilon, \omega, \geq\rangle$ *are mappings from the set of all propositional variables to the powerset of* M *and satisfying the following heredity condition: for any propositional variable* p *and any* $x, y \in M$*,*

1. $x \in \models^+ (p)$ *and* $y \geq x$ *imply* $y \in \models^+ (p)$*,*
2. $x \in \models^- (p)$ *and* $y \geq x$ *imply* $y \in \models^- (p)$*.*

We will write $x \models^+ p$ *(or* $x \models^- p$*) for* $x \in \models^+ (p)$ *(or* $x \in \models^- (p)$ *respectively). These valuations* $\models^+$ *and* $\models^-$ *can be extended to mappings from the set of all formulas to the powerset of* M *by*

1. $x \models^+ \mathbf{1}$ *iff* $x \geq \varepsilon$*,*
2. $x \models^+ \alpha\to\beta$ *iff* $y \models^+ \alpha$ *implies* $x \cdot y \models^+ \beta$ *for all* $y \in M$*,*
3. $x \models^+ \alpha*\beta$ *iff* $y \models^+ \alpha$ *and* $z \models^+ \beta$ *for some* $y, z \in M$ *with* $x \geq y \cdot z$*,*

4. $x \models^+ {\sim}\alpha$ iff $x \models^- \alpha$,

5. $x \models^- \mathbf{1}$ iff $x = \omega$,

6. $x \models^- \alpha \to \beta$ iff $y \models^+ \alpha$ and $z \models^- \beta$ for some $y, z \in M$ with $x \geq y \cdot z$,

7. $x \models^- \alpha * \beta$ iff $y \models^- \alpha$ and $z \models^- \beta$ for some $y, z \in M$ with $x \geq y \cdot z$,

8. $x \models^- {\sim}\alpha$ iff $x \models^+ \alpha$.

Definition 10.5.3 *A* Kripke model *is a structure* $\langle M, \cdot, \varepsilon, \omega, \geq, \models^+, \models^- \rangle$ *such that*

1. $\langle M, \cdot, \varepsilon, \omega, \geq \rangle$ *is a Kripke frame,*

2. $\models^+$ *and* $\models^-$ *are valuations on* $\langle M, \cdot, \varepsilon, \omega, \geq \rangle$.

A formula α *is* true *in a Kripke model* $\langle M, \cdot, \varepsilon, \omega, \geq, \models^+, \models^- \rangle$ *if* $\varepsilon \models^+ \alpha$, *and* valid *in a Kripke frame* $\langle M, \cdot, \varepsilon, \omega, \geq \rangle$ *if it is true for any valuations* $\models^+$ *and* $\models^-$ *on the Kripke frame. A sequent* $\alpha_1, \cdots, \alpha_n \Rightarrow \beta$ *(or* $\Rightarrow \beta$*) is* true *in a Kripke model if so is the formula* $\alpha_1 * \cdots * \alpha_n \to \beta$ *(or* β*), and* valid *in a Kripke frame if so is* $\alpha_1 * \cdots * \alpha_n \to \beta$ *(or* β*).*

Theorem 10.5.4 (Completeness) *Let* C *be a class of Kripke frames,* $L := \{S \mid \mathrm{WL} \vdash S\}$ *and* $L(C) := \{S \mid S$ *is valid in all frames of* $C\}$. *Then* $L = L(C)$.

To prove this theorem, we use the following canonical model.

Definition 10.5.5 *We define a canonical model* $\langle M, \cdot, \varepsilon, \omega, \geq, \models^+, \models^- \rangle$ *such that*

1. $M := \{[\Gamma] \mid [\Gamma]$ *is a finite multiset of formulas*$\}$,

2. *for any* $[\Gamma], [\Delta] \in M$, $[\Gamma] \cdot [\Delta] := [\Gamma, \Delta]$ *(the multiset union),*

3. ε *is the empty multiset* $[\ ]$,

4. ω *is the multiset* $[{\sim}\mathbf{1}]$,

5. *for any* $[\Gamma], [\Delta] \in M$, $[\Gamma] \geq [\Delta]$ *is defined by* $\vdash \Gamma \Rightarrow \Delta^\star$ *where* $\Delta^\star \equiv \gamma_1 * \cdots * \gamma_n$ *if* $\Delta \equiv \langle \gamma_1, \cdots, \gamma_n \rangle$ $(0 < n)$, *and* $\Delta^\star \equiv \mathbf{1}$ *if* Δ *is empty* ($[\Gamma] \doteq [\Delta]$ *is defined by* $\vdash \Gamma \Rightarrow \Delta^\star$ *and* $\vdash \Delta \Rightarrow \Gamma^\star$),

6. *valuations* $\models^+$ *and* $\models^-$ *on* $\langle M, \cdot, \varepsilon, \omega, \geq \rangle$ *are mappings from the set PROP of all propositional variables to the power set of* M *defined by, for any* $p \in PROP$ *and any* $[\Gamma] \in M$,

a) $[\Gamma] \in \models^{+} (p)$ *iff* $\vdash \Gamma \Rightarrow p$

b) $[\Gamma] \in \models^{-} (p)$ *iff* $\vdash \Gamma \Rightarrow {\sim}p$.

Next, we consider a Petri net interpretation for WL.

Definition 10.5.6 (Petri net) *A* Petri net *is a structure* $\langle P, T, \Uparrow, \Downarrow \rangle$ *such that*

1. P *is a set of* places,

2. T *is a set of* transitions,

3. $\Uparrow$ *and* $\Downarrow$ *are mappings from* T *to the set* N *of all multisets over* P.

For $t \in T$, $\Uparrow t$ *and* $\Downarrow t$ *are called the* pre-multiset *and the* post-multiset *of* t *respectively. Each element of* N *is called a* marking.

Definition 10.5.7 *Let* $+$ *be the multiset union operation. A* firing relation $[t\rangle$ *for* $t \in T$ *on* N *is defined as follows: for any* $m_1, m_2 \in N$,

$m_1\ [t\rangle\ m_2$ *iff* $m_1 = m_3 + \Uparrow t$ *and* $\Downarrow t + m_3 = m_2$ *for some* $m_3 \in N$.

A reachability relation $\gg$ *on* N *is defined as follows: for any* $m, m' \in N$,

$m \gg m'$ *iff* $m\ [t_1\rangle\ m_1\ [t_2\rangle \cdots [t_n\rangle\ m_n = m'$ *for some* $t_1, ..., t_n \in T$, $m_1, ..., m_n \in N$ *and* $n \geq 0$.

Definition 10.5.8 *A* Petri net structure *is a structure* $\langle N, +, \varnothing, \gg \rangle$ *such that*

1. N *is the set of all markings,*

2. $+$ *is the multiset union operation on* N,

3. $\varnothing$ *is the empty multiset,*

4. $\gg$ *is a reachability relation on* N.

A Petri net structure $\langle N, +, \varnothing, \gg \rangle$ is in fact a *partially ordered commutative monoid*, that is, we have the following:

1. $\langle N, +, \varnothing \rangle$ is a commutative monoid,

2. $\langle N, \gg \rangle$ is a partially ordered set,

3. $x_1 \gg x_2$ and $y_1 \gg y_2$ imply $x_1+y_1 \gg x_2+y_2$ for all $x_1, x_2, y_1, y_2 \in N$.

We then have the following main proposition.

Proposition 10.5.9 (Correspondence: Petri net structure and Kripke frame) *A Petri net structure $\langle N, +, \varnothing, \gg \rangle$ is just an ω-free reduct $\langle M, \cdot, \varepsilon, \geq \rangle$ of a Kripke frame.*

By this proposition and the canonical model presented in Definition 10.5.5, we can obtain a Petri net interpretation for WL: the reachability in a Petri net corresponds to provability in WL, a place name (or token) in a Petri net corresponds to formula in WL, and a marking in a Petri net corresponds to an antecedent (or consequent) of a sequent in WL, i.e. $[\Gamma] \gg [\Delta]$ corresponds to WL $\vdash \Gamma \Rightarrow \Delta^{\star}$.

In the Petri net interpretation of WL, we can express a Petri net with an infinite reachability tree.

Example 10.5.10 *We give a Petri net $N := \langle P, T, \Uparrow, \Downarrow \rangle$ with $P := \{\alpha, \sim\sim\alpha\}$, $T := \{t_1, t_2\}$, $\Uparrow t_1 := [\alpha]$, $\Downarrow t_1 := [\sim\sim\alpha]$, $\Uparrow t_2 := [\sim\sim\alpha]$ and $\Downarrow t_2 := [\alpha]$. Graphically this becomes the following.*

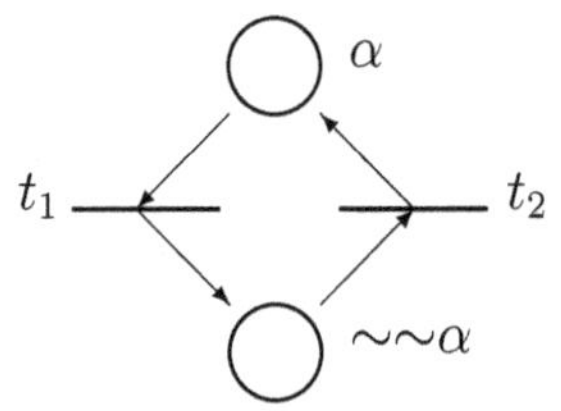

In this example, this net corresponds to the fact that $\vdash \alpha \Rightarrow \sim\sim\alpha$ and $\vdash \sim\sim\alpha \Rightarrow \alpha$. In this net, we consider the following example. If the place α has a token then the transition t_1 is enabled. Moreover, if t_1 fires then the place $\sim\sim\alpha$ gets the token, and then t_2 is enabled. Further, if t_2 fires then the place α gets the token. We can continue the movement of the token infinitely often. This situation is expressed formally as follows:

$$[\alpha]\ \ [t_1\rangle\ \ [\sim\sim\alpha]\ \ [t_2\rangle\ \ [\alpha]\ \ [t_1\rangle\ \ [\sim\sim\alpha]\ \ \cdots.$$

It is known that this net is very useful as a component in a number of applied compound nets. For example, using this net, we can describe a Petri net expressing the producer/consumer problem, and a Petri net solving the mutual exclusion problem (see, e.g., [171], pp. 16–18).

Next we explain the inhibitor arcs. An inhibitor arc is an arc which has, roughly speaking, the meaning "not" in a logic-gate. A transition in a Petri net with inhibitor arc is enabled when tokens are in all of their input places of non-inhibitor arcs and no tokens in all of their input places of inhibitor arcs. It is known that Petri nets with inhibitor arcs are applicable to various kinds of examples such as communication protocols, performance analysis [166], and non-monotonic reasoning in medical diagnosis [177].

The following is a simple example of using an inhibitor arc.

Example 10.5.11 *We consider a Petri net* $N := \langle P, T, \Uparrow, \Downarrow \rangle$ *with the inhibitor arc* i, *where* $P := \{\alpha, {\sim}\alpha, {\sim\sim}\alpha\}$, $T := \{t\}$, $\Uparrow t := [\alpha, {\sim}\alpha]$, $\Downarrow t := [{\sim\sim}\alpha]$. *Graphically this becomes the following.*

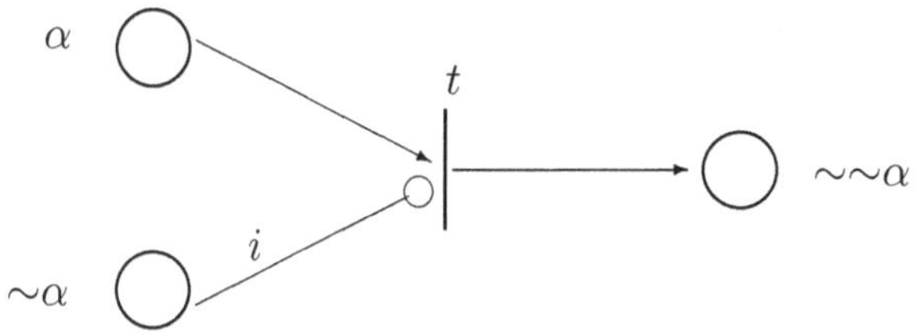

In this example, the net uses the fact that $\alpha, {\sim}\alpha \Rightarrow {\sim\sim}\alpha$ is not always provable in WILL, because WILL has no weakening rule, i.e., WILL is a system of relevance logic, cf., for example, [7, 8]. The failure of the provability of $\alpha, {\sim}\alpha \Rightarrow {\sim\sim}\alpha$ may also be closely related to paraconsistency (Corollary 10.2.5), i.e., the unprovability fact is an instance of the fact that $\alpha, {\sim}\alpha \Rightarrow \beta$ is not always provable. If the place α has a token and the place ${\sim}\alpha$ has no token, then the transition t is enabled, i.e., this fact corresponds to the fact that $\vdash \alpha \Rightarrow {\sim\sim}\alpha$.

The following example [151] expresses the XOR transition by using inhibitor arcs.

Example 10.5.12 *We consider a Petri net* $N := \langle P, T, \Uparrow, \Downarrow \rangle$ *with the inhibitor arcs* i_1 *and* i_2, *where* $P := \{\alpha, {\sim}\alpha, {\sim\sim}\alpha, {\sim\sim\sim}\alpha\}$, $T := \{t_1, t_2\}$, $\Uparrow t_1 := [\alpha, {\sim}\alpha]$, $\Downarrow t_1 := [{\sim\sim}\alpha]$, $\Uparrow t_2 := [\alpha, {\sim}\alpha]$ *and* $\Downarrow t_2 := [{\sim\sim\sim}\alpha]$. *Graphically this becomes the following.*

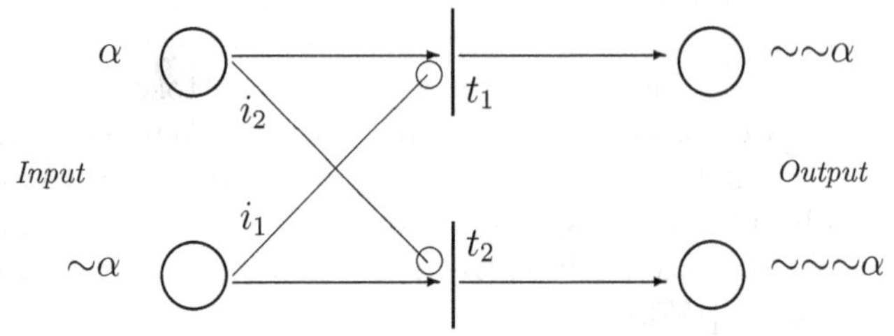

For example, the inhibitor arc i_2 means that if the place α has no token and the place $\sim\alpha$ has some tokens, then the transition t_2 is enabled, but otherwise it is not. Hence this net represents the XOR transition in the following sense. Suppose the places α and $\sim\alpha$ are a piece of input and the places $\sim\sim\alpha$ and $\sim\sim\sim\alpha$ are an output (i.e. both $\sim\sim\alpha$ and $\sim\sim\sim\alpha$ are considered to be an output of the XOR gate). Then, for example, if the place α has a token, but $\sim\alpha$ has no token, then t_1 is enabled, and hence an output place $\sim\sim\alpha$ can obtain such a token. This net corresponds to the following fact: for any formula α, $\vdash \alpha \Rightarrow \sim\sim\alpha$ and $\vdash \sim\alpha \Rightarrow \sim\sim\sim\alpha$, but the sequents $(\alpha, \sim\alpha \Rightarrow \sim\sim\sim\alpha)$, $(\alpha, \sim\alpha \Rightarrow \sim\sim\alpha)$, $(\alpha \Rightarrow \sim\sim\sim\alpha)$ and $(\sim\alpha \Rightarrow \sim\sim\alpha)$ are not always provable.

Next, we give a controlled-NOT transition, which is an example of using the XOR transition. First, we explain the controlled-NOT gate, which is the most fundamental logic-gate in circuits for quantum computation. A set of quantum gates (called a basis) is said to be universal for quantum computation if any unitary operator can be approximated with arbitrary precision by a circuit involving only those gates. An example of such universal basis is the controlled-NOT gate plus the set of all single-qubit gate (see [176]). Thus, the controlled-NOT gate is very important to realize quantum computers, and has been studied by many researchers (see e.g., [128, 19]). The controlled-NOT gate is a reversible logic-gate operation on two bits α_1 and α_2, where α_1 is called the *control bit* and α_2 the *target bit*. The value of α_1 is negated if $\alpha_1 = 1$, otherwise α_2 is left unchanged. In both cases, the control bit α_1 remains unchanged [19]. That is, the truth table is as follows:

Input		Output	
α_1	α_2	α'_1	α'_2
0	0	0	0
0	1	0	1
1	0	1	1
1	1	1	0

We note that, in this table, the part of α_1, α_2 and α'_2 is the XOR gate.

Example 10.5.13 *We consider a Petri net* $N := \langle P, T, \Uparrow, \Downarrow \rangle$ *with the inhibitor arcs* i_1 *and* i_2*, where* $P := \{\alpha, {\sim}\alpha, {\sim}{\sim}\alpha, {\sim}{\sim}{\sim}\alpha, {\sim}{\sim}{\sim}{\sim}\alpha\}$*,* $T := \{t_1, t_2, t_3\}$*,* $\Uparrow t_1 := [\alpha]$*,* $\Downarrow t_1 := [{\sim}{\sim}{\sim}{\sim}\alpha]$*,* $\Uparrow t_2 := [\alpha, {\sim}\alpha]$*,* $\Downarrow t_2 := [{\sim}{\sim}\alpha]$ $\Uparrow t_3 := [\alpha, {\sim}\alpha]$ *and* $\Downarrow t_3 := [{\sim}{\sim}{\sim}\alpha]$*. In the truth table above, we have the following correspondence:* $\alpha = \alpha_1$ *(the control bit),* ${\sim}\alpha = \alpha_2$ *(the target bit),* ${\sim}{\sim}{\sim}{\sim}\alpha = \alpha'_1$ *(an output),* ${\sim}{\sim}\alpha + {\sim}{\sim}{\sim}\alpha = \alpha'_2$ *(an output). Graphically this becomes the following.*

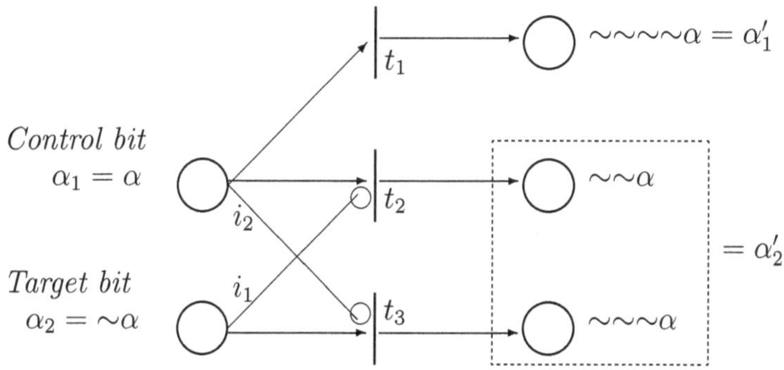

This net represents the controlled-NOT transition in the following sense. Suppose that the places α and ${\sim}\alpha$ are a piece of input, and the places ${\sim}{\sim}\alpha$, ${\sim}{\sim}{\sim}\alpha$ and ${\sim}{\sim}{\sim}{\sim}\alpha$ are a piece of output (i.e. both $[{\sim}{\sim}\alpha, {\sim}{\sim}{\sim}\alpha]$ and $[{\sim}{\sim}{\sim}{\sim}\alpha]$ are considered to be a piece of output of the controlled-NOT gate). Then, for example, if the place α has two tokens, but ${\sim}\alpha$ has no token, then t_1 and t_2 are enabled and hence two output places ${\sim}{\sim}\alpha$ and ${\sim}{\sim}{\sim}{\sim}\alpha$ can obtain such a token.

Other examples of using inhibitor arcs will be discussed in the next section.

10.6. Medical diagnosis

WILL can be used in reasoning for medical diagnosis based on the fact that $\sim$ can express inexact predicates [6, 194] and can allow paraconsistency.

An *inexact predicate* is an incomplete predicate in an empirical domain. An example of an inexact predicate is a symptom (or disease) predicate such as "*depression*(x)", which means that "a person x suffers from first-stage depression or melancholia".[5] This predicate is in-

[5] Another typical example of an inexact predicate is a color predicate such as "*green*(x)", which means "x is green" [194].

complete in the sense that we can not determine exactly that the formula $\sim depression(x) \vee depression(x)$ is true, because, for example, one pathologist may see that the person is suffering from depression, while another pathologist may refute such a diagnosis. This formula can be unprovable in WILL, and hence an expression using $\sim$ is appropriate.

Logics having *paraconsistency* can handle inconsistency-tolerant reasoning more appropriately (see, e.g., [41, 131]). For example, if a large medical knowledge-base KB of diseases and symptoms, such as an expert system based on a logic having paraconsistency, is inconsistent (i.e. there is a symptom predicate $s(x)$ such that $\sim s(x), s(x) \in KB$), then KB does not derive arbitrary disease d, because paraconsistency ensures the fact that for some formulas α and β, the sequent $\sim\alpha, \alpha \Rightarrow \beta$ is not provable. In fact, if α and β are atomic formulas, then such a sequent is not provable in WILL. Of course, in both the classical and intuitionistic logics, we have $\vdash \sim s(x), s(x) \Rightarrow d(x)$ for any formula $d(x)$.

The weak contraposition rule (w-con) was discussed in Section 2. An intuitive meaning of (w-con) is presented in terms of medical reasoning as follows. For example, assuming the fact:

$$\vdash cancer(x) \Rightarrow depression(x),$$

the following fact is true:

$$\vdash \sim depression(x) \Rightarrow \sim cancer(x).$$

However, the latter fact is considered to be not true in the real world. Thus, (w-con) is not very natural in the expression of such medical issues.

Next, we consider the diagnosis of two diseases d_1 and d_2 based on four symptoms s_1, s_2, s_3 and s_4. This problem is a slight modification of the example presented in [131]. Two doctors DOC$_1$ and DOC$_2$ are consulted for a diagnosis of diseases d_1 and d_2. The following expresses the opinion of DOC$_1$.

$$\vdash s_1(x), s_2(x), \sim d_2(x) \Rightarrow d_1(x),$$

$$\vdash s_1(x), s_3(x), \sim d_1(x) \Rightarrow d_2(x),$$

$$\vdash d_1(x), s_2(x) \Rightarrow \sim d_2(x).$$

This situation is expressed using a Petri net with inhibitor arcs as follows.

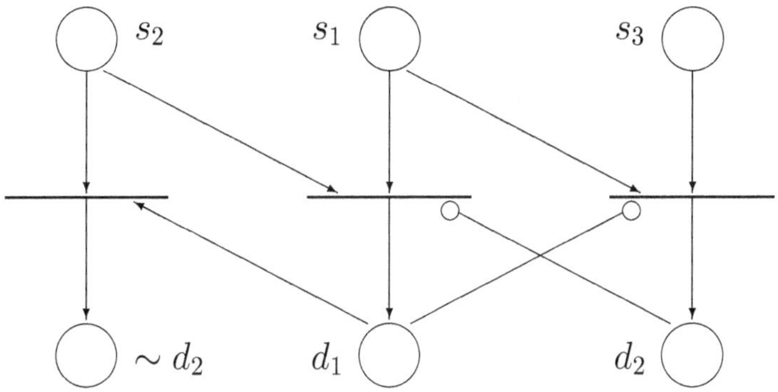

The opinion of DOC_2 is expressed as

$$\vdash s_1(x)^2, s_4(x) \Rightarrow d_1(x),$$
$$\vdash \sim s_1(x), s_3(x) \Rightarrow d_2(x)$$

where the expression $s_1(x)^2$ means $s_1(x) * s_1(x)$. This situation is illustrated below.

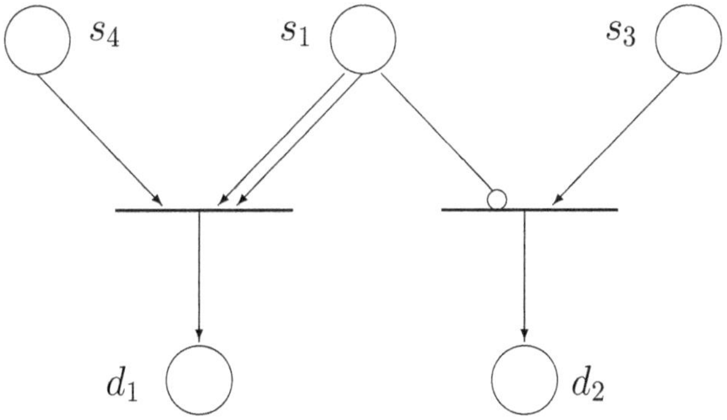

It is assumed that the following information is obtained by pathologists in connection with tests conducted for an individual called "John".

$$\vdash \Rightarrow s_1(john)^2,$$
$$\vdash \Rightarrow s_2(john),$$
$$\vdash \Rightarrow \sim s_3(john),$$
$$\vdash \Rightarrow s_4(john).$$

We can then ask the question "Does John have d_1 or d_2?". We can answer that he has d_1, but does not have d_2:

$$\dfrac{\Rightarrow s_4(john) \qquad \dfrac{\Rightarrow s_1(john)^2 \quad s_1(john)^2, s_4(john) \Rightarrow d_1(john)}{s_4(john) \Rightarrow d_1(john)}}{\Rightarrow d_1(john)},$$

$$\dfrac{\dfrac{\Rightarrow d_1(john) \quad \Rightarrow s_2(john)}{\Rightarrow d_1(john) * s_2(john)} \qquad \dfrac{d_1(john), s_2(john) \Rightarrow {\sim} d_2(john)}{d_1(john) * s_2(john) \Rightarrow {\sim} d_2(john)}}{\Rightarrow {\sim} d_2(john)}.$$

Finally, in this section, resource-conscious medical treatment using linear negation $\cdot^{\perp}$[6] is considered. While strong negation is defined or expressed directly by the connective $\sim$ (thus, called "*direct established falsity*"), linear negation $\cdot^{\perp}$ can be defined indirectly using the multiplicative falsum constant $\mathbf{0}$ as $\alpha^{\perp} := \alpha \rightarrow \mathbf{0}$ (thus, called "*indirect established falsity*"), where the linear implication $\rightarrow$ means the consumption of a resource. The expression of linear negation means that resource consumption induces contradiction, i.e. if a resource α is consumed, then we have contradiction. Here, the word "contradiction", corresponding to $\mathbf{0}$, means that the constant $\mathbf{0}$ in linear logic represents the existence of a "width" for contradiction. The phase space of linear logic has the corresponding constant $\mathbf{0}$ with some width within the domain. This fact is a feature of substructural logic without the weakening rule.

Consider a medicine m as a resource. The expression $m(x) \rightarrow \mathbf{0}$ means that if a person x uses medicine m to recover from a disease, then x makes no recovery from the disease with the medicine. In other words, to use the medicine does not derive a medically beneficial effect. In this expression, $\mathbf{0}$ represents a broad range of facts, such as x taking a turn for the worse, death, the onset of a critical condition, side-effects, or less consciousness (i.e. $\mathbf{0}$ represents a width of contradiction). This interpretation of medicine consumption is also syntactically natural. For example, since linear logic has no contraction rule, the sequent $m(x) * m(x) \rightarrow \mathbf{0} \Rightarrow m(x) \rightarrow \mathbf{0}$ ("if a person x uses two medicines, then x takes a turn for the worse" implies "if x uses a medicine, then x takes a turn for the worse") is not provable. On the other hand, since linear logic has the exchange rule, the sequent $m_1(x) * m_2(x) \rightarrow \mathbf{0} \Rightarrow m_2(x) * m_1(x) \rightarrow \mathbf{0}$ (where m_1 and m_2 represent distinct medicines) is provable, i.e. the order of using medicines does not change the effect of the medicine. These two examples are very natural in realistic medical treatment.

[6] Linear negation cannot be used in the present setting of phase semantics.

Consider a situation in which John takes medicines m_1 (one pill) and m_2 (two pills and one powder, which have the same ingredient). This situation is expressed as follows.

$$\vdash \Rightarrow m_1(john),$$
$$\vdash \Rightarrow m_2(john)^2,$$
$$\vdash \Rightarrow m_2(john).$$

Then, consider the following hypothesis for recovering from disease d_1.

H1: $\vdash m_1(x) * m_2(x)^2 \Rightarrow recover\_d_1(x)$,

H2: $\vdash \Rightarrow (m_2(x)^3)^{\perp}$.

H2 (i.e. $\vdash m_2(x)^3 \Rightarrow \mathbf{0}$) means that taking more than three units of medicine m_1 at a time results in a turn for the worse. We can then pose the question "Can John recover from d_1?". Our answer is "Yes":

$$\cfrac{\cfrac{\Rightarrow m_1(john) \quad \Rightarrow m_2(john)^2}{\Rightarrow m_1(john) * m_2(john)^2} \quad \text{H1}}{\Rightarrow recover\_d_1(john)}\,.$$

However, if the use of medicine m_2 is inappropriate, then John has no recovery from d_1:

$$\cfrac{\cfrac{\Rightarrow m_2(john)^2 \quad \Rightarrow m_2(john)}{\Rightarrow m_2(john)^3} \quad m_2(john)^3 \Rightarrow \mathbf{0}}{\Rightarrow \mathbf{0}}\,.$$

10.7. Electric circuits

An interpretation for a fragment of WILL can be used to discuss resource-sensitive reasoning for direct-current circuit theory. This theory is based on *Ohm's law* and *Kirchhoff's laws* in general. Ohm's law is expressed as $V = I \cdot R$, where V is voltage or electric potential difference, I is electric current, and R is resistance. This means intuitively that if current I [A] flows in resistance R [Ω] , then the voltage across R is V [V]. As a consequence of this fact, electric power $P = V \cdot I$ [W] is *consumed* as *Joule heat* by R. Of Kirchhoff's laws (of current and of voltage), the voltage law is regarded as an energy (resource) conservation law.

In this context, I and V are considered as (energy) resources in an electric circuit. Thus, I and V can be formalized using WILL. First, we introduce a sequent expression of the form $\Gamma : \Delta \Rightarrow \gamma$ for any multisets Γ and Δ of formulas and formula γ. In this expression, Γ, Δ

and γ are called the *electric current context*, the *voltage context*, and the *resultant resistance*, respectively. For example, $\vdash I : V \Rightarrow R$ for some formulas I, V and R, intuitively means that $V = I \cdot R$ holds, i.e., current I and voltage V are consumed as Joule heat by resistance R. Similarly, $\vdash I_1, I_2 : V_1, V_2, V_3 \Rightarrow R$ means that $V_1 + V_2 + V_3 = (I_1 + I_2) \cdot R$ holds. Next, we introduce two resultant resistance expressions of the forms $R_1 \parallel R_2$ (parallel connection of resistance) and $R_1 \mid R_2$ (series connection of resistance), which mean that R_1 and R_2 are connected in parallel, and R_1 and R_2 are connected in series, respectively. The value of current flowing in the resultant resistance can be calculated by the logical rules for the multiplicative conjunction $*$ (corresponds to parallel connection) and additive conjunction $\wedge$ (corresponds to series connection). The value of voltage across the resultant resistance can also be calculated in a similar way. However, calculation of parallel connection and series connection for voltage is the converse of the calculation for current. Thus, we introduce the following inference rules as combined versions of the (right introduction) logical rules for $*$ and $\wedge$.

$$\frac{\Gamma : \Delta_1 \Rightarrow \alpha \quad \Gamma : \Delta_2 \Rightarrow \beta}{\Gamma : \Delta_1, \Delta_2 \Rightarrow \alpha \mid \beta} (\mid), \qquad \frac{\Gamma_1 : \Delta \Rightarrow \alpha \quad \Gamma_2 : \Delta \Rightarrow \beta}{\Gamma_1, \Gamma_2 : \Delta \Rightarrow \alpha \parallel \beta} (\parallel).$$

Next, we consider the role of strong negation in this context. The strong negation connective $\sim$ expresses the minus sign "$-$" of the values of current and voltage in equations formulated based on Kirchhoff's laws and Ohm's law. For example, $\vdash {\sim}I_1, {\sim}I_2 : {\sim}V \Rightarrow R_1 \parallel R_2$ means that the equation $-V = (-I_1 - I_2) \cdot (R_1 \parallel R_2)$ holds. Thus, for example, ${\sim}I$ means the direction of flow of electric current I. This interpretation is very natural, because we have the axiom schemes ${\sim}(\alpha * \beta) \leftrightarrow {\sim}\alpha * {\sim}\beta$ and ${\sim}{\sim}\alpha \leftrightarrow \alpha$, which correspond to the equations $-(I_1 + I_2) = (-I_1) + (-I_2)$ and $-(-I) = I$, respectively.[7]

In this framework, we can calculate the values of current and voltage of any resultant (composite) resistance. Consider the circuit in Figure A. In this circuit, we have the following equations by Kirchhoff's laws: $V = V_1 + V_2 + V_3 + V_4 + V_5$ and $I = I_1 + I_2$, where V is the value of electromotive force, V_i is the value of voltage drop (i.e. electric potential difference) across resistance R_j, and I and I_k are the values of electric current. The situation in Figure A can be formulated as follows.

C1: $\vdash I : V_1 \Rightarrow R_1$,

C2: $\vdash I : V_2 \Rightarrow R_2$,

C3: $\vdash I_1 : V_3 \Rightarrow R_3$,

[7] $\alpha \leftrightarrow \beta$ is defined as $\vdash \alpha \Rightarrow \beta$ and $\vdash \beta \Rightarrow \alpha$.

C4: $\vdash I_2 : V_3 \Rightarrow R_4$,

C5: $\vdash I : V_4 \Rightarrow R_5$,

C6: $\vdash I : V_5 \Rightarrow R_6$.

The requirement is then "Calculate the values of current and voltage for the resultant resistance $((R_1 \mid R_2) \mid (R_3 \parallel R_4)) \mid R_5$". This can be verified using C1–C5 as follows.

$$\dfrac{\dfrac{\dfrac{I : V_1 \Rightarrow R_1 \quad I : V_2 \Rightarrow R_2}{I : V_1, V_2 \Rightarrow R_1 \mid R_2}(\mid) \quad \dfrac{I_1 : V_3 \Rightarrow R_3 \quad I_2 : V_3 \Rightarrow R_4}{I : V_3 \Rightarrow R_3 \parallel R_4}(\parallel)}{I : V_1, V_2, V_3 \Rightarrow (R_1 \mid R_2) \mid (R_3 \parallel R_4)}(\mid) \quad I : V_4 \Rightarrow R_5}{I : V_1, V_2, V_3, V_4 \Rightarrow ((R_1 \mid R_2) \mid (R_3 \parallel R_4)) \mid R_5}(\mid).$$

That is, the answer is "The value of required current is I $(= I_1 + I_2)$ [A], and the value of required voltage is $V_1 + V_2 + V_3 + V_4$ [V]". In addition, by checking the proof figure, we can obtain the values of current and voltage for any resultant resistance.

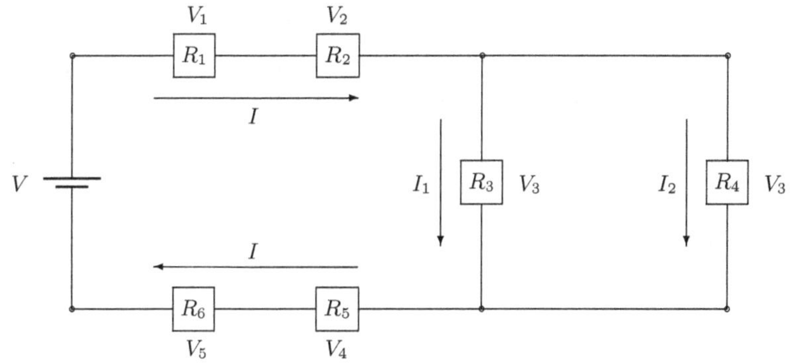

Figure A: An electric circuit.

10.8. Remarks

A logic QILL (*quantized intuitionistic linear logic*), which is closely related to involutive quantales by Mulvey and Pelletier, is studied in [85]. The relationship between QILL and WILL will be discussed in the next chapter. With respect to this investigation, QILL is very similar to WILL. Hence, the present results of phase semantic completeness and cut-elimination, cut-free twist calculus, and Petri net interpretation, can also be applied to QILL with some appropriate modifications.

Finally, other applications of strong negation to concurrency theory and computational linguistics are mentioned below.

In [82], strong negation in ILL$^{\sim}$ of extended WILL with the exponential operator ! was interpreted as a *stop action* of processes in concurrency theory. In this example, the expression $\sim\alpha$ means "process α is stopped or suspended", the rule ($*$left) means "stop-action $\sim(\alpha * \beta)$ halts two processes α and β in parallel", and the rule ($\sim$left) means "suspended process α is restored". Properties such as the *non-existence of a dead-lock process schedule* and *computational relevance principle* were also investigated in this context.

As mentioned in [105, 201], a kind of strong negation can be used to interpret *negative affixes*, which are regarded as expressing *lexical item negation* in natural language sentences, also called predicate-term negation in Aristotelian term logic. Typical examples using a negative affix (such as "in–", "un–", and "non–") include "illogical", "incoherent", "impolite", "non-selfish", and "unhappy". This type of negation is very natural and appropriate for applications of strong negation in the sense that

1. by using a negative affix, the lexicon may represent inexact predicates such as $unhappy(x)$ ($= \sim happy(x)$) or $happy(x)$, where $happy(x) \vee \sim happy(x)$ is not always true,

2. the negative affix is considered to represent directly established falsity,

3. by using a negative affix, the lexicon provides the same part of speech (syntactic part) for its corresponding non-negative part, e.g., both *unhappy* and *happy* are adjectives.

The fact (3) is justified by Proposition 10.2.9, i.e., any formula of the form $\sim p$, where p is a propositional variable such as $happy(x)$, can be translated into another propositional variable p' such as $unhappy(x)$. Another application of stronmg negation in linguistics is presented in Chapter 12. For a comprehensive review on the concept on negation and its history, see e.g., [201, 78].

11. Paraconsistent logics based on involutive quantales

A new logic, quantized intuitionistic linear logic (QILL), is introduced. This logic is closely related to the logic which corresponds to Mulvey and Pelletier's (commutative) involutive quantales. Some cut-free sequent calculi with a new property "quantization principle" and some complete semantics such as an involutive quantale model and a quantale model are obtained for QILL. The relationship between QILL and the extended intuitionistic linear logic with strong negation WILL is also observed using such syntactical and semantical frameworks.[1]

11.1. Introduction

In this chapter, a new logic QILL (quantized intuitionistic linear logic) is introduced in order to obtain a simple perspective for the relationship between involutive quantales and logics with strong negation. Before the precise discussion, quantales and strong negation logics are briefly explained below.

Quantales were first introduced by Mulvey in an attempt to cast light on the connections between C^*-algebras and quantum mechanics. It is well-known that quantales are one of the semantics of intuitionistic linear logic [64, 218] and have a number of applications to computer science such as Petri net specifications [55, 79, 116, 117, 118], process semantics [3, 167] and semantics of logic programs [54].

Involutive quantales were originally introduced by Mulvey and Pelletier [129] in order to *quantize* Hoare and He's calculus of relations [74], which expresses the weakest pre-specification of the specification for a given program. For more information about involutive quantales, see e.g., [130, 150]. The logics which correspond to the involutive quantales and to their neighbors such as Gelfand quantales were proposed and studied by MacCaull [122] for involutive quantales and by Allwein and MacCaull [5] for Gelfand quantales. In [122], some Kripke-type semantics, a sequent calculus and a relational proof system were given for

[1] This chapter is predicated on [85].

the logic of involutive quantales, and the completeness theorems with respect to such Kripke-type semantics were also proved.

An intuitionistic (non-modal) linear logic with strong negation, called here WILL, was studied in [196, 197] and has been discussed in Chapter 10. An extension of WILL with an exponential operator ! was also presented by Kamide [82], and a number of sequent calculi, a complete Kripke semantics and some applications such as applications to process calculus and logic programming language were obtained for this logic in [82].

The contents of this chapter can then be summarized as follows. In Section 11.2, the new logic QILL, which is regarded as a commutative version of the logic corresponding to involutive quantales, is introduced, and the logic WILL is reviewed. It is observed that the involution modal operator occurring in QILL is analogous to the strong negation operator occurring in WILL. Cut-elimination, decidability, conservativity and a new characteristic property, named the "quantization principle", are shown for QILL.

In Section 11.3, two kinds of quantale models for QILL and WILL are introduced. These models have two kinds of valuations v^+ and v^-, which are called "strong valuations" in the case of models for WILL and "involutive valuations" in the case of models for QILL. It is observed that the difference between the models for WILL and those for QILL is only the setting of the (negative) valuation v^- in these models. Involutive quantale models for QILL, which are based on involutive quantales with an involution operator, are also introduced. It is observed that introducing the involution operator and using v^- are essentially the same thing. The soundness theorems (w.r.t. these models) for WILL and QILL are proved.

In Section 11.4, the completeness theorems (w.r.t. the quantale and involutive quantale models) for WILL and QILL are proved using a canonical model construction method based on the MacNeille completion technique discussed in [79, 116, 117, 145]. These completeness results for WILL and QILL are the main contribution of this chapter.

11.2. Sequent calculus

Prior to a detailed discussion, the language and notation used in this chapter are introduced below. *Formulas* are constructed from propositional variables, propositional constants $\mathbf{1}, \top, \bot$, and the connectives $\rightarrow$ (implication), $\wedge$ (conjunction), $*$ (fusion), $\vee$ (disjunction), $\sim$ (strong negation) and $(\cdot)^{\bullet}$ (a modal operator called "involution"). This chapter adopts the notation for the constants $\mathbf{1}, \top$ and $\bot$ in [187], which differs from that in [64], and also uses $\wedge, \vee$, and $*$ which correspond to $\&, \oplus$ and $\otimes$ in [64]. Lower case letters $p, q, ...$ are used to denote propositional variables, lower case Greek letters $\alpha, \beta, ...$ are used to denote formulas, and Greek capital letters $\Gamma, \Delta, ...$ are used to represent finite (possibly empty) sequences of formulas. $[\Gamma]$ is used to denote the multiset which consists of all elements of a sequence Γ of formulas, and $\Gamma^\bullet$ ($\sim\Gamma$) is used to denote the sequence $\langle \gamma^\bullet \mid \gamma \in \Gamma \rangle$ ($\langle \sim\gamma \mid \gamma \in \Gamma \rangle$ resp.). A *sequent* is an expression of the form $\Gamma \Rightarrow \alpha$ where α is non-empty (i.e. a single formula). The symbol $\equiv$ is used to denote the equality of sequences (or multisets) of symbols. $\Delta^\star$ denotes $\delta_1 * \cdots * \delta_n$ if $\Delta \equiv \langle \gamma_1, ..., \gamma_n \rangle$ $(0 \leq n)$, and also denotes $\mathbf{1}$ if Δ is empty. If a sequent S is provable in a sequent system L, then we write $L \vdash S$, and sometimes write $\vdash S$ for $L \vdash S$ by omitting L. Since all logics discussed in this chapter are formulated as sequent calculi, we will occasionally identify a sequent calculus with the logic determined by it.

First, we introduce ILL (intuitionistic linear logic without the exponential !).

Definition 11.2.1 (ILL) *The initial sequents of* ILL *are of the form:*

$$\alpha \Rightarrow \alpha \qquad \Rightarrow \mathbf{1} \qquad \Gamma \Rightarrow \top \qquad \bot, \Delta \Rightarrow \gamma.$$

The structural rules of ILL *are of the form:*

$$\frac{\Gamma \Rightarrow \alpha \quad \alpha, \Sigma \Rightarrow \gamma}{\Gamma, \Sigma \Rightarrow \gamma} \text{ (cut)} \qquad \frac{\Gamma, \beta, \alpha, \Delta \Rightarrow \gamma}{\Gamma, \alpha, \beta, \Delta \Rightarrow \gamma} \text{ (ex)}.$$

The inference rules of ILL *are of the form:*

$$\frac{\Gamma \Rightarrow \gamma}{\mathbf{1}, \Gamma \Rightarrow \gamma} \text{ (}\mathbf{1}\text{we)}$$

$$\frac{\Gamma \Rightarrow \alpha \quad \beta, \Sigma \Rightarrow \gamma}{\alpha \rightarrow \beta, \Gamma, \Sigma \Rightarrow \gamma} \text{ (}\rightarrow\text{left)} \qquad \frac{\Gamma, \alpha \Rightarrow \beta}{\Gamma \Rightarrow \alpha \rightarrow \beta} \text{ (}\rightarrow\text{right)}$$

$$\frac{\alpha, \beta, \Delta \Rightarrow \gamma}{\alpha * \beta, \Delta \Rightarrow \gamma} \text{ (}*\text{left)} \qquad \frac{\Gamma \Rightarrow \alpha \quad \Delta \Rightarrow \beta}{\Gamma, \Delta \Rightarrow \alpha * \beta} \text{ (}*\text{right)}$$

$$\frac{\alpha, \Delta \Rightarrow \gamma}{\alpha \wedge \beta, \Delta \Rightarrow \gamma} \text{ (}\wedge\text{left1)} \qquad \frac{\beta, \Delta \Rightarrow \gamma}{\alpha \wedge \beta, \Delta \Rightarrow \gamma} \text{ (}\wedge\text{left2)}$$

$$\frac{\Gamma \Rightarrow \alpha \quad \Gamma \Rightarrow \beta}{\Gamma \Rightarrow \alpha \wedge \beta} \text{ (}\wedge\text{right)} \qquad \frac{\alpha, \Delta \Rightarrow \gamma \quad \beta, \Delta \Rightarrow \gamma}{\alpha \vee \beta, \Delta \Rightarrow \gamma} \text{ (}\vee\text{left)}$$

$$\frac{\Gamma \Rightarrow \alpha}{\Gamma \Rightarrow \alpha \vee \beta} \text{ (}\vee\text{right1)} \qquad \frac{\Gamma \Rightarrow \beta}{\Gamma \Rightarrow \alpha \vee \beta} \text{ (}\vee\text{right2)}.$$

Definition 11.2.2 (QILL) QILL *(quantized* ILL*) is obtained from* ILL *by adding the initial sequents and inference rules of the form:*

$$\Rightarrow \mathbf{1}^{\bullet} \qquad \Gamma \Rightarrow \top^{\bullet} \qquad \bot^{\bullet}, \Gamma \Rightarrow \gamma$$

$$\frac{\Gamma \Rightarrow \alpha}{\Gamma \Rightarrow \alpha^{\bullet\bullet}}\ (\bullet\text{right}) \qquad \frac{\alpha, \Delta \Rightarrow \gamma}{\alpha^{\bullet\bullet}, \Delta \Rightarrow \gamma}\ (\bullet\text{left})$$

$$\frac{\Gamma \Rightarrow \gamma}{\mathbf{1}^{\bullet}, \Gamma \Rightarrow \gamma}\ (\bullet\mathbf{1}\text{we})$$

$$\frac{\Gamma \Rightarrow \alpha^{\bullet} \quad \beta^{\bullet}, \Sigma \Rightarrow \gamma}{(\alpha \rightarrow \beta)^{\bullet}, \Gamma, \Sigma \Rightarrow \gamma}\ (\bullet{\rightarrow}\text{left}) \qquad \frac{\Gamma, \alpha^{\bullet} \Rightarrow \beta^{\bullet}}{\Gamma \Rightarrow (\alpha \rightarrow \beta)^{\bullet}}\ (\bullet{\rightarrow}\text{right})$$

$$\frac{\alpha^{\bullet}, \beta^{\bullet}, \Delta \Rightarrow \gamma}{(\alpha * \beta)^{\bullet}, \Delta \Rightarrow \gamma}\ (\bullet * \text{left}) \qquad \frac{\Gamma \Rightarrow \alpha^{\bullet} \quad \Delta \Rightarrow \beta^{\bullet}}{\Gamma, \Delta \Rightarrow (\alpha * \beta)^{\bullet}}\ (\bullet * \text{right})$$

$$\frac{\alpha^{\bullet}, \Delta \Rightarrow \gamma}{(\alpha \wedge \beta)^{\bullet}, \Delta \Rightarrow \gamma}\ (\bullet \wedge \text{left1}) \qquad \frac{\beta^{\bullet}, \Delta \Rightarrow \gamma}{(\alpha \wedge \beta)^{\bullet}, \Delta \Rightarrow \gamma}\ (\bullet \wedge \text{left2})$$

$$\frac{\Gamma \Rightarrow \alpha^{\bullet} \quad \Gamma \Rightarrow \beta^{\bullet}}{\Gamma \Rightarrow (\alpha \wedge \beta)^{\bullet}}\ (\bullet \wedge \text{right}) \qquad \frac{\alpha^{\bullet}, \Delta \Rightarrow \gamma \quad \beta^{\bullet}, \Delta \Rightarrow \gamma}{(\alpha \vee \beta)^{\bullet}, \Delta \Rightarrow \gamma}\ (\bullet \vee \text{left})$$

$$\frac{\Gamma \Rightarrow \alpha^{\bullet}}{\Gamma \Rightarrow (\alpha \vee \beta)^{\bullet}}\ (\bullet \vee \text{right1}) \qquad \frac{\Gamma \Rightarrow \beta^{\bullet}}{\Gamma \Rightarrow (\alpha \vee \beta)^{\bullet}}\ (\bullet \vee \text{right2}).$$

Definition 11.2.3 (WILL) WILL *(an extension of* ILL *with strong negation) is obtained from* ILL *by adding the initial sequents and inference rules of the form:*

$$\sim\mathbf{1}, \Gamma \Rightarrow \gamma \qquad \sim\top, \Gamma \Rightarrow \gamma \qquad \Gamma \Rightarrow \sim\bot$$

$$\frac{\Gamma \Rightarrow \alpha}{\Gamma \Rightarrow \sim\sim\alpha}\ (\sim\text{right}) \qquad \frac{\alpha, \Delta \Rightarrow \gamma}{\sim\sim\alpha, \Delta \Rightarrow \gamma}\ (\sim\text{left})$$

$$\frac{\sim\beta, \alpha, \Delta \Rightarrow \gamma}{\sim(\alpha \rightarrow \beta), \Delta \Rightarrow \gamma}\ (\sim{\rightarrow}\text{left}) \qquad \frac{\Gamma \Rightarrow \sim\beta \quad \Delta \Rightarrow \alpha}{\Gamma, \Delta \Rightarrow \sim(\alpha \rightarrow \beta)}\ (\sim{\rightarrow}\text{right})$$

$$\frac{\sim\alpha, \sim\beta, \Delta \Rightarrow \gamma}{\sim(\alpha * \beta), \Delta \Rightarrow \gamma}\ (\sim * \text{left}) \qquad \frac{\Gamma \Rightarrow \sim\alpha \quad \Delta \Rightarrow \sim\beta}{\Gamma, \Delta \Rightarrow \sim(\alpha * \beta)}\ (\sim * \text{right})$$

$$\frac{\sim\alpha, \Delta \Rightarrow \gamma \quad \sim\beta, \Delta \Rightarrow \gamma}{\sim(\alpha \wedge \beta), \Delta \Rightarrow \gamma}\ (\sim \wedge \text{left})$$

$$\frac{\Gamma \Rightarrow \sim\alpha}{\Gamma \Rightarrow \sim(\alpha \wedge \beta)}\ (\sim \wedge \text{right1}) \qquad \frac{\Gamma \Rightarrow \sim\beta}{\Gamma \Rightarrow \sim(\alpha \wedge \beta)}\ (\sim \wedge \text{right2})$$

$$\frac{\sim\alpha, \Delta \Rightarrow \gamma}{\sim(\alpha \vee \beta), \Delta \Rightarrow \gamma}\ (\sim \vee \text{left1}) \qquad \frac{\sim\beta, \Delta \Rightarrow \gamma}{\sim(\alpha \vee \beta), \Delta \Rightarrow \gamma}\ (\sim \vee \text{left2})$$

$$\frac{\Gamma \Rightarrow \sim\alpha \quad \Gamma \Rightarrow \sim\beta}{\Gamma \Rightarrow \sim(\alpha \vee \beta)}\ (\sim \vee \text{right}).$$

The proof of the following theorem is the same as that of the theorem for ILL, except for obvious modifications.

Theorem 11.2.4 (Cut-elimination) *Let L be* QILL *or* WILL. *The rule* (cut) *is admissible in cut-free L.*

Using Theorem 11.2.4, we can show the following.

Corollary 11.2.5 *Let L be* QILL *or* WILL. *Then L is decidable and a conservative extension of* ILL.

We remark that the cut-elimination theorem and decidability of WILL were shown in [196, 197]. These properties for QILL can be shown in a similar way.

Next we consider a characteristic property. Let L be QILL or WILL. L is called *explosive* if for any formulas α and β, the sequent $\alpha, {\sim}\alpha \Rightarrow \beta$ or $\alpha, \alpha^{\bullet} \Rightarrow \beta$ is provable in L.

Corollary 11.2.6 QILL *and* WILL *are not explosive.*

We remark that, roughly speaking, such a property for a logic with negation is called *paraconsistency*, that is, WILL is regarded as paraconsistent. We also remark that for any propositional variable p, the sequents $p^{\bullet} \Rightarrow p$, $p \Rightarrow p^{\bullet}$, ${\sim}p \Rightarrow p$ and $p \Rightarrow {\sim}p$ are not provable in QILL and WILL.

We remark that the following inference rules are admissible in cut-free QILL:

$$\frac{\alpha^{\bullet\bullet}, \Gamma \Rightarrow \gamma}{\alpha, \Gamma \Rightarrow \gamma}\ (\bullet\text{left}^{-1}), \qquad \frac{\Gamma \Rightarrow \alpha^{\bullet\bullet}}{\Gamma \Rightarrow \alpha}\ (\bullet\text{right}^{-1}).$$

We consider the following inference rule for QILL:

$$\frac{\Gamma \Rightarrow \alpha}{\Gamma^{\bullet} \Rightarrow \alpha^{\bullet}}\ (\bullet\text{regu}).$$

We then have the following.

Lemma 11.2.7 *The rule* ($\bullet$regu) *is admissible in cut-free* QILL.

Proof. We show that the rule

$$\frac{\Gamma \Rightarrow \alpha}{\Gamma^{\bullet} \Rightarrow \alpha^{\bullet}}\ (\bullet\text{regu})$$

is admissible in cut-free QILL. We prove this by induction on the cut-free proofs P of $\Gamma \Rightarrow \alpha$ in QILL. We distinguish the cases according to the last inference of P. We show only the following case.

Case (•→left): The last inference of P is of the form:

$$\frac{\Gamma_1 \Rightarrow \beta^\bullet \quad \gamma^\bullet, \Gamma_2 \Rightarrow \alpha}{(\beta{\to}\gamma)^\bullet, \Gamma_1, \Gamma_2 \Rightarrow \alpha}\ (\bullet{\to}\text{left})$$

where $\Gamma \equiv \langle (\beta{\to}\gamma)^\bullet, \Gamma_1, \Gamma_2 \rangle$. By induction hypothesis, we have $\vdash \Gamma_1^\bullet \Rightarrow \beta^{\bullet\bullet}$ and $\vdash \gamma^{\bullet\bullet}, \Gamma_2^\bullet \Rightarrow \alpha^\bullet$, and hence

$$\dfrac{\dfrac{\dfrac{\Gamma_1^\bullet \Rightarrow \beta^{\bullet\bullet}}{\Gamma_1^\bullet \Rightarrow \beta}\ (\bullet\text{right}^{-1}) \quad \dfrac{\gamma^{\bullet\bullet}, \Gamma_2^\bullet \Rightarrow \alpha^\bullet}{\gamma, \Gamma_2^\bullet \Rightarrow \alpha^\bullet}\ (\bullet\text{left}^{-1})}{\beta{\to}\gamma, \Gamma_1^\bullet, \Gamma_2^\bullet \Rightarrow \alpha^\bullet}\ ({\to}\text{left})}{(\beta{\to}\gamma)^{\bullet\bullet}, \Gamma_1^\bullet, \Gamma_2^\bullet \Rightarrow \alpha^\bullet}\ (\bullet\text{left}).$$

∎

The converse of (•regu):

$$\frac{\Gamma^\bullet \Rightarrow \alpha^\bullet}{\Gamma \Rightarrow \alpha}\ (\bullet\text{regu}^{-1})$$

is also derivable in QILL+(•regu). For example, the following is a proof of the case $\Gamma^\bullet \equiv \langle \gamma_1^\bullet, \gamma_2^\bullet \rangle$:

$$\dfrac{\gamma_2 \Rightarrow \gamma_2^{\bullet\bullet} \quad \dfrac{\gamma_1 \Rightarrow \gamma_1^{\bullet\bullet} \quad \dfrac{\dfrac{\gamma_1^\bullet, \gamma_2^\bullet \Rightarrow \alpha^\bullet}{\gamma_1^{\bullet\bullet}, \gamma_2^{\bullet\bullet} \Rightarrow \alpha^{\bullet\bullet}}\ (\bullet\text{regu}) \quad \alpha^{\bullet\bullet} \Rightarrow \alpha}{\gamma_1^{\bullet\bullet}, \gamma_2^{\bullet\bullet} \Rightarrow \alpha}}{\gamma_1, \gamma_2^{\bullet\bullet} \Rightarrow \alpha}}{\gamma_1, \gamma_2 \Rightarrow \alpha}.$$

Then we have the following theorem.

Theorem 11.2.8 QILL *and* QILL + (•regu) + (•regu^{-1}) *are theorem-equivalent.*

The statement of this theorem will be used to prove the completeness theorem (w.r.t. the involutive quantale models) for QILL.

We can show a new characteristic property for QILL, which is named the "*quantization principle*".[2] We call the logic having such a property, a *quantized logic*, and also call the way of constructing the quantized logic from a logic, the *quantization* of a logic.

Theorem 11.2.9 (Quantization principle) *For any formula α,*

[2] The term "quantization" is from [129].

$$\text{QILL} \vdash \Rightarrow \alpha \textit{ iff } \text{QILL} \vdash \Rightarrow \alpha^{\bullet}.$$

This theorem means intuitively that the existence of parallel worlds in the sense of (the philosophy of) the theory of quantum mechanics, i.e. there are a number of worlds (including our real-world) with coherence in parallel. This interpretation of quantum mechanics is called the "many worlds interpretation". There are some alternative interpretations such as the "Copenhagen interpretation". The worlds of electrons are a typical example of such parallel worlds (e.g., the famous two-slit experiment). In these parallel worlds of electrons, we have the problem of observation, i.e., an observation induces a choice of worlds: if we observe the existence of an electron, then we must choose or fix one of the worlds. In this context, "$\vdash \Rightarrow \alpha$" means "$\alpha$ is true in our (chosen) real-world", and "$\vdash \Rightarrow \alpha^{\bullet}$" means "$\alpha$ is true in another (unchosen) parallel world". In fact, if we have the fact "$\vdash \Rightarrow \alpha$", then we can not derive the proof of $\Rightarrow \alpha^{\bullet}$ by proof of $\Rightarrow \alpha$, because the rules ($\bullet$regu) and ($\bullet$regu^{-1}) are admissible, but not derivable. In this context, this theorem particularly represents the case of two parallel worlds, which are called here 1-state and 0-state worlds. This quantum $\{0,1\}$-analogy will be presented semantically in Section 3. We thus believe that the quantization principle is analogous to a phenomenon of the $\{0,1\}$-parallel worlds in the theory of quantum mechanics. More precisely, it is behind the phenomenon that if the physical quantity of a particle for an electron is *quantized*, then the quantum such as a photon in the field has appeared.

Next, we show the Hilbert-style axiomatizations HQILL and HWILL of the sequent calculi QILL and WILL respectively. Let HILL be a Hilbert-style axiomatization of ILL (see e.g., [83, 187]), and $\alpha \leftrightarrow \beta$ be the abbreviation of both $\alpha \rightarrow \beta$ and $\beta \rightarrow \alpha$.

Then HQILL is obtained from HILL by adding the following axiom schemes:

1. $\mathbf{1}^{\bullet} \leftrightarrow \mathbf{1}$,
2. $\top \rightarrow \top^{\bullet}$,
3. $\bot^{\bullet} \rightarrow \bot$,
4. $\alpha^{\bullet\bullet} \leftrightarrow \alpha$,
5. $(\alpha \rightarrow \beta)^{\bullet} \leftrightarrow \alpha^{\bullet} \rightarrow \beta^{\bullet}$,
6. $(\alpha \wedge \beta)^{\bullet} \leftrightarrow \alpha^{\bullet} \wedge \beta^{\bullet}$,
7. $(\alpha \vee \beta)^{\bullet} \leftrightarrow \alpha^{\bullet} \vee \beta^{\bullet}$,

8. $(\alpha * \beta)^\bullet \leftrightarrow \alpha^\bullet * \beta^\bullet$.

It can be observed that the situation of adding the axiom schemes concerning $\cdot^\bullet$ to a logic is analogous to the situation of adding the axiom schemes concerning $\sim$ to a logic.

Introducing the involution connective into an object language is not quite new: This idea was discussed in the papers by Allwein and MacCaull [5, 122]. In [5], the involution connective is explicitly called a "permutation connective" which task is to switch formulas connected by multiplicative conjunction, since the base logic is non-commutative. Thus, the setting of the involution connective in the present chapter and that in [5] are slightly different.

HWILL is obtained from HILL by adding the following axiom schemes:

1. $\sim\mathbf{1}\rightarrow\alpha$,
2. $\sim\top\rightarrow\alpha$,
3. $\alpha\rightarrow\sim\bot$,
4. $\sim\sim\alpha\leftrightarrow\alpha$,
5. $\sim(\alpha\rightarrow\beta)\leftrightarrow\alpha*\sim\beta$,
6. $\sim(\alpha\wedge\beta)\leftrightarrow\sim\alpha\vee\sim\beta$,
7. $\sim(\alpha\vee\beta)\leftrightarrow\sim\alpha\wedge\sim\beta$,
8. $\sim(\alpha*\beta)\leftrightarrow\sim\alpha*\sim\beta$.

11.3. Quantale and involutive quantale models

The original quantales are non-commutative, but in this chapter, we call the commutative versions, quantales.

Definition 11.3.1 (Quantale) *A* quantale *is a structure* $\mathbf{Q} := \langle Q, \bigcup, \cdot, \dot{\mathrm{i}}\rangle$ *satisfying the following conditions:*

1. $\langle Q, \bigcup\rangle$ *is a complete lattice (the least element and the greatest element are respectively denoted by* $\dot{\bot}$ *and* $\dot{\top}$*, and the binary versions of the lattice operations are denoted by* $\cup$ *and* $\cap$*),*
2. $\langle Q, \cdot, \dot{\mathrm{i}}\rangle$ *is a commutative monoid with the identity* $\dot{\mathrm{i}}$*,*
3. $(\bigcup x_i)\cdot y = \bigcup(x_i \cdot y)$ *for all* $x_i, y \in Q$.

We define an operation $\dot{\rightarrow}$ *on* Q *as follows:*

$$y \dot{\rightarrow} z := \bigcup\{x \mid x \cdot y \leq z\}$$

where $\leq$ is defined as $x \leq y$ iff $x \cup y = y$ for all $x, y \in Q$. Then the following condition on Q holds using Condition 3 above:

$$x \leq y \dot{\rightarrow} z \text{ iff } x \cdot y \leq z \text{ for all } x, y, z \in Q.$$

We call the quantale equipped with $\dot{\bot}, \dot{\top}, \cup, \cap$ and $\dot{\rightarrow}$, quantale in the following.

We remark that the following monotonicity condition on a quantale $\mathbf{Q}$ holds:

> $x \leq x'$ and $y \leq y'$ imply $x \cdot y \leq x' \cdot y'$ and $x' \dot{\rightarrow} y \leq x \dot{\rightarrow} y'$ for all $x, x', y, y' \in Q$.

For the purpose of showing the completeness theorems for QILL and WILL, the structure $\mathbf{Q}$ can be modified as follows: the infinite join $\bigcup$ is replaced by the binary $\cup$, and the operation $\dot{\rightarrow}$ is independently defined as the conditions which are described above.

Next we consider *strong valuations* for a quantale model for WILL.

Definition 11.3.2 Strong valuations v^+ *and* v^- *on a quantale* $\mathbf{Q} := \langle Q, \bigcup, \cdot, \dot{\mathbf{1}} \rangle$ *are mappings from the set of all propositional variables to* Q. v^+ *and* v^- *are extended to mappings from the set* Φ *of all formulas to* Q *by*

1. $v^+(\mathbf{1}) := \dot{\mathbf{1}}$,
2. $v^+(\top) := \dot{\top}$,
3. $v^+(\bot) := \dot{\bot}$,
4. $v^+(\alpha \wedge \beta) := v^+(\alpha) \cap v^+(\beta)$,
5. $v^+(\alpha \vee \beta) := v^+(\alpha) \cup v^+(\beta)$,
6. $v^+(\alpha * \beta) := v^+(\alpha) \cdot v^+(\beta)$,
7. $v^+(\alpha \rightarrow \beta) := v^+(\alpha) \dot{\rightarrow} v^+(\beta)$,
8. $v^+(\sim\alpha) := v^-(\alpha)$,
9. $v^-(\mathbf{1}) := \dot{\bot}$,
10. $v^-(\top) := \dot{\bot}$,
11. $v^-(\bot) := \dot{\top}$,

12. $v^-(\alpha \wedge \beta) := v^-(\alpha) \cup v^-(\beta)$,

13. $v^-(\alpha \vee \beta) := v^-(\alpha) \cap v^-(\beta)$,

14. $v^-(\alpha * \beta) := v^-(\alpha) \cdot v^-(\beta)$,

15. $v^-(\alpha \rightarrow \beta) := v^+(\alpha) \cdot v^-(\beta)$,

16. $v^-(\sim\alpha) := v^+(\alpha)$.

An intuitive meaning of the strong valuations is that v^+ and v^- respectively correspond to *provability* and *refutability.*

We can also consider *involutive valuations* for a quantale model for QILL.

Definition 11.3.3 Involutive valuations v^+ *and* v^- *on a quantale* $\mathbf{Q} := \langle Q, \bigcup, \cdot, \dot{\mathbf{1}} \rangle$ *are mappings from the set of all propositional variables to* Q. v^+ *and* v^- *are extended to mappings from the set* Φ *of all formulas to* Q *by 1—7 in Definition 11.3.2 and*

8. $v^+(\alpha^\bullet) := v^-(\alpha)$,

9. $v^-(\mathbf{1}) := \dot{\mathbf{1}}$,

10. $v^-(\top) := \dot{\top}$,

11. $v^-(\bot) := \dot{\bot}$,

12. $v^-(\alpha \wedge \beta) := v^-(\alpha) \cap v^-(\beta)$,

13. $v^-(\alpha \vee \beta) := v^-(\alpha) \cup v^-(\beta)$,

14. $v^-(\alpha * \beta) := v^-(\alpha) \cdot v^-(\beta)$,

15. $v^-(\alpha \rightarrow \beta) := v^-(\alpha) \dot{\rightarrow} v^-(\beta)$,

16. $v^-(\alpha^\bullet) := v^+(\alpha)$.

An intuitive meaning of the involutive valuations is that, for the quantum $\{0,1\}$-analogy discussed, v^+ and v^- respectively correspond to *provability in the* 1*-state* and *provability in the* 0*-state.* More precisely, "$\dot{\mathbf{1}} \leq v^+(\alpha)$" means "$\alpha$ is true in a chosen world (1-state world)", and "$\dot{\mathbf{1}} \leq v^-(\alpha)$", i.e., "$\dot{\mathbf{1}} \leq v^+(\alpha^\bullet)$", means "$\alpha$ is true in another unchosen world (0-state world)".

Definition 11.3.4 (Quantale models) *A* quantale model *for* WILL *(*QILL*) is a structure* $\langle \mathbf{Q}, v^+, v^- \rangle$ *such that* $\mathbf{Q}$ *is a quantale, and* v^+ *and* v^- *are strong valuations (involutive valuations resp.). A formula* α *is* true *in a quantale model* $\langle \mathbf{Q}, v^+, v^- \rangle$ *for* WILL *(*QILL*) if* $\dot{\mathbf{i}} \leq v^+(\alpha)$ *holds, and* strongly valid *(*involutive valid*) in a quantale* $\mathbf{Q}$ *if it is true for any strong valuations (involutive valuations resp.)* v^+ *and* v^- *on the quantale. A sequent* $\alpha_1, \cdots, \alpha_n \Rightarrow \beta$ *(or* $\Rightarrow \beta$*) is* true *in a quantale model* $\langle \mathbf{Q}, v^+, v^- \rangle$ *for* WILL *(*QILL*) if the formula* $\alpha_1 * \cdots * \alpha_n \rightarrow \beta$ *(or* β*) is true in it, and strongly valid (involutive valid) in a quantale if so is* $\alpha_1 * \cdots * \alpha_n \rightarrow \beta$ *(or* β*).*

Theorem 11.3.5 (Soundness) *Let* C *be the class of all quantales,* $L_1(C) := \{S \mid$ *a sequent* S *is strongly valid in all quantales of* $C\}$, $L_2(C) := \{S \mid$ *a sequent* S *is involutive valid in all quantales of* $C\}$, $L_1 := \{S \mid \text{WILL} \vdash S\}$ *and* $L_2 := \{S \mid \text{QILL} \vdash S\}$*. Then* $L_1 \subseteq L_1(C)$ *and* $L_2 \subseteq L_2(C)$*.*

Proof. The proof is straightforward and similar to that for ILL. ∎

Definition 11.3.6 (Involutive quantale) *An* involutive quantale *is a structure* $\mathbf{Q}^\bullet := \langle \mathbf{Q}, \cdot^\circ \rangle$ *satisfying the following conditions:*

1. $\mathbf{Q}$ *is a quantale* $\langle Q, \bigcup, \cdot, \dot{\mathbf{i}} \rangle$ *equipped with* $\dot{\bot}, \dot{\top}, \cup, \cap$ *and* $\dot{\rightarrow}$ *(see Definition 11.3.1),*
2. $\cdot^\circ$ *is a unary operation on* Q *such that*

   C1: $x^{\circ\circ} = x$,

   C2: $(\bigcup x_i)^\circ = \bigcup (x_i)^\circ$,

   C3: $(x \cdot y)^\circ = x^\circ \cdot y^\circ$,

   C4: $(x \cap y)^\circ = x^\circ \cap y^\circ$,

   C5: $(x \dot{\rightarrow} y)^\circ = x^\circ \dot{\rightarrow} y^\circ$,

   C6: $\dot{\mathbf{i}}^\circ = \dot{\mathbf{i}}$,

   C7: $\dot{\top}^\circ = \dot{\top}$,

   C8: $\dot{\bot}^\circ = \dot{\bot}$.

We can derive the following condition on $\mathbf{Q}$ by using C2:

C2′: $x \leq y$ implies $x^\circ \leq y^\circ$ for all $x, y \in Q$.

We remark that the original involutive quantales [129] are extensions of non-commutative quantales. The involutive quantales do not satisfy the conditions C4, C5, C7, C8 (and do not have the operations $\cap, \dot{\rightarrow}$ and the constants $\dot{\bot}, \dot{\top}$) and satisfy the condition

C3′: $(x \cdot y)^\circ = y^\circ \cdot x^\circ$.

We can construct the logic which corresponds to the non-commutative version of the quantales presented (but not the original version). Roughly speaking, the logic of the non-commutative involutive quantales is obtained from QILL by deleting the exchange rule (ex), (•∗left) and (•∗right), and by adding (•regu), (•regu^{-1}) and the following modified versions of (•∗left) and (•∗right):

$$\frac{\Gamma, \beta^\bullet, \alpha^\bullet, \Delta \Rightarrow \gamma}{\Gamma, (\alpha * \beta)^\bullet, \Delta \Rightarrow \gamma} \qquad \frac{\Gamma \Rightarrow \beta^\bullet \quad \Delta \Rightarrow \alpha^\bullet}{\Gamma, \Delta \Rightarrow (\alpha * \beta)^\bullet}$$

and by modifying the sequents of multiset version to that of sequence version.[3] The logic of original involutive quantales was presented by MacCaull [122]. Our sequent calculi differ from the sequent calculus of MacCaull.

Definition 11.3.7 *A* single valuation v *on an involutive quantale* $\mathbf{Q}^\bullet$ *is a mapping from the set of all propositional variables to* Q. *A single valuation* v *is extended to a mapping from the set* Φ *of all formulas to* Q *by*

1. $v(\mathbf{1}) := \dot{\mathbf{1}}$,
2. $v(\top) := \dot{\top}$,
3. $v(\bot) := \dot{\bot}$,
4. $v(\alpha \wedge \beta) := v(\alpha) \cap v(\beta)$,
5. $v(\alpha \vee \beta) := v(\alpha) \cup v(\beta)$,
6. $v(\alpha * \beta) := v(\alpha) \cdot v(\beta)$,
7. $v(\alpha \rightarrow \beta) := v(\alpha) \dot{\rightarrow} v(\beta)$,
8. $v(\alpha^\bullet) := v(\alpha)^\circ$.

Definition 11.3.8 Involutive quantale model) *An* involutive quantale model *is a structure* $\langle \mathbf{Q}^\bullet, v \rangle$ *such that* $\mathbf{Q}^\bullet$ *is an involutive quantale, and* v *is a single valuation. A formula* α *is* true *in an involutive quantale model* $\langle \mathbf{Q}^\bullet, v \rangle$ *if* $\dot{\mathbf{1}} \leq v(\alpha)$ *holds, and* valid *in an involutive quantale* $\mathbf{Q}^\bullet$ *if it is true for any valuation* v *on the involutive quantale. A sequent* $\alpha_1, \cdots, \alpha_n \Rightarrow \beta$ *(or* $\Rightarrow \beta$*) is* true *in an involutive quantale model* $\langle \mathbf{Q}^\bullet, v \rangle$ *if the formula* $\alpha_1 * \cdots * \alpha_n \rightarrow \beta$ *(or* β*) is true in it, and valid in an involutive quantale if so is* $\alpha_1 * \cdots * \alpha_n \rightarrow \beta$ *(or* β*).*

[3] Of course, we must modify all the inference rules by adapting the sequent of sequence version.

Roughly speaking, considering $v^-(\alpha)$ for $v(\alpha)^\circ$, a quantale model with the involutive valuations v^- and v^+ is essentially the same as an involutive quantale model with the single valuation v.

Theorem 11.3.9 (Soundness) *Let C be the class of all involutive quantales, $L(C) := \{S \mid$ a sequent S is valid in all involutive quantales of $C\}$ and $L := \{S \mid \text{QILL} \vdash S\}$. Then $L \subseteq L(C)$.*

11.4. Completeness

In the following, the proofs of the completeness theorems are mainly based on [79], and also based on [55, 116, 117, 145, 187]. The differences from the completeness proofs for the $\{\sim, \bullet\}$-free part in [55, 79, 116, 117, 145, 187] are only the use of the negative valuation v^- and the case for the involution operator. First, we will prove the completeness theorems (for QILL and WILL) using quantale models with strong and involutive valuations. We will prove this result only for WILL since that for QILL is proved similarly. Second, we will prove the completeness theorem for the logic QILL+($\bullet$regu)+($\bullet$regu^{-1}), which is equivalent to QILL by Theorem 11.2.8, using an involutive quantale model with a single valuation.

We will construct a canonical model in the following. First we construct a structure $\mathbf{M} := \langle M, \cdot, [\,], \leq \rangle$ such that

1. $M := \{[\Gamma] \mid [\Gamma]$ is a finite multiset of formulas$\}$,

2. $[\Gamma] \cdot [\Delta] := [\Gamma, \Delta]$ (the multiset union),

3. $[\,]$ is the empty multiset,

4. $[\Gamma] \leq [\Delta]$ iff $\vdash \Gamma \Rightarrow \Delta^\star$ where $\Delta^\star \equiv \gamma_1 * \cdots * \gamma_n$ if $\Delta \equiv \langle \gamma_1, \cdots, \gamma_n \rangle$ $(0 < n)$, and $\Delta^\star \equiv \mathbf{1}$ if Δ is empty.

$[\Gamma] \doteq [\Delta]$ is defined as $[\Gamma] \leq [\Delta]$ and $[\Delta] \leq [\Gamma]$.

$\mathbf{M}$ is called a *pre-ordered commutative monoid* which satisfies the following conditions:

1. $\langle M, \cdot, [\,] \rangle$ is a commutative monoid with the identity $[\,]$,

2. $\langle M, \leq \rangle$ is a pre-ordered set,

3. $x_1 \leq x_2$ and $y_1 \leq y_2$ imply $x_1 \cdot y_1 \leq x_2 \cdot y_2$ for all $x_1, x_2, y_1, y_2 \in M$.

Next we construct the power set structure $\mathbf{P}(\mathbf{M}) := \langle P(M), \bigcup, \circ, \{[\,]\} \rangle$ of $\mathbf{M}$ such that

1. $P(M)$ is the power set of M,

2. $\bigcup$ is the usual set theoretic infinite union on $P(M)$ (we also assume the usual set theoretic operations $\cup$ and $\cap$),

3. $\circ$ is defined as

$$X \circ Y := \{x \cdot y \mid x \in X \text{ and } y \in Y\} \text{ for all } X, Y \in P(M).$$

Moreover we define an operation $\dot{\to}$ as follows:

$$Y \dot{\to} Z := \{x \mid \forall y \in Y \ (x \cdot y \in Z)\} \text{ for all } Y, Z \in P(M).$$

We can assume M (the greatest element) and $\varnothing$ (the least element) as the constants in $P(M)$. We can derive the following condition:

$$X \subseteq Y \dot{\to} Z \text{ iff } X \circ Y \subseteq Z \text{ for all } X, Y, Z \in P(M).$$

Then we have the following.

Proposition 11.4.1 $\mathbf{P(M)}$ *is a quantale.*

Next we construct a structure $\mathbf{C(P(M))} := \langle C(P(M)), \bigcup_c, \circ_c, C\{[\,]\}\rangle$ such that

1. C is a closure operation[4] on $P(M)$ called a *MacNeille closure* such that $CX := (X^{\rightarrow})^{\leftarrow}$ where $X^{\rightarrow} := \{y \mid \forall x \in X \ (x \leq y)\}$ and $X^{\leftarrow} := \{y \mid \forall x \in X \ (y \leq x)\}$,

2. $C(P(M))$ is the set of all C-closed elements of $P(M)$ where X is called a *C-closed element* of $P(M)$ if $CX = X \in P(M)$,

3. $\bigcup_c$ is defined as $\bigcup_c X_i := C(\bigcup X_i)$ for all $X_i \in P(M)$,

4. $\circ_c$ is defined as $X \circ_c Y := C(X \circ Y)$ for all $X, Y \in P(M)$.

Moreover we can assume the elements M (the greatest element) and $C\varnothing$ (the least element) of $C(P(M))$.

We remark that $C(P(M))$ is closed under the operations $\cap, \bigcup_c, \circ_c$ and $\dot{\to}$. Moreover remark that this closure operation C has the following properties: for all $X, Y \in P(M)$,

1. $C(CX \cup CY) = C(X \cup Y)$,

2. $C(CX \circ CY) = C(X \circ Y)$,

[4] That is, C has the properties: for all $X, Y \in P(M)$, $X \subseteq CX$, $CCX \subseteq CX$, $CX \circ CY \subseteq C(X \circ Y)$, $X \subseteq Y$ implies $CX \subseteq CY$.

3. $CCX = CX$.

Then we can show the following.

Proposition 11.4.2 $\mathbf{C}(\mathbf{P}(\mathbf{M}))$ *is a quantale.*

Lemma 11.4.3 *Let C be the MacNeille closure on $P(M)$. Then:*

1. $C\{[\Gamma]\} = \{[\Delta] \mid \vdash \Delta \Rightarrow \Gamma^\star\}$ *for any* $[\Gamma] \in M$,

2. $C\{[\Gamma]\} \subseteq C\{[\Delta]\}$ *iff* $\vdash \Gamma \Rightarrow \Delta^\star$ *for any* $[\Gamma], [\Delta] \in M$,

3. $C\{[(\Gamma^\star) \vee (\Delta^\star)]\} = C\{[\Gamma], [\Delta]\}$ *for any* $[\Gamma], [\Delta] \in M$.

Proof. (1): First, we show $C\{[\Gamma]\} \subseteq \{[\Delta] \mid \vdash \Delta \Rightarrow \Gamma^\star\}$. Suppose $[\Sigma] \in C\{[\Gamma]\}$. Then

$$[\Sigma] \in (\{[\Gamma]\}^{\rightarrow})^{\leftarrow}$$

$$\text{iff } \forall[\Pi] \in \{[\Gamma]\}^{\rightarrow} \ (\vdash \Sigma \Rightarrow \Pi^\star)$$

$$\text{iff } \forall[\Pi] \ (\forall[\Lambda] \in \{[\Gamma]\} \ (\vdash \Lambda \Rightarrow \Pi^\star) \text{ implies} \vdash \Sigma \Rightarrow \Pi^\star)$$

$$\text{iff } \forall[\Pi] \ (\vdash \Gamma \Rightarrow \Pi^\star \text{ implies} \vdash \Sigma \Rightarrow \Pi^\star).$$

Taking Γ for Π, we obtain $\vdash \Sigma \Rightarrow \Gamma^\star$. This means $[\Sigma] \in \{[\Delta] \mid \vdash \Delta \Rightarrow \Gamma^\star\}$. The converse is obvious using (cut) and ($*$left).

(2): First, we show that $C\{[\Gamma]\} \subseteq C\{[\Delta]\}$ implies $\vdash \Gamma \Rightarrow \Delta^\star$. Suppose $C\{[\Gamma]\} \subseteq C\{[\Delta]\}$ holds. Then we have $[\Gamma] \in \{[\Gamma'] \mid \vdash \Gamma' \Rightarrow \Gamma^\star\} \subseteq \{[\Delta'] \mid \vdash \Delta' \Rightarrow \Delta^\star\}$ by Lemma 11.4.3 (1). Therefore $\vdash \Gamma \Rightarrow \Delta^\star$. Next we show the converse. We show that, for any $[\Pi]$, if $\vdash \Pi \Rightarrow \Gamma^\star$ then $\vdash \Pi \Rightarrow \Delta^\star$. Suppose $\vdash \Pi \Rightarrow \Gamma^\star$ and $\vdash \Gamma \Rightarrow \Delta^\star$. Then we obtain $\vdash \Pi \Rightarrow \Delta^\star$ by ($*$left) and (cut).

(3): First we show $C\{[(\Gamma^\star) \vee (\Delta^\star)]\} \subseteq C\{[\Gamma], [\Delta]\}$. Suppose $[\Sigma] \in C\{[(\Gamma^\star) \vee (\Delta^\star)]\}$. Then ($*$): $\vdash \Sigma \Rightarrow (\Gamma^\star) \vee (\Delta^\star)$ by Lemma 11.4.3 (1). We show $[\Sigma] \in C\{[\Gamma], [\Delta]\}$, that is, if $\vdash \Gamma \Rightarrow \Pi^\star$ and $\vdash \Delta \Rightarrow \Pi^\star$ then $\vdash \Sigma \Rightarrow \Pi^\star$ for any $[\Pi] \in M$, because we have that

$$[\Sigma] \in (\{[\Gamma], [\Delta]\}^{\rightarrow})^{\leftarrow}$$

$$\text{iff } \forall[\Pi] \in \{[\Gamma], [\Delta]\}^{\rightarrow} \ (\vdash \Sigma \Rightarrow \Pi^\star)$$

$$\text{iff } \forall[\Pi] \ (\forall[\Lambda] \in \{[\Gamma], [\Delta]\}(\vdash \Lambda \Rightarrow \Pi^\star) \text{ implies} \vdash \Sigma \Rightarrow \Pi^\star).$$

Suppose $\vdash \Gamma \Rightarrow \Pi^\star$, $\vdash \Delta \Rightarrow \Pi^\star$ and (*). We obtain $\vdash \Sigma \Rightarrow \Pi^\star$ by ($\vee$left), ($*$left) and (cut). Next we show $C\{[\Gamma], [\Delta]\} \subseteq C\{[(\Gamma^\star)\vee(\Delta^\star)]\}$. Suppose $[\Sigma] \in C\{[\Gamma], [\Delta]\}$, that is, for any $[\Pi] \in M$, if $\vdash \Gamma \Rightarrow \Pi^\star$ and $\vdash \Delta \Rightarrow \Pi^\star$ then $\vdash \Sigma \Rightarrow \Pi^\star$. Taking $(\Gamma^\star)\vee(\Delta^\star)$ for $\Pi^\star$, we obtain $\vdash \Sigma \Rightarrow (\Gamma^\star) \vee (\Delta^\star)$. Therefore $[\Sigma] \in C\{[(\Gamma^\star) \vee (\Delta^\star)]\}$ by Lemma 11.4.3 (1). ∎

Next we define strong valuations on $\mathbf{C}(\mathbf{P}(\mathbf{M}))$. Strong valuations v^+ and v^- on $\mathbf{C}(\mathbf{P}(\mathbf{M}))$ are mappings from the set of all propositional variables to $C(P(M))$ such that

1. $v^+(p) := C\{[p]\}$,
2. $v^-(p) := C\{[{\sim}p]\}$.

Proposition 11.4.4 *Strong valuations v^+ and v^- on $\mathbf{C}(\mathbf{P}(\mathbf{M}))$ are extend to the mappings from the set Φ of all formulas to $C(P(M))$, i.e., we have the following: for any formula α,*

1. $v^+(\alpha) = C\{[\alpha]\}$,
2. $v^-(\alpha) = C\{[{\sim}\alpha]\}$.

Proof. By induction on the complexity of $\alpha \in \Phi$. The cases for v^+ (except the case $\alpha \equiv {\sim}\beta$) are the same as that in [79]. We show some cases for v^-.

Case $\alpha \equiv {\sim}\beta$ for v^-: We show $v^-({\sim}\beta) = C\{[{\sim}{\sim}\beta]\}$. We can prove (*): $C\{[{\sim}{\sim}\beta]\} = C\{[\beta]\}$ by Lemma 11.4.3 (2) and the fact that $\vdash {\sim}{\sim}\beta \Rightarrow \beta$ and $\vdash \beta \Rightarrow {\sim}{\sim}\beta$. Thus we have the following by the definition of the valuations, the induction hypothesis and (*):

$$v^-({\sim}\beta) = v^+(\beta) = C\{[\beta]\} = C\{[{\sim}{\sim}\beta]\}.$$

Case $\alpha \equiv \beta \wedge \gamma$ for v^-: We show $v^-(\beta \wedge \gamma) = C\{{\sim}(\beta \wedge \gamma)\}$. We can prove (*): $C\{[{\sim}(\beta \wedge \gamma)]\} = C\{[{\sim}\beta \vee {\sim}\gamma]\}$ by Lemma 11.4.3 (2) and the fact that $\vdash {\sim}(\beta \wedge \gamma) \Rightarrow {\sim}\beta \vee {\sim}\gamma$ and $\vdash {\sim}\beta \vee {\sim}\gamma \Rightarrow {\sim}(\beta \wedge \gamma)$. We can show (**): $C\{[{\sim}\beta \vee {\sim}\gamma]\} = C(\{[{\sim}\beta]\} \cup \{[{\sim}\gamma]\})$ by Lemma 11.4.3 (3). Then we have

$$\begin{aligned} v^-(\beta \wedge \gamma) &= v^-(\beta) \cup_c v^-(\gamma) = C\{[{\sim}\beta]\} \cup_c C\{[{\sim}\gamma]\} \\ &= C(C\{[{\sim}\beta]\} \cup C\{[{\sim}\gamma]\}) = C(\{[{\sim}\beta]\} \cup \{[{\sim}\gamma]\}) \\ &= C\{[{\sim}\beta \vee {\sim}\gamma]\} = C\{[{\sim}(\beta \wedge \gamma)]\} \end{aligned}$$

by the definition of the valuations, the induction hypothesis, the definition of $\cup_c$, (**) and (*).

Case $\alpha \equiv \beta \rightarrow \gamma$ for v^-: We will show $v^-(\beta \rightarrow \gamma) = C\{[\sim(\beta \rightarrow \gamma)]\}$. We can prove (*): $C\{[\sim(\beta \rightarrow \gamma)]\} = C\{[\beta * \sim\gamma]\}$ by Lemma 11.4.3 (2) and the fact that $\vdash \sim(\beta \rightarrow \gamma) \Rightarrow \beta * \sim\gamma$ and $\vdash \beta * \sim\gamma \Rightarrow \sim(\beta \rightarrow \gamma)$. Moreover we can show the fact (**): $C\{[\beta * \sim\gamma]\} = C\{[\beta, \sim\gamma]\}$. Then we obtain

$$v^-(\beta \rightarrow \gamma) = v^+(\beta) \circ_c v^-(\gamma) = C\{[\beta]\} \circ_c C\{[\sim\gamma]\} = C(C\{[\beta]\} \circ C\{[\sim\gamma]\}) = C(\{[\beta]\} \circ \{[\sim\gamma]\}) = C\{[\beta] \cdot [\gamma]\} = C\{[\beta, \gamma]\} = C\{[\beta * \sim\gamma]\} = C\{[\sim(\beta \rightarrow \gamma)]\}$$

by the definition of valuations, the induction hypothesis, the definition of $\circ_c$, (**) and (*).

Case $\alpha \equiv \beta \vee \gamma$ for v^-: We show $v^-(\beta \vee \gamma) = C\{[\sim(\beta \vee \gamma)]\}$. We can prove (*): $C\{[\sim(\beta \vee \gamma)]\} = C\{[\sim\beta \wedge \sim\gamma]\}$ by Lemma 11.4.3 (2) and the fact that $\vdash \sim(\beta \vee \gamma) \Rightarrow \sim\beta \wedge \sim\gamma$ and $\vdash \sim\beta \wedge \sim\gamma \Rightarrow \sim(\beta \vee \gamma)$. Also we can prove (**): $C\{[\sim\beta \wedge \sim\gamma]\} = C\{[\sim\beta]\} \cap C\{[\sim\gamma]\}$. We prove this in the following. First we show $C\{[\sim\beta \wedge \sim\gamma]\} \subseteq C\{[\sim\beta]\} \cap C\{[\sim\gamma]\}$, that is, $C\{[\sim\beta \wedge \sim\gamma]\} \subseteq C\{[\sim\beta]\}$ and $C\{[\sim\beta \wedge \sim\gamma]\} \subseteq C\{[\sim\gamma]\}$. This can be proved by Lemma 11.4.3 (2) and the fact that $\vdash \sim\beta \wedge \sim\gamma \Rightarrow \sim\beta$ and $\vdash \sim\beta \wedge \sim\gamma \Rightarrow \sim\gamma$. Next we show $C\{[\sim\beta]\} \cap C\{[\sim\gamma]\} \subseteq C\{[\sim\beta \wedge \sim\gamma]\}$. Suppose $[\Gamma] \in C\{[\sim\beta]\} \cap C\{[\sim\gamma]\}$. Then we have that $\vdash \Gamma \Rightarrow \sim\beta$ and $\vdash \Gamma \Rightarrow \sim\gamma$ by Lemma 11.4.3 (1), and hence $\vdash \Gamma \Rightarrow \sim\beta \wedge \sim\gamma$. This means $[\Gamma] \in C\{[\sim\beta \wedge \sim\gamma]\}$ by Lemma 11.4.3 (1). Now we have the required fact:

$$v^-(\beta \vee \gamma) = v^-(\beta) \cap v^-(\gamma) = C\{[\sim\beta]\} \cap C\{[\sim\gamma]\} = C\{[\sim\beta \wedge \sim\gamma]\} = C\{[\sim(\beta \vee \gamma)]\}$$

by the definition of the valuation, the induction hypothesis, (**) and (*).

Case $\alpha \equiv \top$ for v^-: We show $v^-(\top) = C\{[\sim\top]\}$. We can show (*): $C\{[\sim\top]\} = C\{[\bot]\}$ by Lemma 11.4.3 (2) and the fact that $\vdash \sim\top \Rightarrow \bot$ and $\vdash \bot \Rightarrow \sim\top$. Also we can show (**): $C\{[\bot]\} = C\varnothing$. We prove this in the following. First we show $C\{[\bot]\} \subseteq C\varnothing$. Suppose $[\Gamma] \in C\{[\bot]\}$, that is, (1): $\vdash \Gamma \Rightarrow \bot$ by Lemma 11.4.3 (1). We show $[\Gamma] \in C\varnothing$, that is, $[\Gamma] \in (\varnothing^{\rightarrow})^{\leftarrow}$. Since $\varnothing^{\rightarrow} = M$, we show $\forall[\Delta] \in M\ (\vdash \Gamma \Rightarrow \Delta^\star)$. By the fact $\vdash \bot \Rightarrow \Delta^\star$, (1) and (cut), we obtain $\vdash \Gamma \Rightarrow \Delta^\star$. Second we show $C\varnothing \subseteq C\{[\bot]\}$. Suppose $[\Gamma] \in C\varnothing$, that is, $\forall[\Delta] \in M\ (\vdash \Gamma \Rightarrow \Delta^\star)$. Then taking $\bot$ for Δ, we have $\vdash \Gamma \Rightarrow \bot$. Therefore $[\Gamma] \in C\{[\bot]\}$. Then we have the following:

$$v^-(\top) = C\varnothing = C\{[\bot]\} = C\{[\sim\top]\}$$

by the definition of v^-, (**) and (*). ∎

We then have the following.

Theorem 11.4.5 (Completeness) *Let C be the class of all quantales, $L_1(C) := \{S \mid$ a sequent S is strongly valid in all quantales of $C\}$, $L_2(C) := \{S \mid$ a sequent S is involutive valid in all quantales of $C\}$, $L_1 := \{S \mid \text{WILL} \vdash S\}$ and $L_2 := \{S \mid \text{QILL} \vdash S\}$. Then $L_1 = L_1(C)$ and $L_2 = L_2(C)$.*

Next we prove the completeness theorem (w.r.t. the involutive quantale models) for QILL. In this proof, we will use the rules (•regu) and (•regu^{-1}).

First we introduce a structure $\mathbf{M}^{\bullet} := \langle M, \cdot, [\ ], \leq, \cdot^{\circ}\rangle$ (called a pre-ordered commutative monoid with involution) such that

1. $\langle M, \cdot, [\ ], \leq\rangle$ is $\mathbf{M}$, a pre-ordered commutative monoid,
2. $\cdot^{\circ}$ is a unary operation on M such that
$$[\Gamma]^{\circ} := [\Gamma^{\bullet}] = [\gamma^{\bullet} \mid \gamma \in \Gamma].$$

We construct the powerset structure $\mathbf{P}(\mathbf{M}^{\bullet}) := \langle P(M), \bigcup, \circ, \{[\ ]\}, \cdot^{\circ_p}\rangle$ such that

1. $\langle P(M), \bigcup, \circ, \{[\ ]\}\rangle$ is $\mathbf{P}(\mathbf{M})$,
2. $\cdot^{\circ_p}$ is a unary operation such that
$$X^{\circ_p} := \{[\Gamma]^{\circ} \mid [\Gamma] \in X\} \text{ for all } X \in P(M).$$

Proposition 11.4.6 *$\mathbf{P}(\mathbf{M}^{\bullet})$ is an involutive quantale.*

Proof. We only verify the conditions C2, C3, C5, and C7.

Case C2: We only consider the binary case: $(X \cup Y)^{\circ_p} = X^{\circ_p} \cup Y^{\circ_p}$ for any $X, Y \in P(M)$.

$$\begin{aligned}
& (X \cup Y)^{\circ_p} \\
= \ & \{[\Sigma]^{\circ} \mid [\Sigma] \in \{[\Pi] \mid [\Pi] \in X \cup Y\}\} \\
= \ & \{[\Pi^{\bullet}] \mid [\Pi] \in X \text{ or } [\Pi] \in Y\} \\
= \ & \{[\Gamma^{\bullet}] \mid [\Gamma] \in X\} \cup \{[\Delta^{\bullet}] \mid [\Delta] \in Y\} \\
= \ & X^{\circ_p} \cup Y^{\circ_p}.
\end{aligned}$$

Case C3: We show $(X \circ Y)^{\circ_p} = X^{\circ_p} \circ Y^{\circ_p}$ for any $X, Y \in P(M)$.

$$\begin{aligned}
& (X \circ Y)^{\circ_p} \\
= \ & \{[\Pi]^{\circ} \mid [\Pi] \in X \circ Y\}
\end{aligned}$$

$= \{[\Gamma^\bullet, \Delta^\bullet] \mid [\Gamma] \in X \text{ and } [\Delta] \in Y\}$

$= \{[\Gamma^\bullet] \cdot [\Delta^\bullet] \mid [\Gamma^\bullet] \in X^{\circ_p} \text{ and } [\Delta^\bullet] \in Y^{\circ_p}\}$

$= X^{\circ_p} \circ Y^{\circ_p}.$

Case C5: We show $(X \dot{\rightarrow} Y)^{\circ_p} = X^{\circ_p} \dot{\rightarrow} Y^{\circ_p}$ for any $X, Y \in P(M)$.

$(X \dot{\rightarrow} Y)^{\circ_p}$

$= \{[\Pi]^\circ \mid [\Pi] \in \{[\Delta] \mid \forall[\Gamma] \in X\ ([\Gamma, \Delta] \in Y)\}\}$

$= \{[\Pi^\bullet] \mid \forall[\Gamma] \in X\ ([\Gamma, \Pi] \in Y)\}.$

On the other hand, we have:

$X^{\circ_p} \dot{\rightarrow} Y^{\circ_p}$

$= \{[\Pi'] \mid \forall[\Gamma'] \in X^{\circ_p}\ ([\Gamma', \Pi'] \in Y^{\circ_p})\}$

$= \{[\Pi'] \mid \forall[\Gamma']\ ([\Gamma'] = [\Gamma]^\circ \text{ and } [\Gamma] \in X)\ ([\Gamma', \Pi'] = [\Delta]^\circ \text{ and } [\Delta] \in Y)\}.$

Then we take $\Gamma' \equiv \Gamma^\bullet$ and $\Pi' \equiv \Pi^\bullet$, and hence $X^{\circ_p} \dot{\rightarrow} Y^{\circ_p} = \{[\Pi^\bullet] \mid \forall[\Gamma] \in X\ ([\Gamma, \Pi] \in Y)\}$.

Case C7: We show $M^{\circ_p} = M$.

M°_p}

$= \{[\Gamma]^\circ \mid [\Gamma] \in M\}$

$= \{[\Gamma^\bullet] \mid [\Gamma] \in M\}$

$= \{[\Gamma^\bullet] \mid [\Gamma^{\bullet\bullet}] \in M\}$ (by $[\Gamma^{\bullet\bullet}] \doteq [\Gamma]$)

$= \{[\Pi] \mid [\Pi^\bullet] \in M\}.$

We have that $[\Pi^\bullet] \in M$ iff $[\Pi] \in M$. Then $\{[\Pi] \mid [\Pi^\bullet] \in M\} = \{[\Pi] \mid [\Pi] \in M\} = M$. ∎

Next we construct $\mathbf{C}(\mathbf{P}(\mathbf{M}^\bullet)) := \langle C(P(M)), \bigcup_c, \circ_c, C\{[\ ]\}, \cdot^{\circ_c}\rangle$ such that

1. $\langle C(P(M)), \bigcup_c, \circ_c, C\{[\ ]\}\rangle$ is $\mathbf{C}(\mathbf{P}(\mathbf{M}))$,
2. $\cdot^{\circ_c}$ is a unary operation such that

   $X^{\circ_c} := C(X^{\circ_p})$ for all $X \in P(M)$.

Lemma 11.4.7 *Let C be the MacNeille closure on $P(M)$. Then $(C\{[\Gamma]\})^{\circ_c} = C\{[\Gamma^{\bullet}]\}$ for any $[\Gamma] \in M$.*

Proof.

$$
\begin{aligned}
& (C\{[\Gamma]\})^{\circ_c} \\
= \; & C((C\{[\Gamma]\})^{\circ_p}) \\
= \; & C(\{[\Delta] \mid \vdash \Delta \Rightarrow \Gamma^{\star}\}^{\circ_p}) \text{ (by Lemma 11.4.3 (1))} \\
= \; & C\{[\Pi]^{\circ} \mid [\Pi] \in \{[\Delta] \mid \vdash \Delta \Rightarrow \Gamma^{\star}\}\} \\
= \; & C\{[\Delta^{\bullet}] \mid \vdash \Delta \Rightarrow \Gamma^{\star}\} \\
= \; & C\{[\Delta^{\bullet}] \mid \vdash \Delta^{\bullet} \Rightarrow \Gamma^{\star\bullet}\} \text{ (by (}\bullet\text{regu) and (}\bullet\text{regu}^{-1}\text{))} \\
= \; & C\{[\Delta^{\bullet}] \mid \vdash \Delta^{\bullet} \Rightarrow \Gamma^{\bullet\star}\} \text{ (by } [\Gamma^{\bullet\star}] \doteq [\Gamma^{\star\bullet}] \text{ and (cut))} \\
= \; & C\{[\Sigma] \mid \vdash \Sigma \Rightarrow \Gamma^{\bullet\star}\} \\
= \; & C(C\{[\Gamma^{\bullet}]\}) \text{ (by Lemma 11.4.3 (1))} \\
= \; & C\{[\Gamma^{\bullet}]\}.
\end{aligned}
$$

∎

By using Lemma 11.4.3 (2) and Lemma 11.4.7, we show that the following monotonicity condition for $\cdot^{\circ_c}$:

$$\text{if } X \subseteq Y \text{ then } X^{\circ_c} \subseteq Y^{\circ_c} \text{ for any } X, Y \in C(P(M)).$$

For this condition, it is sufficient to prove the following:

$$C\{[\Gamma]\} \subseteq C\{[\Delta]\} \text{ implies } (C\{[\Gamma]\})^{\circ_c} \subseteq (C\{[\Delta]\})^{\circ_c},$$

because we have Lemma 11.4.3 (3). Now we show this. Suppose $C\{[\Gamma]\} \subseteq C\{[\Delta]\}$. Then we have $\vdash \Gamma \Rightarrow \Delta^{\star}$ by Lemma 11.4.3 (2). Thus we obtain $\vdash \Gamma^{\bullet} \Rightarrow \Delta^{\star\bullet}$ by ($\bullet$regu), and hence obtain $\vdash \Gamma^{\bullet} \Rightarrow \Delta^{\bullet\star}$ by the fact $[(\Delta^{\star})^{\bullet}] \doteq [(\Delta^{\bullet})^{\star}]$ and (cut). This means $(C\{[\Gamma]\})^{\circ_c} \subseteq (C\{[\Delta]\})^{\circ_c}$ by Lemma 11.4.3 (2) and Lemma 11.4.7.

In the following, we show that $C(P(M))$ is closed under the operation $\cdot^{\circ_c}$. Suppose $X \in C(P(M))$, that is, $X = CX$. Then by the monotonicity condition for $\cdot^{\circ_c}$, we have: $X^{\circ_c} = (CX)^{\circ_c} = C((CX)^{\circ_p}) = C(X^{\circ_p}) = C(C(X^{\circ_p})) = C(X^{\circ_c})$. Therefore $X^{\circ_c} \in C(P(M))$.

We then have the following.

Proposition 11.4.8 $\mathbf{C(P(M^{\bullet}))}$ *is an involutive quantale.*

Proof. We only verify the conditions C2—C5 and C8. It is sufficient to consider that all the elements of $C(P(M))$ are of the form $C\{[\Gamma]\}$ (i.e. $\{[\Delta] \mid \vdash \Delta \Rightarrow \Gamma^{\star}\}$), because we have the fact $C\{[\Pi_1], [\Pi_2], ..., [\Pi_n]\} = C\{[(\Pi_1^{\star}) \vee (\Pi_2^{\star}) \vee \cdots \vee (\Pi_n^{\star})]\}$ by Lemma 11.4.3 (3).

• Case C2: We show $(C\{[\Gamma]\} \cup_c C\{[\Delta]\})^{\circ_c} = (C\{[\Gamma]\})^{\circ_c} \cup_c (C\{[\Delta]\})^{\circ_c}$. We can verify $[((\Gamma^{\star}) \vee (\Delta^{\star}))^{\bullet}] \doteq [(\Gamma^{\star})^{\bullet} \vee (\Delta^{\star})^{\bullet}] \doteq [(\Gamma^{\bullet})^{\star} \vee (\Delta^{\bullet})^{\star}]$ for any $[\Gamma], [\Delta] \in M$. Then we have:

$$
\begin{aligned}
& (C\{[\Gamma]\} \cup_c C\{[\Delta]\})^{\circ_c} \\
= \ & (C(C\{[\Gamma]\} \cup C\{[\Delta]\}))^{\circ_c} \\
= \ & (C(\{[\Gamma]\} \cup \{[\Delta]\}))^{\circ_c} \\
= \ & (C\{[\Gamma], [\Delta]\})^{\circ_c} \\
= \ & (C\{[(\Gamma^{\star}) \vee (\Delta^{\star})]\})^{\circ_c} \text{ (by Lemma 11.4.3 (3))} \\
= \ & C\{[((\Gamma^{\star}) \vee (\Delta^{\star}))^{\bullet}]\} \text{ (by Lemma 11.4.7)} \\
= \ & C\{[(\Gamma^{\star})^{\bullet} \vee (\Delta^{\star})^{\bullet}]\} \\
= \ & C\{[(\Gamma^{\bullet})^{\star} \vee (\Delta^{\bullet})^{\star}]\} \\
= \ & C(\{[\Gamma^{\bullet}]\} \cup \{[\Delta^{\bullet}]\}) \text{ (by Lemma 11.4.3 (3))} \\
= \ & C(C\{[\Gamma^{\bullet}]\} \cup C\{[\Delta^{\bullet}]\}) \\
= \ & C\{[\Gamma^{\bullet}]\} \cup_c C\{[\Delta^{\bullet}]\} \\
= \ & (C\{[\Gamma]\})^{\circ_c} \cup_c (C\{[\Delta]\})^{\circ_c} \text{ (by Lemma 11.4.7).}
\end{aligned}
$$

• Case C3: We show $(C\{[\Gamma]\} \circ_c C\{[\Delta]\})^{\circ_c} = (C\{[\Gamma]\})^{\circ_c} \circ_c (C\{[\Delta]\})^{\circ_c})$.

$$
\begin{aligned}
& (C\{[\Gamma]\} \circ_c C\{[\Delta]\})^{\circ_c} \\
= \ & (C(C\{[\Gamma]\} \circ C\{[\Delta]\}))^{\circ_c} \\
= \ & (C(\{[\Gamma]\} \circ \{[\Delta]\}))^{\circ_c} \\
= \ & (C\{[\Gamma, \Delta]\})^{\circ_c} \\
= \ & C\{[\Gamma^{\bullet}, \Delta^{\bullet}]\} \text{ (by Lemma 11.4.7)} \\
= \ & C(\{[\Gamma^{\bullet}]\} \circ \{[\Delta^{\bullet}]\}) \\
= \ & C(C\{[\Gamma^{\bullet}]\} \circ C\{[\Delta^{\bullet}]\}) \\
= \ & C\{[\Gamma^{\bullet}]\} \circ_c C\{[\Delta^{\bullet}]\}
\end{aligned}
$$

$$= (C\{[\Gamma]\})^{\circ_c} \circ_c (C\{[\Delta]\})^{\circ_c} \text{ (by Lemma 11.4.7).}$$

• Case C4: We show $(C\{[\Gamma]\} \cap C\{[\Delta]\})^{\circ_c} = (C\{[\Gamma]\})^{\circ_c} \cap (C\{[\Delta]\})^{\circ_c}$. Before we give the proof, we show (*): $C\{[\Gamma]\} \cap C\{[\Delta]\} = C\{[(\Gamma^\star) \wedge (\Delta^\star)]\}$. Suppose $[\Pi] \in C\{[\Gamma]\} \cap C\{[\Delta]\}$. Then we have $\vdash \Pi \Rightarrow \Gamma^\star$ and $\vdash \Pi \Rightarrow \Delta^\star$ by Lemma 11.4.3 (1). Thus we have $\vdash \Pi \Rightarrow (\Gamma^\star) \wedge (\Delta^\star)$ by ($\wedge$right), and hence $[\Pi] \in C\{[(\Gamma^\star)\wedge(\Delta^\star)]\}$ by Lemma 11.4.3 (1). We can show the converse by using (cut) and the fact that $\vdash (\Gamma^\star) \wedge (\Delta^\star) \Rightarrow \Gamma^\star$ and $\vdash (\Gamma^\star) \wedge (\Delta^\star) \Rightarrow \Delta^\star$. We can verify the fact that $[((\Gamma^\star) \wedge (\Delta^\star))^\bullet] \doteq [(\Gamma^\star)^\bullet \wedge (\Delta^\star)^\bullet] \doteq [(\Gamma^\bullet)^\star \wedge (\Delta^\bullet)^\star]$. Next we show the following required fact by using (*):

$$\begin{aligned}
&(C\{[\Gamma]\} \cap C\{[\Delta]\})^{\circ_c} \\
&= (C\{[(\Gamma^\star) \wedge (\Delta^\star)]\})^{\circ_c} \text{ (by (*))} \\
&= C\{[((\Gamma^\star) \wedge (\Delta^\star))^\bullet]\} \text{ (by Lemma 11.4.7)} \\
&= C\{[(\Gamma^\star)^\bullet \wedge (\Delta^\star)^\bullet]\} \\
&= C\{[(\Gamma^\bullet)^\star \wedge (\Delta^\bullet)^\star]\} \\
&= C\{[\Gamma^\bullet]\} \cap C\{[\Delta^\bullet]\} \text{ (by (*))} \\
&= (C\{[\Gamma]\})^{\circ_c} \cap (C\{[\Delta]\})^{\circ_c} \text{ (by Lemma 11.4.7).}
\end{aligned}$$

• Case C5: We show $(C\{[\Gamma]\} \dot{\rightarrow} C\{[\Delta]\})^{\circ_c} = (C\{[\Gamma]\})^{\circ_c} \dot{\rightarrow} (C\{[\Delta]\})^{\circ_c}$. Before the proof, we show (*): $C\{[(\Gamma^\star) \rightarrow (\Delta^\star)]\} = C\{[\Gamma]\} \dot{\rightarrow} C\{[\Delta]\}$. Suppose $[\Lambda] \in C\{[(\Gamma^\star) \rightarrow (\Delta^\star)]\}$, that is, $\vdash \Lambda \Rightarrow (\Gamma^\star) \rightarrow (\Delta^\star)$, and hence, $\vdash \Lambda, \Gamma^\star \Rightarrow \Delta^\star$. We will show $[\Lambda] \in C\{[\Gamma]\} \dot{\rightarrow} C\{[\Delta]\}$, that is, $[\Lambda] \in \{\Psi_1 \mid \vdash \Psi_1 \Rightarrow \Gamma^\star\} \dot{\rightarrow} \{\Psi_2 \mid \vdash \Psi_2 \Rightarrow \Delta^\star\}$ by Lemma 11.4.3 (1), and hence (**): $\forall [\Pi]$ ($\vdash \Pi \Rightarrow \Gamma^\star$ implies $\vdash \Pi, \Lambda \Rightarrow \Delta^\star$). By $\vdash \Lambda, \Gamma^\star \Rightarrow \Delta^\star$, $\vdash \Pi \Rightarrow \Gamma^\star$ and (cut), we obtain $\vdash \Pi, \Lambda \Rightarrow \Delta^\star$. Next we show the converse. Suppose $[\Lambda] \in C\{[\Gamma]\} \dot{\rightarrow} C\{[\Delta]\}$, that is, (**). We will show $[\Lambda] \in C\{[(\Gamma^\star) \rightarrow (\Delta^\star)]\}$, that is, $\vdash \Lambda, \Gamma^\star \Rightarrow \Delta^\star$. Taking $\Gamma^\star$ for Π in (**), we have $\vdash \Lambda, \Gamma^\star \Rightarrow \Delta^\star$. Moreover we can verify the fact $[((\Gamma^\star) \rightarrow (\Delta^\star))^\bullet] \doteq [(\Gamma^\star)^\bullet \rightarrow (\Delta^\star)^\bullet] \doteq [(\Gamma^\bullet)^\star \rightarrow (\Delta^\bullet)^\star] \in M$. Next we show the required fact by using (*):

$$\begin{aligned}
&(C\{[\Gamma]\} \dot{\rightarrow} C\{[\Delta]\})^{\circ_c} \\
&= (C\{[(\Gamma^\star) \rightarrow (\Delta^\star)]\})^{\circ_c} \text{ (by (*))} \\
&= C\{[((\Gamma^\star) \rightarrow (\Delta^\star))^\bullet]\} \text{ (by Lemma 11.4.7)} \\
&= C\{[(\Gamma^\star)^\bullet \rightarrow (\Delta^\star)^\bullet]\}
\end{aligned}$$

$= C\{[(\Gamma^\bullet)^\star \rightarrow (\Delta^\bullet)^\star]\}$

$= C\{[\Gamma^\bullet]\} \dot{\rightarrow} C\{[\Delta^\bullet]\}$ (by (*))

$= (C\{[\Gamma]\})^{\circ_c} \dot{\rightarrow} (C\{[\Delta]\})^{\circ_c}$ (by Lemma 11.4.7).

- Case C8: We have:

$(C\varnothing)^{\circ_c}$

$= C((C\varnothing)^{\circ_p})$

$= C((C\{[\bot]\})^{\circ_p})$

$= C(\{[\Delta] \mid \vdash \Delta \Rightarrow \bot\}^{\circ_p})$ (by Lemma 11.4.3 (1))

$= C\{[\Delta]^\circ \mid \vdash \Delta \Rightarrow \bot\}$

$= C\{[\Delta^\bullet] \mid \vdash \Delta^\bullet \Rightarrow \bot^\bullet\}$ (by ($\bullet$regu) and ($\bullet$regu^{-1}))

$= C\{[\Pi] \mid \vdash \Pi \Rightarrow \bot^\bullet\}$

$= C(C\{[\bot^\bullet]\})$ (by Lemma 11.4.3 (1))

$= C\{[\bot^\bullet]\}$

$= C\{[\bot]\}$

$= C\varnothing.$

∎

Next we define a *single valuation* on $\mathbf{C}(\mathbf{P}(\mathbf{M}^\bullet))$. A single valuation v on $\mathbf{C}(\mathbf{P}(\mathbf{M}^\bullet))$ is a mapping from the set of all propositional variables to $C(P(M))$ such that

$$v(p) := C\{[p]\}.$$

We can extend to the mapping from the set Φ of all formulas to $C(P(M))$, that is, we can prove the following by induction on the complexity of $\alpha \in \Phi$:

$$v(\alpha) = C\{[\alpha]\}.$$

It is sufficient to prove the fact that $v(\beta^\bullet) = C\{[\beta^\bullet]\}$ because the other cases are similar to that for the quantale model discussed. This fact is proved using the induction hypothesis and Lemma 11.4.7 as follows:

$$v(\beta^\bullet) = v(\beta)^{\circ_c} = (C\{[\beta]\})^{\circ_c} = C\{[\beta^\bullet]\}.$$

Theorem 11.4.9 (Completeness) *Let C be the class of all involutive quantales, $L(C) := \{S \mid$ a sequent S is valid in all involutive quantales of $C\}$ and $L := \{S \mid \mathrm{QILL} \vdash S\}$. Then $L = L(C)$.*

12. Paraconsistent Lambek logics

It would also be interesting to find interpretations of negation other than quasi-Boolean complement, more compatible with the linguistic sense of negative information.

Wojciech Buszkowski [30, p. 124]

12.1. Introduction

In non-commutative substructural logics and in Categorial Grammar, a distinction is drawn between two implication connectives, often denoted by $\backslash$ and $/$, reflecting the linear order of syntactic expressions in most formal and natural languages. These implications satisfy two directional residuation laws with respect to a non-commutative, conjunction connective called "fusion" or "multiplicative conjunction" (cf. p. 81), which in the following will be denoted by $\times$:

$$x \times y \vdash z \text{ iff } y \vdash (x \setminus z), \qquad x \times y \vdash z \text{ iff } x \vdash (z/y),$$

where x, y, z are formulas that stand for syntactic types (also called syntactic categories).

In [196, Section 7.6], [198] the idea of introducing negation into Categorial Grammar has briefly been ventilated by remarking that directional versions of intuitionistic negation do not make sense linguistically. If the constant $\bot$ is seen as a symbol standing for the syntactic type of ungrammatical expressions, then an expression of type $x \backslash \bot$ is an expression that upon combination with an expression of type x from the left results in an ungrammatical expression. If s is the syntactic category of sentences, for instance, then every adjective should belong to the syntactic type $(s \setminus \bot)$. Clearly, the type $(s \setminus \bot)$ does not characterize the complement of the set of sentences, since adverbs in English fail to be sentences but nevertheless may be combined from the left with a sentence to form new sentences. Also intuitively, it seems to be implausible to characterize the negation of some basic syntactic category dynamically as a so-called functor type $(x \setminus y)$ or (y/x).

In [30], W. Buszkowski explores three ways of using negative information in Categorial Grammar: (a) admitting negative data about type assignments in a unification based procedure for learning a categorial grammar, (b) using restrictions of language models, and (c) admitting negated syntactic types. As to introducing a negation connective into Categorial Grammar, Buszkowski [30, p. 118] suggests that "[n]egative postulates can be written $X \to -A$ instead of $X \not\to A$", where the latter means that expression X is not of syntactic type A. Since negative information may be available in the process of defining or learning a categorial grammar's type assignment for some natural language, the introduction of negated syntactic types is of interest.[1]

Buszkowski defines the associative Lambek calculus with negation, **LN**, as the result of enriching the associative Lambek calculus as initially axiomatized in [114] by the double negation laws and the classical contraposition rule (notation adjusted):

$$\text{(DN)} \quad x \to --x, \qquad --x \to x$$

$$\text{(TR)} \quad x \to y \vdash -y \to -x$$

An operation $-$ satisfying (DN) and (TR) is called a De Morgan negation (see [48], [49], [50]), because if $\to$ is a lattice ordering, then $-$ satisfies the De Morgan laws with respect to the lattice meet and the lattice join.

According to Buszkowski, the linguistically intended models for the associative Lambek calculus with negation are so-called powerset residuated semigroups with Boolean complement over semigroups. He observes that **LN** fails to be complete with respect to these models and proves that **LN** is characterized by the class of all residuated semigroups of cones with quasi-Boolean complement (over partially-ordered semigroups), see [30, p. 122 f.].

In [30], the problem of axiomatizing the theory of the class of intended models, the exploration of cut-elimination in a suitable Gentzen-style sequent system for **LN**, and the question whether **LN** is decidable are left for future research.[2]

We introduce negated types into Categorial Grammar and obtain a form of *strong*, non-classical negation that turns the functor-type forming directional implications of Categorial Grammar into *connexive implications*. The introduction of negation into Categorial Grammar thereby

[1] It must be admitted, however, that the significance of negative information in language acquisition is contentious.

[2] The present chapter contains the material from [208] with some notational adjustments and some additional remarks. Moreover, in this chapter we use the arrow $\to$ as a sequent arrow, whereas the symbols $\Rightarrow$ and $\Leftarrow$ are used to denote operations in residuated groupoids and semigroups.

provides additional motivation for systems of connexive logic presented along the lines of [204], [206]. The starting point of our considerations are powerset residuated groupoids and semigroups, see Section 12.2.

The negation $\sim (x\backslash y)$ of the functor-type symbol $(x\backslash y)$ will be understood as follows: an expression e has type $\sim (x\backslash y)$ iff for any expression e' of type x, the phrase structure $(e'e)$ is definitely not of type y. Similarly, the negation $\sim (y/x)$ of (y/x) will be interpreted as follows: an expression e has type $\sim (y/x)$ iff for any expression e' of type x, the phrase structure (ee') is definitely not of type y.

12.2. Negative syntactic types

Recall that a groupoid is a structure $(M, \cdot)$, where M is a non-empty set, and $\cdot$ is a binary operation on M. If $\cdot$ is associative, then $(M, \cdot)$ is called a semigroup. If $(M, \cdot)$ is a groupoid (semigroup), then the structure $(\mathcal{P}(M), \subseteq, \circ, \Rightarrow, \Leftarrow)$ is called the powerset residuated groupoid (semigroup) over $(M, \cdot)$, where $\subseteq$ is the subset relation and for $x, y \subseteq M$ the operations $\Leftarrow$, $\Rightarrow$, and $\circ$, are defined as follows:

$$\begin{aligned} y \Leftarrow x &= \{b \in M \mid (\forall a \in x)\, b \cdot a \in y\} \\ x \Rightarrow y &= \{b \in M \mid (\forall a \in x)\, a \cdot b \in y\} \\ x \circ y &= \{c \in M \mid (\exists a \in x)(\exists b \in y)\, c = a \cdot b\}^3 \end{aligned}$$

The non-associative (associative) Lambek calculus NL (L) (see [114], [115], and Section 12.3) is complete with respect to the class of all powerset residuated groupoids (semigroups), see [29], [31]. Given this semantics, one might try to introduce negated syntactic types by considering functions $-$ on $\mathcal{P}(M)$ satisfying the following conditions:

$$\begin{aligned} -(y \Leftarrow x) &= \{b \in M \mid \neg(\forall a \in x)\, b \cdot a \in y\} \\ -(x \Rightarrow y) &= \{b \in M \mid \neg(\forall a \in x)\, a \cdot b \in y\} \\ -(x \circ y) &= \{c \in M \mid \neg(\exists a \in x)(\exists b \in y)\, c = a \cdot b\} \\ --x &= x \end{aligned}$$

Thus, the defining properties of $(y \Leftarrow x)$, $(x \Rightarrow y)$, and $(x \circ y)$ are classically negated to obtain $-(y \Leftarrow x)$, $-(x \Rightarrow y)$, and $-(x \circ y)$. Clearly,

[3] The notion of a quasi-Boolean complement mentioned in Section 12.1 is defined as follows. If $(M, \cdot)$ is a semigroup and f is a one-one and onto function on M, then for all $x \subseteq M$, the quasi-Boolean complement $-$ in the powerset residuated semigroup $(\mathcal{P}(M), \subseteq, \circ, \Rightarrow, \Leftarrow)$ is defined by:

$$-x := M \setminus f[x]$$

where $f[x] = \{f(a) \mid a \in x\}$ and $\setminus$ is set-theoretical difference.

such one-place functions – exist, since Boolean complement satisfies the above equations. By classical logic we have:

$$\begin{array}{lrcl}
\text{(i)} & -(y \Leftarrow x) & = & \{b \in M \mid (\exists a \in x)\, b \cdot a \notin y\} \\
\text{(ii)} & -(x \Rightarrow y) & = & \{b \in M \mid (\exists a \in x)\, a \cdot b \notin y\} \\
\text{(iii)} & -(x \circ y) & = & \{c \in M \mid (\forall a \in x)(\forall b \in y)\, c \neq a \cdot b\}
\end{array}$$

If $a \notin x$ ($a \neq b$) is taken to give the information that $a \in x$ ($a = b$) is definitely false (and not just not true), then we obtain (i) – (iii) not only in classical, but also in constructive logic with strong negation, see Chapter 2 and referencs given there, such as [6, 73, 82, 83, 92, 113, 134, 141, 170, 186, 196].

12.2.1. Negated functor types

Let us first consider equations (i) and (ii). There seems to be an underlying and usually tacit assumption in Categorial Grammar that expressions behave uniformly with respect to type membership insofar as functor types are total functions.

Uniformity with respect to type membership:

All functor-types are total functions: If an expression forms an expression of type y with some *arguments of type x, then it forms an expression of type y with* every *argument of type x.*

In fact, Buszkowski [30] observes that introducing De Morgan negation gives rise to certain powerset operations defined by existential conditions, operations which are of interest in contexts where uniformity with respect to type membership is violated. The system **LN** can be presented using dual operations $\backslash_d$, $/_d$, $\times_r$, and $\times_l$ that can be defined from $\backslash$, $/$, $\times$ (product), and $-$ such that $/_d$ and $\backslash_d$ are interpreted by the following powerset operations (notation adjusted):

$$\begin{array}{rcl}
(y \Leftarrow_d x) & = & \{b \in M \mid (\exists a \in x)\, b \cdot a \in y\} \\
(x \Rightarrow_d y) & = & \{b \in M \mid (\exists a \in x)\, a \cdot b \in y\}
\end{array}$$

Buszkowski [30, p. 121] explains (notation adjusted) that "[i]n mathematical linguistics, these dual residuals have been considered more often than $\Rightarrow$, $\Leftarrow$, and they seem also to be of interest for categorial grammar in relation to *partial functors*: expressions which form an expression of category y with some (not necessarily all) arguments of category x." In a purely categorial setting, based solely on a type assignment and a syntactic calculus, the linguistic phenomena that might be responded to by postulating partial functors are phenomena that may as well be dealt with by introducing new atomic syntactic types and thereby securing total functionality.[4]

Example 12.2.1 *The definite article* `la` *in French is genus femininum. Assuming that* `femme` *is a noun* N *and* `sourit` *is an intransitive verb* $(n\backslash s)$*, learning that* `((La femme) sourit)` *is a sentence, leads to typing* `la` *as* $((s/(n\backslash s))/N)$*:*

$$\begin{array}{lll} ((\texttt{La} & \texttt{femme}) & \texttt{sourit}) \\ \underline{(s/(n\backslash s))/N} & \underline{N} & (n\backslash s) \\ \multicolumn{2}{l}{\underline{\quad s/(n\backslash s) \qquad\qquad\qquad}} & \\ \quad s & & \end{array}$$

Since `garçon` *is also a noun, and* `((La garçon) sourit)` *is ungrammatical, one might regard* `la`*'s type as that of a partial function accepting only feminine nouns. Assuming uniformity with respect to type membership, however, a more discriminating type assignment accounts for the data, too:*

$$\begin{array}{lll} ((\texttt{La} & \texttt{femme}) & \texttt{sourit}) \\ \underline{(s/(n\backslash s))/N_f} & \underline{N_f} & (n\backslash s) \\ \multicolumn{2}{l}{\underline{\quad s/(n\backslash s) \qquad\qquad\qquad}} & \\ \quad s & & \end{array}$$

$$\begin{array}{lll} *\,((\texttt{La} & \texttt{garçon}) & \texttt{sourit}) \\ (s/(n\backslash s))/N_f & N_m & (n\backslash s) \end{array}$$

Example 12.2.2 *Also the singular versus plural distinction can be accounted for either by stipulating partial functor types or by introducing new syntactic categories. In view of simple sentences like* `Philosophers read books`*, the expression* `read books` *might be assigned the type* $(N \setminus s)$*, and since* `Philosopher read books` *is ungrammatical,* `read books` *might be regarded as belonging to a partial functor type. Alternatively, introducing additional types avoids violating uniformity with respect to type membership:* `read books` *is of type* $(N_p \setminus s)$ *(a plural noun is expected from the left) and* `philosopher` *is of type* N_s *(of singular nouns).*

In the present authors's view, uniformity with respect to type membership is an attractive assumption. In our context it is an *interesting*

[4] The following examples present the standard case for introducing features into Categorial Grammar, see, for example, [46].

assumption, because it justifies the following understanding of $-$ not justified by using classical logic:

$$\begin{aligned} -(y \Leftarrow x) &= \{b \in M \mid (\forall a \in x)\, b \cdot a \notin y\} \\ -(x \Rightarrow y) &= \{b \in M \mid (\forall a \in x)\, a \cdot b \notin y\} \end{aligned}$$

Taking up the idea underlying Buszkowski's proposal concerning notation, we obtain:

$$\begin{aligned} -(y \Leftarrow x) &= \{b \in M \mid (\forall a \in x)\, b \cdot a \in -y\} \\ -(x \Rightarrow y) &= \{b \in M \mid (\forall a \in x)\, a \cdot b \in -y\} \end{aligned}$$

In the language based on $\{\Rightarrow, \Leftarrow, -\}$, $-$ is a rewrite operation in the sense that every term $-x$ can be rewritten as a term in which for every subterm $-y$, y is atomic.

As we shall see, the above algebraic interpretation of type negation turns the directional implications of Categorial Grammar into *connexive implications*, see [204], [206]. Under this reading, the following sequents are correct:

$$\begin{array}{llll} 1. & \sim (x\backslash y) \to (x\backslash \sim y), & 2. & (x\backslash \sim y) \to \sim (x\backslash y), \\ 3. & \sim (y/x) \to (\sim y/x), & 4. & (\sim y/x) \to \sim (y/x). \end{array}$$

For example, any expression of type $(\sim y/x)$ is also of type $\sim (y/x)$, and vice versa. Moreover, we shall have double-negation laws with the effect that an expression e belongs to type $\sim\sim x$ iff e belongs to type x.

12.2.2. Negated product types

Let us now inspect equation (iii). The considerations of this subsection will be negative: we shall assemble some reasons for *not* introducing negated product types.

A reflection on (iii) reveals that this interpretation of the negation of product types $(x \times y)$ seems to provide no information that may be useful in discovering or defining the type assignment of a categorial grammar. Suppose we are informed that an expression e is not of type $(x \times y)$. According to the above interpretation, we learn that e is not a phrase structure $(e_1 e_2)$, where e_1 is of type x and e_2 is of type y. But no particular further conclusion may be drawn from this information. Since $x \circ y = \{c \in M \mid (\exists a \in x)(\exists b \in y)\, c = a \cdot b\} = \{a \cdot b \in M \mid a \in x \text{ and } b \in y\}$, one might alternatively think of setting

$$\text{(iv)} \quad -(x \circ y) \quad = \quad \{a \cdot b \mid (a \notin x \text{ or } b \notin y)\}.$$

Then, an expression e is of type $\sim (x_1 \times x_2)$ iff e is a pair (e_1e_2) such that (i) e_1 is of type x_1 and e_2 is of type $\sim x_2$, or (ii) e_1 is of type $\sim x_1$ and e_2 is of type x_2, or (iii) e_1 is of type $\sim x_1$ and e_2 is of type $\sim x_2$. But if, say, e_1 is of type x and e_2 is of type $\sim y$, concluding that (e_1e_2) is of type $\sim (x \times y)$ is certainly much less informative than concluding that (e_1e_2) is of type $(x \times \sim y)$. And suppose that p, q, r are distinct atomic types and that we learn that p is of type $\sim (q \times r)$. Then, assuming (iv), every expression of type p is of a product type $(\sim q \times r)$ or $(q \times \sim r)$ or $(\sim q \times \sim r)$, which is certainly unwanted.

In addition to the previous appeal to intuition, there is another, slightly more technical reason for refraining from introducing negated product types. Once $\sim (x_1 \times x_2)$ is interpreted in terms of extensional (alias additive) disjunction $\vee$ in the metalanguage, it is natural to add $\vee$ to the object language of the syntactic calculus, too. And if both additive disjunction $\vee$ and negation $\sim$ are in the object language, then additive conjunction $\wedge$ will probably also show up in the metalanguage in order to interpret types $\sim (x \vee y)$, and a natural move is including $\wedge$ in the object language of the syntactic calculus. In algebraic terms this means that the syntactic calculus extended by $\wedge$ and $\vee$ is interpreted in extended powerset structures over residuated groupoids containing an operation $-$ satisfying:

$$\begin{array}{rcl} (x \sqcap y) & = & \{b \in M \mid b \in x \text{ and } b \in y\}, \\ (x \sqcup y) & = & \{b \in M \mid b \in x \text{ or } b \in y\}, \\ -(x \sqcap y) & = & \{b \in M \mid b \in -x \text{ or } b \in -y\}, \\ -(x \sqcup y) & = & \{b \in M \mid b \in -x \text{ and } b \in -y\}. \end{array}$$

Thereby one obtains:

$$\text{(v)} \quad -(x \circ y) \quad = \quad \{b \mid b \in ((-x \circ -y) \sqcup (x \circ -y) \sqcup (-x \circ y))\}$$

and $-$ is a rewrite operation also in the extended language containing $\circ$, $\sqcap$, and $\sqcup$. However, introducing the powerset operations $\sqcup$ and $\sqcap$ also introduces a well-known problem. These operations satisfy the distribution laws:

$$x \sqcap (y \sqcup z) \subseteq (x \sqcap y) \sqcup (x \sqcap z), \quad (x \sqcup y) \sqcap (x \sqcup z) \subseteq x \sqcup (y \sqcap z).$$

In the absence of the structural rules of contraction and weakening, however, the usual sequent rules for $\wedge$ and $\vee$, the lattice-rules, do not allow one to derive the corresponding distribution laws. Of course, one could work with a kind of semantics other than powerset residuated groupoids. Semilattice-ordered groupoids (see the semilattice-ordered groupoids in [196]) or phase structures [92] may be adapted such that each of the following sequents

$$\sim (x \times y) \to (\sim x \times y) \vee (x \times \sim y) \vee (\sim x \times \sim y)$$
$$\sim (x \wedge y) \to (\sim x \vee \sim y)$$
$$\sim (x \vee y) \to (\sim x \wedge \sim y)$$

and their converses are valid (and not only the last two sequents and their converses). But from a linguistic point of view, these semantics seem much less well-motivated than powerset residuated groupoids.

Since, intuitively and in view of equation (iii), learning that an expression is definitely not of type $(x \times y)$ hardly provides any useful information, in this chapter we decide to restrict the language and to exclude negated product types. Syntactically, this means that in the language of residuated groupoids (semigroups), terms $-(x \circ y)$ are not defined, and $-$ remains a rewrite operation in the *restricted* language based on $\{\Rightarrow, \Leftarrow, \circ, -\}$.

12.2.3. Negated atomic types

Note that we have not defined an operation $-$. Instead we have constrained such an operation by conditions on compound terms $(x \Rightarrow y)$, $(y \Leftarrow x)$, and $-x$. What about elements of $\mathcal{P}(M)$ used to interpret atomic type symbols? In other words, how are we supposed to interpret the negation of atomic types? Like in constructive logic with strong negation, positive and negative information is here taken on a par. Evidence concerning atomic type membership may be positive (say, "Expression e is a name") or negative ("Expression e is definitely not a name"). Then, obviously, nothing in particular must be assumed about the negation of atomic types.

One might postulate that for an atomic type symbol x, the interpretations of x and $\sim x$ are always disjoint. From a realistic point of view it seems even appropriate to assume that the interpretation of x is the set-theoretic complement of the interpretation of $\sim x$: any expression either belongs to type x or not, but not both. From an *informational* point of view, however, this is not appropriate. Information about type membership may be supplied from different sources (say, presumably competent speakers), one indicating that an expression e belongs to type x and another indicating that e is clearly not of type x, i.e., is of type $\sim x$. If we do not want to forestall the possibility of such contradictory information, it is wise not to preclude it by a general definition that covers the interpretation also of atomic type symbols.

Therefore, semantically, we want to treat the negation of atomic type symbols as a paraconsistent negation in the sense that the interpretations of x and $\sim x$ are *not* always disjoint, and it does, of course, not follow that in such a case every sequent $x \circ \sim x \rightarrow y$ is valid under that assignment. This is in line also with an informational reading of sequents. Whereas a sequent $X \rightarrow y$ usually is taken to mean that every phrase structure of type X is of type y, under an informational reading $X \rightarrow y$ means that the information that a phrase structure is of type X gives the information that it is of type y.

Note, however, that we do not have additive syntactic type symbols $(y \wedge \sim y)$ in our language. Hence, we have no reason to countenance the provability of pairs of sequents $X \rightarrow y$ and $X \rightarrow \sim y$. In any case, the multiplicative (intensional) multiplicative product operation $\times$ does not support the provability of such pairs of sequents.

12.3. Proof theory

In this section, we shall define two syntactic calculi with negation and connexive implication, one associative and one non-associative. These sequent systems are shown to enjoy cut-elimination. Cut-elimination has a number of useful consequences including, in this case, decidability. The sequent rule for negated directional implications are inspired by the above clauses $-(y \Leftarrow x) = \{b \in M \mid (\forall a \in x)\; b \cdot a \in -y\}$ and $-(x \Rightarrow y)$ $= \{b \in M \mid (\forall a \in x)\; a \cdot b \in -y\}$.

12.3.1. Lambek calculi with connexive implication

We assume an infinite set *At* of atomic type symbols. The set of type symbols and the set of Gentzen terms (G-terms) are inductively defined as follows:

Definition 12.3.1 *1. Every atomic type symbol is a type symbol.*

2. *If x and y are type symbols, then so are $(x \times y)$, $(x\backslash y)$, and (y/x).*
3. *If x is a type symbol and x does not contain (and is different from) any type symbol of the form $(y \times z)$, then $\sim x$ is a type symbol.*
4. *Nothing else is a type symbol.*
5. *Every type symbol is a G-term.*
6. *If X and Y are G-terms, then (XY) is a G-term.*
7. *Nothing else is a G-term.*

A type symbol is also called a formula, and the notion of a subformula is defined in the usual way. We shall now use x, y, z, possibly with subscripts or superscripts, as variables for type symbols, and X, Y, Z, possibly with subscripts or superscripts, as variables for G-terms. Instead of (XY) we shall also write XY. Moreover, we shall write $Z[X]$ to express that Z is a G-term with factor X, i.e., $Z[X] = X$ or $Z[X] = Z'[XX']$ or $Z[X] = Z'[X'X]$, and $Z'[X]$ is a G-term that contains X as a factor and is shorter than $Z[X]$. Note that there is nothing like an "empty G-term".

Definition 12.3.2 *The non-associative (associative) Lambek calculus with negation and connexive implication,* $\mathsf{NL}^{\sim}$ $(\mathsf{L}^{\sim})$, *in the language based on* $\{/, \backslash, \times, \sim\}$ *results from the non-associative (associative) Lambek calculus* NL (L) *upon the addition of the following sequent rules:*

$(\sim\sim \rightarrow)$ $X[x] \rightarrow y \vdash X[\sim\sim x] \rightarrow y$
$(\rightarrow \sim\sim)$ $X \rightarrow x \vdash X \rightarrow \sim\sim x$
$(\sim\backslash \rightarrow)$ $X \rightarrow x, \quad Y[\sim y] \rightarrow z \vdash Y[X \sim (x\backslash y)] \rightarrow z$
$(\rightarrow \sim\backslash)$ $xX \rightarrow \sim y \vdash X \rightarrow \sim (x\backslash y)$
$(\sim/ \rightarrow)$ $X \rightarrow x, \quad Y[\sim y] \rightarrow z \vdash Y[\sim (y/x)X] \rightarrow z$
$(\rightarrow \sim/)$ $Xx \rightarrow \sim y \vdash X \rightarrow \sim (y/x)$

The system NL *in the language based on* $\{/, \backslash, \times\}$ *is given by the following rules:*

(id) $\vdash x \rightarrow x$
(cut) $X \rightarrow x, \quad Y[x] \rightarrow z \vdash Y[X] \rightarrow z$
$(\backslash \rightarrow)$ $X \rightarrow x, \quad Y[y] \rightarrow z \vdash Y[X(x\backslash y)] \rightarrow z$
$(\rightarrow \backslash)$ $xX \rightarrow y \vdash X \rightarrow (x\backslash y)$
$(/ \rightarrow)$ $X \rightarrow x, \quad Y[y] \rightarrow z \vdash Y[(y/x)X] \rightarrow z$
$(\rightarrow /)$ $Xx \rightarrow y \vdash X \rightarrow (y/x)$
$(\times \rightarrow)$ $X[(x_1x_2)] \rightarrow y \vdash X[(x_1 \times x_2)] \rightarrow y$
$(\rightarrow \times)$ $X \rightarrow x, \quad Y \rightarrow y \vdash XY \rightarrow (x \times y)$

The system L *consists of* NL *together with:*

$(assoc_1)$ $X[x(x_1x_2)] \rightarrow y \vdash X[(xx_1)x_2] \rightarrow y$
$(assoc_2)$ $X[(xx_1)x_2] \rightarrow y \vdash X[x(x_1x_2)] \rightarrow y$

In extensions of $\mathsf{NL}^{\sim}$ and $\mathsf{L}^{\sim}$, the earlier sequents 1.–4. have simple proofs, see, for example, the following proof of 4.:

$$\dfrac{\dfrac{x \to x \quad \sim y \to \sim y}{(\sim y/x)\, x \to \sim y}}{(\sim y/x) \to \sim (y/x)}$$

Also the following sequents are easily provable:

$$5.\quad (y/x) \to \sim (\sim y/x), \quad 6.\quad (x\backslash y) \to \sim (x\backslash \sim y).$$

We shall use uppercase Greek letters Δ, Γ, Θ, possibly with sub- or superscripts as schematic letters for finite sets of sequents and s, s_1, s_2 to denote arbitrary single sequents. For $S \in \{\mathsf{NL}^{\sim}, \mathsf{L}^{\sim}\}$, $\Delta \vdash_S s$ means that sequent s is derivable from Δ in system S. If the context is clear, we just write $\Delta \vdash s$, and if Δ is empty, we shall write $\vdash s$. If $\Delta \vdash x \to y$ and $\Delta \vdash y \to x$, we shall also write $\Delta \vdash x \leftrightarrow y$.

The sequents 2. and 4.–6. are directional first-degree versions of the so-called Boethius' Theses. Moreover, if the restriction to non-empty antecedents of sequents is given up, directional versions of the so-called Aristotle's Theses are provable: $\sim (\sim x/x)$, $\sim (x/ \sim x)$, $\sim (\sim x\backslash x)$, and $\sim (x\backslash \sim x)$. Aristotle's Theses and Boethius' Theses are characteristic of systems of *connexive* logic, see [204] and references therein. Therefore, in the presence of $\sim$, the implications $/$ and $\backslash$ are connexive implications. Note that connexive logic is supraclassical: Aristotle's theses and Boethius' Theses are not theorems of classical logic. Since classical logic is Post complete, these principles cannot be consistently added to classical logic. In [203], it has been suggested to extend NL by rules such that the sequents characteristic of quasi-groups become provable: $y \to (x\backslash(x \times y))$, $y \to ((y \times x)/x)$, $(x\backslash(x \times y)) \to y)$, and $((y \times x)/x) \to y$. As in the present case, the result of this addition is a supraclassical system, if the structural rules of (intuitionistic and) classical logic are added.

A type symbol x is in negation normal form if for every subformula $\sim y$ of x, y is atomic. The calculi $\mathsf{NL}^{\sim}$, and $\mathsf{L}^{\sim}$ satisfy a negation normal form theorem.

Definition 12.3.3 *The function $'$ on the set of type symbols is induc-*

tively defined as follows:

$$
\begin{array}{rcl}
x' & = & x, \textit{ for atomic } x \\
(\sim x)' & = & \sim x, \textit{ for atomic } x \\
(\sim\sim x)' & = & x' \\
(x \sharp y)' & = & (x' \sharp y'), \textit{ for } \sharp \in \{\backslash, /, \times\} \\
(\sim (y/x))' & = & ((\sim y)'/x') \\
(\sim (x\backslash y))' & = & (x'\backslash(\sim y)')
\end{array}
$$

If X is a G-term, the G-term X' is inductively defined in the obvious way: x' is defined as above, and $(XY)' = (X'Y')$.

By induction on the complexity of type symbols, one can establish the following:

Observation 12.3.4 *For every type symbol x, x' is in negation normal form and $\vdash_S x \leftrightarrow x'$, for $S \in \{\mathsf{NL}^{\sim}, \mathsf{L}^{\sim}\}$.*

Corollary 12.3.5 *For every G-term X and type symbol x, in $\mathsf{NL}^{\sim}$, $\mathsf{L}^{\sim}$: $\vdash X \rightarrow x$ iff $\vdash X' \rightarrow x'$.*

Let x_y denote a type symbol that contains a certain occurrence of y as a subformula. Let then x_z denote the result of replacing this occurrence of y in x by z. We define the degree $d(y)$ of a type symbol y as the number of separate occurrences of connectives in y.

Observation 12.3.6 *In $\mathsf{NL}^{\sim}$ and $\mathsf{L}^{\sim}$, the relation $\vdash x \leftrightarrow y$ fails to be a congruence relation.*

Proof. In both systems we have $\vdash (x/y) \leftrightarrow (x/((x/y)\backslash x))$, but nevertheless it does not hold that $\vdash \; \sim (x/y) \leftrightarrow \; \sim (x/((x/y)\backslash x))$. ∎

Definition 12.3.7 *The type symbols x and y are provably strongly equivalent in $\mathsf{NL}^{\sim}$ and $\mathsf{L}^{\sim}$ ($\vdash x \rightleftharpoons y$) iff $\vdash x \leftrightarrow y$ and $\vdash \sim x \leftrightarrow \sim y$.*

Observation 12.3.8 (Replacement) *In $\mathsf{NL}^{\sim}$ and $\mathsf{L}^{\sim}$, $\vdash x_y \rightleftharpoons x_z$ if $\vdash y \rightleftharpoons z$.*

Proof. By induction on $l = d(x_y) - d(y)$. For $l = 0$, the claim is obvious. Suppose that $l = m + 1$ and that the claim holds for every $j \leq m$. We consider here only two cases. Case 1: x is a negated atom. Then $x_y = \sim y$ and $x_z = \sim z$. By assumption we have $\vdash \; \sim y \leftrightarrow \; \sim z$, and by $(\sim\sim \rightarrow)$, $(\rightarrow \sim\sim)$ and (cut), $\vdash \; \sim\sim y \leftrightarrow \; \sim\sim z$. Case 2: x_y

$= \sim (x'\backslash z'_y)$. Then $d(x'\backslash z'_y) - d(y) = m$ and $d(z'_y) - d(y) \leq m$. By the induction hypothesis, $\vdash z'_y \rightleftharpoons z'_z$, and thus we obtain

$$\cfrac{\cfrac{\cfrac{x' \to x' \quad \sim z'_y \to \sim z'_y}{x' \sim (x'\backslash z'_y) \to \sim z'_y} \quad \sim z'_y \to \sim z'_z}{x' \sim (x'\backslash z'_y) \to \sim z'_z}}{\sim (x'\backslash z'_y) \to \sim (x'\backslash z'_z)}$$

Similarly, we obtain $\vdash \sim (x'\backslash z'_z) \to \sim (x'\backslash z'_y)$. ∎

12.3.2. Cut-elimination and some of its implications

Theorem 12.3.9 *Applications of (cut) can be eliminated from derivations in* $\mathsf{NL}^{\sim}$ *and* $\mathsf{L}^{\sim}$.

Proof. The proof is a straightforward extension of Lambek's cut-elimination proof for L in [114]. For readers not familiar with this proof we briefly rehearse it in outline here, although these remarks are rather elementary. Consider $\mathsf{NL}^{\sim}$. It is shown that a proof of a sequent $X \to x$ with a single application of (cut) can be transformed into a proof of $X \to x$ not containing any application of (cut). (Then applications of (cut) in a proof can be eliminated successively.) The aim is to replace the single subproof Π' ending in an application of (cut)

$$\cfrac{\begin{array}{cc}\Pi_1 & \Pi_2 \\ X \to x & Y[x] \to z\end{array}}{Y[x] \to z}$$

by a proof Π'' of $Y[x] \to z$ such that every subproof of Π'' ending in an application of (cut) has a complexity smaller than the complexity of Π'. The type symbol x is called the cut-formula. To define a complexity measure, the height $h(\Pi)$ of a proof Π is inductively defined as follows. If Π is an axiomatic sequent $x \to x$, then $h(\Pi) = 0$. If Π is a proof

$$\frac{\Pi_1 \ldots \Pi_n}{X \to x}$$

then $h(\Pi) = max\{h(\Pi_i) \mid 1 \leq i \leq n\} + 1$. The cut-degree $d(\Pi)$ of a proof

$$\Pi = \frac{\Pi_1 \quad \Pi_2}{s} \; (cut)$$

is defined as $(d(x), h(\Pi_1) + h(\Pi_2))$, where x is the cut-formula. It is assumed that the cut-degrees are lexicographically ordered.

Then cases are distinguished as to the possible proofs of the premise sequents of the proof Π' ending in an application of (cut). If one of the premises is an axiom, then Π' is replaced by a (cut)-free proof.

If in the last step of the proof of the left premise sequent of (cut) the main connective of the cut-formula is introduced by a right introduction rule, but in the last step of the proof of the right premise sequent the corresponding left introduction rule has not been applied, a permuting reduction is carried out. The proof of the left premise sequent is used to apply the (cut)-rule in a premise of the right premise sequent of (cut). This gives a height-reducing replacement; the cut-formula remains the same. There is an analogous set of subcases, when in the last step of the proof of the right premise sequent of (cut) the main connective of the cut-formula is introduced by a left introduction rule, but in the last step of the proof of the left premise sequent the corresponding right introduction rule has not been applied. For example:

$$\dfrac{\dfrac{\begin{array}{c}\Pi_1\\ X_1[x_1]\to\sim(x\backslash y)\end{array}}{X_1[\sim\sim x_1]\to\sim(x\backslash y)} \qquad \dfrac{\begin{array}{c}\Pi_2\\ X\to x\end{array}\quad\begin{array}{c}\Pi_3\\ Y[\sim y]\to z\end{array}}{Y[X\sim(x\backslash y)]\to z}}{Y[XX_1[\sim\sim x_1]]\to z}$$

is replaced by

$$\dfrac{\dfrac{\begin{array}{c}\Pi_1\\ X_1[x_1]\to\sim(x\backslash y)\end{array} \qquad \dfrac{\begin{array}{c}\Pi_2\\ X\to x\end{array}\quad\begin{array}{c}\Pi_3\\ Y[\sim y]\to z\end{array}}{Y[X\sim(x\backslash y)]\to z}}{Y[XX_1[x_1]]\to z}}{Y[XX_1[\sim\sim x_1]]\to z}$$

Finally, for all pairs of right and left rules, it may happen that the main connective of the cut-formula is introduced by these rules in both premise sequents of (cut). These are the principal cases, and we shall display here once case of these reductions of the degree of the cut-formula. The proof

$$\dfrac{\dfrac{\begin{array}{c}\Pi_1\\ xX_1\to\sim y\end{array}}{X_1\to\sim(x\backslash y)} \qquad \dfrac{\begin{array}{c}\Pi_2\\ X\to x\end{array}\quad\begin{array}{c}\Pi_3\\ Y[\sim y]\to z\end{array}}{Y[X\sim(x\backslash y)]\to z}}{Y[XX_1]\to z}$$

is replaced by

$$\dfrac{\begin{array}{c}\Pi_2\\ X\to x\end{array} \qquad \dfrac{\begin{array}{c}\Pi_1\\ xX_1\to\sim y\end{array}\quad\begin{array}{c}\Pi_3\\ Y[\sim y]\to z\end{array}}{Y[xX_1]\to z}}{Y[XX_1]\to z}$$

For $\mathsf{L}^{\sim}$, G-terms may be considered as finite *lists* of type symbols, so that the rules $(assoc_1)$ and $(assoc_2)$ can be neglected. ∎

Corollary 12.3.10 $\mathsf{NL}^{\sim}$ $(\mathsf{L}^{\sim})$ *is a conservative extension of* NL (L).

Note that this result also follows from the fact that deleting all negation symbols in the rules of Definition 12.3.2 results in sequent rules of L, so that every sequent provable in $\mathsf{L}^{\sim}$ and $\mathsf{NL}^{\sim}$ remains provable after deleting all negation signs.

Corollary 12.3.11 $\mathsf{NL}^{\sim}$ *and* $\mathsf{L}^{\sim}$ *are decidable.*

Proof. For $\mathsf{NL}^{\sim}$ decidability follows from bottom-up proof search. For $\mathsf{L}^{\sim}$, G-terms again may be considered as finite lists of type symbols, so that the rules $(assoc_1)$ and $(assoc_2)$ can be neglected, and bottom-up proof search terminates. ∎

Definition 12.3.12 *Let* $At^* := \{x^* \mid x \in At\}$ *be a set of new atomic type symbols, and let* $\mathbf{At} := At \cup At^*$. *Let* X^+ *be the result of replacing in the G-term* X *every occurrence of a negated atomic type symbol* $\sim p$ *by* p^*. *For every G-term* X, *the G-term* X^* *is defined as* X'^+.

Corollary 12.3.13 *For every G-term* X *and type symbol* x *in the language based on* At: $\vdash X \to x$ *in* $\mathsf{NL}^{\sim}$ $(\mathsf{L}^{\sim})$ *iff* $\vdash X^* \to x^*$ *in* NL (L) *in the language based on* $\mathbf{At}$.

An atomic type symbol p occurs positively in the scope of an even number of occurrences of $\sim$, and it occurs negatively in the scope of an uneven number of occurrences of $\sim$. A positive (negative) occurrence of p in formula x is positive (negative) in formulas $x \times y$, $y \times x$ and in sequents $X \to x$. A positive (negative) occurrence of p in x is negative (positive) in formulas $\sim x$, $(x\backslash y)$, (y/x) and in sequents $Y[x] \to y$. Let $\times X$ be the result of replacing every sub-G-term (YZ) in X by $(Y \times Z)$. Clearly, $\times X$ is a formula. An atomic type symbol occurs positively (negatively) in X iff it occurs positively (negatively) in $\times X$. Let $pos(X)$ $(neg(X))$ denote the set of all atomic type symbols that occur positively (negatively) in X.

Corollary 12.3.14 (Interpolation) *For* $\mathsf{NL}^{\sim}$ *and* $\mathsf{L}^{\sim}$, *the following holds: if* $\vdash X \to x$, *then there exists a formula* y *such that* $\vdash X \to y$ *and* $\vdash y \to x$, *where* $pos(y) \subseteq (pos(X) \cap pos(x))$ *and* $neg(y) \subseteq (neg(X) \cap neg(x))$

Proof. In view of Corollaries 1 and 12.3.13, the proof in [196, pp. 42, 43] suffices. ∎

Suppose we would want to interpret negated product types $(x \times y)$ by a powerset operation – that satisfies

$$-(x \circ y) \quad = \quad \{b \mid b \in ((-x \circ -y) \sqcup (x \circ -y) \sqcup (-x \circ y))\}$$

We could then introduce the following sequent rules for negated product types:

$$\begin{array}{ll} (\sim\times\rightarrow) & X[(x\sim y)] \rightarrow z,\ \ X[(\sim xy)] \rightarrow z \\ & X[(\sim x\sim y)] \rightarrow z \vdash X[\sim (x \times y)] \rightarrow z \\ (\rightarrow\sim\times) & X \rightarrow x,\ \ Y \rightarrow \sim y \vdash XY \rightarrow \sim (x \times y) \\ & X \rightarrow \sim x,\ \ Y \rightarrow y \vdash XY \rightarrow \sim (x \times y) \\ & X \rightarrow \sim x,\ \ Y \rightarrow \sim y \vdash XY \rightarrow \sim (x \times y) \end{array}$$

With these rules, applications of cut would still be eliminable.

12.4. Semantics

We shall observe that $\mathsf{NL}^{\sim}$ ($\mathsf{L}^{\sim}$) is complete with respect to powerset residuated groupoids (semigroups) but fails to be complete with respect to structures called negative powerset residuated groupoids (semigroups).

Definition 12.4.1 *Let $(\mathcal{P}(M), \subseteq, \circ, \Rightarrow, \Leftarrow)$ be the powerset residuated groupoid (semigroup) over the groupoid (semigroup) $(M, \cdot)$, and let v be a function from* **At** *into $\mathcal{P}(M)$. The interpretation function I_v from negation-free G-terms based on* **At** *into $\mathcal{P}(M)$ is defined as follows:*

$$\begin{array}{rcl} I_v(x) & = & v(x),\ \ x \in \mathbf{At} \\ I_v((x \times y)) & = & I_v(x) \circ I_v(y) \\ I_v((y/x)) & = & I_v(y) \Leftarrow I_v(x) \\ I_v((x\backslash y)) & = & I_v(x) \Rightarrow I_v(y) \\ I_v((X, Y)) & = & I_v(X) \circ I_v(Y) \end{array}$$

Definition 12.4.2 *A sequent $X \rightarrow x$ is valid in a powerset residuated groupoid (semigroup) $(\mathcal{P}(M), \subseteq, \circ, \Rightarrow, \Leftarrow)$ under the valuation function v from* **At** *into $\mathcal{P}(M)$ iff $I_v(X^*) \subseteq I_v(x^*)$. A sequent $X \rightarrow x$ is valid in a powerset residuated groupoid (semigroup) $(\mathcal{P}(M), \subseteq, \circ, \Rightarrow, \Leftarrow)$ iff $X \rightarrow x$ is valid under every valuation function v from* **At** *into $\mathcal{P}(M)$, and $X \rightarrow x$ is valid (simpliciter) iff it is valid in every powerset residuated groupoid (semigroup).*

Theorem 12.4.3 *For all G-terms X, x based on* **At**, $\vdash X \to x$ *in* $\mathsf{NL}^{\sim}$ $(\mathsf{L}^{\sim})$ *iff* $X \to x$ *is valid.*

Proof. By Corollary 12.3.13, $\vdash X \to x$ in $\mathsf{NL}^{\sim}$ $(\mathsf{L}^{\sim})$ iff $\vdash X^* \to x^*$ in NL (L). Since NL (L) is characterized by the class of all powerset residuated groupoids (semigroups), see [29], [31], $\vdash X^* \to x^*$ iff $I_v(X^*) \subseteq I_v(x^*)$ for every valuation function v from **At** into $\mathcal{P}(M)$. ∎

The faithful embedding of $\mathsf{NL}^{\sim}$ $(\mathsf{L}^{\sim})$ into NL (L) under the translation $(\cdot)^*$ shows that after introducing $\sim$ we still have, in a sense, basically the same syntactic calculus. Positive and negative information about atomic type membership is treated on an equal footing, and the rest is rewriting. Negated types are nevertheless useful in approximating a correct type assignment for a language under investigation, because for some expressions only negative information about their membership in certain atomic or functor types may be available.

Definition 12.4.4 *The structure $(\mathcal{P}(M), \subseteq, \circ, \Rightarrow, \Leftarrow, -)$ is called a negative powerset residuated groupoid (semigroup) over the groupoid (semigroup) $(M, \cdot)$, if $(\mathcal{P}(M), \subseteq, \circ, \Rightarrow, \Leftarrow)$ is the powerset residuated groupoid (semigroup) over $(M, \cdot)$, and $-$ is a function on $\mathcal{P}(M)$ satisfying:*

$$\begin{aligned} --x &= x \\ -(y \Leftarrow x) &= \{b \in M \mid (\forall a \in x)\, b \cdot a \in -y\} \\ -(x \Rightarrow y) &= \{b \in M \mid (\forall a \in x)\, a \cdot b \in -y\}. \end{aligned}$$

Clearly such functions exist, since identity satisfies the above equations.

Definition 12.4.5 *Let $(\mathcal{P}(M), \subseteq, \circ, \Rightarrow, \Leftarrow, -)$ be a negative powerset residuated groupoid (semigroup) over the groupoid (semigroup) $(M, \cdot)$, let v be a function from* **At** *into $\mathcal{P}(M)$, and let T be the set of all G-terms based on At. Then the interpretation function I^*_v from $T \cup At^*$ into $\mathcal{P}(M)$ is defined as follows:*

$$\begin{aligned} I^*_v(x) &= v(x), \quad x \in \mathbf{At} \\ I^*_v(\sim x) &= v(x^*), \quad x \in At \\ I^*_v((x \times y)) &= I^*_v(x) \circ I^*_v(y) \\ I^*_v((y/x)) &= I^*_v(y) \Leftarrow I^*_v(x) \\ I^*_v((x\backslash y)) &= I^*_v(x) \Rightarrow I^*_v(y) \\ I^*_v(\sim\sim x) &= I^*_v(x) \\ I^*_v(\sim (x \sharp y)) &= -I^*_v((x \sharp y)), \quad \sharp \in \{/, \backslash\} \\ I^*_v((X, Y)) &= I^*_v(X) \circ I^*_v(Y) \end{aligned}$$

Observation 12.4.6 The following does not hold: For all G-terms X and x based on At, $\vdash X \rightarrow x$ in $\mathsf{NL}^{\sim}$ ($\mathsf{L}^{\sim}$) iff $I_v^*(X) \subseteq I_v^*(x)$ for every negative powerset residuated groupoid (semigroup) $(\mathcal{P}(M), \subseteq, \circ, \Rightarrow, \Leftarrow, -)$ and every valuation function v from **At** into $\mathcal{P}(M)$.

Proof. By Observation 12.3.6, $\sim$ is not extensional. ∎

12.5. Remarks

In this chapter, we have introduced a certain non-classical negation into Categorial Grammar. Buszkowski [30] adds the axioms characteristic of a well-understood basic type of negation to an axiom system for the associative Lambek calculus and then investigates the semantical effects of this addition. In the present chapter, another approach to introducing a negation connective into Categorial Grammar is ventured, namely considering the effects of falsifying the defining properties of the powerset operations used in the semantics of the Lambek calculi. While in the case of the powerset operation used to interpret product types the result casts doubt on the negation of product types, the ideal of uniformity with respect to type membership leads to a notion of negation that gives rise to connexive implications. This is interesting, because connexive implication has been previously discussed completely independently of Categorial Grammar. It may be difficult to judge whether the new kind of negation is more compatible with the linguistic sense of negative information than the quasi-Boolean negation considered by Buszkowski. Under the assumption of uniformity with respect to type membership, the sequent rules for negated directed implications seem justified, and under the translation $(\cdot)^*$, a characterization of $\mathsf{NL}^{\sim}$ and $\mathsf{L}^{\sim}$ is obtained with respect to linguistically intended models. Moreover, the sequent systems $\mathsf{NL}^{\sim}$ and $\mathsf{L}^{\sim}$ have the advantage of admitting cut-elimination and, as a result, being decidable.

Whether the operation $\sim$ in $\mathsf{NL}^{\sim}$ and $\mathsf{L}^{\sim}$ is in fact some kind of *negation* is, perhaps, a matter of debate, as the answer to the question "What is negation?" is contentious, see [61], [201]. Although Nelson's constructive four-valued logic can be faithfully embedded into positive intuitionistic logic, strong negation in Nelson's logic is nevertheless widely accepted as a genuine negation operation. Therefore, the faithful embedding of $\mathsf{NL}^{\sim}$ ($\mathsf{L}^{\sim}$) into NL (L) under the translation $(\cdot)^*$ does not necessarily exclude $\sim$ in $\mathsf{NL}^{\sim}$ and $\mathsf{L}^{\sim}$ from being a negation.

Bibliography

[1] J.M. Abe and S. Akama, Annotated temporal logics $\Delta^*\tau$, in: *Lecture Notes in Artificial Intelligence* 1952, Springer-Verlag, Berlin, pp. 217–226, 2000.

[2] S. Abramsky, Proofs as processes, *Theoretical Computer Science* 135, pp. 5–9, 1994.

[3] S. Abramsky and S. Vickers, Quantales, observational logic and process semantics, *Mathematical Structures in Computer Science* 3, pp. 161–227, 1993.

[4] S. Akama, Tableaux for logic programming with strong negation, in: *Lecture Notes in Computer Science* 1227, Springer-Verlag, Berlin, pp. 31–42, 1997.

[5] G. Allwein and W. MacCaull, A Kripke semantics for the logic of Gelfand quantales, *Studia Logica* 68, pp. 173–228, 2001.

[6] A. Almukdad and D. Nelson, Constructible falsity and inexact predicates, *Journal of Symbolic Logic* 49, pp. 231–233, 1984.

[7] A.R. Anderson and N.D. Belnap, *Entailment: The Logic of Relevance and Necessity, Vol. I*, Princeton University Press, Princeton, 1975.

[8] A.R. Anderson, N.D. Belnap, and J.M. Dunn, *Entailment: The Logic of Relevance and Necessity, Vol. 2*, Princeton University Press, Princeton, 1992.

[9] K.R. Apt and R.N. Bol, Logic programming and negation: a survey, *Journal of Logic Programming* 19/20, pp. 9–71, 1994.

[10] O. Arieli, Paraconsistent declarative semantics for extended logic programs, *Annals of Mathematics and Artificial Intelligence* 36, pp. 381–417, 2002.

[11] O. Arieli and A. Avron, Logical bilattices and inconsistent data, in: *Proceedings of the 9th IEEE Annual Symposium on Logic in Computer Science*, IEEE Press, pp. 468–476, 1994.

[12] O. Arieli and A. Avron, Reasoning with logical bilattices, *Journal of Logic, Language and Information* 5, pp. 25–63, 1996.

[13] O. Arieli and A. Avron, The value of the four values, *Artificial Intelligence* 102, pp. 97-141, 1998.

[14] O. Arieli and A. Avron, Bilattices and paraconsistency, in: D. Batens et al. (eds.), *Frontiers of Paraconsistent Logic*, Research Studies Press, Baldock, Hertfordshire, pp. 11-27, 2000.

[15] A. Arnon, The structure of interlaced bilattices, *Mathematical Structures in Computer Science* 6, pp. 287-299, 1996.

[16] A. Avron, Negation: two points of view, in: D. Gabbay and H. Wansing (eds.), *What is Negation?*, Kluwer Academic Publishers, Dordrecht, pp. 3–22, 1999.

[17] A. Avron, A non-deterministic view on non-classical negations, *Studia Logica* 80, pp. 159–194, 2005.

[18] S. Baratella and A. Masini, A proof-theoretic investigation of a logic of positions, *Annals of Pure and Applied Logic* 123, pp. 135–162, 2003.

[19] A. Barenco, D. Deutsch and A. Ekert, Conditional quantum dynamics and logic gates, *Physical Review Letters* 74, No. 20, pp. 4083–4086, 1995.

[20] G. Bellin and P.J. Scott, On the π-calculus and linear logic, *Theoretical Computer Science* 135, pp. 11–65, 1994.

[21] N.D. Belnap, A useful four-valued logic, in: G. Epstein and J. M. Dunn (eds.), *Modern Uses of Multiple-Valued Logic*, Reidel, Dordrecht, pp. 7–37, 1977.

[22] N.D. Belnap, How computer should think, in: G. Ryle (ed.), *Contemporary Aspects of Philosophy*, Oriel Press, Stocksfield, pp. 30–56, 1977.

[23] N.D. Belnap, Display logic, *Journal of Philosophical Logic* 11, pp. 375–417, 1982. Reprinted with minor changes as §62 of A.R. Anderson, N.D. Belnap, and J.M. Dunn, *Entailment: The Logic of Relevance and Necessity, Vol. 2*, Princeton University Press, Princeton, 1992.

[24] A. Biere, A. Cimatti, E.M. Clarke, O. Strichman and Y. Zhu, Bounded model checking, *Advances in Computers* 58, pp. 118–149, 2003.

[25] H.A. Blair and V.S. Subrahmanian, Paraconsistent logic programming, *Theoretical Computer Science* 68, pp. 135–154, 1989.

[26] B. Brown, On Paraconsistency, in. L. Goble (ed.), *A Companion to Philosophical Logic*, Blackwell Publishers, Oxford, pp. 628–650, 2002.

[27] A. Brunner and W. Carnielli, Anti-intuitionism and paraconsistency, *Journal of Applied Logic* 3, pp. 161–184, 2005.

[28] L. Buisman and R. Goré, A cut-free sequent calculus for bi-intuitionistic logic, in: *Lecture Notes in Computer Science* 4548, Springer-Verlag, Berlin, pp. 90-106, 2007.

[29] W. Buszkowski, Completeness results for Lambek syntactic calculus, *Zeitschrift für mathematische Logik und Grundlagen der Mathematik* 32, pp. 13–28, 1986.

[30] W. Buszkowski, Categorial grammars with negative information, in: H. Wansing (ed.), *Negation. A Notion in Focus*, de Gruyter, Berlin, pp. 107–126, 1996.

[31] W. Buszkowski, Mathematical linguistics and proof theory, in: A. ter Meulen and J. van Benthem (eds.), *Handbook of Logic and Language*, North-Holand, Amsterdam, pp. 683–736, 1997.

[32] P. Cabalar, S. Odintsov, and D. Pearce, Strong negation in well-founded and partial stable semantics for logic programs, in: *Lecture Notes in Artificial Intelligence* 4140, Springer-Verlag, Berlin, pp. 592–601, 2006.

[33] W. Carnielli and J. Marcos, Limits for paraconsistent calculi, *Notre Dame Journal of Formal Logic* 40, pp. 375–390, 1999.

[34] W. Carnielli, M. Coniglio, D. Gabbay, P. Gouveia and C. Sernadas, *Analysis and synthesis of logics. How to cut and paste reasoning systems*, Springer-Verlag, Berlin, 2008.

[35] S. Cerrito, M.C. Mayer and S. Prand, First order linear temporal logic over finite time structures, in: *Lecture Notes in Computer Science* 1705, Springer-Verlag, Berlin, pp. 62–76, 1999.

[36] A. Chagrov and M. Zakharyashev, *Modal Logic*, Oxford University Press, Oxford, 1997.

[37] E.M. Clarke, O. Grumberg, and D.A. Peled, *Model Checking*, The MIT Press, 1999.

[38] T. Crolard, Subtractive logic, *Theoretical Computer Science* 254, pp. 151–185, 2001.

[39] J. Czermak, A remark on Gentzen's calculus of sequents, *Notre Dame Journal of Formal Logic* 18, pp. 471–474, 1977.

[40] N.C.A. da Costa, On the theory of inconsistent formal systems, *Notre Dame Journal of Formal Logic* 15, pp. 497–510, 1974.

[41] N.C.A. da Costa, J.-Y. Béziau, and O.A.S. Bueno, Aspects of paraconsistent logic, *Bulletin of the IGPL* 3, pp. 597–614, 1995.

[42] D. van Dalen, *Logic and Structure*, Fourth Edition, Springer-Verlag, Berlin, 2008.

[43] M. Dam, Process-algebraic interpretations of positive linear and relevant logics, *Journal of Logic and Computation* 4, pp. 939-973, 1994.

[44] C.V. Damásio and L.M. Pereira, A survey of paraconsistent semantics for logic programs, in: D. Gabbay and P. Smets (eds.), *Handbook of Defeasible Reasoning and Uncertainty Management Systems, Vol. 2*, pp. 241–320, Kluwer Academic Publishers, Dordrecht, 1998.

[45] R. Davies, A temporal logic approach to binding-time analysis, in: E. Clarke (ed.), *Proceedings of the Eleventh Annual Symposium on Logic in Computer Science*, IEEE Computer Society, pp. 184-195, 1996.

[46] J. Dörre and S. Manandhar, On constraint-based Lambek calculi, in: P. Blackburn and M. de Rijke (eds.), *Specifying Syntactic Structures*, Studies in Logic, Language and Information, CSLI, Stanford, pp. 25–44, 1997.

[47] J.M. Dunn, Intuitive semantics for first-degree entailment and 'coupled trees', *Philosophical Studies* 29, pp. 149–168, 1976.

[48] J.M. Dunn, Perp and star: Two treatments of negation, in: J. Tomberlin (ed.), *Philosophical Perspectives (Philosophy of Language and Logic)* 7, pp. 331–357, 1993.

[49] J.M. Dunn, Generalized ortho negation, in: H. Wansing (ed.), *Negation. A Notion in Focus*, de Gruyter, Berlin, pp. 3–26, 1996.

[50] J.M. Dunn, A comparative study of various model-theoretic treatments of negation: A History of Formal Neagtion, in: D. Gabbay and H. Wansing (eds.), *What is Negation?*, Kluwer Academic Publishers, Dordrecht, pp. 23–51, 1999.

[51] J.M. Dunn, Partiality and its dual, *Studia Logica* 66, pp. 5–40, 2000.

[52] R. Dyckhoff and S. Negri, Admissibility of structural rules for contraction-free systems of intuitionistic logic, *Journal of Symbolic Logic* 65, pp. 1499–1515, 2000.

[53] S. Easterbrook and M. Chechik, A framework for multi-valued reasoning over inconsistent viewpoints, in: *Proceedings of the 23rd International Conference on Software Engineering (ICSE'01)*, Toronto, pp. 411-420, 2001.

[54] G. Elbl, A declarative semantics for depth-first logic programs, *The Journal of Logic Programming* 41, pp. 27–66, 1999.

[55] U. Engberg and G. Winskel, Completeness results for linear logic on Petri nets, *Annals of Pure and Applied Logic* 86, pp. 101–135, 1997.

[56] M. Fitting, Bilattices in logic programming, in: G. Epstein (ed.), *The Twentieth International Symposium on Multiple-Valued Logic*, IEEE-Press, pp. 238-246, 1990.

[57] M. Fitting, Bilattices and the semantics of logic programming, *Journal of Logic Programming* 11, pp. 91–116, 1991.

[58] M. Fitting, Bilattices are nice things, in: T. Bolander, V. Hendricks, and S.A. Pedersen (eds.), *Self-Reference* , CSLI Publications, Stanford, pp. 53-77, 2006.

[59] J.M. Font, Belnap's four-valued logic and de Morgan lattices, *Logic Journal of the IGPL* 5, pp. 413–440, 1997.

[60] D. Gabbay, What is negation in a system?, in: F. Drake and J. Truss (eds.), *Logic Colloquium '86*, Elsevier, Amsterdam, pp. 95–112, 1988.

[61] D. Gabbay and H. Wansing (eds.), *What is Negation?*, Kluwer Academic Publishers, Dordrecht, 1999.

[62] G. Gargov, Knowledge, uncertainty and ignorance in logic: bilattices and beyond, *Journal of Applied Non-Classical Logics* 9, pp. 195–283, 1999.

[63] M.L. Ginsberg, Multivalued logics: a uniform approach to reasoning in AI, *Computer Intelligence* 4, pp. 256–316, 1988.

[64] J.-Y. Girard, Linear logic, *Theoretical Computer Science* 50, pp. 1–102, 1987.

[65] J.-Y. Girard, A new constructive logic: Classical logic, *Mathematical Structures in Computer Science* 1, pp. 255–296, 1991.

[66] K. Gödel, Zur intuitionistischen Arithmetik und Zahlentheorie, *Ergebnisse eines mathematischen Kolloquiums* 4, pp. 34–38, 1933.

[67] R. Goré, Intuitionistic logic redisplayed, Technical Report TR-ARP-1-1995, Australian National University, Canberra, 1995.

[68] R. Goré, Dual intuitionistic logic revisited, in: *Lecture Notes in Artificial Intelligence* 1847, Springer-Verlag, Berlin, pp. 252–267, 2000.

[69] R. Goré and L. Postniece, Combining derivations and refutations for cut-free completeness in bi-intuitionistic logic, *Journal of Logic and Computation* 20, pp. 233–260. 2010.

[70] R. Goré, L. Postnice and A. Tiu, Cut-elimination and proof-search for bi-intuitionistic logic using nested sequents, in: C. Areces and R. Goldblatt (eds.), *Advances in Modal Logic. Vol. 7*, College Publications, London, pp. 43–66, 2008, see also ⟨http://www.aiml.net/volumes/volume7/⟩.

[71] N.D. Goodman, The logic of contradiction, *Zeitschrift für mathematische Logik und Grundlagen der Mathematik* 27, pp. 119–126, 1981.

[72] A. Grzegorczyk. A philosophically plausible formal interpretation of intuitionistic logic, *Indagationes Mathematicae* 26, pp. 596–601, 1964.

[73] Y. Gurevich, Intuitionistic logic with strong negation, *Studia Logica* 36, pp. 49–59, 1977.

[74] C.A.R. Hoare and J. He, A weakest pre-specification, *Information Processing Letter* 24, pp. 127–132, 1987.

[75] J. Hodas and D. Miller, Logic programming in a fragment of intuitionistic linear logic, *Information and Computation* 110, pp. 327–365, 1994.

[76] I. Hodkinson, F. Wolter and M. Zakharyaschev, Decidable fragments of first-order temporal logics, *Annals of Pure and Applied Logic* 106, pp. 85–134, 2000.

[77] G.J. Holzmann, *The SPIN Model Checker: Primer and Reference Manual*, Addison-Wesley, Boston, 2006.

[78] L. Horn, *A Natural History of Negation*, Chicago University Press, Chicago, 1989, CSLI Publications, Stanford, 2001.

[79] K. Ishihara and K. Hiraishi, The completeness of linear logic for Petri net models, *Logic Journal of the IGPL* 9, pp. 549–567, 2001.

[80] C. Kalicki, Infinitary propositional intuitionistic logic, *Notre Dame Journal of Formal Logic* 21, pp. 216–228, 1980.

[81] N. Kamide, Relevance principle for substructural logics with mingle and strong negation, *Journal of Logic and Computation* 12, pp. 1–16, 2002.

[82] N. Kamide, Sequent calculi for intuitionistic linear logic with strong negation, *Logic Journal of the IGPL* 10, pp. 653–678, 2002.

[83] N. Kamide, A canonical model construction for substructural logics with strong negation, *Reports on Mathematical Logic* 36, 95–116, 2002.

[84] N. Kamide, A note on dual-intuitionistic logic, *Mathematical Logic Quarterly* 49, pp. 519–524, 2003.

[85] N. Kamide, Quantized linear logic, involutive quantales and strong negation, *Studia Logica* 77, pp. 355–384, 2004.

[86] N. Kamide, Combining soft linear logic and spatio-temporal operators, *Journal of Logic and Computation 14*, pp. 625–650, 2004.

[87] N. Kamide, A relationship between Rauszer's H-B logic and Nelson's logic, *Bulletin of the Section of Logic 33*, pp. 237–249, 2004.

[88] N. Kamide, Gentzen-type methods for bilattice negation, *Studia Logica 80*, pp. 265–289, 2005.

[89] N. Kamide, A cut-free system for 16-valued reasoning, *Bulletin of the Section of Logic 34*, pp. 213–226, 2005.

[90] N. Kamide, Natural deduction systems for Nelson's paraconsistent logic and its neighbors, *Journal of Applied Non-Classical Logics 15*, pp. 405–435, 2005.

[91] N. Kamide, An equivalence between sequent calculi for linear-time temporal logic, *Bulletin of the Section of Logic* 35, pp. 187–194, 2006.

[92] N. Kamide, Phase semantics and Petri net interpretations for resource-sensitive strong negation, *Journal of Logic, Language and Information 15*, pp. 371–401, 2006.

[93] N. Kamide, Strong normalization of a typed lambda calculus for intuitionistic bounded linear-time temporal logic, *Reports on Mathematical Logic* 47, pp. 29–61, Jagiellonian University Press, 2012.

[94] N. Kamide, Embedding linear-time temporal logic into infinitary logic: application to cut-elimination for multi-agent infinitary epistemic linear-time temporal logic, in: *Lecture Notes in Artificial Intelligence* 5405, Springer-Verlag, Berlin, pp. 57–76, 2009.

[95] N. Kamide, Proof systems combining classical and paraconsistent negations, *Studia Logica 91*, pp.217–238, 2009.

[96] N. Kamide, An embedding-based completeness proof for Nelson's paraconsistent logic, *Bulletin of the Section of Logic* 39, pp. 1–10, 2010.

[97] N. Kamide, Automating and computing paraconsistent reasoning: Contraction-free, resolution and type systems, *Reports on Mathematical Logic* 45, 3–21, 2010.

[98] N. Kamide, Paraconsistent description logics revisited, in: *Proceedings of the 23rd International Workshop on Description Logics (DL 2010)*, Sun SITE Central Europe (CEUR) Electronic Workshop Proceedings, Technical University of Aachen (RWTH), 12 pp., 2010.

[99] N. Kamide and H. Wansing, Sequent calculi for some trilattice logics, *The Review of Symbolic Logic* 2, pp. 374–395, 2009.

[100] N. Kamide and H. Wansing, Combining linear-time temporal logic with constructiveness and paraconsistency, *Journal of Applied Logic* 6, pp. 33–61, 2010.

[101] N. Kamide and H. Wansing, Symmetric and dual paraconsistent logics, *Logic and Logical Philosophy* 19, pp. 7–30, 2010.

[102] N. Kamide and H. Wansing, A paraconsistent linear-time temporal logic, *Fundamenta Informaticae 106*, pp. 1–23, 2011.

[103] N. Kamide and H. Wansing, Completeness and cut-elimination theorems for trilattice logics, *Annals of Pure and Applied Logic* 162, pp. 816–835, 2011.

[104] N. Kamide and H. Wansing, Proof theory of Nelson's paraconsistent logic: A uniform perspective, *Theoretical Computer Science* 415, pp. 1–38, 2012.

[105] K. Kaneiwa and S. Tojo, An order-sorted logic with implicitly negative sorts (in Japanese), *Journal of Information Processing Society of Japan* 43, pp. 1505–1517, 2002.

[106] I.M. Kanovich, Petri nets, Horn programs, linear logic and vector games, *Annals of Pure and Applied Logic* 75, pp. 107–135, 1995.

[107] I.M. Kanovich, Linear logic as a logic of computations, *Annals of Pure and Applied Logic* 67, pp. 183–212, 1994.

[108] H. Kawai, Sequential calculus for a first order infinitary temporal logic, *Zeitschrift für mathematische Logik und Grundlagen der Mathematik* 33, pp. 423–432, 1987.

[109] H.C.M. Kleijn and M. Koutny, Causality semantics of Petri nets with weighted inhibitor arcs, in: *Lecture Notes in Computer Science* 2421, Springer-Verlag, Berlin, pp. 531–546, 2002.

[110] S. Kosaraju, Limitations of Dijkstra's semaphore primitives and Petri nets, *Operating Systems Review* 7, pp. 122–126, 1973.

[111] M. Kracht, On extensions of intermediate logics by strong negation, *Journal of Philosophical Logic* 27, pp. 49–73, 1998.

[112] F. Kröger, LAR: a logic of algorithmic reasoning, *Acta Informaticae* 8, pp. 243–266, 1977.

[113] F. von Kutschera, Ein verallgemeinerter Widerlegungsbegriff für Gentzenkalküle, *Archiv für Mathematische Logik und Grundlagenforschung* 12, pp. 104–118, 1969.

[114] J. Lambek, The mathematics of sentencs structure, *American Mathematical Monthly* 65, pp. 154–170, 1958.

[115] J. Lambek, On the calculus of syntactic types, in: R. Jacobson (ed.), *Structure of Language and Its Mathematical Aspects*, American Mathematical Society, Providence R.I., pp. 166–178, 1961.

[116] D. Larchey-Wendling and D. Galmiche, Provability in intuitionistic linear logic from a new interpretation on Petri nets — extended abstract —, *Electronic Notes in Theoretical Computer Science* 17, 18 pp., 1998.

[117] D. Larchey-Wendling and D. Galmiche, Quantales as completions of ordered monoids: revised semantics for intuitionistic linear logic, *Electronic Notes in Theoretical Computer Science* 35, 15 pp., 2000.

[118] J. Lilius, High-level nets and linear logic, in: *Lecture Notes in Computer Science* 616, Springer-Verlag, Berlin, pp. 310–327, 1992.

[119] E.G.K. López-Escobar, Refutability and elementary number theory, *Indagationes Mathematicae* 34, pp. 362–374, 1972.

[120] P. Łukowski, A deductive-reductive form of logic: general theory and intuitionistic case, *Logic and Logical Philosophy* 10, pp. 59–78, 2002.

[121] P. Łukowski, A deductive-reductive form of logic: intuitionistic S4 modalities, *Logic and Logical Philosophy* 10, pp. 79–91, 2002.

[122] W. MacCaull, Relational proof system for linear and other substructural logics, *Logic Journal of the IGPL* 5, pp. 673–697, 1997.

[123] S. McCall, Connexive implication and the syllogism, *Mind* 76, pp. 346–356, 1967.

[124] P. Maier, Intuitionistic LTL and a new characterization of safety and liveness, in: *Lecture Notes in Computer Science* 3210, Springer-Verlag, Berlin, pp. 295-309, 2004.

[125] S. Martini and A. Masini, A computational interpretation of modal proofs, in: H. Wansing (ed.), *Proof Theory of Modal Logic*, Kluwer Academic Publishers, Dordrecht, pp. 213–241, 1996.

[126] N. Martí-Oliet and J. Meseguer, From Petri nets to linear logic, *Mathematical Structures in Computer Science* 1, pp. 69–101, 1991.

[127] G. Mints, Gentzen-type systems and resolution rules. Part I. Propositional logic, in: *Lecture Notes in Computer Science* 417, Springer-Verlag, Berlin, pp. 198–231, 1988.

[128] C. Monroe, D.M. Meekhof, B.E. King, W.M. Itano and D.J. Wineland, Demonstration of a fundamental quantum logic gate, *Physical Review Letters* 75, pp. 4712–4717, 1995.

[129] C.J. Mulvey and J.W. Pelletier, A quantisation of the calculus of relations, in: *Category Theory 1991, CMS Conference Proceedings* 13, pp. 345–360, American Mathematical Society, Providence, RI, 1992.

[130] C.J. Mulvey and J.W. Pelletier, On the quantization of points, *Journal of Pure and Applied Algebra* 159, pp. 231–295, 2001.

[131] T. Murata, V.S. Subrahmanian and T. Wakayama, A Petri net model for reasoning in the presence of inconsistency, *IEEE Transactions on Knowledge and Data Engineering* 3, pp. 281–292, 1991.

[132] S. Negri and J. von Plato, Sequent calculus in natural deduction style, *Journal of Symbolic Logic* 66 (4), pp. 1803–1816, 2001.

[133] S. Negri and J. von Plato, *Structural Proof Theory*, Cambridge University Press, Cambridge, 2001.

[134] D. Nelson, Constructible falsity, *Journal of Symbolic Logic* 14, pp.16–26, 1949.

[135] D. Nelson, Negation and separation of concepts in constructive systems, in: A. Heyting (ed.), *Constructivity in Mathematics*, North-Holland Publishing Company, Amsterdam, pp. 208-225, 1959.

[136] S.P. Odintsov, On the embedding of Nelson's logics, *Bulletin of the Section of Logic* 31, pp. 241–250, 2002.

[137] S.P. Odintsov, Algebraic Semantics for Paraconsistent Nelson's Logic, *Journal of Logic and Computation* 13, pp. 453–468, 2003.

[138] S.P. Odintsov, *Constructive Negations and Paraconsistency*, Springer-Verlag, Dordrecht, 2008.

[139] S.P. Odintsov, On axiomatizing Shramko-Wansing's logic, *Studia Logica* 91, pp. 407–428, 2009.

[140] S.P. Odintsov and H. Wansing, Inconsistency-tolerant description logic. Motivation and basic systems, in: V. Hendricks and J. Malinowski (eds.), *Trends in Logic. 50 Years of Studia Logica*, Kluwer Academic Publishers, Dordrecht, pp. 301-335, 2004.

[141] S.P. Odintsov and H. Wansing, Constructive predicate logic and constructive modal logic. Formal duality versus semantical duality, in: V. Hendricks et al. (eds.) *Proceedings of First-Order Logic 75 (FOL 75)*, Logos Verlag, Berlin, pp. 269–286, 2004.

[142] S.P. Odintsov and H. Wansing, Inconsistency-tolerant description Logic. Part II: Tableau algorithms, *Journal of Applied Logic 6*, pp. 343-360, 2008.

[143] M. Okada, An introduction to linear logic: expressiveness and phase semantics, *MSJ Memoirs* 2, pp. 255–295, 1998.

[144] M. Okada, Phase semantic cut-elimination and normalization proofs of first- and higher-order linear logic, *Theoretical Computer Science* 227, pp. 333–396, 1999.

[145] H. Ono, Semantics for substructural logics, in: P. Schroeder-Heister and K. Došen (eds.), *Substructural Logics*, Oxford University Press, Oxford, pp. 259–291, 1993.

[146] H. Ono, *Logic for Information Science* (in Japanease), Nihon Hyouron shya, 297 pp., 1994.

[147] H. Ono and Y. Komori, Logics without the contraction rule, *Journal of Symbolic Logic* 50, 169–201, 1985.

[148] D. Pearce, Reasoning with negative information, II: hard negation, strong negation and logic programs, in: *Lecture Notes in Computer Science* 619, Springer-Verlag, Berlin, pp. 63–79, 1992.

[149] D. Pearce, Sixty years of stable models. Summary, in: *Springer Lecture Notes in Computer Science* 5366, Springer-Verlag, Berlin, p. 52, 2008.

[150] J.W. Pelletier and J. Rosický, Simple involutive quantales, *Journal of Algebra* 195, pp. 367–386, 1997.

[151] J.L. Peterson, *Petri Net Theory and the Modeling of Systems*, Prentice-Hall, Inc., 1981.

[152] A. Pnueli, The temporal logic of programs, in: *Proceedings of the 18th IEEE Symposium on Foundations of Computer Science*, pp. 46–57, 1977.

[153] D. Prawitz, *Natural Deduction: A Proof-theoretical Study*, Almqvist and Wiksell, Stockholm, 1965.

[154] G. Priest, Paraconsistent logic, in: D.M. Gabbay and F. Guenthner (eds.), *Handbook of Philosophical Logic, 2nd edition, Vol. 6*, Kluwer Academic Publishers, Dordrecht, pp. 287–393, 2002.

[155] G. Priest, *Doubt Truth to Be a Liar*, Oxford University Press, Oxford, 2006.

[156] G. Priest and R. Routly, Introduction: paraconsistent logics, *Studia Logica* 43, pp. 3–16, 1982.

[157] G. Priest and K. Tanaka, Paraconsistent logic, in: Edward N. Zalta (ed.), *The Stanford Encyclopedia of Philosophy* (Summer 2009 Edition), URL = ⟨http://plato.stanford.edu/archives/sum2009/entries/logic-paraconsistent/⟩, 2009.

[158] A.P. Pynko, Characterizing Belnap's logic via de Morgan's laws, *Mathematical Logic Quarterly* 41, pp. 442–454, 1995.

[159] A.P. Pynko, Functional completeness and axiomatizability within Belnap's four-valued logic and its expansions, *Journal of Applied Non-Classical Logics* 9, pp. 61–105, 1999.

[160] A.R. Raggio, Propositional sequence-calculi for inconsistent systems, *Notre Dame Journal of Formal Logic* 9, pp. 359–366, 1968.

[161] C. Rauszer, A formalization of the propositional calculus of H-B logic, *Studia Logica* 33, pp. 23–34, 1974.

[162] C. Rauszer, Applications of Kripke models to Heyting-Brouwer logic, *Studia Logica* 36, pp. 61–72, 1977.

[163] C. Rauszer, An algebraic and Kripke-style approach to a certain extension of intuitionistic logic, *Dissertations Mathematicae* CLXVII, PWN, Warszawa, 1980.

[164] W. Rautenberg, *Klassische und nicht-klassische Aussagenlogik*, Vieweg, Braunschweig, 1979.

[165] W. Reisig and G. Rozenberg (Eds.), *Lectures on Petri Nets I: Basic Models (Advances in Petri Nets)*, Lecture Notes in Computer Science 1491, 1998.

[166] W. Reisig and G. Rozenberg (Eds.), *Lectures on Petri Nets II: Applications (Advances in Petri Nets)*, Lecture Notes in Computer Science 1492, 1998.

[167] P. Resende, Quantales, finite observations and strong bisimulation, *Theoretical Computer Science* 254, pp. 95–149, 2001.

[168] G. Restall, Displaying and deciding substructural logics 1: Logics with contraposition, *Journal of Philosophical Logic* 27, pp. 179–216, 1998.

[169] G. Restall, Multiple conclusions, in: P. Hájek, L. Valdes-Villanueva, and D. Westerstahl (eds.), *Logic Methodology and Philosophy of Science. Proceedings of the Twelfth International Congress*, King's College Publications, London, pp. 189–205, 2005.

[170] R. Routley, Semantical analysis of propositional systems of Fitch and Nelson, *Studia Logica* 33, pp. 283–298, 1974.

[171] G. Rozenberg and J. Engelfriet, Elementary net systems, in: [165], 12–121, 1998.

[172] A. Schöter, Evidental bilattice logic and lexical inference, *Journal of Logic, Language, and Information* 5, pp. 65–105, 1996.

[173] P. Schroeder-Heister, Generalized rules, direct negation, and definitional reflection, in: *Proceedings of the 1st World Congress on Universal Logic (UNILOG 2005)*, 2 pages, 2005.

[174] P. Schroeder-Heister, Schluß und Umkehrschluß: Ein Beitrag zur Definitionstheorie, in: C.F. Gethmann (ed.), *Lebenswelt und Wissenschaft. Akten des XXI. Deutschen Kongresses für Philosophie 2008* (*Deutsches Jahrbuch Philosophie*, Vol. 3), Felix Meiner Verlag, Hamburg, pp. 1065–1092.

[175] G.D.M. Serugendo, D. Mandrioli, D. Buchs and N. Guelfi, Real-time synchronised Petri nets, in: *Lecture Notes in Computer Science* 2360, pp. 142–162, 2002.

[176] Y. Shi, Both Toffoli and controlled-NOT need little help to do universal quantum computation, Los Alamos National Laboratory (category: quantum circuit 0205115 v2), http://xxx.laul.gov/archive/quant-ph, 2002.

[177] T. Shimura, J. Lobe and T. Murata, An extended Petri net model for normal logic programs, *IEEE Transaction on Knowledge and Data Engineering* 7, pp. 150–162, 1995.

[178] Y. Shramko, Dual intuitionistic logic and a variety of negations: The logic of scientific research, *Studia Logica* 80, pp. 347-367, 2005.

[179] Y. Shramko, J.M. Dunn and T. Takenaka, The trilattice of constructive truth values, *Journal of Logic and Computation* 11, pp. 761–788, 2001.

[180] Y. Shramko and H. Wansing, Some useful 16-valued logics: how a computer network should think, *Journal of Philosophical Logic* 34, pp. 121-153, 2005.

[181] Y. Shramko and H. Wansing, Hyper-contradictions, generalized truth values and logics of truth and falsehood, *Journal of Logic, Language and Information* 15, pp. 403–424, 2006.

[182] Y. Shramko and H. Wansing, Truth values, in: Edward N. Zalta (ed.), *The Stanford Encyclopedia of Philosophy* (Summer 2014 Edition), URL =
⟨http://plato.stanford.edu/archives/sum2014/entries/truth-values/⟩, first published 2010.

[183] Y. Shramko and H. Wansing, *Truth and Falsehood. An Inquiry into Generalized Logical Values*, Springer-Verlag, Dordrecht, 2011.

[184] A.M. Tamminga and K. Tanaka, A natural deduction system for first degree entailment, *Notre Dame Journal of Formal Logic* 40, pp. 258–272, 1999.

[185] G. Takeuti, *Proof Theory*, North-Holland Publishing Company, Amsterdam, 1975.

[186] R.H. Thomason, A semantical study of constructible falsity, *Zeitschrift für mathematische Logik und Grundlagen der Mathematik* 15, pp. 247–257, 1969.

[187] A.S. Troelstra, *Lectures on Linear Logic*, CSLI Lecture Notes 29, CSLI, Stanford, 1992.

[188] A.S. Troelstra and H. Schwichtenberg, *Basic Proof Theory*, Cambridge University Press, Cambridge, 1996.

[189] I. Urbas, Dual-intuitionistic logic, *Notre Dame Journal of Formal Logic* 37, pp. 440–451, 1996.

[190] D. Vakarelov, Notes on N-lattices and constructive logic with strong negation, *Studia Logica* 36, pp. 109–125, 1977.

[191] D. Vakarelov, Constructive negation on the basis of weaker versions of intuitionistic negation, *Studia Logica* 80, pp. 393–430, 2005.

[192] D. Vakarelov, Non-classical negation in the works of Helena Rasiowa and their impact on the theory of negation, *Studia Logica* 84, pp. 105–127, 2006.

[193] N.N. Vorob'ev, A constructive propositional calculus with strong negation (in Russian), *Doklady Akademii Nauk SSR* 85, pp. 465–468, 1952.

[194] G. Wagner, Logic programming with strong negation and inexact predicates, *Journal of Logic and Computation* 1, pp. 835–859, 1991.

[195] H. Wansing, Formulas-as-types for a hierarchy of sublogics of intuitionistic propositional logic, in: *Lecture Notes in Computer Science* 619, pp. 125–145, 1992.

[196] H. Wansing, *The Logic of Information Structures*, Lecture Notes in Artificial Intelligence 681, 1993.

[197] H. Wansing, Informational interpretation of substructural propositional logics, *Journal of Logic, Language and Information* 2, pp. 285-308, 1993.

[198] H. Wansing, On the expressiveness of categorial grammar, in: V. Sinsini and J. Woleński (eds.), *The Heritage of Kazimierz Ajdukiewicz*, Rudopi, Amsterdam, pp. 337–351, 1995.

[199] H. Wansing, *Displaying Modal Logic*, Trends in Logic Vol. 3, Kluwer Academic Publishers, Dordrecht, 1998.

[200] H. Wansing, Higher-arity Gentzen systems for Nelson's logics, in: J. Nida-Rümelin (ed.), *Prodeedings of the 3rd international congress of the Society for Analytical Philosophy*, de Gruyter, Berlin, pp. 105–109, 1999.

[201] H. Wansing, Negation, in L. Goble (ed.), *The Blackwell Guide to Philosophical Logic*, Blackwell Publishers, Oxford, pp. 415–436, 2001.

[202] H. Wansing, Diamonds are a philosopher's best friends. The knowability paradox and modal epistemic relevant logic, *Journal of Philosophical Logic* 31, pp. 591–612, 2002.

[203] H. Wansing, A rule-extension of the non-associative Lambek calculus, *Studia Logica* 71, pp. 443–451, 2002.

[204] H. Wansing, Connexive modal logic, in: R. Schmidt et al. (eds.), *Advances in Modal Logic. Volume 5*, King's College Publications, London pp. 367-383, 2005, see also ⟨http://www.aiml.net/volumes/volume5/⟩.

[205] H. Wansing, On the negation of action types: Constructive concurrent PDL, in: P. Hájek, L. Valdes-Villanueva, D. Westerstahl (eds), *Logic Methodology and Philosophy of Science. Proceedings of the Twelfth International Congress*, King's College Publications, London, pp. 207–225, 2005.

[206] H. Wansing, Connexive logic, in: Edward N. Zalta (ed.), *The Stanford Encyclopedia of Philosophy (Fall 2014 Edition)*, URL = ⟨http://plato.stanford.edu/archives/fall2014/entries/logic-connexive/⟩, first published 2006.

[207] H. Wansing, Logical connectives for constructive modal logic, *Synthese* 150, pp. 459–482, 2006.

[208] H. Wansing, A note on negation in categorial grammar, *Logic Journal of the IGPL* 15, pp. 271–286, 2007.

[209] H. Wansing, Constructive negation, implication, and co-implication, *Journal of Applied Non-Classical Logics* 18, pp. 341–364, 2008.

[210] H. Wansing, The power of Belnap: sequent systems for $SIXTEEN_3$, *Journal of Philosophical Logic* 39, pp. 369–393, 2010.

[211] H. Wansing, Proofs, disproofs, and their duals, in: L. Beklemishev, V. Goranko and V. Shehtman (eds.), *Advances in Modal Logic. Vol. 8*, College Publications, London, pp. 483–505, 2010, see also ⟨http://www.aiml.net/volumes/volume8/⟩.

[212] H. Wansing and N. Kamide, Intuitionistic trilattice logics, *Journal of Logic and Computation* 20, pp. 1201–1229, 2010.

[213] H. Wansing and Y. Shramko, Harmonious many-valued propositional logics and the logic of computer networks, in: C. Dégremont,

L. Keiff and H. Rückert (eds.), *Dialogues, Logics and Other Strange Things. Essays in Honour of Shahid Rahman,* College Publications, London, pp. 491–516, 2008.

[214] H. Wansing and Y. Shramko, Suszko's Thesis, inferential many-valuedness, and the notion of a logical system, *Studia Logica* 88, pp. 405–429, 89, p. 147, 2008.

[215] D. Wijesekera and A. Nerode, Tableaux for constructive concurrent dynamic logic, *Annals of Pure and Applied Logic* 135, pp. 1–72, 2005.

[216] T. Williamson, *Knowledge and Its Limits,* Oxford University Press, Oxford, 2000.

[217] F. Wolter, On logics with coimplication, *Journal of Philosophical Logic* 27, pp. 353–387, 1998.

[218] D.N. Yetter, Quantales and (noncommutative) linear logic, *Journal of Symbolic Logic* 55, pp. 41–64, 1990.

Index

www.ingramcontent.com/pod-product-compliance
Lightning Source LLC
Chambersburg PA
CBHW060114200326
41518CB00008B/823
* 9 7 8 1 8 4 8 9 0 1 6 7 4 *